MW00844842

MOLECULAR DESCRIPTORS IN QSAR/QSPR

MOLECULAR DESCRIPTORS IN QSAR/QSPR

MATI KARELSON
Department of Chemistry
University of Tartu

A JOHN WILEY & SONS, INC., PUBLICATION
New York • Chichester • Weinheim • Brisbane • Singapore • Toronto

This book is printed on acid-free paper. ♾

Published simultaneously in Canada.

For ordering and customer service, call 1-800-CALL-WILEY.

Library of Congress Cataloging-in-Publication Data:

Karelson, Mati.
Molecular descriptors in QSAR/QSPR / by Mati Karelson.
p. cm.
"A Wiley-Interscience publication."
Includes bibliographical references and index.
ISBN 0-471-35168-7 (alk. paper)
1. QSAR (Biochemistry)—Mathematical models. I. Title.
QP517.S85K37 2000
572′.4—dc21 99-38911

Printed in the United States of America.

10 9 8 7 6 5 4 3 2 1

To Tiia

CONTENTS

PREFACE

The revolutionary progress in computer technology has created an entirely new environment for efficient use of the theoretical constructions of natural science in many areas of applied research. The theoretical approach has proven to be most beneficial in chemistry and bordering sciences, where the experimental study and synthetic development of new compounds and materials is often time consuming, expensive, or even hazardous. The contemporary quantum theory of molecular matter and the respective ab initio computational methods can predict the properties of isolated small molecules within the experimental error. However, the majority of industrially and environmentally important chemical processes, and all biochemical transformations in living organisms, take place in heterogeneous condensed media. The extreme complexity of such media is often prohibitive for the ab initio theory to be used, and thus the relationship between the chemical activity and the molecular structure in these systems is often poorly described and understood.

The direct development of empirical equations that are usually referred to as the quantitative structure–activity/property relationships (QSAR/QSPR) represents an attractive alternative approach to predict the molecular properties in complex environments. Notably, the QSAR methodology has been extremely productive in pharmaceutical chemistry and in computer-assisted drug design. Probably thousands of new medications have been first developed on computer screens before their implementation in synthetic laboratories. In analytical chemistry, the QSPR equations are commonly used to predict the spectroscopic, chromatographic, and other analytical properties of compounds. In recent years, the QSPR approach is rapidly expanding to various areas of industrial and environmental chemistry. In most contemporary applications, *empirical* molecular descriptors that rely on some experimental data have been used in the development of QSAR/QSPR equations. Such descriptors, starting from the original Hammett substituent σ constants to the most popular partition coefficients between water and octanol (log P) are, strictly speaking, restricted to the compounds for which the necessary experimental data are available. Another shortcoming of experimental descriptors evolves from the fact that many of them reflect a complicated combination of different physical interactions. Alternatively, the molecular descriptors can be derived using only the information encoded in the chemical structure of the compound. Importantly, such *theoretical* descriptors can be developed for the compounds that have never been synthesized or experimentally explored.

The motivation for this book is to provide a comprehensive overview of theoretical molecular descriptors used in chemistry and related sciences. During the last half of the twentieth century, much attention has been paid to the interrelation between the topology and chemical properties of molecules in QSAR/QSPR equations. The respective molecular descriptors have been particularly successfully used in prediction of the pharmaceutical activity of compounds. On the other hand, the semiempirical quantum chemical calculations for large molecules have become accessible even using small personal computers. Correspondingly, many descriptors of well-defined physical nature can be acquired from the results of such calculations. The simplest molecular descriptors can be defined just as the counts of different atoms and chemical bonds in a compound, whereas others proceed from the geometry of the molecule. In the present book, a systemic and critical overview of all main classes of theoretical molecular descriptors is given.

The critical analysis of descriptors is accompanied with a review of computational methods that can be employed in the development of QSAR/QSPR equations. The text is supplemented with software that allows to calculate many molecular descriptors for almost any chemical structure and includes basic methods for the development of QSAR/QSPR equations.

The present book is intended to be a helpful tool for both the academic and industrial chemists, from students to advanced researchers, for anyone who is interested in relating the properties of biological activity of compounds with their basic chemical structure.

Finally, I wish to thank my co-workers Dr. Uko Maran, Dr. Rein Hiob, Dr. Jaan Leis, Ms. Helle Kuura, and Mr. Tarmo Tamm for help in writing this book. In particular, I would like to acknowledge the efforts by Mr. Sulev Sild for help in creating the software.

Mati Karelson
Tartu, Estonia

MOLECULAR DESCRIPTORS IN QSAR/QSPR

1

INTRODUCTION

The development of quantum theory in the twentieth century has resulted in methods and techniques that, in principle, should enable us to predict the physical properties and chemical affinity of molecular matter with experimental precision. However, most *real-life* systems, including the processes in chemical reactors, drug–receptor interactions in biological systems, and the degradation of environmental pollutants, are characterized by such a complexity of intra- and intermolecular interactions that their theoretical description from first principles is impossible, even using the most powerful foreseeable computers. Different "model approximations" have therefore been used in ab initio quantum theory to make it applicable for larger chemical systems. In most cases, such simplifications are introduced on the basis of chemical intuition or proceed from various mathematical considerations.

Furthermore, because of a lack of an analytical solution to the so-called many-body problem, ab initio quantum theory of molecules is approximate by its very fundamental mathematical nature. The respective methods for molecular electronic structure calculations are thus often reduced to exercises in applied mathematics, with indirect connection to basic physical phenomena and introduction of various pseudo- or quasiphysical effects. Therefore, the applicability of approximate and semiempirical theories has to be verified for a given property and often even for a given restricted set or class of chemical compounds. In the case of loosely controlled theoretical approximations, the causal relationship between the molecular property and its structure can be obscured.

Over the past several decades, the quantitative structure–activity/property relationships (QSAR/QSPR) have become an alternative powerful theoretical tool for the description and prediction of properties of complex molecular systems in different environments. The QSAR/QSPR approach proceeds from the assumption of the one-to-one correspondence between any physical property, chemical affinity, or biological activity of a chemical compound and its molecular structure. The latter can be represented by the chemical composition, connectivity of atoms, potential energy surface, and electronic wave function of a compound (cf. Fig. 1.1). Various physico-chemical molecular descriptors reflecting the structure can be determined empirically or by using theoretical and computational methods of different complexity. It must be underlined that a necessary requirement for the application of the QSAR/QSPR approach is the knowledge of the exact chemical constitution and/or the three-dimensional molecular structure of the chemical compounds studied.

The QSAR/QSPR relationships are usually derived using the (multiple) linear least-squares regression of the experimentally measured property values P against a preselected set of molecular descriptors (D_1, D_2, D_3, . . .):

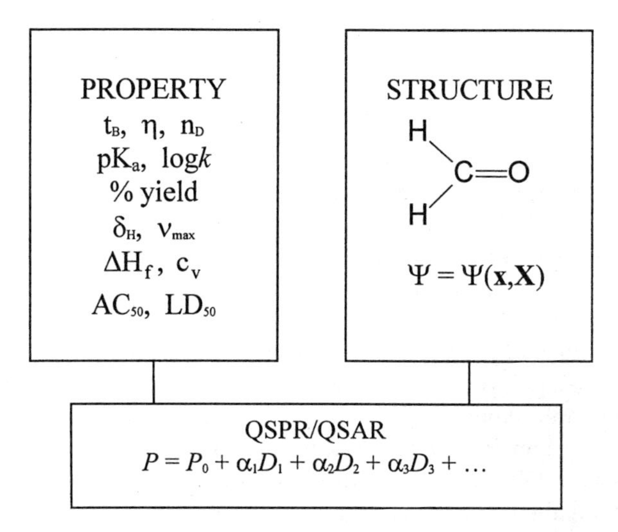

Figure 1.1 QSPR equations relate the property of a molecular system to its structure presented by the chemical constitution, atomic connectivity, and molecular electronic-nuclear wave function $\Psi(\mathbf{x}, \mathbf{X})$, where $\mathbf{x}$ and $\mathbf{X}$ denote the electronic and nuclear coordinates of the system, respectively. In many cases, the QSPR relationship is derived using the (multiple) linear least-squares fitting of the experimentally measured property values P with a selected set of molecular descriptors (D_1, D_2, D_3, . . .). The molecular property P can be a physical constant of the bulk of a compound (such as the normal boiling point t_B, viscosity η, or refractive index n_D at normal temperature), some measure of chemical reactivity (such as the acidic dissociation constant pK_a, rate constant k, or the yield of a chemical reaction), spectroscopic characteristic of a molecule [such as the proton chemical shift δ_H in the nuclear magnetic resonance (NMR) spectrum or the transition frequency ν_{max} in the electronic or vibrational spectrum of a compound], a thermodynamic function (such as the enthalpy of formation ΔH_f or heat capacity c_v), or the biological activity [such as the active concentration (AC_{50}) or the half-lethal dose (LD_{50})] of the compound.

$$P = P_0 + \alpha_1 D_1 + \alpha_2 D_2 + \alpha_3 D_3 + \cdots \qquad (1.1)$$

In this expansion, the property P is a dependent variable whereas the descriptors D_i represent the independent variables. The coefficients α_i measure the weight of each descriptor involved in the relationship. Equation (1.1) is valid only if certain requirements are fulfilled for a given property and a given set of descriptors. First, the property value for a given compound has to be divisible into additive terms, each of which corresponds to a single molecular descriptor. Second, in the case of the variation of chemical structure, such terms should depend linearly on the respective descriptor values. In Figure 1.2, two possible situations are demonstrated. Apparently, the use of Eq. (1.1) is justified if the variation of the chemical structure corresponds to the interval of a descriptor $[D_0, D_1]$. Within the descriptor interval $[D_1, D_2]$, the linear QSAR/QSPR will not be valid.

In such case it is, of course, still possible to find a functional relationship between molecular property and descriptors. Instead of linear regression with molecular descriptors, some nonlinear form of the quantitative structure–property relationship has to be used. The simplest approach involves some nonlinear transformation of the descriptor, for example, the square, the square root, or the logarithm of the natural descriptor can be applied as an independent variable in Eq. (1.1). More systematically, the first few terms of the polynomial expansion of the appropriately rescaled molecular descriptor can be used as the independent variables in multiple linear regression. Various techniques are available for the direct nonlinear least-squares fitting of the property with the suitably chosen mathematical function on the descriptor. The application of artificial neural networks to develop dependence between the property and molecular descriptors of a chemical system also allows accounting for the intrinsically nonlinear relationship between them. Thus, in general, the absence of the (multiple) linear relationship between the observable property and molecular descriptor(s) of a system does not necessarily imply the absence of the causal or functional dependence between them.

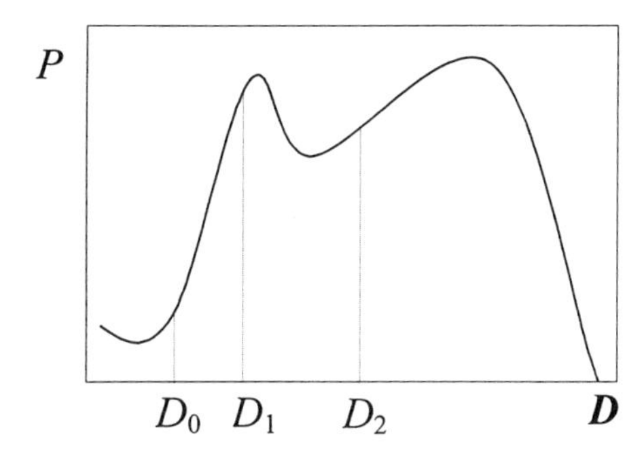

Figure 1.2 Dependence of property P of a molecular system on the molecular descriptor D, presented as a continuous function of the chemical structure.

The first requirement for the validity of Eq. (1.1) that demands the additivity of the different linear terms in the QSAR/QSPR equation may also be violated. In general, this additivity is strictly valid only when the intra- or intermolecular interactions, corresponding to different descriptors, are not affected by each other. In real-life situations, this may often not be the case. A well-known example of the interdependence of different intermolecular interactions is the steric restriction to the resonance effect in the conjugated π-electron molecular systems. One possible way to account for the nonadditive effects is the use of the cross terms between the respective molecular descriptors in the QSAR/QSPR regression:

$$P = P_0 + \alpha_1 D_1 + \alpha_2 D_2 + \alpha_{12} D_1 D_2 + \cdots \qquad (1.2)$$

The addition of the cross terms decreases the number of statistical degrees of freedom for a given regression equation. However, it is acceptable when leading to substantial improvement of the respective QSAR/QSPR description.

The success of the QSAR/QSPR approach is critically dependent on the accurate definition and appropriate use of molecular descriptors. In this book, we attempt to provide a systematic and unified overview of the whole variety of *empirical* and *theoretical* molecular descriptors. This differentiation, while somewhat arbitrary, evolves from the possible limits of the applicability of QSAR/QSPR relationships using the descriptors from every individual group. The QSAR/QSPR equations obtained using solely the theoretically derived descriptors can be used, in principle, for the prediction of the respective properties of any molecular structure. Therefore, such equations can be expanded to the compounds for which the experimental information necessary for the definition of empirical descriptors is missing or which have not yet even been synthesized. The use of empirical descriptors restricts the predictive power of QSAR/QSPR equations to the compounds for which this experimental information is available.

The empirical descriptors can be divided into two general classes (Table 1.1). The first reflects the intramolecular electronic interactions (*structural descriptors*) whereas the second accounts for the intermolecular interactions in

TABLE 1.1 General Classification of Empirical Molecular Descriptors

Class	Subclass
Structural descriptors	Induction constants
	Resonance constants
	Steric constants
Solvational descriptors	Polarity scales
	Polarizability scales
	Acidity scales
	Basicity scales
	Mixed scales

condensed media such as liquids and solutions (*solvational descriptors*). The most widespread structural descriptors are defined to quantify the induction, the mesomeric or resonance, and the steric effects in chemical compounds. The solvational descriptors reflect the interactions of the solute with the bulk of the surrounding solvent (*macroscopic or nonspecific solvent effects*) and the specific bonding, mostly hydrogen bonding between the solute and individual solvent molecules (*microscopic or specific solvent effects*). The macroscopic solvent effects are quantified using various polarity and polarizability scales. The microscopic solvent effect descriptors include general acidity and general basicity scales. Some empirical solvent effect scales (*mixed scales*) may involve both the macroscopic and microscopic effects. A typical representative of such descriptors is the water–octanol partition coefficient, log P.

The theoretical molecular descriptors can be conventionally divided into a number of different classes, proceeding from either their complexity or the method of calculation. The simplest theoretical descriptors are the *constitutional* descriptors that can be constructed from the information about the chemical composition of the chemical compound. The absolute and relative counts of different types of atoms and chemical bonds, the molecular weight, and the number of different rings in the compound are some typical constitutional descriptors. The *topological* descriptors (also called topological indices) describe the atomic connectivity in the molecule. It has been debated that the topological indices may encode more subtle molecular interactions than just the branching of chemical bonds or specific mass distribution in the molecule. The *geometrical* descriptors are derived from the three-dimensional structure of molecules defined by the coordinates of atomic nuclei and the size of the molecule represented, for instance, by the atomic van der Waals radii. For most of the chemical compounds, the molecules possess certain conformational flexibility, and the respective molecular potential surfaces have multiple local minima. Depending on the structure of the molecule, the number of these minima can be very large and, therefore, it may be rather difficult to find the global energy minimum at given experimental conditions. Obviously, the geometrical descriptors may vary significantly depending on the conformation of the molecule. Therefore, it is important to verify the correctness of the conformation used in the calculation of these descriptors. To some extent, the *charge-distribution-related* theoretical descriptors may be also conformation dependent. These descriptors are based on the three-dimensional structure and the charge distribution in the molecule. The latter may be presented by atomic partial charges derived from some empirical schemes or from more sophisticated functions based on the quantum chemically calculated wave function of the molecule. A very interesting and rapidly growing direction in the area of charge-distribution-related descriptors is the use of molecular electrostatic fields. This direction, known primarily as the comparative molecular field analysis (CoMFA), has been successfully applied to the investigation of biological activity of compounds and to the computer-aided molecular design (CAMD) of new materials and pharmaceuticals.

A number of different *molecular-orbital*-based quantum chemical descriptors have been employed in the development of QSAR/QSPR equations. The most widely used are the frontier molecular orbital energies, that is, the calculated energy of the highest occupied molecular orbital (ε_{HOMO}), the energy of the lowest unoccupied molecular orbital (ε_{LUMO}), and the difference between these energies. Also, various reactivity indices derived from Fukui's theory of superdelocalizability or other theoretical constructions have gained popularity among researchers.

Modern quantum chemical program packages include the calculation of the statistical-physical partition function and its derivative thermodynamic functions of molecular systems. These molecular characteristics can be considered as molecular descriptors, particularly appropriate for systems at elevated temperatures where the thermal motion of molecules may substantially influence the process or property studied. It has to be stressed that the possible temperature dependence of molecular properties is not accounted for by traditional QSAR/QSPR descriptors. The respective effects may be, however, of great importance in studying many real-life systems, and thus the application of the theoretical *thermodynamic* or other *temperature-dependent* descriptors would be necessary to obtain a realistic picture of molecular systems. Various possibilities exist to include the temperature dependence of the properties and molecular descriptors in the QSAR/QSPR analysis.

During the last decade, great strides have been made in the development of the quantum theory of solvation. Numerous methods and algorithms introduced calculate the solvent effects on the molecular structure and properties. The theoretically computed individual contributions to the free energy of solvation and other characteristics are attractive as new theoretical *solvational* descriptors of compounds.

Not all theoretical descriptors can be strictly classified according to the above-given scheme (cf. Table 1.2). For example, the topographical indices are derived using the information about both the topology and the geometry of the molecules. The electrotopological indices are based on the topology and charge distribution whereas the charged partial surface area descriptors encode simultaneously the charge distribution and the geometry of compounds. Such descriptors can be classified as the *mixed* or *combined* molecular descriptors.

Molecular descriptors can be defined for the whole molecular system under study or for any part (fragment) of it. For instance, most empirical structural descriptors relate traditionally to molecular fragments called substituents. Accordingly, the molecules in a congeneric series of chemical compounds are formally divided into two or more fragments that correspond to a constant structural unit Y (e.g., the reaction center) and to the variable structural units X_i (substituents). The QSAR/QPSR relationships are thus presented as follows:

$$P = P_0^{(Y)} + \sum_i \sum_k \alpha_i^{(Y)} D_{ik}^{(X)} \tag{1.3}$$

TABLE 1.2 General Classification of Theoretical Molecular Descriptors

Class	Subclass
Constitutional descriptors	Counts of atoms or bonds
	Atomic-weight-based descriptors
Topological descriptors	Topological (connectivity) indices
	Information-theoretical descriptors
	Topochemical descriptors
Geometrical descriptors	Distance-related descriptors
	Surface-area-related descriptors
	Volume-related descriptors
	Molecular steric field descriptors
Charge-distribution-related descriptors	Atomic partial charges
	Molecular electrical moments
	Molecular polarizabilities
	Molecular electrostatic field descriptors
Molecular-orbital-related descriptors	Frontier molecular orbital energies
	Bond orders
	Fukui's reactivity indices
Temperature-dependent descriptors	Thermodynamic functions
	Boltzmann factor-weighted descriptors
Solvational descriptors	Electrostatic energy of solvation
	Dispersion energy of solvation
	Free energy of cavity formation
	Hydrogen bonding descriptors
	Entropy of solvation
	Theoretical linear solvation energy descriptors
Mixed descriptors	Topographical descriptors
	Electrotopological descriptors
	Charged partial surface area descriptors

where $P_0^{(Y)}$ is the intercept corresponding to the constant molecular fragment Y, $D_{ik}^{(X)}$ are the molecular descriptors of type k for the variable fragments X_i, and $\alpha_{ik}^{(Y)}$ are the expansion coefficients characteristic for a given series of compounds X_iY. Many applications, especially the computer-aided molecular design of new compounds and materials, require the description and prediction of properties of compounds with large structural variability. The general QSAR/QSPR equations applicable for compounds involving different chemical functionalities will be useful in these cases. Instead of descriptors referring to structural fragments, the descriptors corresponding to the whole molecule are more appropriate to develop the necessary relationships. Notably, most of the theoretical descriptors listed in Table 1.2 can be calculated either for the whole molecule or for a predefined molecular fragment.

Many QSAR/QSPR applications involve complex multicomponent systems such as solutions and mixtures of chemical compounds, biological objects in vivo and in vitro, chemical reactors and aquifers, and others. Typically, in these cases only one molecular structure is singled out as presumably responsible for the variance of the given property of interest. The quantitative relationship is developed between the property of the whole system and the molecular descriptor(s) of this responsible component. However, the properties of a multicomponent system may depend from the individual contributions from each component and/or from the intermolecular interactions between the different components. In each case, this requires a special execution of the QSAR/QSPR treatment. In the case of negligible intercomponent interaction effects, the QSPR/QSPR expansion for a multicomponent system can be developed using the molar fraction weighted descriptors for individual components. A simple way to introduce the interaction terms between the different components is the use of the cross terms of descriptors belonging to the interacting components.

$$P = P_0 + \sum_i \sum_k \beta_{ik} D_i D'_k \qquad (1.4)$$

where D_i and D'_k are molecular descriptors for components i and k, respectively, and β_{ik} is the specific interaction constant. Depending on the nature of the intermolecular interactions between the different components of a multicomponent system, descriptors D_i and D'_k may be the same or different and reflect the physical mechanism of the interaction. Simple examples of such interactions are the electrostatic repulsion between the similarly charged molecular surfaces and the electrostatic attraction between the oppositely charged surfaces, respectively. In the first case, the same descriptor (fractional negative or positive surface area) is used as the interaction term [Eq. (1.4)]. In the second case, the product of two different descriptors (fractional negative surface area of one component multiplied by the fractional positive surface area of another component) has to be employed. The mathematical form of the dependence of some property of a multicomponent system on the molecular descriptors of different components may be more complex but known. In this case, a least-squares treatment of data according to the respective, possibly nonlinear, equation leads to the physically justified quantitative structure–property relationships.

A large variety of statistical structure–property correlation techniques can be used for the analysis of experimental data in combination with the calculated molecular descriptors. First, the multiple linear regression methods can be applied in the scalar space of the original descriptors, in the principal-component orthogonalized space of the descriptors, or in the target-transformed descriptor space. Different stepwise and stagewise strategies are available for effective search of the best (most informative) multiparameter correlations in large spaces of the natural descriptors. The molecular properties, molecular descriptors, or their combinations can be analyzed using factor-analysis-based pattern recognition methods, including principal component analysis (PCA), partial least squares (PLS), or nonlinear partial least squares (NIPALS). In the case of intrinsically nonlinear dependence between the experimental property of compounds and molecular descriptors, nonlinear regression methods can be applied for the development of QSAR/QSPR equations. The intrinsic nonlinear dependence may also be encoded in the respective artificial neural networks. Notably, the choice of method for the data treatment is largely independent of the descriptors applied. In other words, most molecular descriptors can be applied universally in different QSAR/QSPR treatments. This flexibility makes the molecular descriptors, especially the theoretically derived descriptors, attractive and efficient for the construction of working models to predict the physical, chemical, and biomedical properties of compounds.

Frequently, the concept of molecular similarity is used for the prediction of properties for new and previously unknown molecular systems. It is important to emphasize, however, that there is no absolute similarity measure between any two molecular structures. Let us consider a set of the following six compounds: 2-octanone (**I**), methyl heptanoate (**II**), 2-octanol (**III**), acetone (**IV**), methyl acetate (**V**), and isopropanol (**VI**).

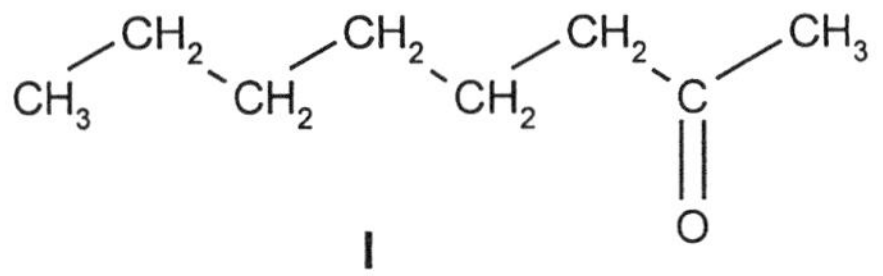

I

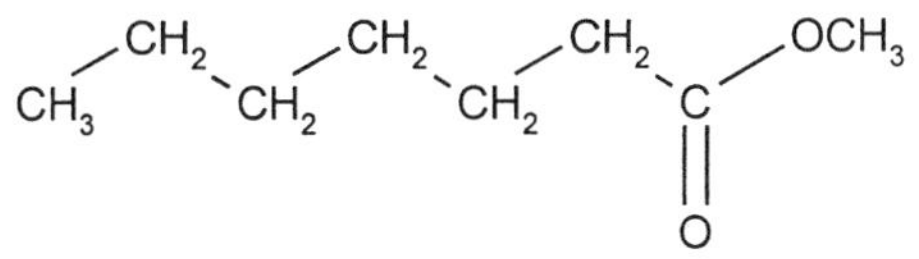

II

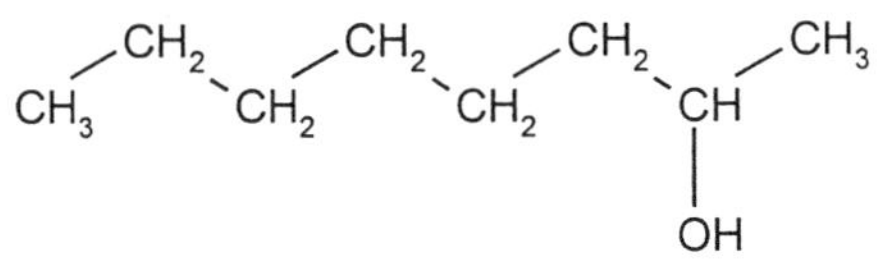

III

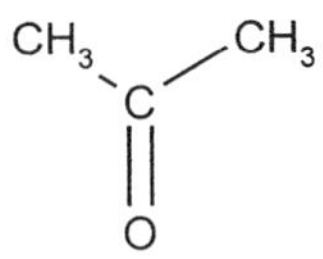

IV

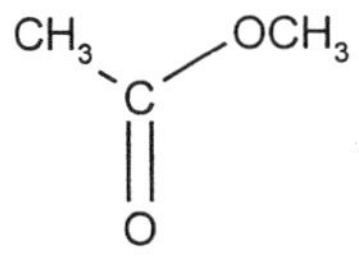

V

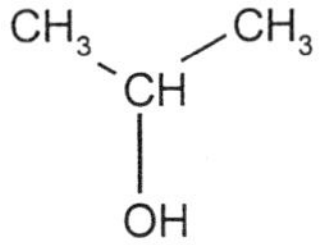

VI

By looking at the simple constitutional descriptors, such as the counts of carbon or hydrogen atoms in the molecule, compounds **I** to **III** belong to one class of structurally similar compounds whereas compounds **IV** to **VI** relate to distinctly different class of compounds (cf. Table 1.3). The same classification is applicable if we compare the molecular volumes or the surface areas of molecules. The theoretically calculated dipole moments of compounds do not show, however, any distinct grouping of molecules. The experimentally determined properties of these compounds exhibit similar diversity. For instance, the boiling point of compounds **I** to **III** is very similar. According to this property, acetone (**IV**) and methyl acetate (**V**) constitute another group of very similar compounds whereas isopropanol (**VI**) stays separately. By comparison of the wavelength of the spectral maxima corresponding to the lowest energy transition in the ultraviolet spectra of compounds, another classification of the six compounds could be obtained. Compounds **I**, **II**, **IV**, and **V** have a carbonyl group in their structure, and the spectral maxima of the accompanying $n - \pi^*$ transitions are located substantially above 200 nm. The maxima for compounds **III** and **VI** are shifted to substantially shorter wavelengths (<200 nm). Thus, according to this property, they form another class of compounds. Finally, the comparison of the chemical reactivity with water results in yet another classification by similarity of the six compounds. Ketones **I** and **IV** undergo the enolization in aqueous solutions whereas esters **II** and **V** are hydrolyzed by water. Alcohols **III** and **VI** do not react with water.

Consequently, the similarity of compounds is related to the specific property investigated. On the other hand, different molecular descriptors also discriminate between the molecular structures differently. Therefore, the multitude of

TABLE 1.3 Numerical Values of Some Molecular Descriptors and Experimental Properties for a Set of Six Compounds: 2-octanone (I), methyl heptanoate (II), 2-octanol (III), acetone (IV), methyl acetate (V), and isopropanol (IV)

Property/Compound	I	II	III	IV	V	VI
Count of C atoms	8	8	8	3	3	3
Count of H atoms	16	16	18	6	6	8
Molecular volume	148.60	157.20	154.71	63.84	72.84	70.34
Molecular surface area	178.15	187.07	179.67	81.72	91.68	85.64
Dipole moment (AM1)	2.742	1.663	1.692	2.922	1.775	1.615
Boiling point (°C)	173	173.5	174	56	57.5	82
λ_{max}(ultraviolet), nm	>200	>200	<200	>200	>200	<200
Reaction with water	E[a]	H[b]	I[c]	E[a]	H[b]	I[c]

[a]Enolization.
[b]Hydrolysis.
[c]Inert.

molecular descriptors available for the development of QSAR/QSPR equations is not really a result of the fantasy of researchers but based on the specific relationships between the properties and molecular characteristics of compounds. It is thus very important to make a correct choice of descriptors for the treatment of a given property. In the following, we proceed with a systematic analysis of the genesis and applicability of the multitude of empirical and theoretical descriptors.

2

EMPIRICAL MOLECULAR DESCRIPTORS

2.1 HISTORICAL BACKGROUND

Investigation of the dependence between the molecular structure and properties of chemical compounds has been an important subject of research throughout modern chemistry. Some authors have noted that the first attempts to relate biological activity with the chemical properties of compounds date back to the middle of nineteenth century. Already in 1863, Cros observed that toxicity of alcohols to mammals increased as the water solubility of the alcohols decreased [1]. At the turn of the last century, Meyer and Overton related the toxicity of organic compounds to their lipophilicity [2–4].

An important step toward the development of quantitative structure–property relationships was made by Brönsted and Pedersen who introduced the equations to correlate the rate constants of different general acid- or base-catalyzed reactions with the respective acidic or basic dissociation constants of the catalytic acids or bases:

$$k_{HA} = g_A (K_{HA})^{\alpha} \tag{2.1}$$

for acids, and

$$k_B = g_B (K_B)^{\beta} \tag{2.2}$$

for bases. The coefficients g_A and g_B and exponents α and β are characteristic for a given catalytic process. In principle, these equations define a linear relationship between the free energy of activation of one process (catalytic reaction) and the free energy of another chemical reaction (acid–base equilibrium) as:

$$\Delta G^{\neq}_{\mathrm{cat}(HA)} = -RT \ln g_{HA} + \alpha \, \Delta G_{HA} \tag{2.3}$$

and

$$\Delta G^{\neq}_{\mathrm{cat}(B)} = -RT \ln g_{B} + \beta \, \Delta G_{B} \tag{2.4}$$

where $\Delta G^{\neq}_{\mathrm{cat}(HA)}$ and $\Delta G^{\neq}_{\mathrm{cat}(B)}$ denote the free energies of activation of the acid- and base-catalyzed reactions, respectively, and ΔG_{HA} and ΔG_B are the free energies of the acid–base equilibria involving the corresponding catalytic acids or bases. The approach that relates the free energies of two processes involving the same molecular structure by equations similar to the Brönsted equations [(2.3) and (2.4)] is known in physical organic chemistry as the method of *linear free energy relationships* (LFER). It has been suggested that this name be replaced by linear Gibbs energy relations (LGER) [5], but the latter is still rarely used.

An important landmark in the development of LFERs was set by Hammett whose classical work (1935) associated the chemical reactivity of meta- and para-substituted benzenes with the acidic dissociation constants of the similarly substituted benzoic acids [6,7]. The well-known Hammett equation has the following form:

$$\log\left(\frac{k_X}{k_H}\right) = \rho\sigma_X \tag{2.5}$$

where k_H and k_X are the rate constants for some chemical reaction involving the unsubstituted benzene and the benzene derivative with the substituent X, respectively. The so-called σ constants

$$\sigma_X = \log\left(\frac{K_X}{K_H}\right) \tag{2.6}$$

were defined as the logarithmic ratio of the acidic dissociation constants of the substituted (by substituent X) and the unsubstituted benzoic acids. The constant ρ is characteristic for a given reaction. The Hammett equation has been extended to other aromatic systems. In addition, σ constants have been developed for the ortho-substituted compounds, but the situation is more complicated in this case because of the spatial closeness of the reaction center and the substituent.

After the first success with the Hammett equation, it was understood that at least two different intermolecular interaction mechanisms are reflected by the σ constants. First, the *induction* effect is originated from the polarization of the chemical bonds by electronegative atoms or atomic groups. Second, depending on the chemical nature and mechanism of the process, direct polar conjugation between the substituent and the reaction center may affect the reactivity in substituted aromatic systems (*mesomeric* or *resonance* effect). A multiparameter extension of the Hammett equation was thus developed to quantify the role of enhanced resonance effects on the reactivity of meta- and para-substituted benzene derivatives as [8–10]:

$$\log\left(\frac{k_X}{k_H}\right) = \rho\sigma + r(\sigma^+ - \sigma) \tag{2.7}$$

or

$$\log\left(\frac{k_X}{k_H}\right) = \rho\sigma + s(\sigma^- - \sigma) \tag{2.8}$$

The second term in Eq. (2.7) reflects the enhanced resonance effect by the electron-donating substituents, quantified by the substituent constant $(\sigma^+ - \sigma)$. The second term in Eq. (2.8) describes the increase of the direct resonance between the reaction center and the electron-accepting substituent in the conjugated para-position using the scale $(\sigma^- - \sigma)$. A more recent generalization of the Hammett equation belongs to Taft and co-workers who introduced the dual substituent parameter (DSP) equation [11,12]:

$$P = P_0 + \rho_I\sigma_I + \rho_R\sigma_R \tag{2.9}$$

which has been applied to the description of chemical, physical, and spectroscopic properties of substituted benzenes. In Eq. (2.9), P denotes the magnitude of a given property for the compound with substituent X, and P_0 is to the property value for the unsubstituted benzene. The terms σ_I and σ_R denote the inductive, or polar, and the resonance substituent constants, respectively. The coefficients ρ_I and ρ_R characterize the relative susceptibility of the property studied to the induction and resonance effect, respectively.

Already in 1953, Taft had extended the LFER approach to the aliphatic organic compounds [13–16]. He noticed that the transition states (TS) of the acid- and base-catalyzed reactions of the ester hydrolysis are geometrically similar, the difference being only in the charge and two additional protons in the TS of the acid-catalyzed process. Therefore, it was plausible to assume that the possible steric effects would cancel if the free energies of activation of these two reactions for the same ester were compared. At the same time, the influence of the possible induction effect by the substituents in the substrate ester molecules has to be significantly different in the case of the acid- and base-catalyzed reaction, respectively. This observation allowed Taft to define a quantitative scale of the induction effect by the substituents in the aliphatic series of compounds as:

$$\sigma^* = \frac{1}{2.48}\left[\log\left(\frac{k_X}{k_0}\right)_B - \log\left(\frac{k_X}{k_0}\right)_A\right] \tag{2.10}$$

where k_X denotes the rate constant for the compound with the substituent X and k_0 is the rate constant for the standard compound, respectively. Subscripts B and A correspond to the base-catalyzed and to the acid-catalyzed reaction, respectively.

The quantitative LFER treatment of steric effects was also pioneered by Taft [14,17]. It was observed that the Hammett ρ constant for the acid hydrolysis of esters was close to zero. Consequently, the induction effect by substituents on this reaction is practically negligible. Thus, the original E_S substituent constants were defined as the logarithmic ratio of the rate constants for the acid hydrolysis of the ester with the substituent X and of the standard compound:

$$E_S = \log\left(\frac{k_X}{k_0}\right)_A \tag{2.11}$$

and employed in the respective general Taft equation as:

$$\log\left(\frac{k_X}{k_H}\right) = \rho^*\sigma^* + \delta E_S \tag{2.12}$$

where δ scales the susceptibility of the given process to the steric effect.

It has to be emphasized that all of these substituent constants were defined using some experimental data; thus, in principle, they are empirical structural molecular descriptors.

In addition to the structural effects on the chemical reactivity and properties of compounds, the solvent effects reflecting the influence of the surrounding environment on the properties of molecules in condensed media have been quantified using the empirical LFER approach. As in the case of LFER constants for structural effects, solvent effect parameters have been introduced proceeding from different physical models of solvation and intermolecular interactions in condensed media. These parameters can be divided into two general groups. The first kind of parameter has been developed using the concept of electrostatic interactions between the solute and the bulk of the solvent. Within this approach, the surrounding medium (i.e., the solvent) is presented as a homogeneous dielectric continuum, polarizable by the electrostatic field created by a discrete solute molecule. Various theoretical constructions predict a linear relationship between the free energy of the observed process and some function of the dielectric constant of the medium [18]. Because of the different characteristic relaxation times of different dielectric relaxation processes in solution (shift of the electron distribution, conformational changes of the molecule), it has been proposed to distinguish between *polarity* and *polarizability* effects [19]. The first refers to the full dielectric polarization of the solvent in the field of the solute molecule. In the respective solvent LFER scales, the macroscopic dielectric permittivity at zero frequency has been employed. The polarizability effects arise from only the short-time electron–nuclear polarization of the solvent, and thus they relate to the dielectric permittivity of solvent at infinite frequency of the external electric field.

Another group of LFER solvent effect scales has been developed to describe the short-range specific solute–solvent interactions. These involve hydrogen bonding and other semichemical interactions between the discrete solute and solvent molecules. The solute–solvent interactions, where the solvent is acting as a hydrogen bond donor and the solute molecule as an acceptor, are referred to as the solvent *acidity*, or *electrophilicity*, effects. The mirror interactions involving the solvent molecules as hydrogen bond acceptors and the solute molecule as the donor of the hydrogen bond are related to the solvent *basicity*, or *nucleophilicity*. It has been recommended to use the first terms in these pairs (acidity and basicity) in the case of equilibrium processes whereas the second terms (electrophilicity and nucleophilicity) are preferable in the case of kinetic or dynamic processes.

The influence of the solvent on the rates and equilibria of chemical reactions was recognized already in the nineteenth century. The solvent effects on the rate of chemical reactions were first reported by Berthelot and Péan de Saint-Gilles [20,21] who noticed that the addition of some solvents to the reaction medium decelerate the esterification of acetic acid by ethanol. The influence of the solvent on the chemical equilibria was discovered in studies of the keto-enol tautomerism of 1,3-dicarbonyl compounds [22–24]. In his classical work on the reaction named after him, Menshutkin related the solvent effects on the rate of reaction to the chemical nature of the solvent [25,26].

One of the first LFER-type solvent effect treatments belongs to Grünwald and Winstein who proceeded from the observation that the S_N1 solvolysis of *tert*-butyl chloride is substantially accelerated by the polar and protic solvents [27].

$$(H_3C)_3C{-}Cl \longrightarrow \left[(H_3C)_3C^{\delta+}\cdots Cl^{\delta-}\right]^{\neq} \longrightarrow (H_3C)_2\overset{+}{C}{-}CH_3 + Cl^- \longrightarrow \text{products}$$

Accordingly, they defined a solvent ionizing power constant, Y, as:

$$Y = \log k_S - \log k_0 \tag{2.13}$$

where k_S is the rate constant of the solvolysis of *tert*-butyl chloride in a given solvent S and k_0 denotes the rate constant of this reaction in the standard solvent (80% aqueous ethanol) at 25°C. The LFER describing the solvent effect on some other reaction (A) can then be presented as:

$$\log k_S^A = \log k_0^A + mY \tag{2.14}$$

where k_S^A and k_0^A are the rate constants of this reaction in a given solvent S and in the standard solvent, and m denotes the sensitivity of the reaction to the solvent effect. It was shown later that the Y parameter involves, apart of the solvent polarity effects, the solvent polarizability and the electrophilicity terms, the latter being related to the electrophilic solvent assistance of the chloride ion elimination during the reaction [19]. Winstein himself extended Eq. (2.14) to account for the solvent nucleophilic assistance effects on the chemical reactions by introducing an additional term $\ell \cdot N$, where N is the nucleophilicity of the solvent and ℓ is the respective sensitivity coefficient. The nucleophilic (basic) solvation effects were also considered within the LFER formalism by Gutmann [28,29]. He provided an empirical scale of the Lewis basicity by defining a donor number (DN) of the electron pair donating solvents as the negative value of the molar enthalpy for adduct formation between antimony pentachloride and a given solvent. Another scale of solvent basicity was defined by Koppel and Paju [30] using the band shifts of the O—H stretching vibration of the phenol $\Delta\tilde{\upsilon}$ in tetrachloromethane, induced by the hydrogen bond formation with the added hydrogen-bonding accepting solvent S:

$$B \equiv \Delta\tilde{\upsilon} = \tilde{\upsilon}_{\mathrm{PhOH}}^{\mathrm{CCl_4}} - \tilde{\upsilon}_{\mathrm{PhOH\cdots S}}^{\mathrm{CCl_4}} \tag{2.15}$$

In general, the LFER equations of solvent effects should include terms accounting both for the nonspecific and specific solute–solvent interactions in solutions. A variety of multiparameter equations has been proposed including different solvent effect scales. One of the first such equations was suggested by Koppel and Palm [31,32] as:

$$P = P_0 + y \cdot Y + p \cdot P + e \cdot E + b \cdot B \tag{2.16}$$

where P and P_0 are the measured property values in a given solvent and in the standard solvent, respectively; $Y = (\varepsilon - 1)/(2\varepsilon + 1)$ is the solvent polarity parameter, $P = (n_D^2 - 1)/(2n_D^2 + 1)$ reflects the solvent polarizability, E is the solvent electrophilicity, and B is the basicity of the solvent. The coefficients y, p, e, and b are determined by the least-squares regression of the experimental property values P against the respective four solvent scales.

An alternative multiparameter approach to the solvent effects was proposed by Taft and Kamlet [33–37] who developed the so-called linear solvation energy relationship (LSER) as:

$$P = P_0 + s(\pi^* + d\delta) + a\alpha + b\beta + h\delta_H^2 + e\xi \tag{2.17}$$

In this equation, P and P_0 express again the measured property values in a given solvent and in the standard solvent, respectively, and π^* is an index of the solvent dipolarity/polarizability. The latter characterizes the solvent ability to stabilize the dipole or the charge of the solute by the dielectric reaction field of the solvent. It was shown that for a series of solvents with a dominant dipolar group, the last parameter is proportional to the total dipole moment of molecules. In Eq. (2.17), δ is a discontinuous polarizability correction term with the value $\delta = 0$ for the nonchlorine substituted aliphatic solvents, $\delta = 0.5$ for polychlorinated aliphatics, and $\delta = 1$ for aromatic solvents. In the same equation, α denotes the measure of the hydrogen bond donating ability of the solvent and corresponds to parameter E in Eq. (2.16). Descriptor β is the measure of the hydrogen bond accepting ability of the solvent similar to parameter B in Eq. (2.16), and δ_H^2 represents the square of the Hildebrandt solubility parameter. Parameter ξ has been introduced to correlate certain types of so-called family-dependent solute basicity properties.

It is, however, doubtful that any of the empirical solvent effect scales would correspond to a single mechanism of physical interactions between the solute and solvent molecules. This has caused an unnecessary confusion and misunderstanding between researchers as the same terms have been used for different experimentally derived solvent effect scales that involve the fundamental physical intermolecular interactions with different weights.

In the following, we proceed with the systematic presentation of different empirical molecular descriptors, together with the analysis of the interrelation between them and the limits of applicability.

2.2 STRUCTURAL DESCRIPTORS

2.2.1 Induction Constants

The induction or inductive effect is one of the fundamental terms in physical organic chemistry that describes the intramolecular interaction between the functional groups or molecular fragments in organic molecules [38]. Two different physical mechanisms have been considered as a cause for this effect [39–44]. No conclusive discrimination between them has been achieved, and thus we briefly describe both. The first mechanism of induction effect was suggested by Kirkwood and Westheimer [45–48], who formulated this effect as resulting from the electrostatic interaction between the charges and/or dipoles in the molecule. The polarization of a chemical bond between the atoms of different electronegativity was conceived as the cause for the formation of partial charges of opposite sign on these atoms and a respective local dipole moment of this bond. Thus, in the model of Kirkwood and Westheimer, the electrostatic interaction between the bond dipoles or the partial charges on atoms is considered the origin of the induction effect.

According to the second physical model, the induction effect is caused by the consecutive polarization of the bonds along the chain of chemically bonded atoms involving at least one bond between the atoms of different electronegativity [49]. The magnitude of this polarization depends on the distance from such bonds and the difference in the electronegativity of the respective atoms. Two groups of substituents can be distinguished according to the direction of the polarization along the chain of carbon atoms. First, the more electronegative substituents X as compared to the carbon atom induce the positive atomic partial charges on the carbon atoms.

$$C^{\delta^{(3)+}}-C^{\delta^{(2)+}}-C^{\delta^{(1)+}}-X^{\Delta+}$$

By convention, these substituents are denoted as the −I groups. The inductive effect fades with distance from the electronegative group and, consequently, $\delta^{(1)+} > \delta^{(2)+} > \delta^{(3)+}$. The majority of functional groups in organic compounds (halogens, carbonyl-, hydroxyl-, amino-, amide-, nitro-, cyano-groups, etc.) possess the −I inductive effect. The alkyl groups, R, are assumed to have an opposite effect by creation of negative atomic partial charges along the carbon chain.

$$C^{\delta^{(3)-}}-C^{\delta^{(2)-}}-C^{\delta^{(1)-}}-R^{\Delta-}$$

The methyl group has been often taken as a standard for which the bond polarization along the chain and the respective inductive effect is postulated to be zero. According to this definition, the longer and more branched hydrocarbon radicals possess the increasing +I effect in the following order:

$$-CH_2-CH_3 < -CH_2-CH_2-CH_3 < -CH(CH_3)_2 < -C(CH_3)_3$$

This definition of the induction effect for alkyl substituents has been disputed by Charton who ascribed the respective change in the free energy of a standard process to the difference in steric repulsion by such substituents [50,51]. Nevertheless, the induction constants for alkyl groups are still used together with the induction constants for electronegative groups in the same LFER correlations.

The unique standardization of the quantitative scale for the induction effect has been also problematic. First, it has to be mentioned that neither the electrostatic nor the bond polarization model of the induction effect can be reduced to fundamental physical interactions. The total energy, E, of a molecule consisting of M nuclei and n electrons is quantum mechanically determined as the solution of the respective Schrödinger equation:

$$\hat{H}\Psi = \left[-\sum_{a=1}^{M} \frac{\nabla_a^2}{2M_a} - \sum_{i=1}^{n} \frac{\nabla_i^2}{2} - \sum_{a=1}^{M} \sum_{i=1}^{n} \frac{Z_a}{r_{ia}} + \frac{1}{2} \sum_{a=1}^{M} \sum_{b=1}^{M} \frac{Z_a Z_b}{r_{ab}} + \frac{1}{2} \sum_{i=1}^{n} \sum_{j=1}^{n} \frac{1}{r_{ij}} \right] \Psi = E\Psi \quad (2.18)$$

This energy can be partitioned into separate contributions, corresponding to individual terms of Hamiltonian $\hat{H}$ in Eq. (2.18). Those include, in the following order, the kinetic energy of nuclei, the kinetic energy of electrons, the electron–nuclear electrostatic attraction, the nuclear–nuclear electrostatic repulsion, and the electron–electron electrostatic repulsion energy, respectively. In Eq. (2.18) the r_{ia}, r_{ab}, and r_{ij} correspond to the distances between the electrons and nuclei, to the internuclear distances and to the interelectronic distances, respectively. The term Ψ denotes the total wave function of the molecule. The induction effect is expected to influence each of the energy terms given in Eq. (2.18), by the perturbation caused with variable substituents in the molecule. The effect on each individual term, however, cannot be scaled out. Therefore, the quantitative measure of the induction effect cannot be determined theoretically, by inspection of the respective terms in the mathematical expression of the quantum mechanical total energy of the molecule. Accordingly, the absolute magnitude of the induction effect is unknown, and thus it has been conventionally defined with respect to some standard level of energy.

The use of different standards for the quantitative definition of induction effect has resulted in numerous empirical scales for this effect. It is evident that the response of different chemical or physical characteristics of a molecule (e.g., the energies of spectral transitions, the rate of chemical reactions, or the position of chemical equilibria) to the inductive effect would be different. In most cases, it is also difficult to prove that other intermolecular (mesomeric, steric) effects do not influence the standard process. Therefore, one reason for the differences in different induction effect scales is the systematic error due to the presence of other effects of unknown extent. In addition, different standard processes take place in different environment, e.g., in different media or at different temperatures. These external factors can also substantially influence the intermolecular induction effects. This influence may be specific for each molecular structure, and thus the substituent effects may be affected differently. An additional systematic deviation of one induction effect scale from another may arise from that reason.

In conclusion, obviously no general empirical induction effect scale is available for every process. Two remarks are essential on this point. First, it is advisable to use the scale of induction effect that is more closely related to the phenomenon studied by the LFER analysis. Second, "bare" induction effect that is not disturbed by surrounding medium should be observed in the case of standard processes occurring in the gas phase at low pressure (e.g., the gas-phase proton affinities).

The original induction effect scales, Hammett's σ and Taft's σ^*, were defined on the basis of standard processes in aqueous solutions [cf. Eqs. (2.6) and (2.10)]. Because of the difficulties with the solubility of many substituted benzoic acids in water, the original σ scale was soon extended using the pK_a of benzoic acids in mixed solvents. In addition, other processes for which the log k or pK is linearly related to the pK_a of benzoic acids in aqueous solutions had been used for the definition of induction constants. For example, it was noticed that the pK_a of acids XCH_2COOH are linearly correlated with the original σ^* constants and thus applicable for the further extension of the induction effect scale in aliphatic and alicyclic systems. In fact, a new scale

$$\sigma_I^{(X)} = pK_a^{(XCH_2COOH)} - pK_a^{(CH_3COOH)} \tag{2.19}$$

was defined on the basis of the acidic dissociation constants of substituted acetic acids [52]. The similar σ_I constants have been developed also for heterocyclic systems [53].

As discussed, the original Hammett σ constants determined from the dissociation constants of substituted benzoic acids included, at least for the electron-donating mesomeric groups in para-position, a contribution related to the direct mesomeric interaction through the aromatic cycle; for example:

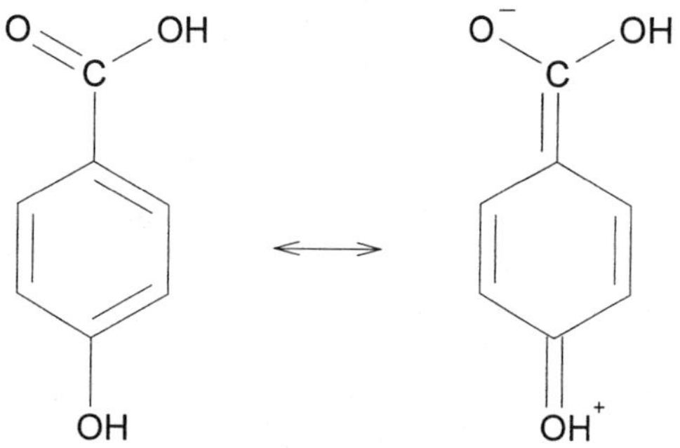

O OH C OH ⟷ O⁻ OH C OH⁺

The comparison of the σ and σ_I constants revealed that even in the case the meta-substituents the two scales are poorly correlated. One possible reason for this discrepancy is the possible mesomeric conjugation between the electronegative substituents and the aromatic phenyl ring. Thus, a new scale for the induction effect in aromatic systems (σ^0) was defined on the basis of the following relation:

$$\sigma^0_{(XC_6H_4)} = \sigma^*_{(XC_6H_4)} - 0.600 \tag{2.20}$$

where $\sigma^*_{(XC_6H_4)}$ is determined from the pK_a of phenyl-substituted acetic acids [54,55]. In result, these constants have the same scaling as the σ^* constants for the aliphatic substituents. The scale of σ^0 constants has been extended using the rate constants of the hydrolysis of substituted phenylacetic acids [56] and by applying a general statistical treatment of similar data on the reactivity of substituted phenyl systems [57]. In each of these systems, the reaction center was separated from the aromatic ring by the methylene group, which disabled the possible mesomeric conjugation with the substituent in the ring.

To eliminate the possible mesomeric effects, several other standard reactions have been employed for the definition of the induction constants in cyclic systems. Those include the pK_a of dissociation 4-substituted bicyclo[2,2,2]octane carboxylic acids (**I**) [58] and the pK_a of protonation of 4-substituted quinuclidines (**II**) [59–61].

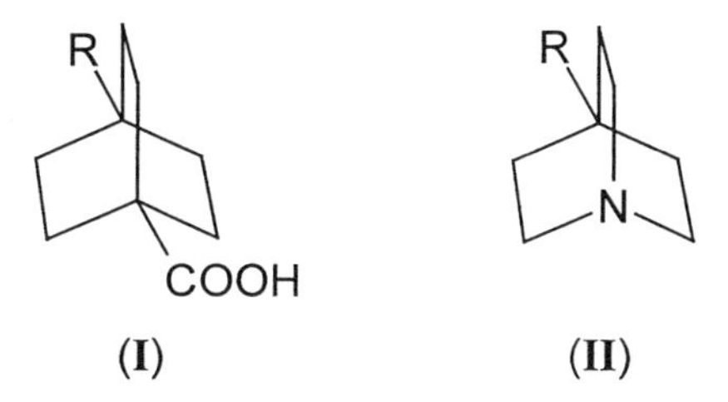

(**I**) (**II**)

In these systems, the substituent R is geometrically rigidly fixed with respect to the reaction center and the possible mesomeric interaction between these two groups has been disabled because of aliphatic —CH_2—CH_2— bridges between them. Table 2.1 lists various substituent induction constants together with reference to the standardization process used. Table 2.2 presents the numerical values of some common induction constants for a selection of substituents [62,63,63a].

Two principles have been employed in the calculation of the induction constants of complex substituents involving several electronegative groups, aliphatic chains, and multiple substitutions in aromatic rings. First, the induction effect is assumed to be additive, that is, the induction constant σ_I for a multiply substituted radical can be added to the σ_I constants of the respective substituent groups [64–66]:

$$\sigma_I^{(-\mathrm{CX_1X_2X_3})} = \sum_{i=1}^{3} \sigma_I^{(-\mathrm{CH_2X}_i)} \tag{2.21}$$

It has been argued, however, that in the case of several strong electron-withdrawing groups (e.g., Cl, F) at the same carbon atom, the additivity of the substituent constants can break down, leading to a "saturation" of the induction effect. Originally, the induction effect by several substituents in an aromatic system (e.g., in the phenyl ring) has been also assumed to be additive [67]. However, it was later shown that this additivity does not hold strictly [68,69]. An alternative "experimental" scale of Hammett σ constants for multiply substituted phenyl groups has been suggested by Hansch and others [70]. The Hammett σ constants are also expected to be position dependent. Nevertheless, a strong collinearity has been reported between the σ constants for the same substituent at the para- and the meta-position of phenyl ring, σ_m and σ_p, respectively [62], expressed by the following equation:

$$\sigma_p = (0.08 \pm 0.02) - (1.19 \pm 0.04)\sigma_m$$
$$R^2 = 0.885 \qquad s = 0.137 \qquad n = 530$$

The second principle applied to the induction interaction asserts that the effect

TABLE 2.1 Substituent Induction Constant Scales in Aromatic and Aliphatic Systems, Defined by Using Different Standard Processes

Scale Notation	Standard Process	Reference
σ	pK_a of substituted benzoic acids in aqueous solutions	*a*
σ^*	Difference in log k of acidic and alkaline hydrolysis of esters	*b*
σ^0	pK_a of substituted phenylacetic acids and log k of hydrolysis of phenylacetates	*c,d*
$\bar{\sigma}^0$	log k of alkaline hydrolysis of substituted aryl tosylates	*e*
σ'	pK_a of substituted bicyclooctane carboxylic acids	*f*
σ_{QN}	pK_a of substituted quinuclidines	*g*
$\sigma_F(\sigma_I)$	Averaging over several processes	*h*
σ_I	pK_a of substituted acetic acids XCH_2COOH	*i*
σ_F	^{19}F NMR chemical shifts of meta-substituted fluorobenzenes	*j*
σ_I	pK_a of *N*-substituted guanidinium ions $XHN{=}C(NH_2)_2$	*k*
F	pK_a of substituted bicyclooctane carboxylic acids	*l*
σ^n	Averaging over several processes	*m*
σ_Φ	pK_a of substituted hypophosphorous acids	*n*
σ_χ	Quantum chemically calculated partial charge on hydrogen atom in compound HX	*o*
i	Group electronegativity scale	*p*
σ_α	Electrostatic polarization potential calculations	*q*
σ^*_{theor}	Additive increments for individual atoms	*r*

[a] L.P. Hammett, *Chem. Rev.* **17,** 125 (1935).
[b] R.W. Taft, in *Steric Effects in Organic Chemistry* (M.S. Newman, ed.) p. 556. Wiley, New York, 1956.
[c] R.W. Taft, S. Ehrenson, I.C. Lewis, and R.E. Glick, *J. Am. Chem. Soc.* **81,** 5352 (1959); R.W. Taft, *J. Phys. Chem.* **64,** 1805 (1960).
[d] Y. Yukawa, Y. Tsuno, and M. Sawada, *Bull. Chem. Soc. Jpn.* **39,** 2274 (1966); **45,** 1198 (1972).
[e] V.M. Maremäe and V.A. Palm, *Reakts. Sposobn. Org. Soedin.* **1,** 85 (1964).
[f] J.D. Roberts and W.T. Moreland, Jr., *J. Am. Chem. Soc.* **75,** 2167 (1953); H.D. Holtz and L.M. Stock, *ibid.* **86,** 5188 (1964).
[g] R.W. Taft and C.A. Grob, *J. Am. Chem. Soc.* **96,** 1236 (1974); C.A. Grob and M.G. Schlageter, *Helv. Chim. Acta.* **59,** 264 (1976); C.A. Grob, *ibid.* **68,** 882 (1985).
[h] R.W. Taft and R.D. Topsom, *Prog. Phys. Org. Chem.* **16,** 1 (1987).
[i] M. Charton, *Prog. Phys. Org. Chem.* **13,** 119 (1981).
[j] R.W. Taft, E. Price, I.R. Fox, I.C. Lewis, K.K. Andersen, and G.T. Davis, *J. Am. Chem. Soc.* **85,** 709 (1963); P.R. Wells, S. Ehrenson, and R.W. Taft, *Prog. Phys. Org. Chem.* **6,** 147 (1968).
[k] P.J. Taylor and A.R. Wait, *J. Chem. Soc., Perkin Trans. 2,* p. 1765 (1986).
[l] C.G. Swain and E.C. Lupton, *J. Am. Chem. Soc.* **90,** 4328 (1968); C.G. Swain, S.H. Unger, N.R. Rosenquist, and M.S. Swain, *ibid.* **105,** 492 (1983).
[m] H. van Bekkum, P.E. Verkade, and B.M. Wepster, *Recl. Trav. Chim.* **78,** 815 (1959).
[n] M.I. Kabachnik, *Dokl. Akad. Nauk SSSR* **110,** 393 (1956).
[o] S. Marriott, W.F. Reynolds, R.W. Taft, and R.D. Topsom, *J. Org. Chem.* **49,** 959 (1984).
[p] N. Inamoto and S. Masuda, *Tetrahedron Lett.,* p. 3237 (1977); N. Inamoto, K. Tori, and Y. Yoshimura, *ibid.*, p. 4547 (1978).
[q] W.J. Hehre, C.-F. Pau, A.D. Headley, R.W. Taft, and R.D. Topsom, *J. Am. Chem. Soc.* **108,** 1711 (1986); P. Čarsky, P. Nauš, and O. Exner, *J. Phys. Org. Chem.* **11,** 485 (1998).
[r] A.R. Cherkasov, V.I. Galkin, and R.A. Cherkasov, *J. Phys. Org. Chem.* **11,** 437 (1998).

TABLE 2.2 Numerical Values for Selection of Induction Effect Substituent Constants in Aliphatic and Aromatic Molecular Systems[a] [62,63,63a]

Substituent	σ_m	σ_p	σ^*	σ_I	F
H	0.00	0.00	0.49	0.00	0.03
CH_3	−0.07	−0.17	0.00	−0.04	0.01
CH_2CH_3	−0.07	−0.15	−0.10	−0.01	0.00
$CH_2CH_2CH_3$	−0.06	−0.13	−0.12	−0.01	0.01
$CH(CH_3)_2$	−0.04	−0.15	−0.19	0.01	0.04
Cyclopropyl	−0.07	−0.21	0.04	0.01	0.02
$(CH_2)_3CH_3$	−0.08	−0.16	−0.13	−0.04	−0.01
$CH_2CH(CH_3)_2$	−0.07	−0.12	−0.13	−0.03	−0.01
$CH(CH_3)C_2H_5$	−0.08	−0.12	−0.21	−0.03	−0.02
$C(CH_3)_3$	−0.10	−0.20	−0.30	−0.07	−0.02
Cyclobutyl	−0.05	−0.14	−0.15		0.02
$(CH_2)_4CH_3$	−0.08	−0.15	−0.16	−0.03	−0.01
$CH_2C(CH_3)_3$	−0.05	−0.17	−0.17	−0.07	0.03
$C(C_2H_5)(CH_3)_2$	−0.06	−0.18	−0.33	−0.08	0.03
Cyclopentyl	−0.05	0.14	−0.20		0.02
Cyclohexyl	−0.05	−0.15	−0.18	−0.02	0.03
$(CH_2)_6CH_3$	−0.77	−0.16	−0.37	−0.04	0.00
$CH{=}CH_2$	0.06	−0.04	0.59	0.11	0.13
$CH_2CH{=}CH_2$	−0.11	−0.14	0.12	0.02	−0.06
trans-$CH{=}CHCH_3$	0.02	−0.09	0.36	0.07	0.09
$C(CH_3){=}CH_2$	0.09	0.05	0.48	0.10	0.13
C_6H_5	0.06	−0.01	0.60	0.12	0.12
$CH_2C_6H_5$	−0.08	−0.09	0.22	−0.08	−0.04
$CH(C_6H_5)_2$	−0.03	−0.05	0.41	0.19	0.01
$CH{=}CHC_6H_5$	0.03	−0.07	0.41	0.02	0.10
$C{\equiv}CC_6H_5$	0.14	0.16	1.35	0.14	0.15
$CH_2CH_2C_6H_5$	−0.07	−0.12	0.08	0.02	−0.01
$C{\equiv}CH$	0.21	0.23	2.15	0.29	0.22
F	0.34	0.06	3.19	0.52	0.45
Cl	0.37	0.23	2.94	0.47	0.42
Br	0.39	0.23	2.80	0.44	0.45
I	0.35	0.18	2.22	0.39	0.42
OH	0.12	−0.37	1.37	0.29	0.33
NH_2	−0.16	−0.66	0.62	0.12	0.08
NH_3^+	0.86	0.60	3.61	0.61	0.92
NO_2	0.71	0.78	4.66	0.64	0.65
SH	0.25	0.15	1.52	0.26	0.30
S^-	−0.36	−1.21	−0.42		0.03
SCl	0.44	0.48	2.50	0.40	0.42
PH_2	0.06	0.05		0.06	0.09
SiH_3	0.05	0.10	0.40	0.09	0.06
$SiCl_3$	0.48	0.56	1.77	0.39	0.44
$SiBr_3$	0.48	0.57	2.40	0.39	0.44
GeH_3	0.00	0.01	0.70		0.03
$GeCl_3$	0.71	0.79	3.90	0.63	0.65

TABLE 2.2 ***(Continued)***

Substituent	σ_m	σ_p	σ^*	σ_I	F
$GeBr_3$	0.66	0.73	3.70	0.59	0.61
OCH_3	0.12	−0.27	1.77	0.27	0.29
OCH_2CH_3	0.10	−0.24	1.68	0.28	0.26
$OCH_2CH_2CH_3$	0.10	−0.25	1.57	0.28	0.26
$OCH(CH_3)_2$	0.10	−0.45	1.51		0.34
$O(CH_2)_3CH_3$	0.10	−0.32	1.55	0.28	0.29
$O(CH_2)_4CH_3$	0.10	−0.34	1.52		0.29
$OCH{=}CH_2$	0.21	−0.09		0.38	0.34
OC_6H_5	0.25	−0.03	2.24	0.40	0.37
OCH_2F	0.20	0.02	2.31	0.37	0.29
OCH_2Cl	0.25	0.08	2.58	0.41	0.33
$OCHF_2$	0.31	0.18	2.81	0.36	0.37
$OCHCl_2$	0.38	0.26	3.06	0.49	0.43
OCF_3	0.38	0.35		0.39	0.39
$OCCl_3$	0.43	0.35	3.19	0.51	0.46
$OCOCH_3$	0.39	0.31	2.45	0.33	0.42
$OCOC_6H_5$	0.21	0.13	2.40	0.43	0.26
OCN	0.67	0.54	5.00	0.80	0.69
ONO_2	0.55	0.70	3.77	0.66	0.48
OSO_2CH_3	0.39	0.36	3.62	0.55	0.40
$OSO_2C_6H_5$	0.36	0.33	3.62		0.37
OSO_2CF_3	0.56	0.53	4.37	0.84	0.56
$OP(O)(C_3F_7)_2$	0.66	0.56		0.77	0.67
$NHCH_3$	−0.21	−0.70	0.94	0.13	0.03
$N(CH_3)_2$	−0.16	−0.83	1.02	0.06	0.15
$N^+(CH_3)_3$	0.88	0.82	4.16	0.93	0.86
$NH(CH_2)_3CH_3$	−0.34	−0.51	1.08		−0.21
$N(C_2H_5)_2$	−0.23	−0.72	1.00	0.02	0.01
NHC_6H_5	−0.02	−0.56		0.02	0.22
$N(C_6H_5)_2$	0.00	−0.22		0.11	0.12
NHOH	−0.04	−0.34	0.30	0.12	0.11
$NHNH_2$	−0.02	−0.55	0.40	0.14	0.22
NHCHO	0.19	0.00	1.62	0.33	0.28
$NHCOC_6H_5$	0.02	−0.19	1.68	0.28	0.13
$NHCOCH_2Cl$	0.17	−0.03	2.06	0.35	0.27
$NHCOOC_2H_5$	0.11	−0.15	1.99	0.28	0.23
$NHCSNH_2$	0.22	0.16	1.80	0.29	0.26
$NHSO_2C_6H_5$	0.16	0.01	1.99	0.33	0.24
$NHSO_3CF_3$	0.44	0.39	3.10	0.49	0.45
$N(COCH_3)_2$	0.35	0.33	2.31	0.37	0.36
$N(CF_3)_2$	0.40	0.53	3.10	0.49	0.35
$N(CH_3)COCH_3$	0.31	0.26	2.25	0.36	0.34
$N(COF)_2$	0.58	0.57	3.60	0.58	0.57
$N(CH_3)SO_2CH_3$	0.21	0.24	2.10	0.34	0.21
$N(CH_3)SO_2CF_3$	0.46	0.44	3.00	0.48	0.46
$N(SO_2CH_3)_2$	0.47	0.49	2.80	0.45	0.45

TABLE 2.2 (*Continued*)

Substituent	σ_m	σ_p	σ^*	σ_I	F
$N(SO_2CF_3)_2$	0.61	0.83	4.40	0.70	0.50
$N(CF_3)C{=}O(F)$	0.50	0.50	3.50	0.50	0.49
$NHP(O)(C_3F_7)_2$	0.28	0.18		0.39	0.33
CHO	0.35	0.42	2.15	0.27	0.33
$COCH_3$	0.38	0.50	1.65	0.30	0.33
COC_2H_5	0.38	0.48	1.61		0.34
$COC(CH_3)_3$	0.27	0.32	1.45	−0.01	0.26
COC_6H_5	0.34	0.43	2.20	0.20	0.31
COF	0.55	0.70	2.44	0.39	0.48
COCl	0.51	0.61	2.37	0.44	0.46
COOH	0.37	0.45	2.94	0.30	0.34
COO^-	−0.10	0.00	0.75	−0.19	−0.10
$COOCH_3$	0.37	0.45	2.00	0.32	0.34
$COOC_2H_5$	0.37	0.45	2.26	0.21	0.34
$COOC_6H_5$	0.37	0.44		0.42	0.34
$CONH_2$	0.28	0.36	1.66	0.28	0.26
$CONHC_6H_5$	0.23	0.41		0.25	0.17
$COCF_3$	0.63	0.80	3.70	0.45	0.54
CN	0.56	0.66	3.64	0.53	0.51
CH_2F	0.12	0.11	1.10	0.12	0.15
CH_2Cl	0.11	0.12	1.01	0.17	0.13
CH_2Br	0.12	0.14	1.00	0.12	0.14
CH_2I	0.10	0.11	0.85	0.17	0.12
CHF_2	0.29	0.32	2.05	0.29	0.29
$CHCl_2$	0.31	0.32	2.01	0.30	0.31
$CHBr_2$	0.31	0.32	1.95	0.30	0.31
CHI_2	0.26	0.26	1.62	0.26	0.27
CF_3	0.43	0.54	2.56	0.40	0.38
CCl_3	0.40	0.46	2.65	0.31	0.38
CBr_3	0.28	0.29	2.43	0.26	0.28
CH_2CF_3	0.12	0.09	0.87	0.16	0.15
CH_2OH	0.00	0.00	0.56	0.11	0.03
CH_2OCH_3	0.08	0.01	0.52	0.11	0.13
$CH_2OC_6H_5$	0.06	0.07	0.85		0.08
CH_2OCOCH_3	0.04	0.05	0.70	0.15	0.07
CH_2NH_2	−0.03	−0.11	0.40	0.00	0.04
$CH_2NH_3^+$	0.59	0.53	2.24	0.36	0.59
$CH_2N(CH_3)_2$	0.00	0.01	0.24	0.05	0.03
$CH_2NH^+(CH_3)_2$	0.40	0.43	1.04		0.39
$CH_2N(CH_3)_3^+$	0.40	0.44	1.90	0.25	0.38
$CH_2NHCOCH_3$	0.05	−0.05	0.43	0.09	0.12
CH_2COO^-	0.07	−0.16	−0.06	0.14	0.19
CH_2CONH_2	0.06	0.07	0.65	0.06	0.08
CH_2CN	0.16	0.18	1.30	0.20	0.17
$CH_2SO_2C_6H_5$	0.15	0.16	1.37		0.17
CH_2SOCF_3	0.25	0.24	1.62	0.26	0.27

TABLE 2.2 ***(Continued)***

Substituent	σ_m	σ_p	σ^*	σ_I	F
$CH_2SO_2CF_3$	0.29	0.31	1.75	0.28	0.29
CH_2SCF_3	0.12	0.15	0.75	0.16	0.13
CH_2SCN	0.12	0.14	1.18		0.14
$CH_2Si(CH_3)_3$	−0.16	−0.21		−0.07	−0.09
$CH_2PO(OC_2H_5)_2$	0.12	0.06	0.78	0.13	0.17
$CH_2Mn(CO)_5$	−0.14	−0.44			0.02
$CH(OH)CH_3$	0.08	−0.07	0.46	0.04	0.16
$CH(OH)C_6H_5$	0.00	−0.03	0.76	0.10	0.05
$CH(SCF_3)_2$	0.44	0.44	2.75	0.44	0.43
$CH(CN)_2$	0.53	0.52	3.40	0.55	0.52
$C(OCH_3)_3$	−0.03	−0.04		−0.12	0.01
$C(CN)_3$	0.97	0.96		0.98	0.92
$C(OH)(CF_3)_2$	0.29	0.30	1.75	0.28	0.29
$C(CH_3)(CN)_2$	0.60	0.57	6.94	0.63	0.59
$C(NO_2)(CH_3)_2$	0.18	0.20	1.10		0.19
$C(CH_3)(NO_2)_2$	0.54	0.61	3.36		0.50
$C(C_2H_5)(NO_2)_2$	0.56	0.64	3.35		0.51
$C(NO_2)_3$	0.72	0.82	4.60		0.65
$C(SCF_3)_3$	0.51	0.53	3.06	0.49	0.49
CF_2CF_3	0.47	0.52	2.56	0.41	0.44
$CF_2CF_2CF_3$	0.44	0.48	2.83	0.39	0.42
$CF(CF_3)_2$	0.37	0.53	3.00	0.48	0.31
$C(CF_3)_3$	0.55	0.55		0.27	0.52
$(CF_2)_3CF_3$	0.47	0.52	2.44	0.39	0.44
CH_2CH_2COOH	−0.03	−0.07	0.35		0.02
$CH_2CH_2Si(CH_3)_3$	−0.16	−0.17	−0.25	−0.04	−0.11
$CH_2CH(OH)CH_3$	−0.12	−0.17	0.16	0.03	−0.06
$CH_2C(OH)(CH_3)_2$	0.20	0.26	1.41		0.19
$CH{=}CHCOCH_3$	0.21	−0.01	1.08		0.31
$CH{=}C(CN)_2$	0.66	0.84		0.43	0.57
trans-$CH{=}CHNO_2$	0.32	0.26	1.75	0.24	0.35
$CH{=}CHSO_2CF_3$	0.31	0.55	1.94	0.34	0.22
$C{\equiv}CCF_3$	0.41	0.51	1.94	0.35	0.37
2-Furyl	0.06	0.02	1.08	0.04	0.10
2-Thienyl	0.09	0.05		0.21	0.13
3-Thienyl	0.03	−0.02	0.65		0.08
Cyclo-C_4F_7	0.48	0.53	2.81	0.45	0.45
C_6F_5	0.26	0.27	2.00	0.31	0.27
C_6Cl_5	0.25	0.24	1.56	0.25	0.27
$N{=}NC_6H_5$	0.32	0.39	1.87	0.25	0.30
$N{=}CCl_2$	0.21	0.13	1.81	0.29	0.26
$N{=}C{=}O$	0.27	0.19	2.25	0.36	0.31
$N{=}C{=}S$	0.48	0.38	2.62	0.42	0.51
$N{=}C(CF_3)_2$	0.29	0.23	2.20	0.35	0.32
SCH_3	0.15	0.00	1.66	0.25	0.23
$S^+(CH_3)_2$	1.00	0.90	5.09	0.89	0.98

TABLE 2.2 (*Continued*)

Substituent	σ_m	σ_p	σ^*	σ_I	F
SC_2H_5	0.18	0.03	1.44	0.25	0.26
$SCH(CH_3)_2$	0.23	0.07	1.56	0.26	0.30
$SCH{=}CH_2$	0.26	0.20	1.31	0.21	0.29
$SCH_2CH{=}CH_2$	0.19	0.12	1.45		0.23
$SC{\equiv}CH$	0.26	0.19	2.00	0.32	0.30
SC_6H_5	0.23	0.07	1.87	0.31	0.30
SCH_2F	0.23	0.2	1.67	0.27	0.25
$SCHF_2$	0.33	0.37	2.06	0.33	0.32
SCF_3	0.40	0.50	2.70	0.42	0.36
$S(OCH_3)$	0.21	0.17	1.56	0.25	0.24
$SN(CH_3)_2$	0.12	0.09	0.94	0.15	0.15
SCN	0.51	0.52	3.43	0.64	0.49
$SCOCH_3$	0.39	0.44	2.29	0.21	0.37
$SCOCF_3$	0.48	0.46	3.19	0.51	0.48
$S(O)CH_3$	0.52	0.49	2.88	0.49	0.52
$S(O)C_6H_5$	0.50	0.44	3.24	0.51	0.51
S(O)F	0.74	0.83	4.12	0.66	0.67
$S(O)CF_3$	0.63	0.69	4.30	0.64	0.58
$S(O)OCH_3$	0.50	0.54	2.84	0.45	0.47
$S(O)CHF_2$	0.54	0.58	4.10	0.65	0.51
$SO_2C_2H_5$	0.66	0.77	3.74	0.60	0.59
$SO_2C_6H_5$	0.62	0.68	3.25	0.56	0.58
SO_2CHF_2	0.75	0.86	3.69	0.75	0.67
SO_2CH_3	0.83	0.96	4.32	0.78	0.74
SO_2NH_2	0.53	0.60	2.61	0.46	0.49
$SO_2N(CH_3)_2$	0.51	0.65	2.62	0.42	0.44
SO_2CN	1.10	1.26	5.90	0.94	0.97
$P(CH_3)_2$	0.03	0.06		0.08	0.05
$P^+(CH_3)_3$	0.74	0.73	2.50	0.56	0.71
$P(C_2H_5)_2$	0.10	0.13		−0.04	0.11
$P(C_6H_5)_2$	0.11	0.19	1.06	0.05	0.10
$P^+(CH_3)(C_6H_5)_2$	1.13	1.18		0.60	1.04
$P(N(CH_3)_2)_2$	0.18	0.25		0.00	0.17
$P(CN)_2$	0.82	0.90	4.60	0.58	0.75
$P(CF_3)_2$	0.60	0.69	3.12	0.50	0.55
PF_4	0.63	0.80	2.80	0.45	0.54
$P(O)(CH_3)_2$	0.43	0.50		0.31	0.40
$P(O)(C_2H_5)_2$	0.37	0.47		0.28	0.33
$P(O)(C_4H_9)_2$	0.35	0.49		0.24	0.30
$P(O)(C_6H_5)_2$	0.38	0.53		0.26	0.32
$P(O)(OCH_3)_2$	0.42	0.53		0.36	0.37
$P(O)(OC_2H_5)_2$	0.55	0.60		0.30	0.52
$P(O)(OC_4H_9)_2$	0.41	0.57	1.77		0.35
$P(O)(N(CH_3)_2)_2$	0.30	0.40		0.16	0.27
$P(O)(C_3F_7)_2$	0.95	1.10		0.79	0.84
$P(S)(C_2H_5)_2$	0.39	0.46		0.31	0.36

TABLE 2.2 (*Continued*)

Substituent	σ_m	σ_p	σ^*	σ_I	F
$P(S)(C_6H_5)_2$	0.29	0.47		0.15	0.23
$P(S)(C_6H_5)(4\text{-}CH_3\text{-}C_6H_4)$	0.09	0.30		0.25	0.03
$SiH(CH_3)_2$	0.01	0.04	−0.60	0.07	0.03
$Si(CH_3)_3$	−0.04	−0.07	−0.81	−0.11	0.01
$Si(CH_3)(C_6H_5)_2$	0.10	0.13	0.19	0.07	0.11
$Si(OCH_3)_3$	0.09	0.13	0.02	0.01	0.10
$Si(OC_2H_5)_3$	0.02	0.08		−0.04	0.03
$Si(CH_3)_2OSi(CH_3)_3$	0.00	−0.01	−0.81	−0.13	0.04
$B(OH)_3^-$	−0.48	−0.44		−0.36	−0.42
$As(C_2H_5)_2$	0.22	0.00		0.28	0.32
$As(C_6H_5)_2$	0.03	0.09		−0.01	0.04
$As(O)(C_2H_5)_2$	0.57	0.44		0.63	0.60
$As(O)(C_6H_5)_2$	0.54	0.64		0.43	0.49
$As(S)(C_2H_5)_2$	0.52	0.44		0.55	0.54
SeCN	0.61	0.66	3.61	0.58	0.57
$SeCF_3$	0.44	0.45	2.62	0.28	0.43
$SeOCF_3$	0.81	0.83	4.75	0.76	0.76
HgF	0.34	0.33	2.10	0.18	0.35
HgCN	0.28	0.34	1.40	0.23	0.27
$HgCF_3$	0.29	0.32	1.70	0.27	0.29
$HgSCF_3$	0.39	0.42	2.30	0.37	0.38
$HgOCOCH_3$	0.39	0.4	2.40	0.38	0.39
$HgOCOCF_3$	0.50	0.52	3.00	0.48	0.48
$Sn(CH_3)_3$	0.00	0.00	−1.29	−0.11	0.03
$Sn(C_2H_5)_3$	0.00	0.00		0.02	0.03
IF_2	0.85	0.83	5.40	0.86	0.82
IF_4	1.07	1.15	6.30	1.00	0.98
$I(OCOCH_3)_2$	0.85	0.88	5.10	0.82	0.80
$I(OCOCF_3)_2$	1.28	1.34	7.60	1.22	1.18
$4\text{-}C_2H_5\text{-}C_6H_4$	0.07	−0.02	0.47	0.10	0.13
$3\text{-}F\text{-}C_6H_4$	0.15	0.10	0.82	0.16	0.19
$4\text{-}F\text{-}C_6H_4$	0.12	0.06	0.62	0.13	0.17
$3\text{-}Cl\text{-}C_6H_4$	0.15	0.10	0.85		0.19
$4\text{-}ClC_6H_4$	0.15	0.12	0.75		0.18
$3\text{-}Br\text{-}C_6H_4$	0.09	0.08	0.86		0.12
$4\text{-}Br\text{-}C_6H_4$	0.15	0.12	0.74		0.18
$3\text{-}I\text{-}C_6H_4$	0.13	0.06	0.90	0.16	0.18
$4\text{-}IC_6H_4$	0.14	0.10	0.87	0.15	0.18
$4\text{-}OCH_3\text{-}C_6H_4$	0.05	−0.08	0.36	0.11	0.13
$3\text{-}NO_2\text{-}C_6H_4$	0.21	0.20	1.09	0.19	0.23
$4\text{-}NO_2\text{-}C_6H_4$	0.25	0.26	1.14	0.23	0.26
$2,4,6\text{-}(NO_2)_3\text{-}C_6H_2$	0.26	0.30			0.26

[a] σ_m, Hammett σ constant for meta-substituted phenyls; σ_p, Hammett σ constant for para-substituted phenyls; σ^*, Taft's σ constant for aliphatic substituents; σ_I, DSP σ_I constant; F, Swain–Lupton field effect parameter.

caused by the terminal group of an atomic chain will fade by a constant factor at each atom of the chain. This transmission factor $z_I(z^*)$ for a carbon atom (more precisely, for a $—CH_2—$ group) was calculated from the data on several experimental series. Typically, the reactivity of compounds $X(CH_2)_{n+1}Y$ was plotted against the reactivity of compounds $X(CH_2)_nY$.

$$\log K[X(CH_2)_{n+1}Y] = \log K_0 + z_I \log K[X(CH_2)_nY] \tag{2.22}$$

As an example, the plot of $pK_a(XCH_2CH_2COOH)$ against the pK_a (XCH_2COOH) is presented on Figure 2.1. It was noticed that the slopes of such linear relationships are different for the electronegative and for the alkyl substituents, respectively. Accordingly, it has been suggested that the mechanisms of the inductive effect have to be different for the alkyl and for the electronegative substituents. The average transmission coefficients $z_I^{CH_2}$(alkyl) = 0.204 and $z_I^{CH_2}$(electronegative) = 0.388 are suggested for these two groups of substituents, respectively [71]. The transmission coefficients have also been calculated for some other, heteroatom and cyclic bridge fragments (cf. Table 2.3). It can be noticed, however, that the transmission coefficient values for some cyclic fragments, obtained from different reaction series, are rather disperse and thus not quite reliable for the estimation of induction constants of these systems [71].

Most physical interactions can be described by the following equation for the energy of interaction of two objects, A and B:

$$E_{AB} = \alpha_i x_A x_B \tag{2.23}$$

where α_i is a universal characteristic for a given type of interaction, and x_A and

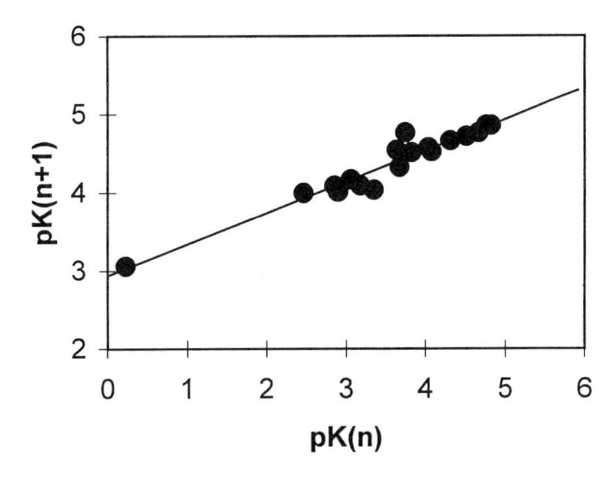

Figure 2.1 Linear relationship between the pK_a of aliphatic carboxylic acids XCH_2CH_2COOH and XCH_2COOH in aqueous solutions at 25°C.

TABLE 2.3 Induction Effect Transmission Coefficients, z_X, for Some Carbon, Heteroatomic, and Cyclic Fragments

Fragment	z_X	Reference
$—CH_2—$	0.41 ± 0.05	*a*
	0.388 ± 0.015	*b*
$—CH_2CH_2—$	0.20 ± 0.02	*a*
	0.205 ± 0.005	*c*
—NH—	0.40	*a*
$—OCH_2—$	0.30 ± 0.04	*a*
	0.28	*c*
$—SCH_2—$	0.40	*a*
	0.32 (0.40[f])	*c*
$—SeCH_2—$	0.35	*a*
	0.39	*c*
$—NHCH_2—$	0.305	*a*
$—^{+}NH_2CH_2—$	0.11	*a*
$—SO_2CH_2—$	0.24	*c*
trans-CH=CH—	0.50 ± 0.02	*a*
	0.49 ± 0.01	*c*
cis-CH=CH—	0.44 ± 0.04	*c*
trans-$(CH=CH)_2$-*trans*	0.24	*c*
cis-$(CH=CH)_2$-*trans*	0.23	*c*
trans-SCH=CH—	0.39	*c*
cis-SCH=CH—	0.37	*c*
trans-S(O)CH=CH—	0.19	*c*
cis-S(O)CH=CH—	0.27	*c*
trans-SO_2CH=CH—	0.22	*c*
cis-SO_2CH=CH—	0.22	*c*
—CH≡CH—	0.36 ± 0.03	*a*
	0.48	*c*
—C(O)NH—	0.31 ± 0.02	*a*
trans 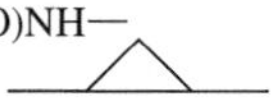	0.28 ± 0.07	*a*
	0.33	*c*
cis 	0.37 ± 0.06	*a*
	0.28	*c*
cis 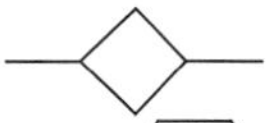	0.16	*a*
trans 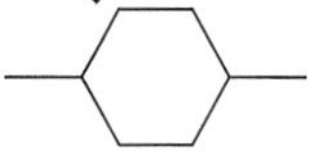	0.145 ± 0.031	*b*
	0.095 ± 0.018	*b*
	0.068 ± 0.007	*b*
	0.058 ± 0.007	*b*
	0.055 ± 0.010	*b*
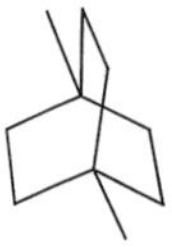	0.158 ± 0.012	*b,d*
	0.163 ± 0.014	*b,d*
	0.160	*b,e*
	0.094 ± 0.012	*b,e*
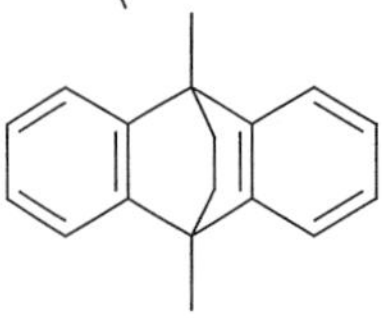	0.112 ± 0.023	*b*
	0.161 ± 0.011	*b*

[a] Yu.A. Zhdanov and V.I. Minkin, *Correlation Analysis in Chemistry* (in Russian). Izd. Rostov Univ., Rostov, 1966.
[b] V.A. Palm, *Foundations of the Quantitative Theory of Organic Reactions* (in Russian). Khimiya, Leningrad, 1967.
[c] C. Hansch and A. Leo, *Exploring QSAR. Fundamentals and Applications in Chemistry and Biology,* ACS Prof. Ref. Book. American Chemical Society, Washington, DC, 1995.
[d] H.D. Holtz and L.M. Stock, *J. Am Chem. Soc.* **86,** 5188 (1964).
[e] J.D. Roberts and W.T. Moreland, *J. Am. Chem. Soc.* **75,** 2167 (1953).
[f] In 50% dioxane/water mixture.

x_B are the characteristics of the intensity of interaction for two given objects, respectively.

Formally, Eq. (2.23) corresponds to the general form of the electrostatic or gravitational interaction energy when the electrical charges or masses of interacting particles are employed as the interaction intensity parameters, x, respectively. The parameters α_i include in that case the respective universal constant of interaction and the reciprocal of the distance between two interacting objects. An analogous equation may be formally assumed for the induction effect by using the σ^* constants (or σ_I constants) as the interaction intensity parameters. Accordingly, for a molecular system that is divided into the substituent X and the reaction center Y, the induction energy between these fragments can be expressed as:

$$E_{XY}^{\text{ind}} = \alpha^* \sigma_X^* \sigma_{Y'}^* \tag{2.24}$$

where α^* is a universal induction effect parameter. The distance dependence may be considered negligible because the substituent and the reaction center are always separated by only one chemical bond that has approximately the same length. Therefore, in the first approximation α^* can be considered a constant. During a chemical reaction, the reaction center changes, that is, $Y \rightarrow Y'$, and the substituent constant σ^* for the reaction center changes as well. The induction interaction energy for the final state of the reaction (XY') can be thus expressed as:

$$E_{XY'}^{\text{ind}} = \alpha^* \sigma_X^* \sigma_Y^* \tag{2.25}$$

The contribution of the induction effect into the energy of the reaction is therefore given as:

$$\Delta E_{XY \rightarrow XY'}^{\text{ind}} = \alpha^* (\sigma_{Y'}^* - \sigma_Y^*) \sigma_X^* \tag{2.26}$$

In the case of a chemical reaction for which the induction substituent constants of the reaction center are known both for the initial and the final (or transition) state, the numerical value of the universal constant α^* can be calculated. By noticing that

$$\rho^* = \alpha^* (\sigma_{Y'}^* - \sigma_Y^*) \tag{2.27}$$

for a given reaction series, the constant α^* can be found as a slope of the linear relationship between the experimental values of ρ^* and the difference $(\sigma_{Y'}^* - \sigma_Y^*)$. Most chemical reactions proceed with the change of the ionic charge on the reaction center. Thus it is necessary to evaluate the σ^* constants for the ionic substituents that exhibit, apart from the induction effect, the direct ion–ion or ion–dipole electrostatic interaction with the reaction center. This has been reflected by confusion in determining induction constants for charged substituents. The use of the data from different standard reaction series has lead, as a rule, to significantly different σ^* constants for the ionic substituents. However, it has been suggested that the direct ion–ion interaction energy between two ionic centers in a molecule can be accounted for using a simple electrostatic equation [72,73]:

$$\Delta E_{\text{ion}} = \frac{Z_X Z_Y}{D r_{XY}} \tag{2.28}$$

where Z_X and Z_Y denote the ionic charges of two interacting molecular fragments, D is the macroscopic dielectric constant of the reaction medium, and r_{XY} is the distance between the respective formally charged atoms. The inductive contribution to the free energy of reaction involving the change of the ionic charge at the reaction center and a substituent with ionic charge is thus given by the following equation:

$$\Delta G_{\text{ind}} = \Delta G_{Y \to Y'} - \Delta E_{\text{ion}} \tag{2.29}$$

where $\Delta G_{Y \to Y'}$ is the total change of the free energy of reaction with the changing reaction center, Y. The induction constants, derived for the substituents with ionic charge from the contribution ΔG_{ind}, are practically independent of the reaction series and thus applicable in Eq. (2.27). The least-squares treatment of the data according to this equation has been carried out for 24 reaction series including the acidic dissociation of aliphatic carboxylic acids, alcohols, thiols and ammonium ions, base-catalyzed hydrolysis of esters, reaction of carboxylic acids with diazodiphenyl methane, enolization of nitromethane catalyzed by carboxylate anions, and other reactions in different solvents (Fig. 2.2) [74,75]. The average value $\alpha^* = 2.62 \pm 0.08$ was obtained as a result of the respective least-squares treatment. This value of α^* can be applicable for the further determination of the induction constants of the reaction center, particularly for the reaction centers in the transition state.

It is important to comment once more that the linear relationship given in Figure 2.2 is valid only if the electrostatic interaction energy between the ionic charges in the molecules is subtracted from the total effect according to Eq. (2.28). Notably, the electrostatic term may have both the positive and negative values, depending on the sign of interacting charges.

Equation (2.27) gives also an explicit interpretation of the ρ^* constant for a reaction series. It reflects the magnitude of the change of the σ^* constants for the reaction center in the initial and in the final (transition) state, respectively. The values $\rho^* > 0$ indicate the increase of the electronegativity, as measured by the σ^* constants, of the reaction center during the reaction. On the other hand, the values $\rho^* < 0$ correspond to the decrease of the electronegativity of the reaction center during the reaction. For instance, $\rho^* = (1.81 \pm 0.08)$ [71] has been obtained for the acidic dissociation of substituted carboxylic acids XCOOH in aqueous solutions at 25°C. Consequently, σ^*(COOH)

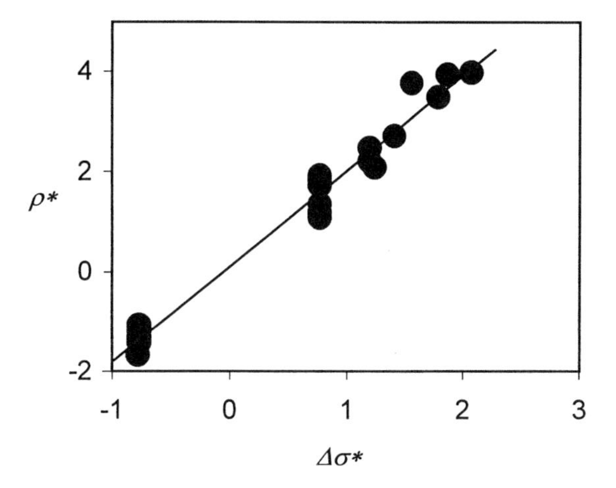

Figure 2.2 Linear relationship between the ρ^* and the difference between the induction substituent constants of the reaction center in the initial and final (transition) state, $(\sigma^*_{y'} - \sigma^*_{y})$, for 24 chemical reactions in different media.

$> \sigma^*(COO^-)$, that is, the carboxyl group (initial state of the reaction center), is more electronegative than the carboxylate group (final state of the reaction center).

These rules can be helpful in the study of the mechanism of a chemical reaction. The bimolecular rate constants of the neutral hydrolysis of substituted carboxylic acid (2-nitrophenyl)esters in aqueous solutions

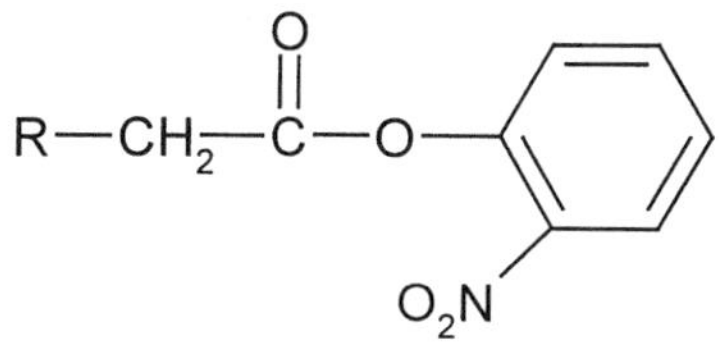

has been correlated with the Taft σ^* constants as follows [63]:

$$\log k_{AB} = (-6.87 \pm 0.05) + (2.26 \pm 0.08)\sigma^*_R$$

$$R^2 = 0.994 \qquad s = 0.114 \qquad n = 7$$

The positive sign of ρ^* for this reaction implies that the reaction center (the ester group) has to be more electronegative in the transition state of the reaction than in the initial compound.

The bimolecular rate constant for the reaction of acid-catalyzed hydration of a double bond in a series of compounds

$$RCH_2\text{—}O\text{—}CH{=}CH_2 + H_2O \xrightarrow{H^+} RCH_2\text{—}O\text{—}CH(OH)\text{—}CH_3$$

has been described by the following Taft equation [63]:

$$\log k_{AB} = (-1.47 \pm 0.07) - (0.31 \pm 0.03)\sigma_R^*$$
$$R^2 = 0.903 \qquad s = 0.132 \qquad n = 11$$

The negative sign of ρ^* for this reaction corresponds to a less electronegative reaction center in the transition state of the reaction as compared to the initial state.

The induction effect may be dependent on the solvent in which the process or property of compounds is measured [76–78]. However, in most cases, the solvent effect on the induction constants is assumed negligible. Notable effects are usually observed in the case of ionic substituents. For instance, the pK_a of acidic dissociation of aliphatic alcohols has been correlated with Taft σ^* constants as:

$$pK_a = (15.54 \pm 0.15) + (2.72 \pm 0.10)\sigma^*$$
$$R^2 = 0.968 \qquad s = 0.035 \qquad n = 17$$

in the case of aqueous solutions and

$$pK_a = (19 \pm 3) + (6.99 \pm 1.29)\sigma^*$$
$$R^2 = 0.853 \qquad s = 5.7 \qquad n = 13$$

in the case of the gas phase [79]. Proceeding from Eq. (2.27), the following values of the σ^* constant for the negatively charged oxy-group can be determined: $\sigma^*_{O^-}$ (gas phase) = -2.4; $\sigma^*_{O^-}$ (water) = 0; σ^*_{OH} (full protonization) = $+1.41$. Obviously, the hydrogen bonding of the O^- group with water molecules increases significantly its electronegativity and the respective σ^* constant.

In the discussion of the intramolecular induction interactions, it has been customary to distinguish between the electronegative (χ), field (F), and polarization (α) contributions to the overall effect [12,80]. Various schemes have been developed to estimate those individual contributions, based on the experimental data or quantum chemical modeling of appropriately chosen model systems [81–85]. In most cases, such divisions have relied on certain, more or less arbitrary, assumptions and thus have not gained much popularity.

Most applications of induction constants involve the reactivity of compounds in chemical reactions. As expected, the Hammett σ constant correlates well with the pK_a of various acid–base equilibria at different conditions. For instance, the pK_a of substituted benzoic acids are still excellently described by the following LFER equations:

$$pK_a = (4.22 \pm 0.03) - (1.01 \pm 0.06)\sigma$$
$$R^2 = 0.996 \qquad s = 0.024 \qquad n = 9$$

in aqueous solutions at the elevated temperature (45°C) [86];

$$pK_a = (3.74 \pm 0.02) - (1.01 \pm 0.04)\sigma$$
$$R^2 = 0.998 \qquad s = 0.013 \qquad n = 7$$

in aqueous solutions at the elevated pressure (3000 bar) [87];

$$pK_a = (5.69 \pm 0.02) - (1.44 \pm 0.05)\sigma$$
$$R^2 = 0.976 \qquad s = 0.076 \qquad n = 86$$

in 50% aqueous ethanol [88,89];

$$pK_a = (20.68 \pm 0.27) - (2.52 \pm 0.27)\sigma$$
$$R^2 = 0.974 \qquad s = 0.213 \qquad n = 9$$

in acetonitrile [90]; and

$$pK_a = (10.99 \pm 0.22) - (2.39 \pm 0.36)\sigma$$
$$R^2 = 0.962 \qquad s = 0.123 \qquad n = 11$$

in dimethyl sulfoxide [91]. These results demonstrate the applicability of the free energy at ionization of meta- and para-substituted benzoic acids in aqueous solutions at normal conditions for the correlation of the free energies of the same process at elevated temperatures and pressures and in different solvents. However, the sensitivity of the process on the variation of the substituent may be rather different at different conditions.

Similarly, the Hammett equation has been applied to the description of ionization constants of phenols and thiophenols in different media. For example, the pK_a of substituted phenols in aqueous solutions has been described by the following equation [92,93]:

$$pK_a = (9.84 \pm 0.07) - (2.00 \pm 0.14)\sigma$$
$$R^2 = 0.982 \qquad s = 0.114 \qquad n = 20$$

whereas the pK_a of the acidic dissociation of thiophenols in 95% aqueous ethanol can be predicted by the following relationship [94]:

$$pK_a = (9.17 \pm 0.13) - (2.35 \pm 0.28)\sigma$$
$$R^2 = 0.962 \qquad s = 0.194 \qquad n = 15$$

The influence of the induction effect on acid–base equilibria involving aliphatic compounds has been described by linear relationships employing the Taft σ^* or σ_I constants. Thus, the following equation is applicable for the pK_a of ionization of aliphatic alcohols ROH in aqueous solutions at 25°C [63]:

$$pK_a = (15.56 \pm 0.29) - (1.27 \pm 0.04)\sigma^*$$
$$R^2 = 0.990 \qquad s = 0.16 \qquad n = 11$$

The ionization of substituted phosphorous acids RPO_3H_2 in aqueous solutions at 25°C has been described by the following LFER equation [63]:

$$pK_a = (1.47 \pm 0.05) - (1.72 \pm 0.08)\sigma^*$$
$$R^2 = 0.966 \qquad s = 0.058 \qquad n = 17$$

and the ionization of substituted thiols RSH in aqueous solutions at 25°C has been described as follows [75]:

$$pK_a = (10.30 \pm 0.09) - (3.98 \pm 0.15)\sigma^*$$
$$R^2 = 0.982 \qquad s = 0.27 \qquad n = 18$$

It has been noticed that the highest sensitivity of the acid–base equilibrium constants toward the induction effect is exhibited in the case of processes in low dielectric constant media, including the gas phase [95,96]. The free energies of these acid–base equilibria are often correlated not with particular σ constants but with the relative substituent-induced changes, $\delta\,\Delta G^0$, in the free energy of some standard process, either in the gas phase or in some other medium.

The induction effect is also one of the main structural factors determining the rate of various chemical reactions. For example, the log k of the alkaline hydrolysis of esters $XC_6H_5COOCH_3$ in aqueous solutions at 25°C correlate with Hammett σ constants as follows [97]:

$$\log k = (1.92 \pm 0.01) - (1.66 \pm 0.04)\sigma$$
$$R^2 = 0.998 \qquad s = 0.022 \qquad n = 14$$

The correlation of log k for the alkaline hydrolysis of the same esters in 85% methanol at 25°C has, however, a significantly larger absolute value of ρ (2.25 instead of 1.66) than the respective reaction in aqueous solutions [98]:

$$\log k = (-3.72 \pm 0.06) - (2.25 \pm 0.16)\sigma$$
$$R^2 = 0.992 \qquad s = 0.076 \qquad n = 10$$

this difference in ρ values in different media has been interpreted as the change of the susceptibility of the reaction center to the substituent inductive effect induced by the solvent. In the above example, water is expected to polarize the carbonyl bond more than methanol. Correspondingly, the transition state of the reaction is more similar to the reaction product, and thus the influence of the substituent is diminished because of a weaker bond between the substituent and the reaction center.

The log k of the alkaline hydrolysis of amides of substituted benzoic acids $XC_6H_5CONH_2$ in 80% aqueous ethanol solutions at 52.8°C has been correlated with the σ^0 constants as follows [63]:

$$\log k = (-5.12 \pm 0.03) - (1.33 \pm 0.07)\sigma^0$$
$$R^2 = 0.994 \qquad s = 0.05 \qquad n = 4$$

A diminished value of ρ^0 as compared to the reaction of alkaline hydrolysis of esters may be interpreted as the result of the additional polarization of the carbonyl group by the adjacent more electronegative and electron-donating NH_2 group. Such polarization should reduce the influence by substituents in the phenyl ring. The mechanistic interpretation of ρ values is customary but not always conclusive. Thus it has been advised to complement the ρ analysis with additional independent experimental studies of the reaction mechanism [99].

Some additional information may be gained from the study of the temperature dependence of the ρ values. For instance, the alkaline hydrolysis of aliphatic amides $RCONH_2$ in aqueous solutions has been described by the following two-parameter equations at different temperatures [63]:

$$\log k = (-2.99 \pm 0.62) - (2.48 \pm 1.36)\sigma^* + (1.30 \pm 0.19)E_S$$
$$R^2 = 0.904 \qquad s = 0.24 \qquad n = 11$$

at 65°C,

$$\log k = (-3.25 \pm 0.54) - (3.19 \pm 1.18)\sigma^* + (1.20 \pm 0.17)E_S$$
$$R^2 = 0.837 \qquad s = 0.24 \qquad n = 17$$

at 75°C, and

$$\log k = (-3.16 \pm 0.57) - (3.71 \pm 1.27)\sigma^* + (1.27 \pm 0.18)E_S$$
$$R^2 = 0.837 \qquad s = 0.24 \qquad n = 17$$

at 85°C, where E_S is the Taft steric constant of substituents. A substantial increase in the value of ρ^* together with an increase in temperature may be caused by the respective change of the physical properties (e.g., dielectric constant) of water. Notably, the steric effect on the reaction remains practically constant at different temperatures.

As described by the few examples above, many LFER equations employing the induction constants describe the chemical reactivity of compounds. However, various inductive σ constants have been also used to correlate the biological activity of compounds. Extensive databases on Hammett-type equations, developed for the description of biological activity of compounds, have been compiled by Hansch and others [100–104]. Frequently, the induction effect exhibits small influence on the biological activity. This has been interpreted as due to the equalization of the intramolecular electronic effects in the substrate molecule exposed to the enzyme field. However, in some cases the induction effect has significant influence on the biological activity. For instance, the inhibitory potency, IC_{50}, of substituted aniline mustards

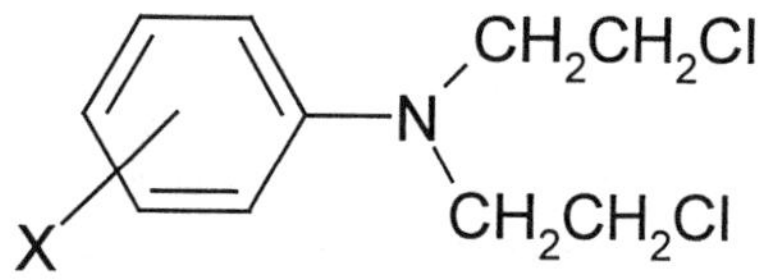

CH_2CH_2Cl
CH_2CH_2Cl
N
X

on the Chinese hamster ovary-derived UV_4 cells has been described by the following Hammett-type equation [105,106]:

$$\log \frac{1}{IC_{50}} = -(0.21 \pm 0.25) - (2.50 \pm 0.41)\sigma$$

$$R^2 = 0.941 \qquad s = 0.35 \qquad n = 13$$

The ρ value for this activity is similar to the ρ values of the nucleophilic substitution of aniline mustards by nitrobenzylpyridine in 50% acetone–water mixtures at 66°C (-3.21 ± 0.84) and in ethanol at 80°C (-2.38 ± 0.99) [107]. This similarity of ρ values indicates the similarity of the respective reaction mechanisms. As a possibility, the aniline mustards are thus expected to destroy the replicative mechanisms of a cell by electrophilic attack and subsequent alkylation on DNA bases [108].

The derivatives of mitomycin C

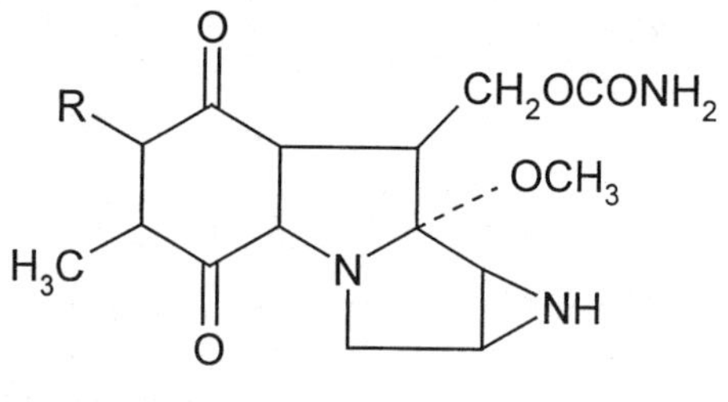

O
R
CH_2OCONH_2
OCH_3
H_3C
N
NH
O

have been proposed as potential clinical antitumor agents [109]. While the activity of some mitomycin derivatives has been correlated with the hydrophilicity of compounds [110], a significant correlation was also found for N7-aryl-substituted mitomycin C analogs with the Hammett σ constants as [111]:

$$\log \frac{1}{IC_{50}} = 8.46 - 0.92\sigma$$

$$R^2 = 0.792 \qquad s = 0.22 \qquad n = 8$$

where IC_{50} denotes the activity of compounds against P388 leukemia in culture. The conclusion about the importance of electronic intramolecular effects of substituents in the 7-position on the antitumor activity of mitomycin C derivatives was later confirmed by adaptive least-squares [112] and fuzzy adaptive least-squares [113] analysis of the respective data [114].

For a series of 1-phenyl-3,3-dimethyltriazenes,

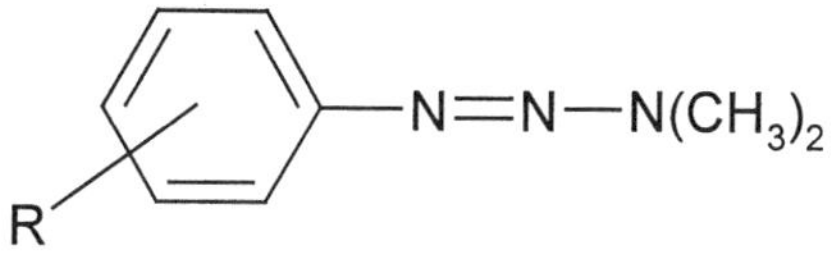

the activity against sarcoma-180 ascitic tumor was correlated with the σ constant of the substituent at phenyl ring [115]:

$$\log \frac{1}{C} = (3.41 \pm 0.03) - (0.69 \pm 0.09)\sigma$$

$$R^2 = 0.846 \qquad s = 0.09 \qquad n = 13$$

A substantially weaker correlation was obtained between the toxicity of 1-phenyl-3,3-dimethyltriazenes and the Hammett σ constants ($R^2 = 0.476$). Thus, the antitumor activity of these compounds is more significantly determined by the intramolecular electronic effects than general toxicity.

In a similar study, no induction effect influence by a substituent in the phenyl ring was established on the cytotoxic activity of rubidazone analogs [116,117]:

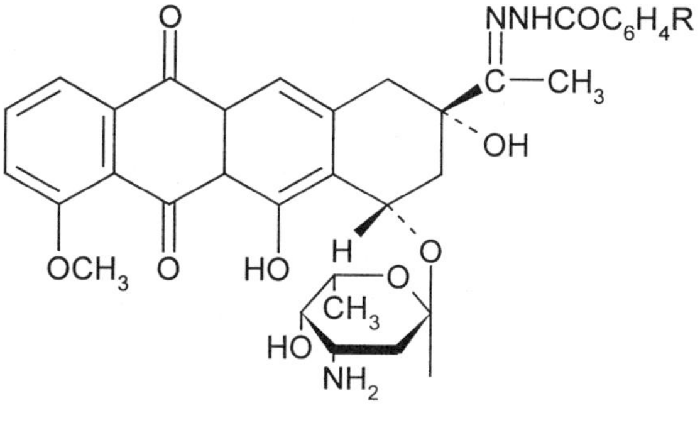

At the same time, a significant correlation with Hammett σ constants of substituents on the phenyl ring was obtained for the cardiotoxicity of these compounds as:

$$\log \mathrm{MCCD} = (1.63 \pm 0.19) - (0.53 \pm 0.07)\sigma$$

$$R^2 = 0.884 \qquad s = 0.065 \qquad n = 9$$

where MCCD stands for minimum cumulative cardiotoxic dose in rat. Thus, it allows to associate the cardiotoxicity of rubidazone analogs with the mechanism of activity influenced by intramolecular inductive effect in the molecules.

In the development of ρ-σ correlations, the representative selection of substituents is, however, very important. Otherwise, the relationships for small sets of data or data with little variance in σ constants may be fictitious and arise from the chance correlations between the induction constants and some other, physically determining factors. For instance, the anticancer activity of a series of [bis(1-aziridinyl)phosphonyl]carbamates defined through the effective dose ED_{50}

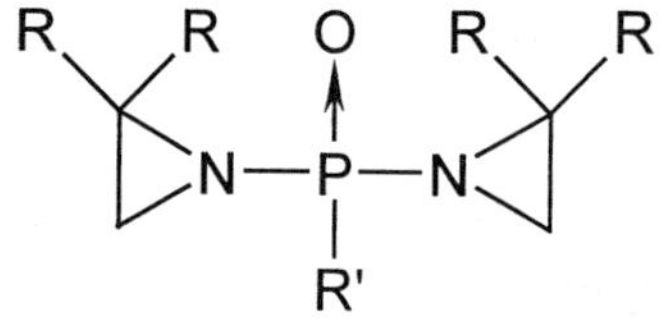

has been correlated with the Taft induction constant σ^* as [118]:

$$\log \frac{1}{\mathrm{ED}_{50}} = 3.34 + 2.74\sigma_R^*$$

$$R^2 = 0.893 \qquad s = 0.26 \qquad n = 10$$

However, it has been pointed out [108] that the substituents applied as R were only the hydrogen and the methyl group. Therefore, a simple use of an indicator variable, accounting for the presence of one of these substituents, would give the same result. Consequently, the above equation cannot be considered as a proof for the importance of induction effect in the biological activity of these compounds.

The empirical induction constants have been also extensively used for the LFER description of physical and spectroscopic properties of compounds. In particular, the shifts in nuclear magnetic resonance (NMR) spectra of compounds have been used for the definition of induction effect scales. The σ_F scale has been defined by Taft and co-workers [119] using the ^{19}F NMR shielding effect, $\int_H^{m-X}$, of meta-substituted benzenes, in dilute hydrocarbon solvents as:

$$\sigma_F = 0.16 + 0.137 \int_H^{m-X} \tag{2.30}$$

The respective scale has been extended to more than 270 substituents [62]. It has been, however, noticed that the σ_F scale correlates poorly with the substituent (Y) induced variations on the NMR chemical shifts of ^{13}C, ^{19}F, ^{29}Si, ^{119}Sn, and $^{207}Pb(X)$ atoms, directly attached to the bridgehead carbon of bicyclo[2,2,2]octane ring system [120–126].

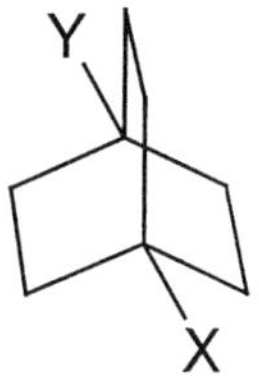

This discrepancy has been explained by the presence of additional "through-space" electrostatic interaction between the closely located C—X and C—Y bonds.

Katritzky and others have correlated the intensity of the ring-stretching vibrational modes of benzenoid derivatives in the 1600 to 1500 cm^{-1} region with the substituent σ^0 constants [127–129]. This approach has been successfully extended to polysubstituted benzenes and to analogous heteroaromatic compounds [130–135]. The infrared spectral transition energies corresponding to the bond-stretching vibrations of various groups (carbonyl-, nitroso-, nitro-, cyano-, etc.) have also been correlated with the σ^* constants of adjacent substituents [136–141].

The substituent induction constants have been employed for the correlation of substituent-induced shifts in the electronic spectra of compounds. For example, the energy of the long-wavelength transition in the electronic spectra of 1-arylazoazulenes

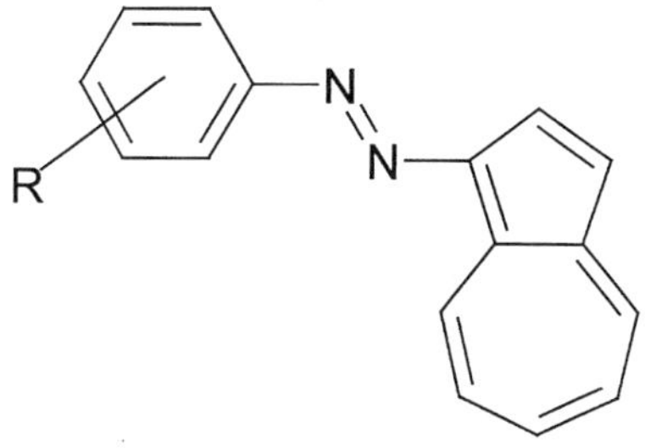

has been successfully correlated with the Hammett σ constants of substituents R [142]. The correlations of the electronic spectral transition energies with the Taft σ^* constants have been established for aliphatic nitro-compounds [143], aliphatic ketones [144], and saturated carboxylate esters [145].

Other physical properties of organic compounds related to induction substituent constants by LFER-type relationships include dipole moments [146–

153], ionization potentials of compounds [154–158], electron affinities of compounds [159], and polarographic half potentials [160]. It has to be admitted that the σ-ρ analysis of spectral and physical properties of compounds is more or less only of historical interest, and direct quantum chemical calculations are now mostly employed for the prediction of these data. Nevertheless, in the case of large molecular systems with complex electronic structure, the quantum chemical methods may fail, and the σ-ρ approach offers thus an alternative for the analysis and prediction of the respective spectroscopic data.

2.2.2 Resonance Constants

The induction effect is not the only "formal" intramolecular interaction mechanism defined in physical organic chemistry. Proceeding from the valence bond theory, Pauling and Wheland pointed out that another type of intramolecular interaction may exist between the atoms or atomic groups in the molecule that is called the resonance or the mesomeric effect [161,162]. They suggest that for a proper representation of the structure and charge distribution of a chemical compound, a single valence formula may be insufficient and a more realistic picture should be given by a combination of formal valence formulas. Within this approach, these classical formulas are called the resonance forms. Two types of resonance interaction have been distinguished. The first, called the nonpolar resonance, relates to the redistribution of the single and multiple chemical bonds in the formula, without the charge separation. A classical example of such resonance is the representation of the benzene structure by different Kekulé and Dewar structures:

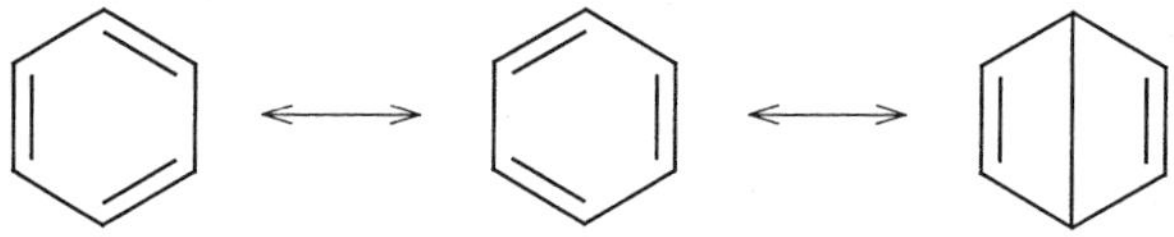

The above resonance forms reflect the equalization of the C—C chemical bonds in the benzene ring and the uniformity of the charge distribution in the molecule.

The second type of the resonance interaction, called the polar resonance, involves, apart from the redistribution of single and multiple bonds, the change of formal ionic charges on atoms in different resonance forms. In the case of benzene, the respective polar resonance forms can be written as follows:

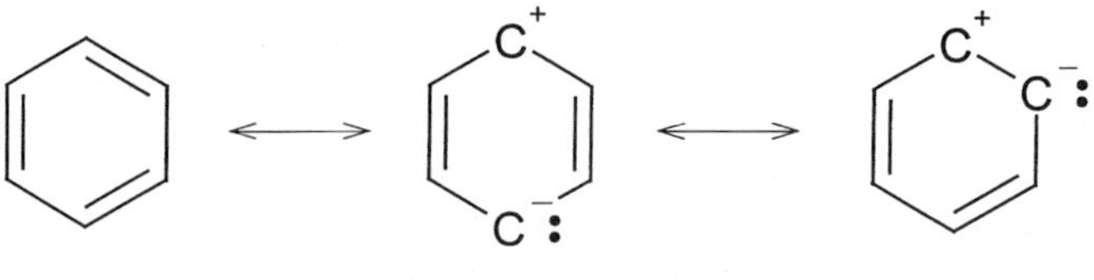

The relative weight of the polar resonance forms into the total structure of benzene is, of course, much smaller than for the Kekulé form. Nevertheless, the presence of these forms predicts accurately the dual reactivity of the carbon atoms of the benzene ring, which can act both as the nucleophiles (having partially free electron pair) or the electrophiles (having partially free orbital).

As a rule, the polar resonance takes place between two or more electronegative groups, connected to each other directly or through a system of conjugated π bonds. Contrary to the induction interaction, the polar resonance is a non-symmetric interaction involving always the interacting electron-acceptor ($-$R) and electron-donor ($+$R) groups. Therefore, separate scales of resonance effect have to be developed in the framework of the LFER approach for these two differently behaving groups. Most electronegative substituents fall into one of the possible types of resonant groups. The typical electron-acceptor ($-$R) substituents are the nitro- ($-NO_2$), cyano- (—CN), and carboxylic (—COOH) groups. The hydroxyl- (—OH), amino- ($-NH_2$), and halogen (—F, —Cl, —Br, —I) groups belong to the electron-donor ($+$R) substituents.

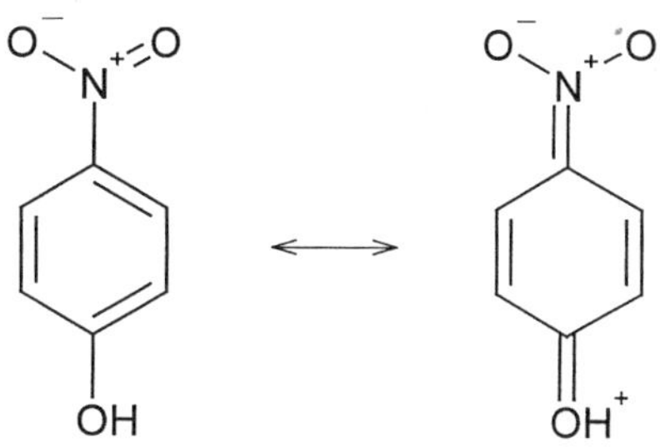

However, some substituents have an amphoteric nature. For instance, depending on the counterpart in the resonance interaction, the nitroso-group may possess either the quality of the $+$R or the $-$R group, respectively.

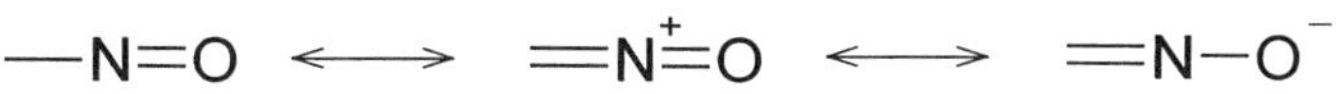

$$—N{=}O \longleftrightarrow {=}\overset{+}{N}{=}O \longleftrightarrow {=}N{-}O^{-}$$

The polar resonance involves always the electronegative groups and, consequently, in such molecular systems, the induction interaction between the same groups has to be effective, too. Therefore, with no intrinsic interrelation between these two interactions, the respective total substituent effect has to be described by two parameters, one corresponding to the induction effect and another to the resonance effect. A respective two-parameter equation has been suggested as:

$$\delta\,\Delta G = fF + rR + h \tag{2.31}$$

where F and R are the field (induction) and the resonance constants for substituents, respectively, f and r are transmission coefficients that depend only on the reaction or property described by the change of the free energy, $\delta\,\Delta G$, while h is the intercept for the correlation equation corresponding to change in $\delta\,\Delta G$ for the standard substituent, the hydrogen atom [163–165]. This equation, however, has been criticized from several points of view [166–168]. The main argument against the use of the two-parameter representation of electronic effects is based on the asymmetry of the polar resonance effect, requesting the use of separate scales for electron-donating and electron-accepting substituents, respectively. Indeed, Eq. (2.31) is derived proceeding from the implicit assumption that in the case of any process, the intensity of the interaction of the +R and −R groups with the changing reaction center should be the same. By introducing a universal interaction constant for the resonance interaction, α_R, similar to the α^* for the induction interaction in the aliphatic systems, an expression can be derived for the contribution of the resonance effect to the total free energy change of the system [cf. Eq. (2.27)]. To involve both the +R and −R resonance effects in the same reaction series, let us assume that the reaction center in one state (e.g., in the initial state of the chemical reaction) has a +R character and in another state (e.g., in the final state of the reaction) a −R character. Thus the −R substituents will have the resonance interaction with the reaction center in the initial state (Y) whereas the +R substituents interact with the reaction center in the final state (Y'). Accordingly, the change of the free energy of the process originating from the resonance effect, will be given as:

$$\delta\, \Delta G_{\text{res}} = \alpha_R(\sigma^+_{Y'} - \sigma^+_Y)\sigma^-_X + \alpha_R(\sigma^-_{Y'} - \sigma^-_Y)\sigma^+_X \tag{2.32}$$

where X denotes the substituent. It is obvious that while in the standard reaction series used to define the σ^+ and σ^- constants they form a common scale, it will not be the case in other reaction series because the terms $(\sigma^+_{Y'} - \sigma^+_Y)$ and $(\sigma^-_{Y'} - \sigma^-_Y)$ are different for a different reaction series. As a result, the ρ constants related to two types of resonance interaction in a given reaction have to be different.

Consequently, different scales are needed for the +R and −R substituent resonance constants. The respective σ^+ and σ^- constants in the Yukawa–Tsuno equations [cf. Eqs. (2.7) and (2.8)] were defined using the different reaction series for different scales. Originally, the σ^- scale was established using the pK_a values of para-substituted phenols [54,169,170]. The direct resonance delocalization of the negative charge by an electron-acceptor substituent in the conjugated para-position of the phenolate anion is substantially larger than the respective direct resonance interaction in the substituted phenol molecule itself.

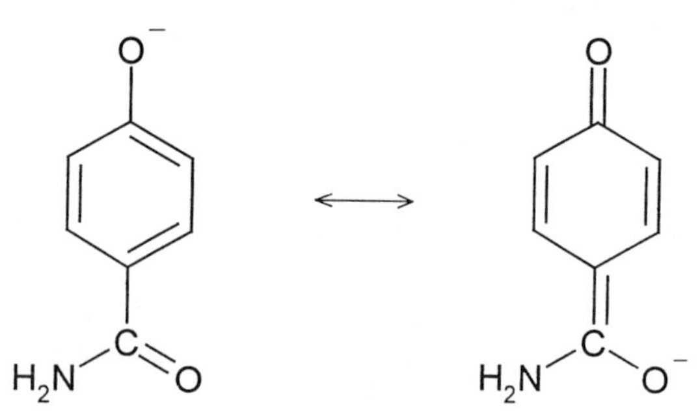

Consequently, the σ^- constants have been related to the enhanced substituent effect in the series of acidic dissociation of para-substituted phenols as compared to the pK_a of acidic dissociation of substituted benzoic acids. Formally, the σ^- constant for a substituent X is defined as:

$$\sigma^- = \frac{\mathrm{p}K_X - \mathrm{p}K_H}{\rho_{\mathrm{PhOH}}} \tag{2.33}$$

where pK_X and pK_H refer to the dissociation of the substituted and unsubstituted phenol, respectively, and ρ_{PhOH} is the constant of reaction determined on the basis of pK_a of nonconjugated substituents.

The σ^+ scale of substituent resonance constants has been defined originally from the rate constants of the solvolysis of cumyl chlorides

HO—C$_6$H$_4$—C(CH$_3$)$_2$—Cl ⟶ [HO—C$_6$H$_4$—C$^+$(CH$_3$)$_2$ ⟷ HO$^+$=C$_6$H$_4$=C(CH$_3$)$_2$]

in 90% acetone aqueous solutions [171,172]. The direct resonance of the reaction center in transition state with the +R groups (here, OH— group) in the para-position of the phenyl ring leads to an additional enhancement of the reaction rate. The σ^+ constant for a substituent X was defined as:

$$\sigma^+ = \frac{\log k_X - \log k_H}{\rho_{\mathrm{solv}}} \tag{2.34}$$

where log k_X and log k_H denote the logarithms of rate constants for the solvolysis reaction of the substituted and unsubstituted cumyl chloride, respectively, and ρ_{solv} is the constant of reaction determined on the basis of data for the nonconjugated substituents.

As discussed above, both the induction and resonance effects may simultaneously affect the reactivity and properties of aromatic molecular systems. In the framework of the LFER approach, a dual-substituent parameter (DSP) equation has been introduced by Taft and others to describe these effects as:

$$\delta\,\Delta G = \rho_I\sigma_I + \rho_R\sigma_R \tag{2.35}$$

where σ_I denotes the induction constant and σ_R the resonance constant for a given substituent. Four different sets of σ_R scales were proposed depending on the extent of the interaction between the substituent and the reaction center. The σ_R^0 constants should be valid in the case of absence of direct resonance between the substituent and the reaction center. The σ_R scale is related to the substituent conjugation with the low electron density reaction centers. The

σ_R^- and σ_R^+ constants apply in the case of the substituent interaction with the high electron density reaction centers. The original σ_I and σ_R scales had been evaluated on the basis of eight reactions with high reliability of data [11] and later extended for a larger set of substituents [62]. Various response constant scales are listed in Table 2.4. The numerical values of some commonly used constants for different types of substituents are presented in Table 2.5. Notably,

TABLE 2.4 Substituent Resonance Constant Scales Defined by Using Different Standard Processes

Scale Notation	Standard Process	Reference
σ^+	log k of hydrolysis of cumyl chlorides	*a*
σ^-	pK_a of substituted phenols and anilinium ions	*b*
$\Delta\sigma^+$	$\sigma^+ - \sigma$	*c*
$\Delta\sigma^-$	$\sigma^- - \sigma$	*d*
σ_R^0	$\sigma_R^0 = \sigma^0 - \sigma_I$	*e*
σ_R	$\sigma_R = \sigma - \sigma_I$	*f*
σ_c^+	$\sigma_c^+ = \sigma^+ - \sigma_I$	*g*
σ_c^-	$\sigma_c^- = \sigma^+ - \sigma_I$	*g*
R	$R = \sigma_{para} - 0.921F$	*h*
R^+	$R^+ = \sigma^+ - 0.921F$	*i*
R^-	$R^- = \sigma^- - 0.921F$	*i*
σ_c	Sensitivity of substituent to electron demand at reaction center	*j*
σ_D	Intrinsic resonance between the substituent and reaction center	*j*
σ_R^+	Difference in the ^{19}F NMR chemical shifts of the para- and meta-substituted fluorobenzenes	*k*
σ_R^-	Difference in the ^{19}F NMR chemical shifts of the para- and meta-substituted fluorobenzenes	*k*
σ_R^0	Ab initio calculation of the π-electron density shift in monosubstituted ethylenes	*l*

[a] Y. Okamoto and H.C. Brown, *J. Org. Chem.* **22,** 485 (1957); H.C. Brown and Y. Okamoto, *J. Am. Chem. Soc.* **80,** 4979 (1958).
[b] S. Ehrenson, R.T.C. Brownlee, and R.W. Taft, *Prog. Phys. Org. Chem.* **1,** 10 (1973).
[c] Y. Tsuno, T. Ybata, and Y. Yukawa, *Bull. Chem. Soc. Jpn.* **32,** 960 (1959); Y. Yukawa, and Y. Tsuno, *ibid.*, pp. 965, 971.
[d] M. Yoshioka, K. Hamamoto, and T. Kubota, *Bull. Chem. Soc. Jpn.* **35,** 17623 (1962).
[e] R.W. Taft, S. Ehrenson, I.C. Lewis, and R. E. Glick. *J. Am. Chem. Soc.* **81,** 5352 (1959).
[f] R.W. Taft, *J. Am. Chem. Soc.* **79,** 5075 (1957).
[g] O.Exner, *Collect. Czech. Chem. Commun.* **25,** 642 (1960).
[h] C.G. Swain and E.C. Lupton, *J. Am. Chem. Soc.* **90,** 4328 (1968); C.G. Swain, S.H. Unger, N.R. Rosenquist, and M.S. Swain, *ibid.* **105,** 492 (1983).
[i] C. Hansch, A. Leo, and R.W. Taft, *Chem. Rev.* **91,** 165 (1991).
[j] M. Charton, *Prog. Phys. Org. Chem.* **16,** 287 (1987).
[k] R.W. Taft and R.D. Topsom, *Prog. Phys. Org. Chem.* **16,** 1 (1987).
[l] S. Marriott, A. Silvestro, and R.D. Topsom, *J. Chem. Soc., Perkin Trans. 2,* pp. 457 (1988).

TABLE 2.5 Numerical Values for Selection of Empirical Substituent Resonance Constant Scales Applicable in Aromatic Systems

Substituent	σ^+	σ^-	σ_R	R	R^+	R^-
H	0.00	0.00	0.00	0.00	0.00	0.00
CH_3	−0.31	−0.17	−0.14	−0.18	−0.32	−0.18
CH_2CH_3	−0.30	−0.19	−0.10	−0.15	−0.30	−0.19
$CH_2CH_2CH_3$	−0.29	−0.06	−0.11	−0.14	−0.30	−0.07
$CH(CH_3)_2$	−0.28	−0.16	−0.12		−0.32	−0.20
Cyclopropyl	−0.41	−0.09		−0.23	−0.43	−0.11
$(CH_2)_3CH_3$	−0.29	−0.12		−0.15	−0.28	−0.11
$CH_2CH(CH_3)_2$		0.01		−0.11		0.00
$C(CH_3)_3$	−0.26	−0.13		−0.18	−0.17	−0.04
Cyclobutyl	−0.29	−0.07		−0.16	−0.31	−0.09
$(CH_2)_4CH_3$		−0.19		−0.14		−0.18
$CH_2C(CH_3)_3$	−0.31			−0.14	−0.22	
Cyclopentyl	−0.30	−0.18		−0.16	−0.32	−0.20
Cyclohexyl	−0.29	−0.14		−0.18	−0.32	−0.17
1-Adamantyl	−0.38	−0.14	−0.15	−0.06	−0.31	−0.07
$CH{=}CH_2$	−0.16		−0.03	−0.17	−0.29	
$CH_2CH{=}CH_2$			−0.14	−0.08		
$CH{=}CH{-}CH{=}CH^-$	−0.14	0.12		−0.03	−0.33	−0.07
$CH_2CH{=}C(CH_3)_2$	−0.26					
$C{\equiv}CH$	0.18	0.53	−0.04		−0.04	0.31
C_6H_5	−0.18	0.02		−0.13	−0.30	−0.10
$CH_2C_6H_5$	−0.28	−0.09		−0.05	−0.45	−0.26
$CH_2CH_2C_6H_5$	−0.28	−0.12		−0.11	−0.27	−0.11
$CH{=}CHC_6H_5$	−1.00	0.13		−0.17	−1.10	0.03
CCC_6H_5	−0.03	0.30			−0.18	0.15
$CH(C_6H_5)_2$	−0.19			−0.06	−0.18	
$C(C_6H_5)_3$	−0.21			0.01	−0.22	
F	−0.07	−0.03	−0.48	−0.39	−0.52	−0.48
Cl	0.11	0.19	−0.25	−0.19	−0.31	−0.23
Br	0.15	0.25	−0.25	−0.22	−0.30	−0.20
I	0.14	0.27	−0.16	−0.24	−0.28	−0.15
OH	−0.92	−0.37	−0.43	−0.70	−1.25	−0.70
NH_2	−1.30	−0.15	−0.50	−0.74	−1.38	−0.23
NH_3^+		−0.56		−0.32		−1.48
NO		1.63	0.32	0.42		1.14
NO_2	0.79	1.27	0.15	0.13	0.14	0.62
N_3		0.11		−0.40		−0.37
SH	−0.03		−0.15	−0.15	−0.33	
S^-	−2.62		−0.32	−1.24	−2.56	
SCN		0.59		0.03		0.23
SF_5		0.86	0.08	0.12		0.30
SO_2F		1.54	0.23	0.19		0.82
SO_2^-		0.08		−0.08		0.05
SO_3^-		0.58		0.06		0.29
PH_2	0.06		−0.05	−0.04	−0.03	

TABLE 2.5 ***(Continued)***

Substituent	σ^+	σ^-	σ_R	R	R^+	R^-
PCl_2	0.62		0.11	0.11	0.12	
$P(O)Cl_2$	0.38		0.17	0.20	−0.32	
$P(S)Cl_2$	0.33			0.17	−0.30	
$B(OH)_2$	0.38			0.15	0.41	
BCl_2	0.86		0.30			
SiH_3	0.14		0.09	0.04	0.08	
$SiCl_3$	0.57		0.17	0.12	0.13	
$SiBr_3$	0.41			0.13	−0.03	
$GeCl_3$	0.57			0.14	−0.08	
$As(O)(OH)_2$		0.97				0.93
AsO_3H^-		0.46		−0.06		0.42
OCH_3	−0.78	−0.26	−0.43	−0.56	−1.07	−0.55
$OCH(CH_3)_2$	−0.85		−0.43		−1.19	
$OCH_2CH_2CH_3$	−0.83		−0.52	−0.51	−1.09	
OC_6H_5	−0.50	−0.10		−0.40	−0.87	−0.47
$OCHF_2$		0.11	−0.24	−0.19		−0.26
OCF_3		0.27	−0.04	−0.04		−0.12
OCF_2CF_3		0.28	−0.25	−0.27		−0.27
OCF_2CHF_2		0.21		−0.13		−0.17
OCH_2O^-	−0.68			−0.05	−0.57	
OCH_2CH_3	−0.81	−0.28	−0.44	−0.50	−1.07	−0.54
$OCOCH_3$	−0.19		−0.21	−0.11	−0.61	
$OCOC_6H_5$	−0.07			−0.13	−0.33	
OSO_2CH_3	0.16		−0.21	−0.04	−0.24	
OSO_2CF_3		0.49		−0.03		−0.07
$NHCH_3$	−1.81		−0.52	−0.73	−1.78	
$N(C_2H_5)_2$	−2.07	−0.43		−0.73	−2.08	−0.44
$N^+(CH_3)_3$	0.41	0.77	−0.11	−0.04	−0.45	−0.09
$N(CF_3)_2$		0.53	0.01			0.18
NHC_6H_5	−1.40	−0.29	−0.35	−0.78	−1.43	−0.32
$NHCOCH_3$	−0.60	−0.46		−0.31	−0.91	−0.77
$NHCOC_6H_5$	−0.60			−0.32	−0.73	
$NHSO_2C_6H_5$	−0.98			−0.23	−1.22	
CHO	0.73	1.03	0.24	0.09	0.40	0.70
$COCH_3$		0.84	0.20	0.17		0.51
COC_6H_5	0.51	0.83		0.12	0.20	0.52
COCl	0.79	1.24	0.21	0.46	0.33	0.78
COOH	0.42	0.77		0.11	0.08	0.43
$COOCH_3$	0.49	0.75	0.16	0.11	0.15	0.41
$COOC_2H_5$	0.48	0.75	0.16	0.11	0.14	0.41
COO^-	−0.02	0.31		0.10	0.08	
$CONH_2$		0.61	0.01	0.10		0.35
$CON(CH_3)_2$		0.70				
$COCF_3$	0.85	1.09	0.33		0.31	0.55
CN	0.66	1.00	0.08	0.15	0.15	0.49
CH_2Cl	−0.01		−0.03		−0.14	

TABLE 2.5 (*Continued*)

Substituent	σ^+	σ^-	σ_R	R	R^+	R^-
CH_2Br	0.02		−0.10	0.00	−0.12	
CF_3	0.61	0.65	0.10	0.16	0.23	0.27
CH_2CF_3		0.14	−0.04	−0.06		−0.01
CF_2CF_3		0.69		0.08		0.25
CH_2OH	−0.04	0.08	−0.07	−0.03	−0.07	0.05
CH_2OCH_3	−0.05			−0.12	−0.18	
CH_2CN	0.16	0.11	−0.08	0.01	−0.01	−0.06
CH_2COOH	−0.01	0.05				
CH_2COO^-	−0.53	−0.16		−0.35		
CH_2COCH_3	0.03					
CH_2COOCH_3		0.07		−0.02		
$CH_2COOC_2H_5$	−0.16					
$CH_2N^+(CH_3)_3$		0.57		0.06		0.19
$CH_2Si(CH_3)_3$	−0.62	−0.22		−0.12	−0.62	−0.22
$CH_2Ge(CH_3)_3$	−0.61					
$CH_2Sn(CH_3)_3$	−0.92					
$CH_2CH_2CH_2N^+(CH_3)_3$		0.09		−0.13		−0.03
$CF(CF_3)_2$		0.68	0.04	0.22		0.37
$C(CF_3)_3$		0.71		0.02		0.42
$(CF_2)_3CF_3$		0.73		0.08		0.29
$CH_2CH_2N^+(CH_3)_3$		0.09		−0.06		−0.10
$CH_2CH_2Si(CH_3)_3$		−0.16		−0.06		−0.05
$CF{=}CFCF_3$		0.65	0.14	0.10		0.29
$CH{=}CHCF_3$-*trans*		0.34	0.07	0.02		0.10
$CH{=}CHCF_3$-*cis*		0.29		0.01		0.11
$CH{=}CHSO_2CF_3$		0.83		0.33		0.61
$CH{=}CHCOOH$		0.62				
$CH{=}CHCOCH_3$	0.39			−0.32	0.08	
$CH{=}CHNO_2$-*trans*		0.88	0.13			0.52
$CH{=}C(CN)_2$	0.82	1.20		0.27	0.25	0.63
$C(CN){=}C(CN)_2$		1.70		0.33		1.05
$CH{=}NC_6H_5$		0.54		0.09		0.21
C_6F_5	0.23	0.43		0.00	−0.09	0.11
$N{=}CHC_6H_5$		0.22		−0.69		0.08
$N{=}C{=}O$	−0.19		−0.17	−0.12	−0.50	
$N{=}C{=}S$		0.34	−0.06	−0.13		−0.17
$N{=}NC_6H_5$	−0.19	0.45		0.09	−0.49	0.15
$N{=}NCF_3$	0.74			0.18	0.24	
$N{=}N{-}CN$	1.03			0.47	0.47	0.41
$N{=}NN(CH_3)_2$	−0.46		−0.17	−0.01	−0.43	
$N{=}P(C_6H_5)_3$	−1.65	−0.77		−0.67	−1.55	−0.67
SCH_3	−0.60	0.06	−0.17	−0.23	−0.83	−0.17
SC_6H_5	−0.55	0.18		−0.23	−0.85	−0.12
SCF_3		0.64	0.06	0.14		0.28
$S(CF_2)_2CF_3$		0.65	0.08	0.05		0.22
$SCF(CF_3)_2$		0.69	0.09	0.03		0.23

TABLE 2.5 ***(Continued)***

Substituent	σ^+	σ^-	σ_R	R	R^+	R^-
$SC(CF_3)_3$		0.79	0.14	0.11		0.32
$S^+(CH_3)_2$		0.83	0.17	−0.08		−0.15
$S(=O)CH_3$		0.73	0.00	−0.03		0.21
$S(=O)C_6H_5$		0.76		−0.07		0.25
$S(=O)CHF_2$		0.93	0.11			0.42
$S(=O)CF_3$		1.05	0.08	0.11		0.47
$S(=O)OCH_3$		0.89	0.09	0.07		0.65
SO_2CH_3		1.13	0.11	0.19		0.60
$SO_2C_6H_5$		1.21		0.10		0.65
SO_2CH_2F		1.17	0.04			
SO_2CHF_2		1.44		0.19		0.77
SO_2CF_3		1.63		0.22		0.89
$SO_2(CF_2)_2CF_3$		1.75	0.33	0.28		0.94
$SO_2CF(CF_3)_2$		1.76	0.33	0.30		0.96
$SO_2C(CF_3)_3$		1.81	0.34	0.29		0.97
$SO_2OC_6H_5$		1.11		−0.04		−1.48
SO_2NH_2		0.94		0.11		0.45
$SO_2N(CH_3)_2$	0.86	0.99	0.12	0.21	0.42	0.55
$S(C=O)CH_3$		0.46	0.01	0.07		0.09
$P(CH_3)_2$		0.22	0.06	0.01		0.17
$P(C_6H_5)_2$	0.70	0.26		0.09	0.60	0.16
$P^+(CH_3)_3$		0.95	0.08	0.02		0.59
$P(O)(CH_3)_2$		0.74	0.10	0.10		0.34
$P(O)(C_6H_5)_2$	0.52	0.68		0.21	0.20	0.36
$P(S)(CH_3)_2$		0.62				
$PS(C_6H_5)_2$		0.73		0.24		0.50
$SiH(CH_3)_2$	−0.04		0.04		−0.07	
$Si(CH_3)_3$	0.02		0.05	−0.08	0.01	
$Si(C_2H_5)_3$	0.04		0.05			
$Si(CH_3)_2C_6H_5$	0.08				0.02	
$Si(CH_3)(C_6H_5)_2$	−0.04		0.06	0.02	−0.15	
$Si(C_6H_5)_3$	0.12	0.29	0.06	0.14	0.16	0.33
$SiF(CH_3)_2$	0.17		0.11	0.04	0.00	
$SiCl(CH_3)_2$	0.02		0.10	0.05	−0.14	
$SiF_2(CH_3)$	0.23			−0.09	−0.09	
$SiCl_2(CH_3)$	0.08		0.15	0.10	−0.21	
$Si(CH_3)_2OCH_3$	−0.02			−0.11	−0.11	
$SiCH_3(OCH_3)_2$	0.01			0.05	−0.04	
$Si(OCH_3)_3$	0.13		0.08	0.03	0.03	
$Si(OC_2H_5)_3$	0.17		0.08	0.05	0.14	
$Si(CH_3)_2Si(CH_3)_3$	−0.23	0.02	0.04			
SeC_6H_5	−0.47	0.13	−0.19			
$As(C_2H_5)_2$		0.08		−0.32		−0.24
$As(C_6H_5)_2$		0.29	−0.07	0.05		0.25
$Ge(C_6H_5)_3$	−0.15		−0.08	0.01	−0.35	
$Ni(CH_3)_2$	−1.70	−0.12			−1.85	−0.27

TABLE 2.5 (*Continued*)

Substituent	σ^+	σ^-	σ_R	R	R^+	R^-
$Sn(CH_3)_3$	−0.12			−0.03	−0.46	
2-Thienyl	−0.43	0.19		−0.08	−0.56	0.06
3-Thienyl	−0.38	0.13		−0.10	−0.46	0.05
2-Selenieyl		0.22		−0.06		0.12
2-Tellurienyl		0.25		−0.07		0.15
2-Pyridyl		0.55	0.22	−0.23		0.15
3-Pyridyl		0.58		0.01		0.34
4-Pyridyl		0.81		0.23		0.60

for the original set of 24 substituents, highly significant correlations were obtained between the Swain–Lupton F and σ_I constants and the Swain–Lupton R and σ_R constants, respectively, as follows [102]:

$$F = (0.051 \pm 0.029) - (0.852 \pm 0.070)\sigma_I$$

$$R^2 = 0.962 \qquad s = 0.038 \qquad n = 24$$

and

$$R = (0.011 \pm 0.036) - (0.980 \pm 0.011)\sigma_R$$

$$R^2 = 0.943 \qquad s = 0.076 \qquad n = 24$$

However, such correlations are, as expected, much poorer when a wider selection of substituents is involved. The DSP substituent constants have been also predicted using the back-propagation neural networks [173].

In most treatments, the induction and resonance mechanisms of interaction are assumed independent. Apparently, this assumption has been valid in the case of many applications. However, in special cases the intensity of the resonance effect may be affected by the inductive interaction between the substituents and otherwise round. An example of such interdependence between two intramolecular interactions is the so-called positive bridge effect (PBE), which was discovered in the molecular systems with two aromatic rings connected through a bridge group having a lone electron pair. For instance, the ratio of the rate constants of the acylation reaction between the picryl chloride and unsubstituted and para-nitrosubstituted bridge compounds

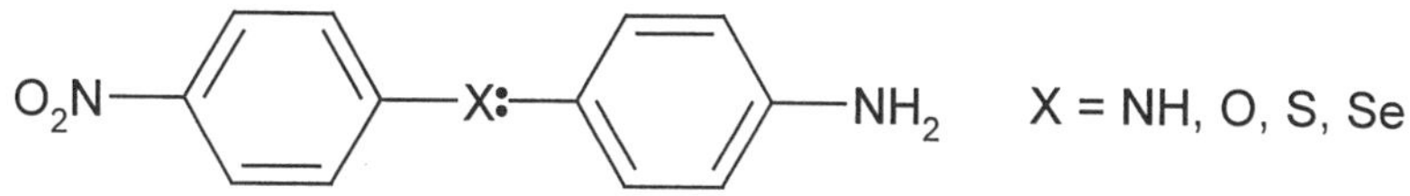

O_2N–C_6H_4–X:–C_6H_4–NH_2 X = NH, O, S, Se

was found to be significantly larger than in the case of the respective biphenyl systems [71]:

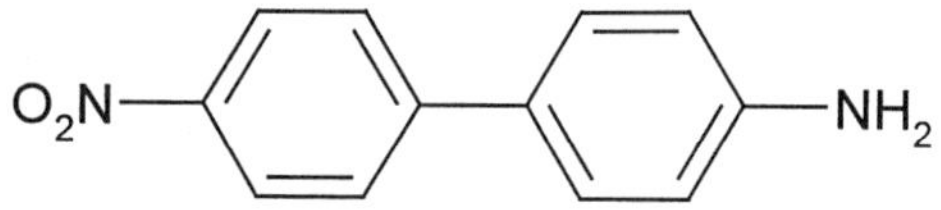

O_2N NH_2

This observation, as well as the observation of higher values of ρ^0 in bridged systems as compared to the biphenyl systems was interpreted as the result of the induction interaction on the availability of the lone electron pair of the bridge group (+R group).

Numerous attempts have been made to develop theoretically various substituent constants related to intramolecular inductive or resonance effects, in terms of the quantum chemical characteristics of the molecules. For instance, Jaffé [174–176] found that the value of the Hammett σ constant was linearly related with π-electron densities in the phenyl ring. Another linear relationship was reported early between the σ constants and π-electron localization energies [177]. An operational value of such relationships lays in the possibility of the extension of known ρ-σ correlations by the inclusion of the substituents that have the empirical electronic substituent constants unknown or that are difficult to measure. The Hammett σ and Taft σ^0 constants have been also examined using the semiempirical modified neglect of diatomic differential overlap (MNDO) quantum chemical characteristics of a series of benzoic acids and benzoate anions [178]. The most significant correlation ($r = 0.921$; $s = 0.17$; $n = 18$) for σ was found with the calculated electronic charge on the oxygen of the anion (q_O). A nearly equally good linear correlation ($r = 0.899$; $s = 0.19$; $n = 18$) was observed between σ and the reciprocal value of E_{HOMO} (energy of highest occupied molecular orbital) of the benzoate anion. Since σ and σ^0 are well correlated with each other, it was not surprising that σ^0 regressed well against the same two parameters, although the preferred parameter was E_{HOMO}^{-1}:

$$\sigma^0 = 34.38E_{HOMO}^{-1} + 7.87$$

$$r = 0.946 \qquad s = 0.12 \qquad n = 18$$

This correlation supports the presumption that the Hammett σ constant and the related equilibrium constant for the ionization of substituted benzoic acids are determined by the electronic effects in anion. The data on charge distribution and the HOMO energies obtained from the ab initio calculations with Slater-type STO-3G basis sets augmented by diffuse functions for all oxygen atoms resulted in similar correlations [179]. The statistical quality of the correlations obtained were surprisingly good, considering the neglect of solvation effects, which are known to be of considerable importance in acid–base equilibria [96]. From the electrostatic considerations, both E_{HOMO} and q_O^- can be considered as the "enthalpic" descriptors. The ionization equilibrium of carboxylic acids, however, is known to be an isoentropic reaction series, and thus the correlation of the pK_a with this type of molecular descriptors is not unexpected. Notably, Austin Model 1 (AM1) calculated partial charges on the oxygen atom q_{O}^{-} of benzoate anions correlate even better with a little more limited set of Hammett σ constants [180].

The ab initio quantum chemical calculations have been also used for the separation of the inductive and mesomeric effects in aromatic systems [181]. However, such separations depend on certain assumptions about the absence or presence of one or another effect in the system that makes the conclusions somewhat arbitrary.

Finally, it should be emphasized that reliable substituent constants have been reported only for the meta- and para-substituted phenyl systems. In the case of ortho-substituted compounds, it has been demonstrated that different standard processes lead to very different substituent effect scales [71]. This may be caused by the cooperative field and steric interactions between the spatially closely fixed substituent and reaction center that are strongly dependent on the nature of the process involved. Thus, no universal inductive, resonance, or steric constant scale will be reported here for the ortho-substituted phenyl compounds.

The resonance constants have been used extensively, both alone and in the framework of multiparameter equations, for the description of chemical reactivity in systems involving the direct mesomeric interaction between substituent and reaction center. For example, the alkaline hydrolysis of tropolone ethers in 60% aqueous dioxane solutions at 30°C correlates well with the σ constants as [182]:

$$\log k = (-3.26 \pm 0.28) - (4.22 \pm 0.57)\sigma^{-}$$

$$R^2 = 0.968 \qquad s = 0.331 \qquad n = 11$$

This reaction is strongly accelerated by $-$R groups in conjugated position to the reaction center, being in a direct resonance with this center in the transition state of the reaction. On the other hand, a substantial direct resonance of $+$R groups with the transition state of the solvolysis of substituted diphenylchloromethanes in ethanol at 25°C has resulted in a highly significant correlation equation involving the σ^{+} constants as:

$$\log k = (-4.16 \pm 0.06) - (4.35 \pm 0.19)\sigma^{+}$$

$$R^2 = 0.996 \qquad s = 0.102 \qquad n = 14$$

In most cases, however, the multiparameter approach is needed for a satisfactory LFER description of data. For instance, the acetolysis rate constants for substituted neophyl brosylates at 75°C have been described by the following equation [183]:

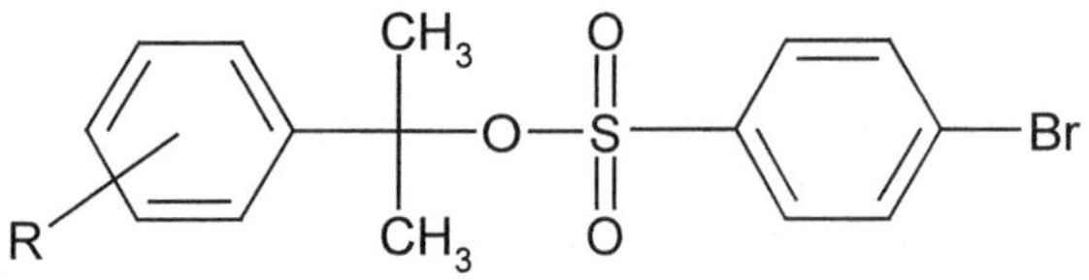

$$\log(k/k_0) = (-3.78 \pm 0.48)(\sigma^0 + 0.578\Delta\sigma_R^+)$$
$$R^2 = 0.9994 \qquad s = 0.038 \qquad n = 29$$

where k_0 denotes the rate constant for the unsubstituted compound and $\Delta\sigma_R^+ = \sigma^+ - \sigma^0$. The coefficient in front of the last term has been fitted using the nonlinear least-squares treatment. A similar equation had been obtained for the rate constants of acetolysis of benzyl tosylates at 25°C as [183]:

$$\log(k/k_0) = -3.444(\sigma^0 + 1.524\Delta\sigma_R^+)$$
$$R^2 = 0.954 \qquad s = 0.401 \qquad n = 25$$

The substituent effects on the ketone-acetal equilibrium constants in a series of substituted acetophenones in methanol at 25°C

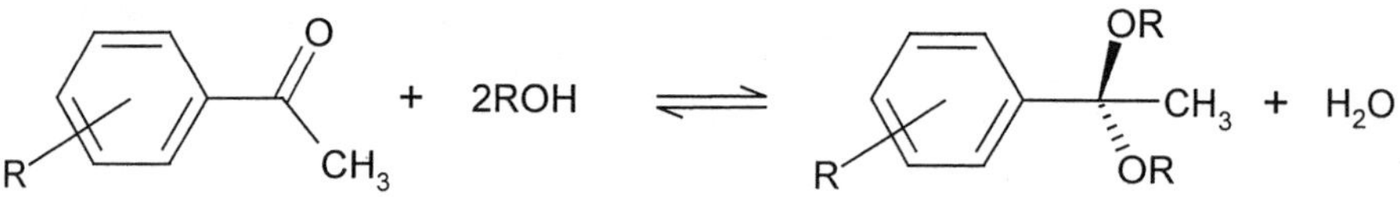

have been correlated using the Taft–Lewis DSP equation as [184]:

$$\log(K/K_0) = (1.73 \pm 0.11)\sigma_I - (1.34 \pm 0.07)\sigma_R^+$$
$$R^2 = 0.998 \qquad s = 0.04 \qquad n = 10$$

where K_0 is the equilibrium constant for the unsubstituted compound. Analogous equations have been developed also for the rate constants of the forward and reverse reactions of dimethyl acetal formation from substituted acetophenones [185]. Thus, the acetal formation rate constants are described by the following DSP equation:

$$\log(k^f/k_0^f) = (-1.45 \pm 0.11)\sigma_I - (0.71 \pm 0.07)\sigma_R^+$$
$$R^2 = 0.966 \qquad s = 0.070 \qquad n = 8$$

whereas the acetal hydrolysis rate constants are presented as follows:

$$\log(k^r/k_0^r) = (-3.17 \pm 0.13)\sigma_I - (2.02 \pm 0.07)\sigma_R^+$$
$$R^2 = 0.992 \qquad s = 0.083 \qquad n = 8$$

The influence of electronic effects is substantially higher for the reverse reaction, the acetal hydrolysis. Therefore, the difference between the electronegativity and the electron donor/acceptor properties of the reaction center in the transition state and in the product of the reaction (acetal), respectively, is larger than the corresponding difference for the reagent (ketone) and the transition state.

The sensitivity of the reactivity on substituent effects may be significantly affected by the reaction medium. For instance, the pK_a of acidic dissociation of meta- and para-substituted benzoic acids in different media has been correlated as [79]:

$$\mathrm{p}K_a = (1.63 \pm 0.46) - (11.00 \pm 0.48)\sigma^0 - (2.24 \pm 0.62)\sigma_R^+$$

$$R^2 = 0.980 \qquad s_0 = 0.07 \qquad n = 16$$

in the gas phase,

$$\mathrm{p}K_a = (12.26 \pm 0.07) - (2.35 \pm 0.08)\sigma^0 - (1.37 \pm 0.48)\sigma_R^+$$

$$R^2 = 0.994 \qquad s_0 = 0.033 \qquad n = 19$$

in dimethyl formamide,

$$\mathrm{p}K_a = (8.63 \pm 0.04) - (1.57 \pm 0.10)\sigma^0 - (0.30 \pm 0.12)\sigma_R^+$$

$$R^2 = 0.970 \qquad s_0 = 0.053 \qquad n = 16$$

in butanol, and

$$\mathrm{p}K_a = (4.21 \pm 0.01) - (1.01 \pm 0.06)\sigma^0 - (0.251 \pm 0.086)\sigma_R^+$$

$$R^2 = 0.953 \qquad s_0 = 0.047 \qquad n = 30$$

in the aqueous solutions. These results demonstrate that the sensitivity of the acid–base equilibrium constants toward both the substituent induction and resonance effects is significantly reduced in high polarity and hydrogen-bonding solvents.

A similar tendency can be observed in the case of the pK_a of acidic dissociation of the para-substituted phenols in different media. The respective DSP equations are as [79]:

$$\mathrm{p}K_a = (1.58 \pm 0.55) - (14.10 \pm 1.13)\sigma_I - (13.10 \pm 0.67)\sigma_R^-$$

$$R^2 = 0.992 \qquad s_0 = 0.038 \qquad n = 9$$

for the data in the gas phase,

$$\mathrm{p}K_a = (14.03 \pm 0.20) - (5.39 \pm 0.52)\sigma_I - (5.24 \pm 0.39)\sigma_R^-$$
$$R^2 = 0.978 \qquad s_0 = 0.057 \qquad n = 12$$

for the data in acetone, and

$$\mathrm{p}K_a = (14.96 \pm 0.09) - (2.78 \pm 0.21)\sigma_I - (2.63 \pm 0.17)\sigma_R^-$$
$$R^2 = 0.993 \qquad s_0 = 0.32 \qquad n = 6$$

for the data in ethanol. Apparently, the influence of possible hydrogen bonding of the reaction center (phenolate anion) by the solvent decreases the intramolecular electronic effects more than the increase of the polarity of the medium.

The decrease of the sensitivity of the induction and resonance effects by high-polarity and hydrogen-bonding solvents is, however, not a general rule and depends on the reaction mechanism. For instance, the rate constants of the monomolecular decomposition of (α-aryl)ethyl chlorides can be described by the following two-parameter equations:

$$\log k_I = (0.18 \pm 0.07) - (1.414 \pm 0.0.015)\sigma^0 - (1.353 \pm 0.121)\sigma_R^+$$
$$R^2 = 0.996 \qquad s_0 = 0.022 \qquad n = 7$$

in the gas phase [79,186], and

$$\log k_I = (0.06 \pm 0.04) - (5.44 \pm 0.16)\sigma^0 - (6.00 \pm 0.14)\sigma_R^+$$
$$R^2 = 0.998 \qquad s_0 = 0.015 \qquad n = 14$$

in the 80% aqueous acetone solutions [186]. A substantial increase in the ρ constants reflects the increase in the difference between the electronegativity and electron-acceptor/donor properties of the reaction center in the initial and transition state, respectively, while transferred from the gas phase to a condensed medium.

Various substituent resonance constants have been applied for the description of the biological activity of compounds. For instance, the 50% inhibitory concentrations, I_{50}, of ethylphosphonic acid esters

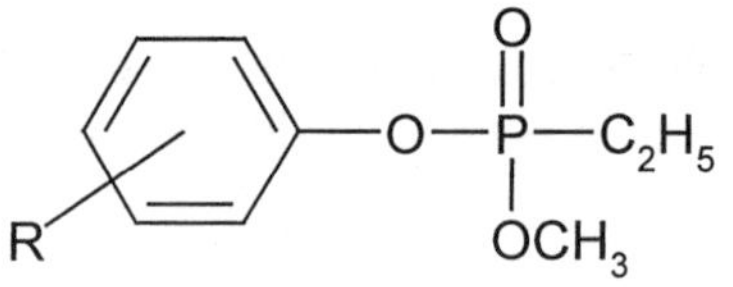

on fly head cholinesterase [187] have been correlated with the resonance constants as follows [188]:

$$\log \frac{1}{C} = 4.31 + (2.42 \pm 0.49)\sigma^-$$

$$R^2 = 0.970 \qquad n = 7$$

The ρ^- for this enzymatic reaction as well as the ρ^- value for the inhibitory activity of phosphorous acid esters ($\rho^- = 2.45 \pm 0.54$) [188,189]

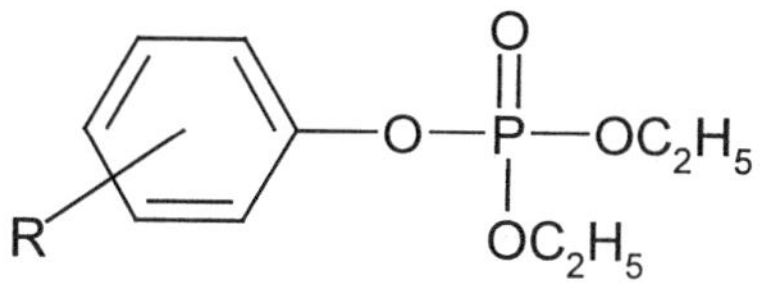

are significantly higher than the ρ^- value for the alkaline hydrolysis for the last series of compounds:

$$\log k = (1.56 \pm 0.07) + (1.25 \pm 0.14)\sigma^-$$

$$R^2 = 0.938 \qquad s = 0.151 \qquad n = 25$$

Consequently, the transition state of the enzymatic reaction has to be more susceptible to electron-acceptor substituents in the phenyl ring as compared to the transition state of alkaline hydrolysis. Comparisons of this kind may be useful in the determination of the mechanism of enzymatic reactions, also in terms of early and late transition states of reaction.

The resonance substituent constants have been also used to correlate the activity of oxidoreductases. The rate constant of the reduction of meta- and para-substituted nitrobenzenes to hydroxylamides by xanthine oxidase has been described by the following one-parameter equation [102]:

$$\log k = (1.73 \pm 0.11) + (1.09 \pm 0.30)\sigma^-$$

$$R^2 = 0.876 \qquad s = 0.192 \qquad n = 21$$

The positive sign of ρ^- implies that the more electronegative and electron-withdrawing substituents stabilize the transition state and accelerate the enzymatic reduction of compounds. The V_{max} of oxidation of substituted phenylglycines

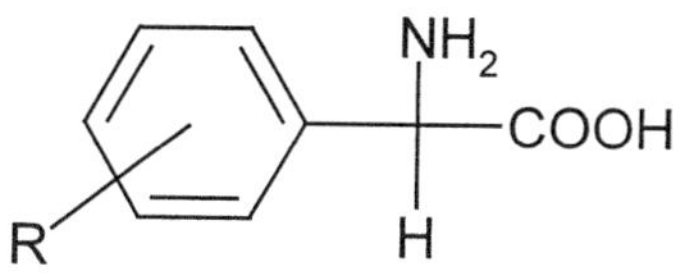

by amino acid oxidase at pH 8.5 has been correlated using the following equation [190]:

$$\log V_{max} = (0.28 \pm 0.31) + (2.05 \pm 0.45)\sigma^+$$

$$R^2 = 0.882 \qquad s = 0.514 \qquad n = 15$$

The relatively large value of ρ^+ has been interpreted as indication of the involvement of radical ions in the transition state of the reaction.

2.2.3 Steric Constants

Induction and resonance interactions are caused by electronic polarization in the molecule. However, molecular properties are also affected by the finite size of the atomic constituents and groups in the molecule and the interactions and geometrical restrictions thereof. The respective effects are known in chemistry as steric effects. In most contemporary applications, steric interactions can be evaluated using different theoretical models spanning from various molecular mechanics techniques to the advanced quantum chemical calculations. However, the steric effects and the respective quantitative substituent constants have been also defined in the framework of the linear free energy relationship formalism. As mentioned [cf. Eqs. (2.11) and (2.12)], the original substituent steric constants within the LFER scheme, E_S, were derived by Taft on the basis of experimental kinetic data on the acidic hydrolysis of esters. It had been debated later that these constants reflect also the possible hyperconjugation effects of substituents. Several "corrected" scales have been proposed for the description of the steric effects in the molecule. Proceeding from the estimate of the hyperconjugation term in the LFER expansion, obtained using molecular orbital (MO) calculations, Hancock et al. [191] proposed new steric constants as:

$$E_S^C = E_S + 0.306(n_H - 3) \tag{2.36}$$

where n_H is the number of hydrogen atoms at the connecting (first) carbon atom of the substituent, with the number 3 corresponding to the standard substituent (—CH_3). Another scale of the steric constants was proposed with the correction accounting for the C—C bond hyperconjugation:

$$E_S^0 = E_S - 0.33(3 - n_H) + 0.13n_C \tag{2.37}$$

for nonalkyl substituents and

$$E_S^0 = E_S - 0.20(3 - n_H) \tag{2.38}$$

for the alkyl substituents [71]. In Eq. (2.37), n_C denotes the number of carbon atoms bonded to the connecting carbon atom of a substituent.

Another parameter related to molecular volume and steric effects is the molar refractivity (MR). Experimentally, it is obtained from

$$\mathrm{MR} = \frac{n_D^2 - 1}{n_D^2 + 2}\frac{MW}{d} \tag{2.39}$$

where n_D is the index of refraction, d is density, and MW is the molecular weight of a compound. Various substituent steric constant scales are listed in Table 2.6. The numerical values of some common empirical substituent steric constants are given in Table 2.7.

The steric constants have been often used in the LFER correlations of the chemical reactivity and biological activity. In most cases, the steric interaction term appears together with other substituent effect (inductive and/or resonance) terms. One of the exceptions is the following nucleophilic substitution reaction at the aromatic carbon atom [192]:

O_2N–$C_6H_3(NO_2)$–Br + RNH_2 ⟶ O_2N–$C_6H_3(NO_2)$–NHR + HBr

for which log k_{AB} correlates with the steric constants alone as [63]:

$$\log k_{AB} = (-2.51 \pm 0.05) + (1.49 \pm 0.08)E_S$$
$$R^2 = 0.986 \qquad s = 0.06 \qquad n = 7$$

TABLE 2.6 Empirical Substituent Steric Constant Scales Defined by Using Different Standard Processes

Scale Notation	Standard Process	Reference
E_s	log k of acidic hydrolysis of aliphatic esters	*a*
E_s^c	$E_S^C = E_S + 0.306(n_H - 3)$	*b*
E_s^0	$E_S^0 = E_S - 0.33(3 - n_H) + 0.13n_C$ for nonalkyl substituents	*c*
E_s^0	$E_S^0 = E_S - 0.20(3 - n_H)$ for the alkyl substituents	*d*
E_s'	log k of acidic hydrolysis of aliphatic esters	*e*
υ_x	van der Waals radii of atoms	*f*
MR	Molecular refraction	*f*

[a] R.W. Taft, *J. Am. Chem. Soc.* **74,** 3120 (1952); W.A. Pavelich and R.W. Taft, *ibid.* **79,** 4935 (1957).
[b] C.K. Hancock, E.A. Meyers, and B.J. Jager, *J. Am. Chem. Soc.* **83,** 4211 (1961).
[c] V.A. Palm, *Foundations of the Quantitative Theory of Organic Reactions* (in Russian). Khimiya, Leningrad, 1967.
[d] J.-A. MacPhee, A. Panaye, and J.-E. Dubois, *Tetrahedron* **34,** 3553 (1978); *J. Org. Chem.* **45,** 1164 (1980).
[e] M. Charton, *J. Am. Chem. Soc.* **91,** 615 (1969); *Prog. Phys. Org. Chem.* **8,** 235 (1971).
[f] R.C. Weast, ed., *CRC Handbook of Chemistry and Physics,* 65th ed. CRC Press, Boca Raton, FL, 1984.

TABLE 2.7 Numerical Values for Selection of Empirical Substituent Steric Constant Scales [63,63a]

Substituent	E_S	E_S^C	E_S^0	MR
H	0.00	0.00	0.00	0.10
CH_3	−1.24	−0.32	−0.32	0.57
C_2H_5	−1.31	−0.70	−0.57	1.03
C_3H_7	−1.60	−0.99	−0.86	1.50
$CH(CH_3)_2$	−1.71	−1.40	−1.14	1.50
C_4H_9	−1.63	−1.02	−0.89	1.96
$CH_2CH(CH_3)_2$	−2.17	−1.56	−1.43	1.96
$CH(CH_3)C_2H_5$	−2.37	−2.06	−1.80	1.94
$C(CH_3)_3$	−2.78	−2.78	−2.39	1.97
Cyclobutyl	−1.30	−0.99	−0.73	1.79
C_5H_{11}	−1.64	−1.03	−0.90	2.43
$CH_2CH_2CH(CH_3)_2$	−1.59	−0.98	−0.98	2.43
$CH_2C(CH_3)_3$	−2.98	−2.37	−2.24	2.43
$CH(C_2H_5)_2$	−3.22	−2.91	−2.65	2.42
Cyclopentyl	−1.75	−1.44	−1.18	2.20
C_6H_{13}	−1.54	−1.23	−0.97	2.89
$(CH_2)_3CH(CH_3)_2$	−1.67	−1.06	−0.93	2.89
$CH_2CH_2C(CH_3)_3$	−1.58	−0.97	−0.84	2.90
$CH(CH_3)C(CH_3)_3$	−4.57	−4.26	−4.00	2.89
Cyclohexyl	−2.03	−1.72	−1.46	2.67
$CH(C_3H_7)_2$	−3.35	−3.04	−2.78	3.36
$CH(CH_3)CH_2C(CH_3)_3$	−3.09	−2.78	−2.52	3.36
$C(C_2H_5)_3$	−5.04	−5.04	−4.65	3.35
$C(CH_3)_2C(CH_3)_3$	−5.14	−5.14	−4.75	3.36
C_8Hl_7	−1.57	−0.96	−0.83	3.81
$C(CH_3)_2CH_2C(CH_3)_3$	−3.81	−3.81	−3.42	3.82
$CH(CH_2CH(CH_3)_2)_2$	−3.71	−3.40	−3.14	4.29
$C(CH_3)_2$-*c*-butyl	−3.73	−3.73	−3.34	4.08
CH_2-*c*-hexyl	−2.22	−1.61	−1.48	3.13
$CH{=}CH_2$	−2.84	−2.53	−2.40	1.10
$CH{=}CHCH_3$	−2.87	−2.56	−2.43	1.56
$C(CH_3){=}CH_2$	−3.32	−3.32	−3.06	1.56
C_6H_5	−1.01	−1.01	−1.01	2.54
$CH_2C_6H_5$	−1.61	−1.00	−0.87	3.00
$CH_2CH_2C_6H_5$	−1.62	−1.01	−0.88	3.47
$(CH_2)_3C_6H_5$	−1.69	−1.08	−0.95	3.93
F	−0.46	−0.46	−0.46	0.10
Cl	−0.97	−0.97	−0.97	0.60
Br	−1.16	−1.16	−1.16	0.89
I	−1.40	−1.40	−1.40	1.39
OH	−0.55	−0.55	−0.55	0.28
NH_2	−0.61	−0.61	−0.61	0.54
NO_2	−1.01	−1.01	−1.01	0.74
SH	−1.07	−1.07	−1.07	0.89
SF_5	−2.91	−2.91	−2.91	1.00

TABLE 2.7 ***(Continued)***

Substituent	E_S	E_S^C	E_S^0	MR
CN	−0.51	−0.51	−0.51	0.63
OCH_3	−0.55	−0.55	−0.55	0.79
SCH_3	−1.07	−1.07	−1.07	1.38
$N(CH_3)_3^+$	−2.84	−2.84	−2.84	
$Si(CH_3)_3$	−3.36	−3.36	−3.36	2.50
CH_2F	−1.48	−0.87	−0.87	0.54
CH_2Cl	−1.48	−0.87	−0.87	1.05
CH_2Br	−1.51	−0.90	−0.90	1.34
CH_2I	−1.61	−1.00	−0.87	1.86
CH_2OH	−1.21	−0.60	−0.60	0.72
CH_2OCH_3	−1.43	−0.82	−0.82	1.20
$CH_2OC_2H_5$	−1.61	−1.00	−1.00	1.67
$CH_2OC_3H_7$	−1.63	−1.02	−1.02	2.14
$CH_2OC_4H_9$	−1.66	−1.05	−1.05	2.60
$CH_2OCH_2CH(CH_3)_2$	−1.71	−1.10	−1.10	2.60
$CH_2OC_6H_5$	−1.57	−0.96	−0.96	3.18
CH_2NO_2	−2.71	−2.10	−2.10	1.20
CH_2COCH_3	−1.99	−1.38	−1.25	1.51
CH_2CN	−2.18	−1.57	−1.44	1.01
CH_2SCH_3	−1.58	−0.97	−0.97	1.84
$CH_2SC_2H_5$	−1.71	−1.10	−1.10	2.41
$CH_2SC_6H_5$	−2.03	−1.42	−1.42	3.79
$CH_2N(CH_3)_3^+$	−4.13	−3.52	−3.52	
CH_2CH_2Cl	−2.14	−1.53	−1.40	1.51
CH_2CH_2Br	−2.24	−1.63	−1.50	1.80
CH_2CH_2I	−2.26	−1.65	−1.52	2.33
$CH_2CH_2OCH_3$	−2.01	−1.40	−1.27	1.67
CH_2CH_2COOH	−2.21	−1.60	−1.47	1.65
$CH_2CH_2COOC_2H_5$	−2.14	−1.53	−1.40	2.58
CH_2CH_2CN	−2.14	−1.53	−1.40	1.48
$CH_2CH_2NH_3^+$	−3.06	−2.45	−2.32	
$CH_2CH_2N(CH_3)_3^+$	−3.23	−2.62	−2.49	
$CH_2CH(OH)CH_3$	−2.31	−1.70	−1.57	1.65
$(CH_2)_3F$	−1.64	−1.03	−0.90	1.48
$(CH_2)_3Cl$	−1.72	−1.11	−0.98	1.98
$(CH_2)_3OCH_3$	−1.66	−1.05	−1.05	2.14
$(CH_2)_3OC_2H_5$	−1.69	−1.08	−1.08	2.60
$(CH_2)_3COOH$	−1.65	−1.04	−0.91	2.12
$(CH_2)_3COOC_2H_5$	−1.74	−1.13	−1.00	3.04
$(CH_2)_3CN$	−1.74	−1.13	−1.00	1.94
$(CH_2)_3N(CH_3)_3^+$	−2.59	−1.98	−1.85	
$(CH_2)_4OCH_3$	−1.58	−1.27	−1.14	2.60
$(CH_2)_4COOH$	−1.56	−0.95	−0.82	2.58
CHF_2	−1.91	−1.60	−1.60	0.52
$CHCl_2$	−2.78	−2.47	−2.47	1.53
$CHBr_2$	−3.10	−2.79	−2.79	1.68

TABLE 2.7 ***(Continued)***

Substituent	E_S	E_S^C	E_S^0	MR
$CH(Cl)CH_3$	−1.74	−1.43	−1.30	1.51
$CH(Br)CH_3$	−1.93	−1.62	−1.49	1.80
$CH(I)CH_3$	−2.60	−2.29	−2.16	2.32
$CH(OH)CH_3$	−1.15	−0.84	−0.71	1.18
$CH(OH)C_2H_5$	−1.58	−1.27	−1.14	1.65
$CH(OH)C_3H_7$	−1.57	−1.26	−1.13	2.11
$CH(OH)C_4H_9$	−1.55	−1.24	−1.24	2.58
$CH(OH)C_5H_{11}$	−1.58	−1.27	−1.14	3.04
$CH(OH)CH_2OH$	−2.05	−1.74	−1.61	1.35
$CH(OCH_3)CH_3$	−1.88	−1.57	−1.44	1.67
$CH(OC_2H_5)CH_3$	−1.93	−1.62	−1.49	2.13
$CH(SC_2H_5)CH_3$	−2.77	−2.46	−2.33	2.75
$CH(SC_2H_5)_2$	−3.55	−3.24	−3.24	4.01
CF_3	−2.40	−2.40	−2.40	0.50
CCl_3	−3.30	−3.30	−3.30	2.01
CBr_3	−3.67	−3.67	−3.67	2.88
$C(OH)(CH_3)_2$	−2.98	−2.98	−2.72	1.64
$C(CN)(CH_3)_2$	−2.00	−2.00	−1.74	1.94

Also, the bimolecular rate constants of the acid-catalyzed esterification of carboxylic acids in methanol at 40°C

$$RCOOH + CH_3OH \longrightarrow RCOOCH_3 + H_2O$$

have been correlated with the steric constants alone as [63]:

$$\log k_{AB} = (0.45 \pm 0.03) + (1.01 \pm 0.01)E_S$$

$$R^2 = 0.998 \qquad s = 0.051 \qquad n = 15$$

This correlation is not unexpected because it is related to the reverse reaction of the acid-catalyzed hydrolysis of esters that was used as the standard reaction for the definition of E_S constants. Consequently, both reactions have the same transition state.

The bimolecular rate constants k_{AB} of the alkaline hydrolysis of carboxylic acid methyl esters $RCOOCH_3$ in aqueous solutions at 25°C have been correlated simultaneously with the induction and steric constants of substituents R [63]:

$$\log k_{AB} = (1.58 \pm 0.07) + (2.45 \pm 0.05)\sigma^* + (0.80 \pm 0.05)E_S$$

$$R^2 = 0.994 \qquad s = 0.137 \qquad n = 24$$

A similar equation was obtained for the bimolecular rate constants k_{AB} of the alkaline hydrolysis of carboxylic acid propyl esters $RCOOC_3H_7$ at the same reaction conditions [63]:

$$\log k_{AB} = (1.35 \pm 0.05) + (1.00 \pm 0.05)\sigma^* + (1.51 \pm 0.05)E_S$$

$$R^2 = 0.994 \qquad s = 0.077 \qquad n = 12$$

As expected, the bulkier propyl group as compared to the methyl group increases the sensitivity of the reaction rate toward the steric interaction between the acid and alcohol residues of the ester in the transition state. Interestingly, this effect is accompanied by a significant decrease of the sensitivity of reaction rate to induction effect.

Apart from the LFER correlations of the chemical reactivity, the empirical steric constants have been widely used in the QSAR studies of the drug–receptor or substrate–enzyme systems, both in vivo and in vitro. For example, the chymotrypsin hydrolysis of para-nitrophenyl esters at pH 5.92 has been characterized by the following LFER equation [193]:

$$\log(k_2/K_m) = (2.23 \pm 0.52) + (1.76 \pm 0.42)E_S + (0.79 \pm 0.52)\pi$$

$$R^2 = 0.962 \qquad s = 0.201 \qquad n = 8$$

where π is the Hansch hydrophobic substituent constant [194]. In the last equation, the main term can be related to the steric hindrance in the active center of enzyme. The antibacterial action of isonicotinic hydrazides has been also correlated with a two-parameter equation as [195]:

$$\log(1/C) = 5.78 + 0.89E_S - 3.70F$$

$$R^2 = 0.835 \qquad s = 0.368 \qquad n = 17$$

where F is the Swain–Lupton field-inductive parameter.

The molar refractivity (MR) is another steric constant widely used in QSAR correlations. For example, the inhibitory activity of 7-substituted 4-hydroxyquinoline-3-carboxylic acids

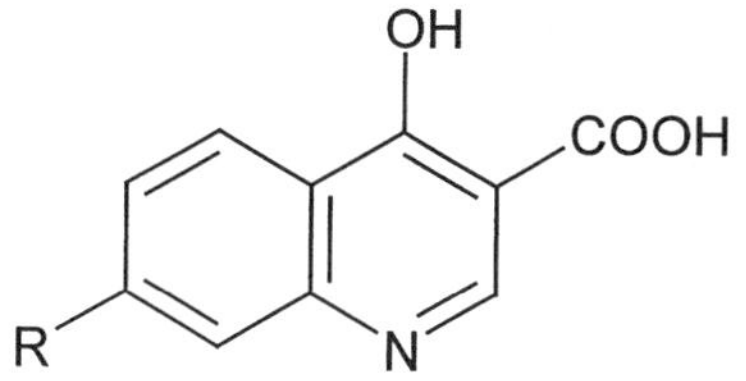

on the mitochondrial malate dehydrogenase (MDH) and skeleton muscle lactate dehydrogenase (LDH) has been correlated by the following one-parameter equations [196]:

$$\log(1/C)_{\mathrm{MDH}} = 2.46 + 0.58\mathrm{MR}$$
$$R^2 = 0.961 \qquad s = 0.34 \qquad n = 29$$

and

$$\log(1/C)_{\mathrm{LDH}} = 2.79 + 0.36\mathrm{MR}$$
$$R^2 = 0.960 \qquad s = 0.20 \qquad n = 11$$

These equations illustrate the importance of steric interactions between the side chain of the substrate and the active center of the enzyme.

The popularity of empirical steric constants has declined because of the development of a variety of theoretically calculable geometrical molecular descriptors. The latter have the advantage of being connected to clearer mathematical (geometrical) models that enable better interpretation of the relationships with the molecular properties studied. An overview of geometrical descriptors is given in Chapter 3.

2.3 SOLVATIONAL DESCRIPTORS

2.3.1 Macroscopic Solvent Effect Scales

The solvation of molecules in condensed disordered media such as liquids or other dense fluids involves a multitude of physically distinct intermolecular interactions. These interactions bring about the changes in the electronic structure and geometry of the solute molecule, the changes in the radial and angular distribution of the solvent molecules, and overall electrical polarization of the solvent around a dissolved solute molecule. It has been customary to distinguish between the *equilibrium* and *nonequilibrium* solvation effects. The equilibrium effects can be described by the difference in the chemical potential of the solute in the gas phase and in solution at thermodynamic equilibrium. Alternatively, these effects can be related to the time-independent changes in the electronic and geometrical structure of the solute–solvent system. The nonequilibrium effects arise from the coupling of electron–nuclear motions in the solute and solvent molecules and incomplete relaxation of the solvent in the case of fast solute-related processes. The theoretical treatment of such effects requires the use of time-dependent quantum mechanics and molecular dynamics.

The Gibbs free energy of solvation is defined as [197]:

$$\Delta G_{\mathrm{sol}} = \Delta G_s - \Delta G_{\mathrm{gas}} \tag{2.40}$$

where both the free energy of the solute in the solution (ΔG_s) and the free energy of the solute in the gas phase (ΔG_{gas}) can be, in principle, rigorously presented by the respective partition functions, Z_s and Z_s, in those two media as follows:

$$\Delta G_s = -RT \ln Z_s \tag{2.41}$$

$$\Delta G_{\text{gas}} = -RT \ln Z_{\text{gas}} \tag{2.42}$$

It should be emphasized that all components of the partition function

$$Z = Z_{\text{tr}} Z_{\text{rot}} Z_{\text{vib}} Z_{\text{el}} \tag{2.43}$$

that is, the respective translational, rotational, vibrational, and electronic partition functions may be affected by the solvation of the molecule. Therefore, a rigorous ad hoc approach to the solvation is very complicated and requires the use of fundamental physical theory for the description of different motions of particles and molecules in different media.

The free energy of solvation, ΔG_{sol}, can be also divided into the respective enthalpic and entropic contributions, ΔH_{sol} and ΔS_{sol}, as follows [198,199]:

$$\Delta G_{\text{sol}} = \Delta H_{\text{sol}} - T\,\Delta S_{\text{sol}} \tag{2.44}$$

This separation brings about the differentiation between the enthalpic and entropic solvent effects, each of which should be treated using different, in principle, theoretical models. Another partitioning of the free energy of solvation can be based on the separation of physically distinct interactions between the solute and solvent molecules. One possible way is to divide the solvation free energy, ΔG_S, into the following contributions [200]:

$$\Delta G_S = \Delta G_{\text{elst}} + \Delta G_{\text{disp}} + \Delta G_{\text{cav}} + \Delta G_{\text{HB}} \tag{2.45}$$

where $\Delta G_{\text{elst}} = E_{\text{elst}}$ is the electrostatic interaction energy of the solute charge distribution with the reaction field it creates in the surrounding polarizable dielectric medium, $\Delta G_{\text{disp}} = E_{\text{disp}}$ is the energy of the dispersion interaction between the solute and adjacent solvent molecules, ΔG_{cav} is the free energy of the cavity formation in the solution, and ΔG_{HB} denotes the energy of the solute–solvent hydrogen bonding. The first three terms of Eq. (2.45) can be related to the size and shape of the molecular cavity in the solvent created in the process of dissolution of the solute molecule. The electrostatic and dispersion interaction energies can be calculated quantum chemically by solving the appropriate Schrödinger equation. The free energy of cavity formation has essentially an entropic nature, and hence it should be calculated using the statistical mechanical approach.

Historically, the complexity and multitude of intermolecular interactions in condensed media has given rise to numerous *empirical* or *parametric* and *physical* models of solvation. The various physical models will be considered in Chapter 3 as related to the respective theoretical solvational descriptors. In this paragraph, we will concentrate on empirical molecular descriptors based on the parametric approach. This approach has emerged from the widespread observance of the solvation-related linear free energy relationships, formally similar to the relationships describing the intramolecular substituent effects [78,201]. According to the parametric approach, the solvation is considered as a composite of physically independent processes, each of which can be described by a single empirical scale of the solvent effect. These processes have been conventionally divided into two groups, reflected partially by the partitioning of the free energy of solvation according to Eq. (2.45). The first is defined as related to the nonspecific or macroscopic solvent effects. These include the solvent dielectric polarization in the field of solute molecule, the isotropic solute–solvent dispersion interactions, and the solute cavity formation in the bulk of the solvent. The second type of solvent effects, the specific or microscopic effects, involve the formation of directed semichemical bonds (e.g., hydrogen bonds) and other anisotropic interactions between the solute and solvent molecules in solution. Notably, each empirical solvent scale used in the development of solvation linear free energy relationships has been derived from some experimental data that presumably correspond to a single solvent effect according to the above-given classification. The empirical approach has, however, a deficiency that no predictions can be made in the case of chemical and physical processes for which the experimental information to define the necessary solvent effect scales is missing.

Most empirical LFER solvent effect molecular descriptors have been related, at the same time, to various quantum mechanical and statistical-physical models and concepts of the intermolecular interactions in condensed media. In other words, the empirical solvent effect scales based on certain macroscopic experimental data are often related to individual, physically distinct mechanisms of solvation at the microscopic molecular level. Unfortunately, different experimentally observable chemical and physical processes are influenced, as a rule, simultaneously by several solute–solvent interactions in solution. For instance, the dissolution process involves always the cavity formation in the solvent and the solute–solvent dispersion interactions at close distances. This created the problem of finding suitable experimental phenomena that is affected only or predominantly by a single mechanism of intermolecular interactions. The different mechanisms of interaction have, in general, different relative intensity of influence on different experimental molecular properties in solution, and thus the "mixed" empirical solvent effect scales cannot be expected to be universally applicable. It is also obvious that, in general, the treatment of solvent effects requires the use of a multiparameter LFER expansion, each term of which corresponds to a certain model interaction between the dissolved molecules and the solvent. Therefore, the selection of appropriate solvent effect scales is nontrivial and should be based on the independent knowledge about the process or molecular property studied.

In the following, we will concentrate on the nonspecific solvent effect scales. As discussed, the respective intermolecular interactions relate to the bulk of the solvent and not to the directional short-distance intermolecular interactions between the solute and the individual solvent molecules. Different model concepts about the physical interactions in the solution have been used in the definition of empirical scales for the nonspecific solvent effects. For instance, Kosower [202–205] and Dimroth and Reichardt [206–209] have suggested empirical solvent polarity scales as related to the solvent-induced shifts in the maxima of the ultraviolet/visible (UV/Vis) spectra of certain compounds. These shifts originate from the difference in the solvation of the light-adsorbing indicator molecule in the ground state and in the excited state, respectively [210,211]. The molecules for which the ground state is more stabilized by the solvent than the excited state exhibit a hypsochromic or blue shift in the UV/Vis spectrum with the increase of the polarity of the surrounding solvent. The respective compounds have been called the negatively solvatochromic indicators [209]. In the absence of hydrogen bonding between the indicator and the solvent molecules, the solvent stabilization is expected to arise from the electrostatic interaction of the charge distribution of indicator molecule with the solvent reaction field. The leading term in the respective electrostatic energy corresponds to the dipole–dipole interaction, and thus the change in the dipole moment of the molecule due to the light adsorption usually determines the size of the shift of the spectrum. In the case of negatively solvatochromic indicators, the dipole moment of the molecule is diminished in the first excited state as compared to the ground state. In the case of positively solvatochromic indicators that exhibit a bathochromic or red shift in the UV/Vis spectrum with the increase of the polarity of surrounding solvent, the dipole moment of the indicator molecule increases in the excited state.

The choice of indicator compounds for the definition of the nonspecific solvent polarity scales has been directed by the necessity to eliminate the possible specific hydrogen-bonding effects on their spectra. Thus, Kosower [202,203] suggested the *Z* scale of the solvent polarity based on the solvent shifts of the UV spectral maxima of pyridinium oxide

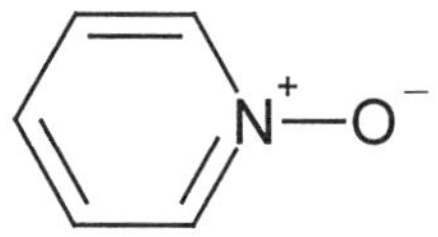

and (4-carboxymethyl)*N*-ethylpyridinium iodide

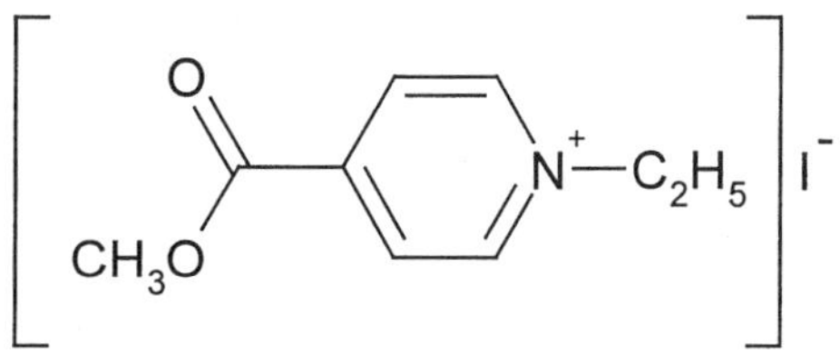

as follows:

$$Z = hv_{\max} \tag{2.46}$$

where υ_{max} is the peak of the maximum in the UV absorption spectrum of the indicator dye in a given solvent. The numerical values of the Z scale were defined by the transition energy (2.46) in kilocalories/mol. Notably, an approximately linear relationship was found between the Z scale and the Y scale by Grünwald and Winstein derived from the data on the solvolysis [Eq. (2.13)] [27]. The use of pyridinium oxide as a universal indicator dye was, however, limited by several unfavorable properties of this compound. First, it is not soluble enough in the low-polarity solvents to measure the spectrum reliably. Second, the long-wavelength charge-transfer band used for the definition of Z values is shifted in high-polarity solvents to shorter wavelengths where it has been hidden underneath the much stronger $\pi \rightarrow \pi^*$ absorption band.

To overcome these difficulties, Dimroth et al. [206] suggested the use of the pyridinium *N*-phenolbetaine as an indicator dye for the determination of another solvent polarity scale:

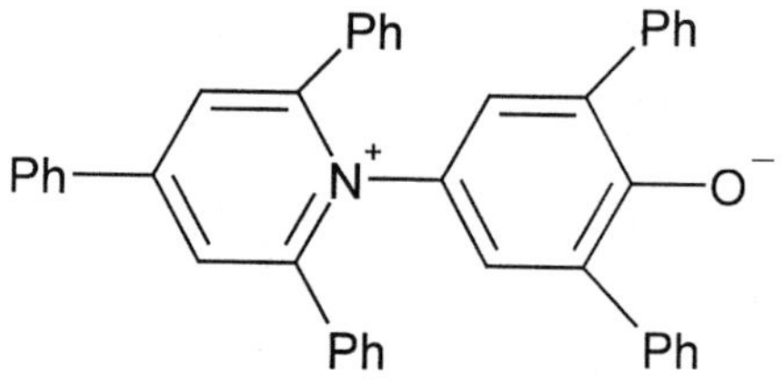

The E_T scale was defined simply as the transition energy for the longest wavelength solvatochromic adsorption band of the above indicator dye (Dimroth's dye) in kilocalories/mole. In some sources, this scale has been also denoted as E_T (30) after the dye sequential order number (no. 30) in the original article. A normalized scale of solvent polarity based on E_T values has been formulated using water and tetramethylsilane (TMS) as extreme reference solvents as follows [212]:

$$E_T^N = \frac{E_T(\text{solvent}) - E_T(\text{TMS})}{E_T(\text{water}) - E_T(\text{TMS})} = \frac{E_T(\text{solvent}) - 30.7}{32.4} \tag{2.47}$$

The advantage of the normalized E_T^N scale is that it is not connected to a specific system of physical units [Systemé International (SI), centimeter-gram-second (CGS), etc.]. The E_T (30) scale has been determined experimentally for many individual solvents [209,213–219], for various binary solvent mixtures [220–238], for liquid organic salts [239–242], and for various microheterogeneous media [243–249].

A number of other solvent polarity scales has been defined as based on the solvatochromic shifts of the electronic spectra of different indicator compounds. For instance, Brooker et al. [250] defined the χ_B and χ_R scales of solvent polarity using a negatively solvatochromic merocyanine dye with a very large blue shift (9500 cm^{-1} between toluene and water) and a positively solvatochromic merocyanine dye with large red shift (−4400 cm^{-1} between iso-octane and 20:80 mixture of lutidine with water),

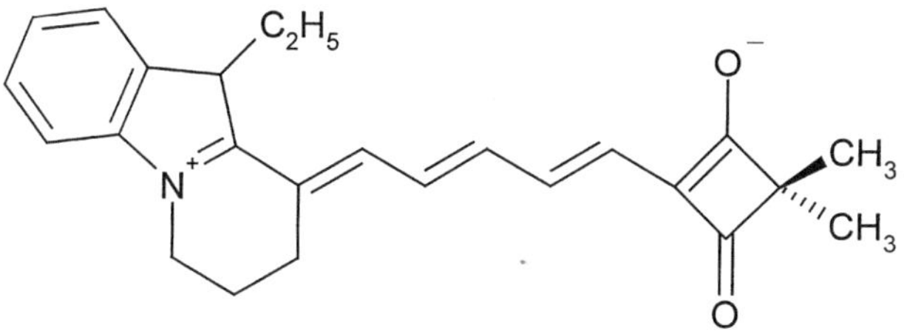

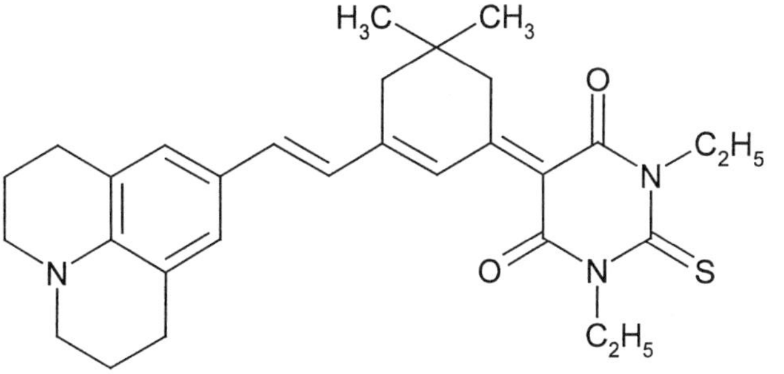

respectively.

The other solvent polarity scales defined from the UV/Vis spectra of negatively solvatochromic compounds include the E_K scale of Walther [251], the E^*_{MLCT} scale of Manuta and Lees [252], the $E_{CT}(\pi)$ scale by Kaim et al. [253], the E_T^{SO} scale [254], the E_B scale [255], and the Φ scale based on the solvent-sensitive $n \rightarrow \pi^*$ transition of selected aliphatic ketones [256].

On the other hand, the π^* scale by Kamlet, Abboud, and Taft [257,258] has been defined on the basis of the positive solvatochromic shifts in the long-wavelength spectrum of the following compounds:

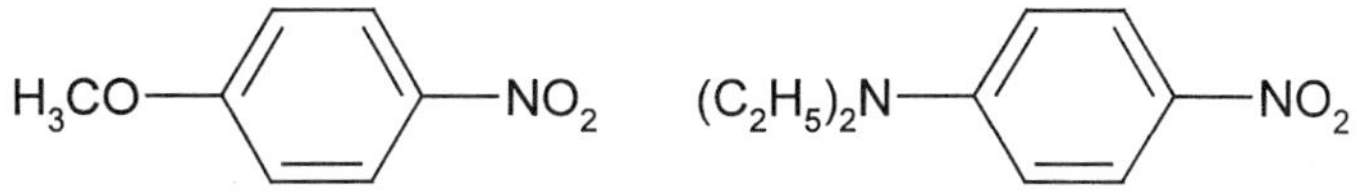

H_3CO–C_6H_4–NO_2 $(C_2H_5)_2N$–C_6H_4–NO_2

These shifts are determined by a substantially more polar excited state of the molecules as compared to the ground state. The excited state is more stabilized by the larger reaction field in high-polarity solvent that leads to the lowering of the transition energy in such a solvent. The π^* scale has been recently revised using the UV/Vis spectroscopic data of *N*,*N*-dimethyl-4-nitroaniline in different solvents [259]. The values of π^* were given for altogether more than 200 solvents.

The other solvent polarity scales defined on the basis of UV/Vis spectra of positively solvatochromic compounds are the π^*_{azo} scale of Buncel and Rajagopal [260], the RPM (relative polarity mass) scale [261], the S scale by Zelinskii's group [262], and the E_{LMCT} scale [263]. Reichardt [209] counted at least 18 different solvent polarity scales based on the solvatochromic shifts in the UV/Vis spectra of indicator dyes. He also listed a large number of other positively or negatively solvatochromic compounds that can be used for the definition of the solvent polarity scales [209].

Another group of nonspecific solvent effect scales is based on experimental data on the chemical reactivity of compounds in different media. The first of such scales was the ionizing power, Y, defined [27] from the rate constants of the *tert*-butyl chloride solvolysis in polar and protic solvents as given by Eqs. (2.13) and (2.14). It was, however, argued that in the nucleophilic solvents of low polarity, the solvolysis of the *tert*-butyl chloride may be influenced by the nucleophilic assistance of the solvent, and the reaction mechanism has a partially S_N2 character [264,265]. In addition, it was pointed out that the electrophilic assistance of solvent by coordinating the leaving group might significantly influence the solvolysis rates of alkyl halides [266,267].

To overcome these problems, Winstein and co-workers proposed to use the solvolysis of the *p*-methoxyneophyl-*p*-toluenesulfonate as the new standard reaction for the definition of solvent polarity [268]. The limiting stage of this reaction is an anchimerically assisted rearrangement that guarantees the existence of the pure S_N1 mechanism even in the low-polarity nucleophilic solvents. The respective Y' scale was defined analogously to the original Y scale as:

$$Y' = \log k_S^{\text{Tos}} - \log k_0^{\text{Tos}} \tag{2.48}$$

where k_S^{Tos} and k_0^{Tos} are the rate constants of the solvolysis of *p*-methoxyneophyl-*p*-toluenesulfonate in a given solvent and in the standard solvent (80% aqueous solution of ethanol), respectively. The new scale (2.48) has been successful in the description of the solvent effect on various reactions proceeding according to the S_N1 mechanism.

Other chemical reactions have been also employed as standards for the definition of solvent polarity scales. For instance, the second-order rate constants of the quaternization of tri-*n*-propylamine by iodomethane have been used for the determination of the S scale of solvent polarity [269]. Additional solvent effect scales have been defined as based on the electrophilic aliphatic substitution reaction between tetramethyltin and bromine (X) [270,271] and using the *exo*/*endo* product ratio in the Diels–Alder addition of cyclopentadiene to methyl acrylate (Ω) [272].

Most LFER approaches to the solvent effects employ several scales, each of which corresponds to a physically distinct solute–solute interaction. One of the first systematic multidimensional approaches to solvent effects was established by Katritzky's group [273]. Subsequently, a four-parameter equation was proposed to describe the solvent effect on a given molecular property, A [274]:

$$A = A_0 + yY + pP + eE + bB \tag{2.49}$$

where Y is the polarity of the solvent, P is the polarizability of the solvent, E is the electrophilicity, and B is the basicity of the solvent. The intercept A_0 corresponds to the standard solvent (in this case, the gas phase) and y, p, e, and b denote the respective regression coefficients. We turn our attention to a possible confusion with the terminology. The term *polarity* has been often used to describe all possible solvent effects or all nonspecific solvent effects. Within the framework of Eq. (2.49), this term is related to a special scale of the nonspecific solvent effect.

Equation (2.49) was successfully applied to the description of the solvent dependence of more than 60 properties on the chemical reactivity, spectral characteristics, and other properties of compounds in a large variety of the condensed media. The first two terms in this equation involving the solvent polarity and polarizability scales were developed proceeding from the Kirkwood–Onsager reaction field theory of solvation. The solvent polarity and polarizability scales may have different mathematical formulations depending on the choice of the physical model for the description of the interaction between the solute charge distribution and the solvent reaction field. For instance, following the Born equation for the free energy of ionic solvation [18],

$$\Delta G_{\mathrm{ion}} = \frac{1 - \varepsilon}{2\varepsilon} \frac{Q^2}{R} \tag{2.50}$$

the factor $Y = 1/(1 - \varepsilon)$ may be used as the definition of the solvent polarity, with the term Q^2/R characterizing the process. Proceeding from the dipolar term of the expansion for Kirkwood electrostatic solvation energy [18],

$$E_{\mathrm{dip}} = \frac{1 - \varepsilon}{2\varepsilon + 1} \frac{\mu^2}{r^3} \tag{2.51}$$

the factor $Y = (1 - \varepsilon)/(2\varepsilon + 1)$ relates to the solvent polarity, and the term μ^2/r^3 is the characteristic of the process (interaction between the dipolar charge distribution in a spherical cavity and the corresponding solvent reaction field). The Kirkwood function has been especially widely used for the description of the solvent effects on various electronic, vibrational, or magnetic spectroscopic phenomena in condensed media [275–278].

The use of the macroscopic dielectric constant in the definition of above-given factors implies full dielectric relaxation of the solvent in the electrostatic field of the solute. However, in the case of processes and species characterized by short lifetimes (spectroscopic transitions, transition states of chemical reactions), the solvent may not have enough time to be fully relaxed in the field of instantaneous change of the charge distribution. The respective solvent reaction field will rise only from the electron–nuclear polarization of the solvent whereas the spatial distribution of the solvent molecules in the surroundings of the solute molecule remains unchanged. The latter corresponds to the equilibrium distribution in the electrostatic field of the long-living ground or initial state of the solute. No relation is expected to exist, in general, between the full polarization and the electron–nuclear polarization of the solvent. Therefore, another scale for nonspecific solvent effects has been introduced that accounts only for the instantaneous polarization of the solvent.

Proceeding from the definition of the Kirkwood electrostatic solvation energy, the respective solvent polarizability scale has been defined as:

$$P = \frac{1 - \varepsilon_\infty}{2\varepsilon_\infty + 1} \approx \frac{1 - n_D^2}{2n_D^2 + 1} \tag{2.52}$$

where ε_∞ is the dielectric permittivity of the solvent at the infinite frequency of the external electric field. The latter can be approximated with the square of the index of refraction of the solvent, n_D^2. A list of various nonspecific solvent effect scales is given in Table 2.8. The numerical data for some common scales for a variety of solvents are presented in Table 2.9.

The scales of the solvent polarity and polarizability have been applied in a large number of solvent effect LFER treatments. However, we turn our attention to some difficulties related to the interpretation of the regression coefficients (intensity factors) in the respective multiparameter LFER equations. First of all, it follows directly from Eqs. (2.49) to (2.51) that the charge distribution in the solute molecule is required to be independent of the solvent into which the solute molecule is submerged. Unfortunately, it may often not be the case, particularly in systems that involve the intramolecular polar resonance. For instance, the quantum chemical self-consistent reaction field molecular orbital (SCRF MO) calculations on disubstituted ethylenes

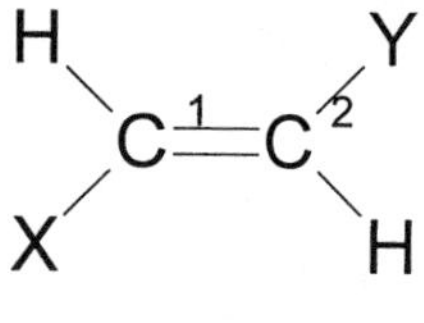

where X is an electron-donating (+R) group and Y denotes an electron-accepting (−R) group, reveal a significant charge redistribution in the molecule in a high dielectric constant medium (cf. Table 2.10).

This, on the other hand, causes substantial change in the dipole moment of the molecule. According to Eq. (2.51), the square of the dipole moment of the molecule is involved in the term e from Eq. (2.49). The variation of the dipole moment in solvents of different dielectric permittivity causes therefore the respective change in the e value. Consequently, the linear equation (2.49) should not be valid any more. The large changes in the dipole moment of some substituted benzoic acids, phenols, and anilines are also demonstrated by data in Table 2.11. It is also obvious from these data that the relative increase in the

TABLE 2.8 Empirical Nonspecific Solvent Effect and Solvent Polarity Scales

Scale Notation	Standard Process	Reference
Y	Kinetics of the solvolysis of *tert*-butyl chloride	*a*
Z	Solvatochromic shifts in the electronic spectra of pyridinium oxide and (4-carboxymethyl)*N*-ethylpyridinium iodide	*b*
E_T	Solvatochromic shifts in the electronic spectra of pyridinium *N*-phenolbetaine	*c*
E_T^N	Normalized E_T scale	*d*
Y	Kirkwood's function on the dielectric constant	*e*
P	Kirkwood's function on the squared index of refraction	*e*
χ_B	Negative solvatochromic shifts in the electronic spectra of a merocyanine dye	*f*
E_K	Negative solvatochromic shifts in the charge-transfer electronic spectra of $Mo^0(CO)_4$ diimine complex	*g*
E^*_{MLCT}	Negative solvatochromic shifts in the charge-transfer electronic spectra of $W^0(CO)_4$ bipyridyl complex	*h*
$E_{\text{CT}}(\pi)$	Negative solvatochromic shifts in the electronic spectra	*i*
E_T^{SO}	Negative solvatochromic shifts in the electronic spectra of *N,N*-(dimethyl)thiobenzamide-*S*-oxide	*j*
E_B	Negative solvatochromic shifts in the electronic spectra 2,2,6,6-tetramethylpiperidine-*N*-oxyl	*k*
Φ	Negative solvatochromic shifts in the electronic spectra of ketones	*l*
MLCT	Negative solvatochromic shifts in the electronic spectra of $Fe^{II}(1,10\text{-phenanthroline})_2(CN)_2$	*m*
Py	Solvent-dependent ratios of emission intensities of selected vibronic fluorescence bands of pyrene and other polycylic aromatic hydrocarbons, and nitrogen-containing aromatic polycyclic compounds	*n*
χ_R	Positive solvatochromic shifts in the electronic spectra of a merocyanine dye	*f*
π^*	Positive solvatochromic shifts in the electronic spectra *N,N*-diethyl-4-nitroaniline, *N,N*-dimethyl-4-nitroaniline and 4-nitroanisole	*o*
π^*_{azo}	Positive solvatochromic shifts in the electronic spectra of six merocyanine dyes	*p*
RPM	Positive solvatochromic shifts in the electronic spectra of $(CH_3)_2N(CH{=}CH)_2CHO$	*q*
S	Solvatofluorochromic shifts in the spectra of *N*-methylphtahalimide	*r*
E_{LMCT}	Positive solvatochromic shifts in the charge-transfer electronic spectra of $M_2[Fe^{III}(CN)_5]L$	*s*
SPP^N	UV/Vis spectra of 2-dimethylamino-7-nitrofluorene and 2-fluoro-7-nitrofluorene	*t*
$E_{\text{CT}}(A)$	UV CT absorption maxima of *tetra-n*-hexylammonium iodide trinitrobenzene	*u*
Y', W	Kinetics of the solvolysis of the *p*-methoxyneophyl-*p*-toluenesulfonate	*v*
S	Kinetics of the quaternization of tri-*n*-propylamine by iodomethane	*w*

TABLE 2.8 ***(Continued)***

Scale Notation	Standard Process	Reference
X	Electrophilic aliphatic substitution reaction between tetramethyltin and bromine	*x*
Ω	*Exo/endo* product ratio in the Diels-Alder addition reaction of cyclopentadiene with methyl acrylate	*y*
V_m	Characteristic volume of the solute	*z*
δ_H^2	Hildebrand's solubility parameter squared	*aa*
$\log L^{16}$	Gas-hexadecane partition coefficient	*bb*
R_2	Excess molar refraction	*cc*
π_2^H	Gas chromatographic retention data	*dd*
S'	Critically analyzed solvatochromic data	*ee*
Y_X	Kinectics of the solvolysis of the 1-adamantyl and 2-adamantyl substrates	*ff*

[a]E. Grunwald and S. Winstein, *J. Am. Chem. Soc.* **70,** 846 (1948).
[b]E.M. Kosower, *J. Am. Chem. Soc.* **80,** 3253 (1958); E.M. Kosower, J.A. Skorcz, W.M. Schwarz, and J.W. Patton, *ibid.* **82,** 2188 (1960).
[c]K. Dimroth, C. Reichardt, T. Siepmann, and F. Bohlmann, *Justus Liebigs Ann. Chem.* **661,** 1 (1963).
[d]C. Reichardt and E. Harbusch-Görnert, *Liebigs Ann. Chem.* p. 721 (1983).
[e]I.A. Koppel and V.A. Palm, in *Advances in Linear Free Energy Relationships* (N.B. Chapman and J. Shorter, eds.), Chapter 5. Plenum, London, 1972.
[f]L.G.S. Brooker, A.C. Craig, D.W. Heseltine, P.W. Jenkins, and L.L. Lincoln, *J. Am. Chem. Soc.* **87,** 2443 (1965).
[g]D. Walther, *J. Prakt. Chem.* **316,** 604 (1974).
[h]D.M. Manuta and A.J. Lees, *Inorg. Chem.* **26,** 3212 (1986).
[i]W. Kaim, B. Olbrich-Deussner, and T. Roth, *Organometallics* **10,** 410 (1991).
[j]W. Walter and O. Bauer, *Liebigs Ann. Chem.*, pp. 407 (1977).
[k]A. Janowski, I. Turowska-Tyrk, and P.K. Wrona, *J. Chem. Soc., Perkin Trans. 2*, p. 821 (1985).
[l]J.-E. Dubois and A. Bienvenüe, *Tetrahedron Lett.*, pp. 1809 (1966).
[m]R.W. Soukup and R. Schmid, *J. Chem. Educ.* **62,** 459 (1985).
[n]D.C. Dong and M.A. Winnik, *Photochem. Photobiol.* **35,** 17 (1982).
[o]M.J. Kamlet, J.-L. Abboud, and R.W. Taft, *J. Am. Chem. Soc.* **99,** 6027, 8325 (1977); C. Laurence, P. Nicolet, M.T. Dalati, J.-L.M. Abboud, and R. Notario, *J. Phys. Chem.* **98,** 5807 (1994).
[p]E. Buncel and S. Rajagopal, *J. Org. Chem.* **54,** 798 (1989).
[q]S. Dähne, F. Shob, K.-D. Nolte, and R. Radeglia, *Ukr. Khim. Zh.* **41,** 1170 (1975).
[r]I.A. Zhmyreva, V.V. Zelinskii, V.P. Kolobkov, and N.D. Kranitskaya, *Dokl. Akad. Nauk SSSR, Ser. Khim.* **129,** 1089 (1959).
[s]F. Armand, H. Sakuragi, and K. Tokumaru, *J. Chem. Soc., Faraday Trans.* **89,** 1021 (1993).
[t]J.A. Paéz, N. Campillo, and J. Elguero, *Gazz. Chim. Ital.* **126,** 307 (1996).
[u]J.G. Davis and H.E. Hallam, *Spectrochim. Acta, Part A* **23A,** 593 (1967).
[v]S.G. Smith, A.H. Fainberg, and S. Winstein, *J. Am. Chem. Soc.* **83,** 618 (1961).
[w]Y. Drougard and D. Decroocq, *Bull. Soc. Chim. Fr.* p. 2972 (1969).
[x]M. Gielen and J. Nasielski, *J. Organomet. Chem* **1,** 173 (1963); **7,** 273 (1967).
[y]J.A. Berson, Z. Hamlet, and W.A. Mueller, *J. Am. Chem. Soc.* **84,** 297 (1962).
[z]M.H. Abraham, and J.C. McGowan, *Chromatographia* **23,** 213 (1987).
[aa]J.H. Hildebrand, *The Solubility of Non-Electrolytes*. Reinhold, New York, 1950.
[bb]M.H. Abraham, P.L. Grellier, and R.A. McGill, *J. Chem. Soc., Perkin Trans. 2*, pp 797 (1987).
[cc]M.H. Abraham and J.C. McGowan, *Chromatographia* **23,** 213 (1987).
[dd]M.H. Abraham, in *Quantitative Treatments of Solute/Solvent Interactions* (P. Politzer and J.S. Murray, eds.), pp. 83–134. Elsevier, Amsterdam, 1994.
[ee]R.S. Drago, *J. Chem. Soc., Perkin Trans. 2,* p. 1827 (1992).
[ff]T.W. Bentley and G.E. Carter, *J. Am. Chem. Soc.* **104,** 5741 (1982); *J. Org. Chem.* **48,** 579 (1983); T.W. Bentley, G.E. Carter, and K. Roberts, *ibid.* **49,** 5183 (1984); D.N. Kevill, M.S. Bahari, and S.W. Anderson, *J. Am. Chem. Soc.* **106,** 2895 (1984); D.N. Kevill and S.W Anderson, *J. Org. Chem.* **50,** 3330 (1985); *J. Am. Chem. Soc.* **108,** 1579 (1986); T.W. Bentley and K. Roberts, *J. Org. Chem.* **50,** 4281 (1985).

dipole moment due to the surrounding polarizable medium is highly individual for a given combination of the substituents at the benzene ring of the molecule. This increase does not correlate with the absolute value of the dipole moment but depends on the relative position of the substituents. The largest increase is observed in the case of molecules having direct polar resonance between the substituents (*p*-nitrophenol and *p*-nitroaniline).

The changes in the electron distribution due to the electrostatic solvation are reflected in the respective calculated Mulliken atomic partial charges. For instance, the polarity of the C=O double bond in the 2-H-oxo form of the 3-hydroxypyrazole, as estimated from the Mulliken charges, is substantially increased by the dielectric medium:

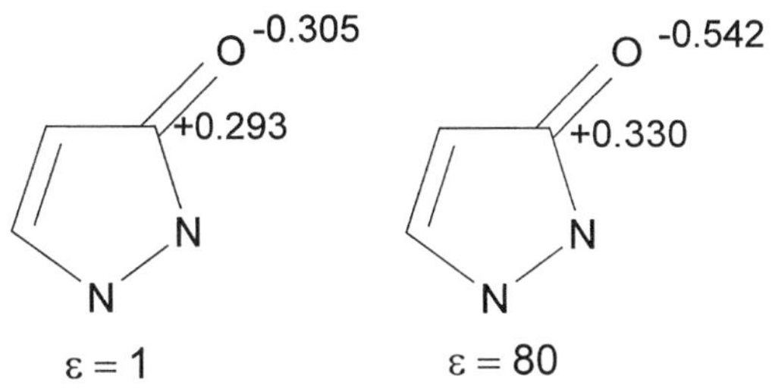

O -0.305 +0.293 N N $\varepsilon = 1$ O -0.542 +0.330 N N $\varepsilon = 80$

In fact, two different resonant forms of this compound may be written as being predominant in each of these media:

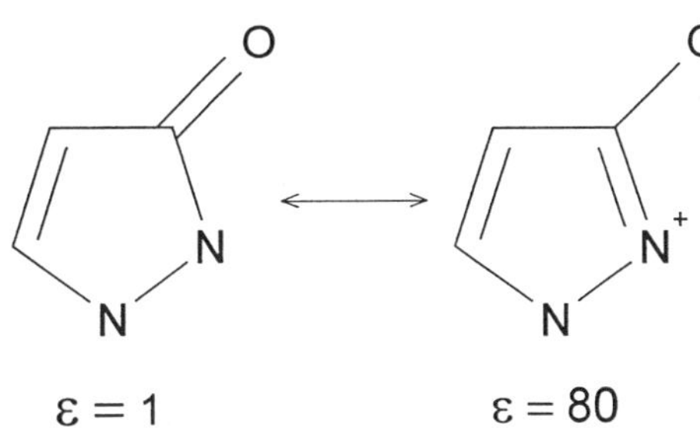

O N N $\varepsilon = 1$ ⟷ O^- N^+ N $\varepsilon = 80$

Notably, the left-hand structure has a dienic π-electron system whereas the right-hand structure, corresponding to the high dielectric constant medium, has a six π-electron aromatic system. Therefore, the electrostatic solvation can interfere with the aromaticity of a cyclic compound [279].

The effect of the solvent reaction field on the solute molecular geometry, conformational transitions, and equilibria in solutions may also have important consequences as related to the use of LFER-type expansions for the solvent effects [280–284]. In most cases, the bond lengths and valence bond angles of the solute are little affected by the surrounding dielectric medium. However, larger effects may be observed in systems with large solvent assistance to the polar resonance. For instance, in the 2-H-oxo form of the 3-hydroxypyrazole,

TABLE 2.9 Numerical Values of Some Common Empirical Nonspecific Solvent Effect Scales for a Variety of Media[a]

Solvent	E_T	Z	Y	P	E^*_{MLCT}	$E_{CT(\pi)}$	E_T^{SO}	E_B	Φ
1,2-Dichloroethane	41.3	63.4	0.432	0.210	0.64	0.065		0.23	
1,2-Dimethoxyethane	38.2	59.1				1.000			
1,2-Ethanediol	56.3	85.1					86.0		0.295
1,3,5-Trimethylbenzene	32.9		0.230	0.227	0.25	0.489		0.08	
1,4-Dioxane	36.0		0.226	0.203		0.476	80.8	0.16	0.115
1-Butanol	49.7	77.7	0.458	0.195	0.55		84.7	0.47	
1-Hexanol	48.8		0.446	0.201				0.48	
1-Octanol	48.1		0.427	0.205					
1-Pentanol	49.1		0.448	0.199	0.45			0.44	
1-Propanol	50.7	78.3	0.464	0.190				0.47	
2-Butanol	47.1		0.456	0.194				0.43	
2-Butanone	41.3		0.460	0.189				0.16	
2-Methyl-1-propanol	48.6		0.459	0.194				0.44	
2-Propanol	48.4	76.3	0.462	0.187	0.46		84.8	0.47	0.190
Acetic acid	51.7	79.2					88.8	0.73	0.370
Acetone	42.2	65.5	0.466	0.180	0.82	0.800	81.8	0.19	
Acetonitrile	45.6	71.3	0.480	0.174	0.98	0.535	82.4	0.28	0.135
Acetophenone	40.6		0.461	0.237				0.19	
Aniline	44.3						82.2		
Anisole	37.1		0.347	0.232		0.28		0.17	
Benzene	34.3	54.0	0.231	0.228	0.34	0.270	80.3	0.09	0.020
Benzonitrile	41.5	65.0	0.471	0.235		0.329		0.20	
Benzyl alcohol	50.4		0.444	0.239	0.67		84.7		
Bromobenzene	36.6	59.2	0.373	0.244					
Carbon disulfide	32.8								
Carbon tetrachloride	32.4		0.226	0.215	0.12	0.058	79.9	0.09	−0.010
Chlorobenzene	36.8	58.0	0.378	0.234	0.46	0.115	80.8		
Chloroform	39.1	63.2	0.359	0.210	0.42	0.000	82.5	0.29	0.095
Cyclohexane	30.9					0.053		0.00	0.035
Cyclohexanol	47.2		0.452	0.216	0.49				

Cyclohexanone	39.8		0.455	0.212	0.69	0.775			
Dichloromethane	40.7	64.7	0.422	0.204		0.030	82.0	0.28	
Diethyl ether	34.5				0.32	0.455		0.08	0.080
Diisopropylether	34.1		0.329	0.184					
Dimethyl sulfoxide	45.1	71.1	0.484	0.221	1.00		81.7	0.23	
Di-*n*-butylether	33.0	60.1	0.289	0.195				0.05	
Ethanol	51.9	79.6	0.470	0.181	0.69		85.1	0.51	0.255
Ethyl acetate	38.1	59.4	0.385	0.185		0.694		0.17	
Fluorobenzene	37.0	60.2	0.374	0.217					
Formamide	55.8	83.3					84.5	0.59	0.245
HMPA	40.9	62.8						0.12	0.115
Methanol	55.4	83.6	0.477	0.169	0.73		85.9	0.55	0.285
Methyl acetate	38.9		0.395	0.181					
N,N-Dimethylacetamide	42.9	66.9	0.480	0.208	0.93			0.19	
N,N-Dimethylformamide	43.2	68.4	0.480	0.204	0.95		81.7	0.21	0.135
n-Heptane	31.1								
n-Hexane	31.0							0.00	0.070
Nitrobenzene	41.2		0.479	0.242		0.272			
Nitromethane	46.3	71.2	0.481	0.189				0.36	
N-Methylacetamide	52.0	77.9							
N-Methylformamide	54.1						83.4		
o-Dichlorobenzene	38.0	60.0	0.428	0.242	0.48	0.056			
Piperidine	35.5				0.41				
Propylene carbonate	46.0	72.4							
Pyridine	40.5	64.0			0.77		81.1	0.20	
Quinoline	39.4								
tert-Butanol	43.3	7.13	0.442	0.191	0.51				0.130
Tetrahydrofuran	37.4	58.8	0.405	0.198	0.59	0.753		0.11	0.100
Tetramethylene sulfone	44.0	77.5							
Toluene	33.9		0.239	0.226	0.30	0.327		0.11	
Triethylamine	32.1				0.23		80.0		
Trimethyl phosphate	43.6							0.2	
Water	63.1	94.6					87.4	1.00	0.545
Reference[a]	*c*	*b*	*e*	*e*	*h*	*i*	*j*	*k*	*l*

TABLE 2.9 ***(Continued)***

Solvent	Py	χ_R	π^*	π^*_{azo}	S	SPP^N	$E_{CT(A)}$	δ	S'
1,2-Dichloroethane	1.46		0.807		−0.151			20.0	2.10
1,2-Dimethoxyethane	1.48								
1,2-Ethanediol	1.64	40.4	0.932		0.068		27.5	29.9	
1,3,5-Trimethylbenzene						0.576			1.54
1,4-Dioxane	1.50	48.4	0.553	0.34	−0.179	0.701		20.5	1.93
1-Butanol	1.06	44.5	0.503		−0.024		25.9	23.3	
1-Octanol	0.92								
1-Pentanol	1.02								
1-Propanol	1.09	44.1	0.534	0.86	−0.016	0.847	26.2		
2-Butanol	1.03								
2-Butanone	1.58					0.881			
2-Methyl-1-propanol	1.02								
2-Propanol	1.08	44.5	0.505		−0.041		25.0		
Acetic acid			0.664		0.005	0.781		20.7	
Acetone	1.64	45.7	0.683	0.53	−0.175	0.881	21.2	20.2	2.58
Acetonitrile	1.79	45.7	0.713	0.63	−0.104	0.895	23.1	24.3	3.00
Acetophenone							20.9		
Aniline								21.1	
Anisole	1.24					0.823	21.6		
Benzene	1.05	46.9	0.535	0.40	−0.215	0.667	21.4	18.8	1.73
Benzonitrile		43.3	0.904	0.82		0.960			2.63
Benzyl alcohol	1.24								
Bromobenzene	1.07	44.6	0.794		−0.164	0.824	21.3		2.10
Carbon disulfide							20.3	20.4	1.51
Carbon tetrachloride		48.7	0.294		−0.245	0.632	21.3	17.6	1.49
Chlorobenzene	1.08	45.2	0.709	0.58	−0.182	0.824		19.4	1.98
Chloroform	1.25	44.2	0.760	0.62	−0.200	0.786		19.0	1.74
Cyclohexane	0.58	50.0	0.000	0.00	−0.324	0.557		16.8	1.11
Cyclohexanol								23.3	

Cyclohexanone	1.47						20.3	2.35	
Dichloromethane		44.9	0.802	0.62	−0.189	0.876	22.6		2.08
Diethyl ether	1.02	48.3	0.273	0.16	−0.277	0.694	20.7		1.73
Diisopropylether	0.93								
Dimethyl sulfoxide	1.95		1.000	1.00		1.000	22.4	24.5	3.00
Di-*n*-butylether	0.84	48.6	0.239		−0.286	0.652			1.58
Ethanol	1.18	43.9	0.540		0.000	0.853	26.7	26.0	
Ethyl acetate	1.37	47.2	0.545	0.37	−0.210	0.795	21.0	18.6	2.15
Formamide	1.57		1.118		0.046	0.833		39.3	
HMPA			0.871						2.52
Methanol	1.35	43.1	0.586	0.89	0.050	0.857	26.8	29.6	
Methyl acetate						0.785			
N,N-Dimethylacetamide		43.0	0.882	0.86		0.970			2.70
N,N-Dimethylformamide	1.81	43.7	0.875	0.86	−0.142	0.954		24.8	2.80
n-Heptane									0.79
n-Hexane	0.58	50.9	−0.181	−0.09	−0.337	0.519		14.9	0.68
Nitrobenzene		42.6	1.029	0.91	−0.218	1.009			2.61
Nitromethane		44.0	0.848	0.70	−0.134	0.907	22.1		3.07
N-Methylacetamide	1.48								
N-Methylformamide								32.9	
o-Dichlorobenzene	1.02								
Piperidine			0.720						
Propylene carbonate							23.1		3.10
Pyridine	1.42	43.9	0.867	0.80	−0.197	0.922		21.9	2.44
tert-Butanol			0.534		−0.105		23.5	21.7	
Tetrahydrofuran	1.35	46.6	0.576			0.838	21.1	18.6	2.08
Tetramethylene sulfone			0.997					27.4	
Toluene	1.04	47.2		0.38	−0.237	0.655	21.3		1.66
Triethylamine		49.3	0.140		−0.285				1.43
Trimethyl phosphate						0.889			2.79
Water	1.87		1.090	1.03	0.154			47.9	
Reference[a]	*n*	*f*	*o*	*p*	*r*	*t*	*u*	*aa*	*ee*

[a]References correspond to those given in Table 2.8.

TABLE 2.10 AM1 SCF MO and SCRF MO Calculated Mulliken Partial Charges δ_C^1 and δ_C^2, on Two Carbon Atoms of Disubstituted Ethylenes and Calculated Dipole Moments (Debye) of Molecules in Media with Dielectric Permittivity $\varepsilon = 1$ and $\varepsilon = 80$

Substituents	$\delta_C^1(1)$	$\delta_C^1(80)$	$\delta_C^2(1)$	$\delta_C^2(80)$	$\mu(1)$	$\mu(80)$
R^1 = —OH R^2 = —OH	−0.124	−0.124	−0.124	−0.125	0.0	0.042
R^1 = —NH_2 R^2 = —NH_2	−0.155	−0.159	−0.155	−0.151	0.0	0.177
R^1 = —NO_2 R^2 = —NO_2	−0.119	−0.113	−0.155	−0.151	0.0	0.256
R^1 = —CN R^2 = —CN	−0.038	−0.034	−0.038	−0.042	0.0	0.175
R^1 = —OH R^2 = —NO_2	0.108	0.135	−0.357	−0.378	4.768	5.686
R^1 = —OH R^2 = —CN	0.061	0.080	−0.237	−0.264	2.981	3.596
R^1 = —NH_2 R^2 = —NO_2	0.109	0.147	−0.366	−0.428	7.237	0.528
R^1 = —NH_2 R^2 = —CN	0.047	0.116	−0.248	−0.344	4.960	7.201

TABLE 2.11 AM1 SCF and SCRF Calculated Total Dipole Moments μ (Debye) for Some Substituted Benzoic Acids, Phenols, and Anilines in Gas Phase ($\varepsilon = 1$) and in Polarizable Dielectric Medium ($\varepsilon = 80$)

Compound	Substituent	$\mu(1)$	$\mu(80)$	$\Delta\%^a$
Benzoic acid	—H	2.040	2.951	45
	m—CN	2.267	2.880	27
	p—CN	1.753	2.342	34
	m—NO_2	4.113	5.399	31
	p—NO_2	3.543	5.201	47
Phenol	—H	1.236	1.535	24
	m—CN	2.096	2.827	35
	p—CN	3.286	4.744	44
	m—NO_2	3.963	5.605	41
	p—NO_2	5.234	8.051	54
Aniline	—H	1.537	1.697	10
	m—CN	4.053	5.486	35
	p—CN	4.916	7.519	53
	m—NO_2	5.839	7.839	34
	p—NO_2	7.244	12.590	74

[a]Relative increase in the dipole moment $\Delta\%\% = 100[\mu(\varepsilon = 80) - \mu(\varepsilon = 1)]/\mu(\varepsilon = 1)$.

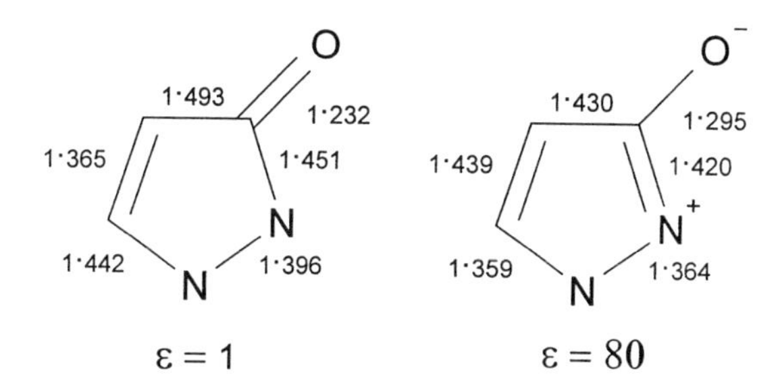

the AM1 SCRF calculated bond lengths are significantly altered in the high dielectric constant medium as compared to those in the gas phase. First, the C—O bond length is substantially increased, from the length typical for a double bond (1.232 Å) to a value close to that for a single bond (1.295 Å). This variation in bond lengths is also in accordance with the above-given formal representation of the predominant resonance structure for this compound in the corresponding media. In the heterocyclic ring system, the bond lengths are significantly more equalized in the high dielectric constant medium as expected for an aromatic cyclic structure.

The solvent effects on the conformational dynamics and equilibria of the molecules in liquids and solutions have been rather thoroughly studied, particularly as related to conformations of biomolecules [285–294]. Even the conformation of such a simple system as the push–pull disubstituted carbon radical $^{\cdot}$CH(CN)OH may be affected by the dielectric medium. Let us define a conformational angle α in this radical between the two planes determined by the sp^2 radical C center and the OH bond:

The AM1 SCF and AM1 SCRF calculated dependence of the relative energy (zero corresponds to the energy at $\alpha = 0°$) on this conformational angle in the $^{\cdot}$CH(CN)OH radical is given in Figure 2.3. It can be seen that the conformational minimum is shifted from $\alpha = 0°$ in the gas phase to the $\alpha = 180°$ in the high dielectric constant medium. This is caused by a large increase in the dipole moment of the $^{\cdot}$CH(CN)OH radical in this conformation and the corresponding large increase in the interaction with the respective reaction field. The AM1 SCRF calculated dipole moment in the medium with the dielectric constant $\varepsilon = 80$ is $\mu = 1.569D$ for the conformation with $\alpha = 0°$ and $\mu = 4.875D$ for the conformation with $\alpha = 180°$. These two conformations correspond to the energy minimums in the gas phase and in the condensed medium, respectively [295]. Proceeding from the SCRF model calculations, a similar electrostatic solvent

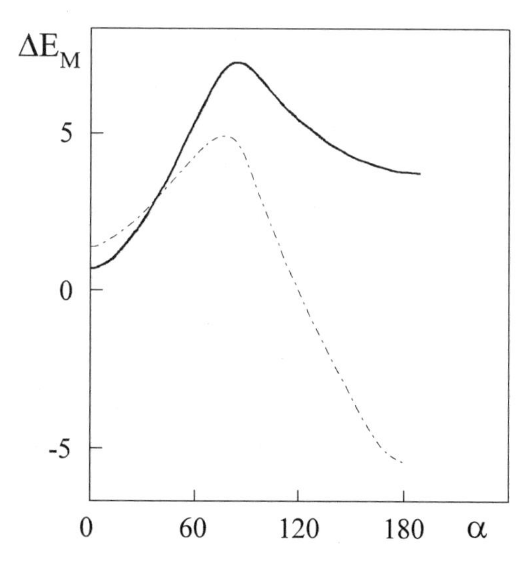

Figure 2.3 Dependence of the merostabilization energy ΔE_{M} (kcal/mol) of $^{\cdot}$CH(CN)OH radical on the conformational angle α (*deg*). The solid line corresponds to the gas phase and the dashed line to the dielectric medium with $\varepsilon = 80$.

effect of approximately 1 kcal/mol has been predicted for the relative energies of different conformational minima of some alkanes [285], and 1,2-dichloroethane and furfural [286] in aqueous solutions and in nonpolar media, respectively.

The solvent effect on the rotational barrier and relative energy of different conformational minima of formamide [293] and other small amides [291,292,294] has been also examined using the SCRF electrostatic solvation model within the ab initio molecular orbital method. The rotation about the C—N bond in amides is of fundamental importance for understanding the essence of conformational preferences in proteins. The development of the respective accurate potentials is important for the molecular dynamics simulation of the structure of these macromolecules.

It has been established that the conformational barriers at the C—N bond of formamide are substantially influenced by the presence of the high dielectric constant medium [293]. Two transition states of the C_s symmetry were found in the potential surface of this molecule, one of which corresponds to the rotation about the C—N bond (TS1) and another to the inversion at the nitrogen atom (TS2). The ab initio 6-31G**//6-31G** calculated rotational barrier relative to the planar structure (TS1) was increased by 1.4 to 4.8 kcal/mol and the inversion barrier reduced by 0.9 to 1.2 kcal/mol according to different reaction field models when the molecule was transferred from the gas phase to a dielectric medium with $\varepsilon = 78.54$ [293]. These trends can be rationalized in terms of the increase of the weight of the more polar resonance form of formamide in the high dielectric constant medium:

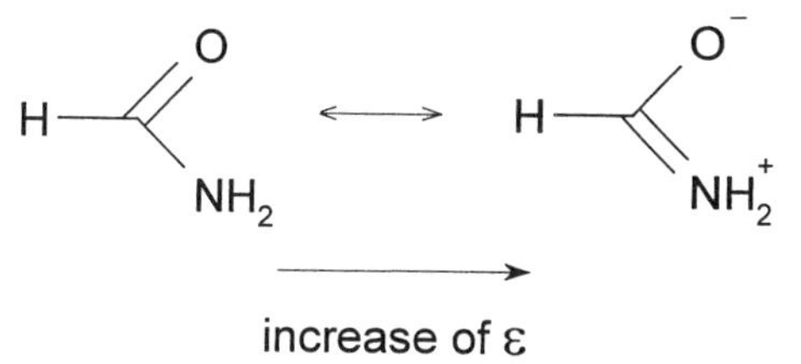

H–C(=O)–NH_2 ⟷ H–C(O^-)=NH_2^+

increase of ε

Obviously, the double-bond character of the C—N bond is increased by the higher dielectric constant of the medium, and thus the respective rotation barrier about this bond is increased, too. At the same time, the nitrogen atom is transformed from the sp^3 valence hybridization given by the left-hand side resonance form to a more sp^2-like valence state presented by the right-hand side resonance form. Accordingly, the inversion barrier at this atom should be reduced by the high dielectric constant medium. As a result of these trends, the two conformational transition states of formamide are separated by about 2.7 kcal/mol in the gas phase, with TS1 of the rotational barrier being lower. In the high dielectric constant media, both the rotation and inversion transition states are predicted to have practically the same energy. This conclusion is of great importance for the prediction of conformational transformations and equilibria in polypeptides and proteins at different biological conditions. From another point of view, these results indicate the dangers related to the mechanical use of LFER solvent effect equations since the molecular geometry may be substantially affected by the surrounding medium. Accordingly, the solvent-effect-related molecular descriptors do not correspond to the same molecular structure in different media.

Another possible misinterpretation of LFER correlations of a solvent effect using the Kirkwood polarity scale has been discussed by us in relation to the calculated merostabilization effects in push–pull substituted carbon radicals [296,297]. This effect [298–301], also called the "captodative" effect [302,303] has been defined as an additional stabilization of the carbon radical with an electron-donor and an electron-acceptor group simultaneously connected to the radical center. The respective merostabilization energy is defined as [295]:

$$\Delta E_M = E_{XY} - \frac{1}{2}(E_{XX} + E_{YY}) \qquad (2.53)$$

where E_{XY} is the energy of the electron-donor and electron-acceptor substituted radical $^{\cdot}CHXY$, E_{XX} is the energy of the doubly electron-donor substituted radical $^{\cdot}CHX_2$, and E_{YY} is the energy of the doubly electron-acceptor substituted radical $^{\cdot}CHY_2$. Our studies have shown that the merostabilization is largely supported by the presence of the polarizable dielectric continuum. In Table 2.12, the results of the AM1 unrestricted Hartree–Focu self-consistent reaction field (UHF SCRF) calculations on the $^{\cdot}CH(ON)_2$, $^{\cdot}CH(CN)_2$, and $^{\cdot}CH(CN)OH$

radicals, corresponding to three dielectric media are presented [295]. In the gas phase ($\varepsilon = 1$), no additional stabilization was observed for the $^{\cdot}CH(CN)OH$ radical. However, even in the low dielectric constant medium, substantial stabilization energy of this radical ΔE_M was predicted by those calculations (3.53 kcal/mol). This stabilization was further enhanced by the high dielectric constant medium (to 12.57 kcal/mol). Also, a dramatic change in the value of the dipole moment was observed for the $^{\cdot}CH(CN)OH$ radical while transferred from the gas phase to the high dielectric constant medium (cf. Table 2.12). Such dependence of the dipole moment on the dielectric constant of the medium leads to a principally nonlinear relationship between the electrostatic solvation energy E_{el} and the Kirkwood–Onsager factor $\Gamma = (1 - \varepsilon)/(2\varepsilon + 1)$ [Eq. (2.51)]. The calculated merostabilization energy ΔE_M was predicted to have a nonlinear dependence from this factor (Fig. 2.4). Notably, the classical Kirkwood–Onsager formula using constant values of the dipole moment of the compounds involved in the gas phase leads to a qualitatively wrong prediction of the slope of this relationship.

The above-given cautions about the use of Kirkwood's dielectric constant function in LFER solvent effect expansions apply also for the solvent polarizability scales defined from the Kirkwood function or other reaction field models. It also means that whenever the electron distribution or the geometrical structure of a solute is substantially distorted by the solvent, the linear relationships for the respective solvent effects will not be valid any more. In Chapter 3, a more throughout analysis of the influence of solvent on the molecular descriptors, including the solvational descriptors themselves, is given.

One of the most popular sets of empirical solvent effect scales was proposed by Kamlet, Taft, and others [33,304–306]. The respective linear solvation energy relationship (LSER) model has the following general form:

$$\begin{aligned}\text{Property} = {} & \text{bulk/cavity term(s)} + \text{dipolarity/polarizability term(s)} \\ & + \text{hydrogen bonding term(s)} + \text{constant}\end{aligned} \tag{2.54}$$

The first two terms in Eq. (2.54) describe the nonspecific solvent effects. Thus, the bulk and cavity terms model the endoergic energy to overcome the solvent cohesive forces to create the hole for the solute in the solvent. This term is related to the cavity formation energy discussed above as one of the physical models of solvation [cf. Eq. (2.45)]. In the first LSER model, the molar volume of the solute (V_M) was suggested as the respective solvent effect descriptor. In later treatments, this descriptor has been replaced by the intrinsic volume of the solute molecule, V_I, or by the characteristic volume of the solute, V_m. For the interaction of solvents with a given solute, it was also suggested to use the Hildebrand solubility parameter δ_H^2 as the bulk term in the LSER equation.

The dipole/polarizability terms in Eq. (2.54) describe the van der Waals interactions and electrostatic interactions between dipoles and dipoles and induced dipoles of the solute and the solvent. In the framework of the LSER

TABLE 2.12 AM1 UHF SCRF Calculated Enthalpies of Formation ΔH_f^0 (kcal/mol), Dipole Moments μ (Debye), and Merostabilization Energies ΔE_M (kcal/mol) in Media of Different Dielectric Constant, ε

Radical	ε	ΔH_f^0	μ	ΔE_M
$^{\cdot}CH(OH)_2$	1	−80.03	1.289	
	2	−84.38	1.411	
	80	−86.56	1.514	
$^{\cdot}CH(CN)_2$	1	82.14	2.247	
	2	79.34	2.583	
	80	76.51	2.915	
$^{\cdot}CH(CN)OH$	1	0.65	1.419	1.09
	2	−6.05	4.875	−3.53
	80	−17.59	6.193	−12.57

formalism, the product of the solute and solvent dipolarity parameters, π^*, has been used as the solvent effect descriptor for the dipole/polarizability term. The dipolarity descriptor π^* has been defined from the solvent-induced shifts in the UV/Vis spectra of positively solvatochromic compounds. In addition, the polarizability term proportional to the excess molar refraction has been introduced into the LSER equation. The various forms of the LFER/LSER equations for the description of solvent effects are given in Table 2.13. Notably, Famini and co-workers [307–312] have developed an LSER equation based solely on theoretically derived molecular solvational descriptors. The respective theoretical linear solvation energy relationship (TLSER) formalism is discussed in Chapter 3.

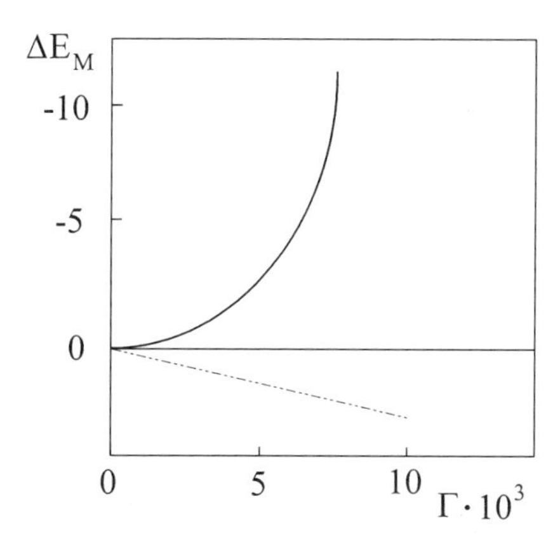

Figure 2.4 The dependence of the AM1 UHF SCRF calculated merostabilization energy ΔE_M (kcal/mol) of the $^{\cdot}CH(CN)OH$ radical on the Kirkwood–Onsager factor Γ. The solid line corresponds to the SCRF theory, the dashed line to the classical theory.

TABLE 2.13 Some Multiparameter LFER/LSER Equations for Description of Solvent Effects on Molecular Properties

Equation	Reference
$P = P_0 + \alpha_1 \delta \, \Delta F_1 + \alpha_2 \delta \, \Delta F_2 + \cdots$	*a*
$P = P_0 + y \cdot Y + p \cdot P + e \cdot E + b \cdot B$	*b*
$P = P_0 + s(\pi^* + d\delta) + a\alpha + b\beta + h\delta_H^2 + e\xi$	*c*
$p_{ij} = c_i + a_i A_j + b_u B_j$	*d*
$\Delta\chi = E_A' E_B^* + C_A' C_B^* + S'P + W$	*e*
$P = P_0 + \alpha E_T(30) + \beta \mathrm{DN}$	*f*
$P = P_0 + a\mathrm{DN} + b\mathrm{AN} + c\Delta G_{\mathrm{vap}}^0$	*g*
$\log SP = c + rR_2 + s\pi_2^H + a \sum \alpha_2^H + b \sum \beta_2^0 + vV_x$	*h*

[a] F.W. Fowler, A.R. Katritzky, and R.J.D. Rutherford, *J. Chem. Soc. B*, p. 460 (1971).
[b] I.A. Koppel and V.A. Palm, in *Linear Free Energy Relationships* (N.B. Chapman and J. Shorter, eds.), Chapter 5. Plenum, London, 1972.
[c] J.-L. M. Abboud, M.J. Kamlet, and R.W. Taft, *Prog. Phys. Org. Chem.* **13,** 485 (1981).
[d] C.G. Swain, M.S. Swain, A.L. Powell, and J. Alunni, *J. Am. Chem. Soc.* **105,** 502 (1983).
[e] S. Joerg, R.S. Drago, and J. Adams, *J. Chem. Soc., Perkin Trans. 2*, p. 2431 (1997).
[f] T.M. Krygowski and W.R. Fawcett, *J. Am. Chem. Soc.* **97,** 2143 (1975).
[g] U. Mayer, *Monatsh. Chem.* **109,** 421 (1978).
[h] M.H. Abraham, H.S. Chadha, J.P. Dixon, C. Rafols, and C. Treiner, *J. Chem. Soc., Perkin Trans. 2*, p. 887 (1995).

An alternative multiparameter LFER approach to the solvent effects was proposed by Abraham [313,314]. His method proceeds with the definition of a new solute descriptor, the gas-hexadecane partition coefficient (Ostwald solubility coefficient) as follows [315]:

$$\log L^{16} = \frac{C_{\mathrm{C_{16}H_{34}}}}{C_{\mathrm{gas}}} \tag{2.55}$$

where $C_{\mathrm{C_{16}H_{34}}}$ and C_{gas} are the equilibrium concentrations of the solute in the hexadecane and in the gas phase at the normal conditions, respectively. The $\log L^{16}$ values have been calculated from the gas chromatographic (GC) retention data using hexadecane as a stationary phase. For the nonvolatile compounds at room temperature, the $\log L^{16}$ values have been also obtained from the GC measurements on nonpolar phases such as aqualane or apiezon at elevated temperatures. The experimental GC data involve the retention volumes, the relative retention times, or the retention indices.

The next descriptor in Abraham's approach is the excess molar refraction, R_2, defined as the difference between the molar refraction values for a solute X and for an alkane of the same characteristic volume, V_X, respectively [316]:

$$R_2 = \mathrm{MR}_X - \mathrm{MR}_{\mathrm{alkane}} \tag{2.56}$$

The molar refraction itself is calculated from the following equation:

$$\mathrm{MR}_X = \frac{n_D^2 - 1}{(n_D^2 + 2)V_X} \tag{2.57}$$

where n_D is the refractive index of the solute for the sodium D line at 25°C. The GC data have been fitted according to the following equation:

$$\log SP = c + rR_2 + l \log L^{16} \tag{2.58}$$

where SP are either the retention volumes or the relative retention times for a series of solutes, and c, r, and l are the regression coefficients. Eq. (2.58), the dependent variable logSP can be substituted by the retention index I. For instance, the following regression equation was obtained for a large number of solutes at 110°C [317]:

$$\frac{I}{10} = 6.669 + 8.918R_2 + 20.0002 \log L^{16}$$

$$n = 138 \qquad r = 0.9995 \qquad s = 0.449 \qquad F = 67{,}950$$

The preceding equation was employed for the prediction of log L^{16} values for about 1500 solutes for which the experimental data on I were available [317a].

The two descriptors—the gas-hexadecane partition coefficient and the excess molar refraction—are designated to account for the interactions of compounds with nonpolar condensed media. These interactions may include, with some fraction, the cavity formation energy in a condensed medium and the solute–solvent dispersion forces. However, the respective LFER terms cannot be put into one-to-one correspondence to the theoretical terms of the free energy of solvation [Eq. (2.45)] or to the individual LSER contributions for the nonspecific solvation [Eq. (2.54)].

The Abraham solvent effect equation includes also a dipolarity/polarizability term. The respective descriptors have been also defined as proceeding from the GC data but using the results obtained for a polar nonacidic stationary phase. In this case, the interaction of the basicity centers of the solute with the stationary phase is avoided. The dipolarity/polarizability descriptors (π_2^H) are derived by from the GC data according to the following equation [318]:

$$\pi_2^H = s^{-1}(\log SP - rR_2 - l \log L^{16} - c) \tag{2.59}$$

where the coefficients r, l, and c are taken from Eq. (2.58). This dipolarity/polarizability descriptor is again of purely empirical nature and not directly comparable with any of the individual terms in the theoretical formula for the free energy of solvation. A three-layer feed-forward neural network has been applied to predict the dipolarity/polarizability descriptors π_2^H for a number of solvents [319]. The Abraham approach involves also the hydrogen-bonding related specific solvation descriptors, which will be reviewed more thoroughly in the next paragraph. A general form of Abraham's equation is given in Table 2.13.

Drago has suggested a scale of solvent polarity that enables us to estimate the influence of nonspecific solvation on a wide variety of physicochemical properties for solutes of widely varying shapes and polarity [320]. His equation has the following appearance:

$$\Delta\chi = S'P + W \tag{2.60}$$

where $\Delta\chi$ is the value of the physicochemical property measured in the specified solvent, S' is the measure of the solvent's polarity, P is the measure of the susceptibility of the solute probe to solvation, and W is the value of $\Delta\chi$ for the standard system (with $S' = 0$). The S' scale was constructed by using the solvatochromic data for a carefully chosen set of solvents and indicator probes that do not exhibit substantial donor–acceptor interactions between them. The Drago equation was later expanded by adding the terms to include also the systems with specific interactions involved (cf. the next paragraph). The original S' scale has been correspondingly refined and extended to donor and acceptor solvents [321,322].

The nonspecific solvent effect scales alone should be sufficient for the description of the solute–solvent systems that do not exhibit substantial donor–acceptor interactions or other specific effects. This can be achieved by the appropriate selection of the solvents or properties to be correlated. Methodologically, it is advisable first to develop an LSER regression equation for such a limited set of solvents, involving only one or two descriptors related to the nonspecific solvent effects. In the next step, the deviations from this master equation would be correlated with appropriate specific solvent effect scales for a wider selection of solvents. It is, however, important to use the scales defined in the framework of a single, consistent multiparameter approach. The experimental (empirical) solvent effect scales do not correspond exactly to distinct physical interactions. Accordingly, the different definitions of the scales contain the contributions by different interactions in a different amount that makes the scales of different origin incompatible.

2.3.2 Microscopic Solvent Effect Scales

The empirical scales for the microscopic or specific solvent effects are also derived on the basis of the appropriately chosen experimental data. In most cases, two types of specific effects have been distinguished. Both of these are related to the donor–acceptor interactions or to the hydrogen bonding between the solute and the solvent molecules. First, the specific solvent effects originating from the hydrogen bonding between the solute as the hydrogen-bonding donor and a solvent molecule as an acceptor have been related to the basicity or the nucleophilicity of the solvent. The same terms may characterize a solvent acting as a Lewis base in the donor–acceptor interaction with the solute molecule. On the other hand, the acidity or the electrophilicity of the solvent has been related to the specific interaction that involves the solute as the hydrogen bond acceptor and a solvent molecule as the hydrogen bond donor. In general, the same terms characterize a solvent acting as a Lewis acid in the interaction with the solute molecule. Following the terminology of physical organic chemistry, the first terms in these pairs (acidity and basicity) are mostly used for the molecular systems in equilibria. The second terms in these pairs (nucleophilicity and electrophilicity) are usually applied in the case of time-dependent processes or kinetic data. However, it should be mentioned that the differently named solvent effect scales have been used for both types of phenomena.

One of the first basicity scales, introduced for the solvent effect LFER treatment, was the Gutmann donor number or donicity (DN). This scale is defined as the negative value of the molar enthalpy for the formation of adducts between antimony pentachloride and the solvent [28,323,324].

$$\mathrm{DN} = -\Delta H_{\mathrm{B:SbCl5}} \tag{2.61}$$

The original DN values were measured in dilute solutions of 1,2-dichloroethane, and thus they are expected to correspond to the spatially separated solvent molecules. The corresponding enthalpy measurements in tetrachloromethane gave similar numbers [325]. This result confirmed the validity of the donicity scale in chemically similar environments. Later, it was suggested to use the bulk donicity scales, which account for the additional cooperative effects in highly structured solvents such as water, alcohols, or amines, by measuring the DN data in a neat solvent [326]. However, the cooperative effects seem to be difficult to quantify as different experimental methods may give substantially different donicities for a given solvent. The donor numbers have been used for the description of a variety of thermodynamic, kinetic, electrochemical, and spectroscopic data [327–329]. However, primarily due to different experimental shortcomings, the DN descriptors have been subject to severe criticism [330,331]. A critical compilation and discussion of donor numbers has been given by Marcus [332].

Another Lewis basicity scale was suggested by Gal and Maria [333–335]. This scale was derived from the standard molar enthalpies of the 1:1 adduct formation of electron pair donor solvents with gaseous boron trifluoride in dilute dichloromethane solutions. A hydrogen bond basicity scale pK_{HB} has been defined for 18 thiocarbonyl bases [335a] and for 65 aromatic six-membered *N*-heterocycles using the equilibrium with 4-fluorophenol as a reference compound in CCl_4 at 298 K [336]. A basicity scale B_P was introduced on the basis of the mean spherical approximation (MSA) model expression for the Gibbs solvation energy of monoatomic monovalent cation [337].

$$\Delta G_S(\mathrm{Cat}^+) = -\frac{N_A e^2}{8\pi\varepsilon_0}\left(1 - \frac{1}{\varepsilon_S}\right)\left(\frac{B_P}{1 + r_i B_P}\right) \tag{2.62}$$

where $\Delta G_S(\mathrm{Cat}^+)$ is the Gibbs free energy of solvation of a given cation, Cat^+, with the radius r_i in a given solvent, N_A is Avogadro's number, e is the electron charge, and ε_0 and ε_S are the dielectric permittivity of the vacuum and the relative dielectric constant of the solvent, respectively. A fair correlation has been established between the B_P parameter and Gutmann's donor number, DN [337]. Persson et al. have designed a solvent donor scale, D_S, based on the shift in the symmetric stretching frequency in $HgBr_2$ in different solvents [338]. This scale was suggested as particularly suitable for the soft acceptors. An alternative scale for hard acceptors, D_H, was proposed as based on the Gibbs energy of transfer of the sodium ion from water to a given solvent [338]. A basicity scale, designated as B_{sc}, has been introduced using the solvatochromic shifts of the maximum wavelength for the absorption band of Cu(II)*N*,*N*,*N'*,*N'*-tetramethylethylenediamine acetoacetonate [339]. It was suggested that this process should be particularly convenient for the experimental definition of the basicity scales for a large variety of solvents, both protic and aprotic. A critical overview of various basicity scales has been presented by Persson [340].

The acidity (electrophilicity) of the solvents has been described by Gutmann and co-workers by using the respective solvent acceptor numbers (AN) [341–344]. The AN scale was defined from the perturbations of the ^{31}P NMR chemical shifts of triethylphosphane oxide due to the complex formation with the electron pair acceptor solvents, S

$$(C_2H_5)P{=}O + S \rightleftharpoons (C_2H_5)\overset{\delta+}{P}{=}\overset{\delta-}{O}\cdots S$$

as follows:

$$\mathrm{AN} = \frac{\delta_{\mathrm{corr}}(S) - \delta_{\mathrm{corr}}(n\text{-hexane})}{\delta_{\mathrm{corr}}((C_2H_5)_3PO\cdots SbCl_5) - \delta_{\mathrm{corr}}(n\text{-hexane})} \times 100 \qquad (2.63)$$

where δ_{corr} denote the ^{31}P NMR chemical shifts of $(C_2H_5)_3PO$ in the solvent S, in the *n*-hexane, and in its 1:1 adduct with the antimony pentachloride in 1,2-dichloroethane, respectively. Thus the acceptor number (or acceptivity) reflects the capability of a solvent to form a Lewis acid–base complex with the probe compound (triethylphosphane oxide) relative to the $SbCl_5$. The AN scale correlates well with the $A(^{14}N)$ solvent parameters determined from the nitrogen hyperfine splitting constants in the electron spin resonance (ESR) spectra of some aminyloxide radicals in different solvents [345–348]. Notably, neither of these scales correlate with the Kirkwood–Onsager functions of the dielectric constant of the solvent and are thus expected to describe the specific solute–solvent interactions [349].

An acidity scale has been introduced on the basis of the MSA model expression for the Gibbs solvation energy of a monoatomic monovalent ion [350]. With only the ion–dipole interactions considered, the MSA equation has in this case the following form:

$$\Delta G_S = -\frac{N_A e^2}{8\pi\varepsilon_0}\left(1 - \frac{1}{\varepsilon_S}\right)\left(\frac{1}{r_i + \delta_S}\right) \tag{2.64}$$

where N_A is Avogadro's number, e is the electron charge, ε_0 and ε_S are the dielectric permittivity of vacuum and the relative dielectric constant of the solvent, respectively, and δ_S is the MSA distance parameter. The δ_S parameter depends on both the solvent and the probe ion. The acidity scale A_P was defined using the linear relationship between the reciprocal of δ_S parameter for halide ions and other solvent acidity parameters (AN, E_T) as:

$$\Delta G_S(\mathrm{Hal}^-) = -\frac{N_A e^2}{8\pi\varepsilon_0}\left(1 - \frac{1}{\varepsilon_S}\right)\left(\frac{A_P}{1 + r_i A_P}\right) \tag{2.65}$$

where $\Delta G_S(\mathrm{Hal}^-)$ is the Gibbs free energy of solvation of a given halide ion Hal^- with radius r_i in a given solvent.

A different approach to describe quantitatively the general acid–base properties of solvents was proposed by Drago [351,352]. He suggested to correlate the standard enthalpy of the reaction between an acceptor A and a donor B with the formation of a 1:1 adduct in an inert solvent by using the following four-parameter equation:

$$-\Delta H^0_{AB} = E_A E_B + C_A C_B \tag{2.66}$$

In Eq. (2.66) parameters E_A and E_B describe the tendency of an acid and a base, respectively, to participate in electrostatic interactions. Parameters C_A and C_B reflect the involvement of an acid and a base, respectively, in covalent interactions. The original set of E_A and E_B parameters was determined primarily from the data on enthalpies of adduct formation of iodine or phenol as electron pair acceptors with amines as electron pair donors. The best set of parameters was obtained by the optimization on the large database on enthalpies. The modification of Eq. (2.66) by the inclusion of the respective term for the nonspecific solvent effects [Eq. (2.60)] has led to a unified solvation model (USM) described by the following equation [321,322,353]:

$$\Delta\chi = E'_A E^*_B + C'_A C^*_B + S'P + W \tag{2.67}$$

where $\Delta\chi$ is the value of the property measured in the specified solvent, S' is the measure of the solvent's polarity, P is the measure of the susceptibility of the solute probe to nonspecific solvation, and W is the value of $\Delta\chi$ for the standard system. The descriptors E'_A and C'_A are the electrostatic and covalent acceptor parameters for one component, respectively, and E^*_B and C^*_B are the corresponding parameters for the response of the another component to the donor electrostatic and covalent interactions, respectively.

Swain et al. developed the following two-parameter equation for the LFER description of the solvent effects [354]:

$$p_{ij} = c_i + a_i A_j + b_i B_j \quad (2.68)$$

In this equation, p_{ij} denotes the value of the given property i in a solvent j, A_j is the solvent acity characterizing the anion-solvating ability of the solvent, B_j is the solvent basity characterizing the cation-solvating ability of the solvent, and a_i, b_i, and c_i are the parameters of regression. The terms *acity* and *basity*, although representing some kind of acidity and basicity, respectively, were chosen to emphasize their relation to the neat solvents. The original A_j and B_j scales were determined by fitting the data on 77 solvent-dependent chemical and physical properties in different solvents. The numerical scaling of the acity and basity was obtained using optional conditions $A_j = B_j = 0$ for *n*-heptane and $A_j = B_j = 1$ for water. The overall correlation with the data from 1080 individual sets was excellent (overall correlation coefficient $R = 0.991$ [354]). An obvious conclusion that only two parameters are sufficient to describe all solvent effects was, however, questioned because of subjective selection of data used in the development of acity and basity scales and in the subsequent validation of Eq. (2.65) [355,356].

Another two-parameter equation for the description of the solvent effects was developed using the combination of Gutmann's donor numbers DN and the solvatochromic $E_T(30)$ scale as [357–360]:

$$P = P_0 + \alpha E_T(30) + \beta \mathrm{DN} \quad (2.69)$$

where P is the property studied and P_0 is the property value in the standard solvent. This equation had been derived on the assumption that the nonspecific solvent effects are negligible or accounted for implicitly by two specific solvent effect parameters involved. The last equation has been successfully applied for a number of solvent-dependent properties. Nevertheless, it is difficult to assume a general validity of the last equation. Accordingly, an extended multiparameter equation was proposed by Mayer [361,362] as:

$$P = P_0 + a\mathrm{DN} + b\mathrm{AN} + c\Delta G^0_{\mathrm{vap}} \quad (2.70)$$

where DN and AN are the Gutmann donor and acceptor numbers, respectively, and $\Delta G^0_{\mathrm{vap}}$ is the standard molar Gibbs energy of vaporization of a given solvent.

Different acidity and basicity scales have been employed in various other multiparameter approaches to the solvent effects [273]. In the original Koppel–Palm multiparameter equation of the solvent effects [Eq. (2.49)], the basicity parameter B was defined using the solvent-induced shifts of the O—D stretching band in the infrared spectra of deuterated methanol, CH_3OD [274].

$$B \equiv \Delta\tilde{\upsilon} = \tilde{\upsilon}_{\mathrm{MeOD}} - \tilde{\upsilon}_{\mathrm{MeOD\cdots S}} \quad (2.71)$$

Later, this scale was refined and substantially extended using the solvent-induced infrared (IR) band shifts of the O—H stretching vibration of phenol in tetrachloromethane [30]. The electrophilicity scale E applied in Eq. (2.49) was derived from the data on solvatochromic shifts in different solvents. The original $E_T(30)$ scale was corrected by subtraction of the terms corresponding to the influence of nonspecific solvent effects and by adjustment of the scale for the gas phase to be the origin:

$$E = E_T(30) - 25.57 - 14.39Y - 9.08B \qquad (2.72)$$

The original Kamlet–Taft equation [35] (2.17) involved also two parameters for the specific solute–solvent interactions, the solvent basicity, and the solvent acidity, respectively. The hydrogen bond donor acidity α was originally determined for each solvent as an average value of this descriptor derived from data for six different processes [363]. These data included (i) enhanced solvatochromic shift for Dimroth's betaine, 4-(2,4,6-triphenylpyridinium)2,6-diphenyloxide relative to 4-nitroanisole [364]; (ii) enhanced solvatochromic shift for Brooker's merocyanine relative to 4-nitroanisole [250]; (iii) augmented value of Brownstein's S scale [365] relative to solvatochromic shift for 4-nitroanisole; (iv) enhanced solvatochromic shift for bis[α-(2-pyridylbenzylidene)-3,4-dimethylaniline]bis(cyano)iron(II) clathrate complex [366] relative to N,N-diethyl-4-nitroaniline; (v) reduced free energy of transfer of tetraethylammonium iodide [367] relative to solvatochromic shift for 4-nitroanisole; and (vi) Y' scale derived from the rate constants of *tert*-butyl chloride solvolysis at 120°C [368], augmented relative to solvatochromic shift for 4-nitroanisole. In all these processes, the hydrogen bonding by solvent S—H as a bond donor is expected to be much more significant compared to the hydrogen-bonding effect on the UV/Vis spectrum of 4-nitroanisole (weak hydrogen bond acceptor).

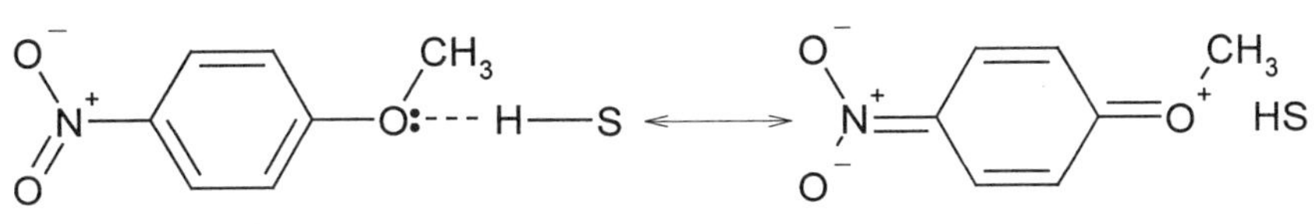

The original basicity scale β [369] was derived using the data for five processes, including: (i) enhanced solvatochromic shift for 4-nitroaniline relative to N,N-diethyl-4-nitroaniline; (ii) enhanced solvatochromic shift for 4-nitrophenol relative to 4-nitroanisole; (iii) pK_{HB} for the formation of hydrogen-bonded complex with 4-fluorophenol; (iv) limiting ^{19}F NMR shift for hydrogen-bonded complex with 4-fluorophenol; (v) pK_{HB} for the formation of hydrogen-bonded complex with phenol. This scale was refined and extended [370] using the data for another three indicator processes: (vi) enhanced solvatochromic shift for 2-nitroaniline relative to N,N-dimethyl-2-nitroaniline; (vii) enhanced solvatochromic shift for 2-nitro-4-toluidine relative to N,N-dimethyl-2-nitrotoluidine; and (viii) enhanced solvatochromic shift for 2-nitro-4-anisidine relative to N,N-dimethyl-2-nitro-4-anisidine. In these cases, two processes are compared, one of which involves the strong bonding with the solvent S as a hydrogen bond acceptor and the second with the absence of that interaction.

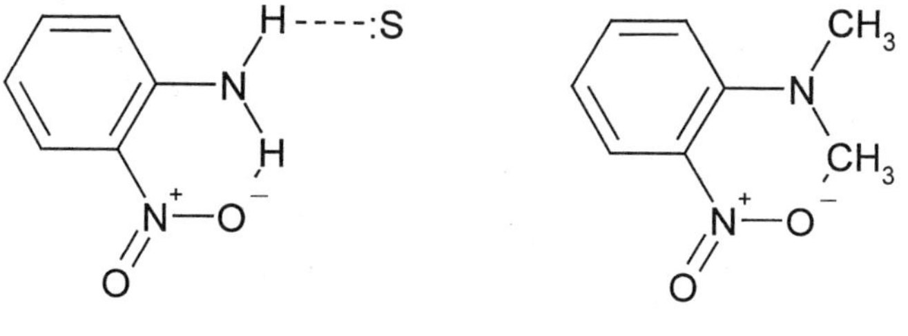

2-nitroaniline

N,N-dimethyl-2-nitroaniline

The spectral transition energy of 2-nitroaniline is additionally red shifted in hydrogen bond acceptor solvents because of enhanced solvation of the excited state as compared to the ground state of this molecule.

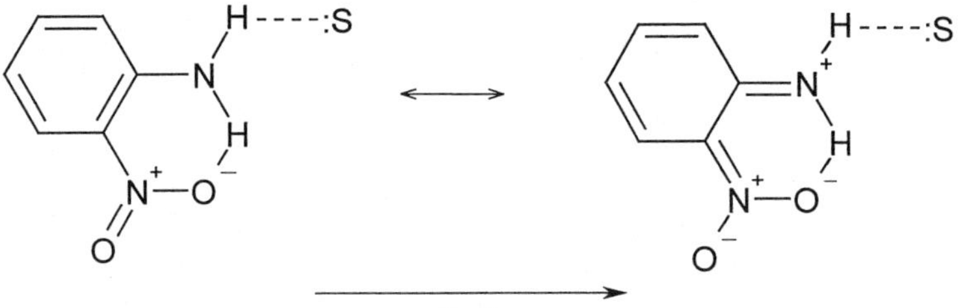

excitation

The relative spectral shifts $\Delta\Delta\upsilon$ in the spectra for processes (vi) relate to β as:

$$\beta = (0.025 - \Delta\Delta\upsilon)/1.017 \tag{2.73}$$

Following up the work by Kamlet and Taft [363,369], Abraham et al. [371] developed the solute scales for hydrogen-bonding acidity (α_2^H) and hydrogen-bonding basicity (β_2^H). These scales were established as the linear functions for the formation of 1:1 complexes between the solute molecule and given reference base or acid, respectively, in the tetrachloromethane. It was found [371] that the equilibrium constant (K_c) for the formation of a hydrogen bond complex in a low-polarity medium (CCl_4) at 25°C can be expressed by the following basic equation:

$$\log K_c = -1.10 + 7.35\alpha_2^H\beta_2^H \tag{2.74}$$

where K_c is in units of liters/mole. The parameters α_2^H and β_2^H in Eq. (2.74) are the scaled descriptors for the relative hydrogen bond donor (HBD) and hydrogen bond acceptor (HBA) strengths of the combining electroneutral molecules. Equation (2.74) was found to hold for over 1300 equilibrium constants with the average precision of 0.1 logarithmic units. The systematic deviations from Eq. (2.74) for certain pairs of acids and bases (e.g., chloroform or diphenylamine as the HBD with the amines, ethers or sulfides as the HBA) has been discussed by several groups [372–374]. The equations similar to Eq. (2.74) have been developed for the equilibrium constants of hydrogen-bonded complexes in 1,1,1-trichloroethane and in the gas phase. It was anticipated that the hydrogen bonding assisted docking of drug molecules and molecular recognition could be within the utility of these equations [306].

Values of α_2^H and β_2^H parameters are available for many compounds, mostly for the singly functionalized molecules. The quantitative family structure–HBA relationships have been determined from the IR spectroscopic data for a series of alcohols, phenols, nitriles, formamidines, amides, ureas, lactones, carbonates, nitro compounds, and lactams [375–380]. By using the empirical constants of substituent induction, resonance, and steric effects, it had been possible to reliably predict the β_2^H values of polyfunctional drug fragments. Some typical values of the descriptors α_2^H and β_2^H are given in Table 2.14.

A further methodology for the determination of the overall HBD solvation energy and the respective LFER parameter $\varepsilon\alpha$ was developed by Taft et al. [306]. This approach was based on the use of the partition coefficient between the octanol and chloroform; log P and the following equation was developed for the calculation of $\varepsilon\alpha$ values for individual compounds:

$$\varepsilon\alpha = \frac{\log P(\text{octanol/chloroform}) + 0.01V_X + 0.03}{3.23} \tag{2.75}$$

where V_X is the characteristic volume of the molecule. It was demonstrated that the new HBD parameters for monofunctional compounds were in good correspondence with the earlier HBD scale (α_2^H). Thus, Eq. (2.75) was applied to calculate the $\varepsilon\alpha$ parameters for the polyfunctional HBD compounds. It has been noted, however, that the applicability of this equation for that purpose depends on several assumptions. For instance, the existence of an LFER between the free energy of hydrogen bonding between the HBD solvent and octanol, and the α_2^H descriptors is required. In addition, the solute dipolarity/polarizability and HBA interactions must be nearly equal in the wet octanol and in the chloroform, to be canceled out in the free energy of the solute transfer between these two solvents. Furthermore, the V_X parameter should adequately describe the difference in the free energy of solute cavity formation in octanol and in chloroform, respectively.

A different approach to the quantitative empirical description of the solvent properties has been based on the concept of hardness and softness of the molecules participating in the acid–base interaction. A simple rule governing the stability of Lewis acid–base complexes states that hard acids prefer to coordinate to hard bases and, correspondingly, soft acids to soft bases. Marcus has

TABLE 2.14 Values of α_2^H and β_2^H for Some Compounds, Obtained from Equilibrium Constants of Hydrogen Bond Formation [306]

Compound	α_2^H	β_2^H
Methanol	0.36	0.41
Ethanol	0.33	0.44
Acetic acid	0.59	0.43
Acetone	0.04	0.49
Acetonitrile	0.09	0.44
N-Methylformamide	0.38	0.66
Benzene	0.00	0.14
Phenylethyne	0.12	0.16
Pyridine	0.00	0.63
Piperidine	0.08	0.74
Morpholine	0.15	0.64
Trifluoraacetic acid	0.95	0.26

defined a scale to describe the hardness/softness of solvents as the difference between the mean of the standard molar Gibbs energies of transfer of sodium and potassium ions from water (W) to a given solvent (S), and the respective transfer energy for silver ions [381]:

$$\mu = \{\tfrac{1}{2}[\Delta G_t^0(\mathrm{Na}^+, W \rightarrow S) + \Delta G_t^0(\mathrm{K}^+, W \rightarrow S)] - \Delta G_t^0(\mathrm{Ag}^+, W \rightarrow S)\} \times \frac{1}{100} \quad (2.76)$$

The sodium and potassium ions represent the hard ions whereas the silver cation is assumed a soft ion. Thus, the former ions should remain in water, which is a typical hard base, whereas the silver ions migrate preferably to softer solvents. The degree of softness among the solvents increases in the following series: O-donors (alcohols, ketones, amides) < N-donors (nitriles, pyridines, amines) < S-donor solvents (thioethers, thioamides).

A multiparameter LFER equation based on the solvent ionization potentials and electron affinities and reflecting the hardness/softness of solvents has been developed by Dougherty [382]. In his approach, two donor–acceptor type of interactions were considered to exist between the solvent and a probe ion. The first was described by the energy difference between the HOMO of the solvent and the LUMO of the probe ion. The second interaction was characterized by the energy difference between the LUMO of the solvent and the HOMO of an ion. Accordingly, the following equation was proposed:

$$\delta E_S = C_1(\mathrm{IP}_S + \mathrm{EA}_S) + C_2(\mathrm{IP}_S + C_3(\mathrm{EA}_S)^2 + C_4 \quad (2.77)$$

In this equation, the sum $(\mathrm{IP}_S + \mathrm{EA}_S)$ has been interpreted as the solvent ionizing power, the IP_S term reflects the solvent nucleophilicity and the term $(EA_S)^2$ the electrophilicity of the solvent.

A list of empirical scales for the description of specific solvent effects is given in Table 2.15. The numerical data on various empirical scales of acidity and basicity of solvents are presented in Tables 2.16 and 2.17.

The empirical solvent effect scales have been widely used for the description of chemical reactivity and physical and spectroscopic properties of compounds in different media. Some compilations of the respective LSER/LFER equations have been published in several reviews [19,36,209,274,383–385]. In the following, we shall give some representative examples of the application of empirical solvent effect scales in the description of various experimental data.

The incompatibility of solvent scales from different sources can be illustrated by the treatment of data on the rate constant of the *tert*-butyl chloride solvolysis. These data were explicitly used for the definition of the single Y scale of solvent polarity [27] [Eq. (2.13)]. However, the LSER treatment of these data using the Kamlet–Taft equation (2.17) has lead to the following four-parameter equation [386]:

$$\log k = (-14.60 \pm 0.29) + (4.8 \pm 0.7)10^{-3}\delta_H^2 + (5.10 \pm 0.37)\pi^* + (4.17 \pm 0.11)\alpha + (0.73 \pm 0.21)\beta$$
$$R^2 = 0.9946 \qquad s = 0.242 \qquad n = 21$$

where k is the rate constant of the *tert*-butyl chloride solvolysis in a given solvent. The β term that measures the nucleophilic assistance to the reaction is, however, small and insignificant in the treatment of $\log k$ for the solvolysis of *tert*-butyl bromide and *tert*-butyl iodide in different solvents. The respective LSER equations have the following form [386]:

$$\log k = (-11.51 \pm 0.25) + (3.5 \pm 0.9)10^{-3}\delta_H^2 + (5.47 \pm 0.43)\pi^* + (3.06 \pm 0.13)\alpha$$
$$R = 0.9870 \qquad s = 0.299 \qquad n = 21$$

for the solvolysis of *tert*-butyl bromide and

$$\log k = (-10.14 \pm 0.24) + (6.30 \pm 0.32)\pi^* + (2.50 \pm 0.12)\alpha$$
$$R = 0.9793 \qquad s = 0.307 \qquad n = 21$$

for the solvolysis of *tert*-butyl iodide. Notably, only two parameters, the dipolarity/polarizability π^* and the solvent acidity α, are sufficient for the description of the solvent effects on the last reaction.

Similar correlations were obtained for the rate constants of the reaction between thiophen-2-sulphonyl chloride and ring-substituted aniline derivatives [387].

TABLE 2.15 Empirical Scales for Description of Specific Solvent Effects

Scale Notation	Standard Process	Reference
Basicity Scales		
DN	Molar enthalpy for the formation of adduct between $SbCl_5$ and the solvent	*a*
B(OH); B(OD)	Band shifts of the O—H stretching vibration of the phenol in tetrachloromethane and O—D stretching vibration of the deuterated methanol in pure solvent	*b*, *u*
$\Delta H^0_{BF_3}$	Molar enthalpies for the formation of adduct between the gaseous BF_3 and solvent	*c*
pK_{HB}	Equilibrium of the complex formation with 4-fluorophenol in CCl_4	*d*
B_P	Gibbs solvation energy of monoatomic monovalent cation	*e*
D_S	Shift in the symmetric stretching frequency of $HgBr_2$	*f*
D_H	Gibbs energy of transfer of the sodium ion from water to a given solvent	*f*
B_{sc}	Solvatochromic shifts for the absorption band of Cu(II) *N,N,N′,N′*-tetramethylethylenediamine acetoacetonate	*g*
B_i	Statistical treatment of multiple data sets	*h*
β	Statistical treatment of multiple data sets	*i*
β_2^H	Complex formation between the solute and given reference acid in CCl_4	*j*
μ	Difference between the mean of the standard molar Gibbs energies of transfer of sodium and potassium ions from water to a given solvent and the respective transfer energy for the silver ions	*k*
IP_S	Ionization potential of solvent (nucleophilicity)	*l*
E_B^*	Electrostatic component of the basicity (multiple data)	*m*
C_B^*	Covalent component of the basicity (multiple data)	*m*
CP	Ni^{II} complex formation with solvents	*n*
D_π	Kinetics of reaction between DDM and TCNE	*o*
$\Delta \upsilon_{CI}$	IR stretching of C—I bonds in iodine cyanide	*p*
N_T	Solvent nucleophilicity	*q*
Acidity Scales		
AN	^{31}P NMR chemical shifts of triethylphosphane oxide complexes with the electron pair acceptor solvents	*r*
$A(^{14}N)$	Nitrogen hyperfine splitting constants in the ESR spectra of aminyloxide radicals in different solvents	*s*
A_P	Gibbs solvation energy of a monoatomic monovalent ion	*t*
A_j	Statistical treatment of multiple data sets	*h*
E	Dimroth's E_T scale with the subtraction of the nonspecific solvation effect	*u*

TABLE 2.15 ***(Continued)***

Scale Notation	Standard Process	Reference
α	Statistical treatment of multiple data sets	*v*
α_2^H	Complex formation between the solute and given reference base in CCl_4	*j*
$\varepsilon\alpha$	Octanol/water partition coefficient and the solute volume	*w*
$(EA_S)^2$	Electron affinity of solvent squared (electrophilicity)	*l*
E'_A	Electrostatic component of the acidity (multiple data)	*m*
C'_A	Covalent component of the acidity (multiple data)	*m*
$\Delta\delta(^{15}N)$	Chemical shifts in the ^{15}N NMR spectrum of pyridine	*x*

[a]V. Gutmann, *Coord. Chem. Rev.* **2,** 239 (1967); *Chimia* **23,** 285 (1969); *Electrochim. Acta* **21,** 661 (1976); Y. Marcus, *J. Solut. Chem.* **13,** 599 (1984).
[b]I.A. Koppel and A.I. Paju, *Org. React.* **11,** 121 (1974).
[c]P.-C. Maria and J.-F. Gal, *J. Phys. Chem.* **89,** 1296 (1985); P.-C. Maria, J.-F. Gal, J. de Franceschi, and E. Fargin, *J. Am. Chem. Soc.* **109,** 483 (1987).
[d]C. Laurence, M. Berthelot, J.-Y. Le Questel, and M.J. El Ghomari, *J. Chem. Soc., Perkin Trans. 2*, p. 2075 (1995); M. Berthelot, C. Laurence, M. Safar, and F. Besseau, *ibid.*, p. 283 (1998).
[e]W.R. Fawcett, in *Quantitative Treatments of Solute/Solvent Interactions* (P. Politzer and J.S. Murray, eds.), pp. 183–212. Elsevier, Amsterdam, 1994.
[f]I. Persson, M. Sandström, and P.L. Goggin, *Inorg. Chim. Acta* **129,** 183 (1987).
[g]M. Sandström, I. Persson, and P. Persson, *Acta Chem. Scand.* **44,** 653 (1990).
[h]C.G. Swain, M.S. Swain, A.L. Powell, and J. Alunni, *J. Am. Chem. Soc.* **105,** 502 (1983).
[i]M.J. Kamlet and R.W. Taft, *J. Am. Chem. Soc.* **98,** 377, 3233 (1976).
[j]M.H. Abraham, P.L. Grellier, D.V. Prior, R.W. Taft, J.J. Morris, P.J. Taylor, C. Lauren, M. Berthelot, R.M. Doherty, M.J. Kamlet, J.-L.M. Abboud, K. Sraidi, and J. Guiheneuf, *J. Am. Chem. Soc.* **10,** 8534 (1988).
[k]Y. Marcus, *J. Phys. Chem.* **91,** 4422 (1987).
[l]R.C. Dougherty, *Tetrahedron Lett.*, p. 385 (1975).
[m]R.S. Drago, M.S. Hirsch, D.C. Ferris, and C.W. Chronister, *J. Chem. Soc., Perkin Trans. 2*, p. 219 (1994); J. George and R.S. Drago, *Inorg. Chem.* **35,** 239 (1996); S. Joerg, R.S. Drago, and J. Adams, *J. Chem. Soc., Perkin Trans. 2*, p. 2431 (1997).
[n]M. Munakata, S. Kitagawa, and M. Miyazima, *Inorg. Chem.* **24,** 1638 (1985).
[o]T. Oshima, S. Arikata, and T. Nagai, *J. Chem. Res., Synop.*, p. 204 (1981).
[p]C. Laurence, M. Queignec-Cabanetos, T. Dziembowska, R. Queignec, and B. Wojtkowiak, *J. Am. Chem. Soc.* **103,** 2567 (1981).
[q]D.N. Kevill, in *Advances in Quantitative Structure-Property Relationships* (M. Charton, ed.), Vol. 1, pp. 81–115. JAI Press, Greenwich, CT.
[r]U. Mayer, V. Gutmann, and W. Gerger, *Monatsh. Chem.* **106,** 1235 (1975); V. Gutmann and G. Resch, in *Ions and Molecules in Solution* (N. Tanaki, H. Ohtaki, and R. Tamamushi, eds.). Elsevier, Amsterdam, 1982; p. 203 ff.
[s]B.R. Knauer and J.J. Napier, *J. Am. Chem. Soc.* **98,** 4395 (1976); O.W. Kolling, *Anal. Chem.* **49,** 591 (1977); **55,** 143 (1983).
[t]W.R. Fawcett, *J. Phys. Chem.* **97,** 9540 (1993).
[u]I.A. Koppel and V.A. Palm, in *Advances in Linear Free Energy Relationships* (N.B. Chapman and J. Shorter, eds.), Chapter 5. Plenum, London, 1972.
[v]M.J. Kamlet and R.W. Taft, *J. Am. Chem. Soc.* **98,** 2886 (1976).
[w]R.W. Taft and J.S. Murray, in *Quantitative Treatments of Solute/Solvent Interactions* (P. Politzer and J.S. Murray, eds.), p. 62. Elsevier, Amsterdam, 1994.
[x]O.W. Kolling, *Anal. Chem.* **51,** 1324 (1979).

TABLE 2.16 Numerical Values of Some Common Empirical Basicity or Neucleophilicity Scales for Description of Specific Solvent Effects[a]

Solvent	DN	B(OH)	B(OD)	ΔH^0_{BF3}	D_S	D_H	B_i
1,2-Dichloroethane	0.0	40	49		7	0.0	0.8
1,2-Dimethoxyethane	20.0	238					0.5
1,2-Ethanediol	20.0	224		74.1	20		0.8
1,3,5-Trimethylbenzene	10.0	77	62				
1,4-Dioxane	14.3	237	128		18		0.7
1-Butanol	29.0	231	119		19	11.0	0.4
1-Hexanol			120				
1-Octanol	32.0		121				
1-Pentanol	25.0		120				
1-Propanol			119			7.9	0.4
2-Butanol		240	122				
2-Butanone	17.4	209	109	76.1			0.7
2-Methyl-1-propanol		230	119				
2-Propanol	36.0	236	122	112.1			0.4
Acetic acid	20.0	131		76.0	14		0.1
Acetone	17.0	224	123	60.4	15	20.7	0.8
Acetonitrile	14.1	160	103		12	9.9	0.9
Acetophenone	15.0	202	108	74.5			0.9
Aniline	35.0	346			34		1.2
Anisole	9.0	155	75		10		0.7
Benzene	0.1	48	52	55.4	9		0.6
Benzonitrile	11.9	155	97		12	−0.1	0.9
Benzyl alcohol	23.0		112				
Bromobenzene	3.0	40	49				0.7
Carbon disulfide	2.0						0.4
Carbon tetrachloride	0.0	0	31				0.3
Chlorobenzene	3.3	38	45				0.7

Chloroform	4.0	14	39				0.7
Cyclohexane	0.0	0		78.6			0.1
Cyclohexanol	25.0	242	124	76.4			
Cyclohexanone	18.0	242	132	105.3			0.8
Dichloromethane	1.0	23	43	78.8	6		0.8
Diethyl ether	19.2	280			12		0.3
Diisopropylether	19.0	293	134	76.6			
Dimethyl sulfoxide	29.8	360	192		28	37.8	1.1
di-*n*-Butylether	19.0	285	129	10.0			0.3
Ethanol	32.0	235	177	75.6	19	14.0	0.5
Ethyl acetate	16.5	181	89				0.6
Fluorobenzene	3.0	38	34				
Formamide	24.0	270		117.5	21	33.0	1.0
HMPA	38.8	471			34		1.1
Methanol	30.0	218	114		18	16.7	0.5
Methyl acetate	16.3	170	88	72.8	12	13.3	
N,*N*-Dimethylacetamide		343	178	110.5	24	36.8	1.0
N,*N*-Dimethylformamide	26.6	291	166		24	34.3	0.9
n-Heptane	0.0	0					0.0
n-Hexane	0.0	0		35.8			0.0
Nitrobenzene	4.4	67	63	37.6	9	−9.0	0.9
Nitromethane	2.7	65	58	128.1	9	−1.0	0.9
N-Methylacetamide	27.8						
N-Methylformamide	27.0	287			22		
o-Dichlorobenzene	3.0	28	43	64.2			
Piperidine	40.0				48		
Propylene carbonate	15.1				12		
Pyridine	33.1	472			38	8.7	1.0
Quinoline	32.0	494					
tert-Butanol	38.0	247	125	90.4			0.5
Tetramethylene sulfone		157			15	28.0	

TABLE 2.16 ***(Continued)***

Solvent	DN	B(OH)	B(OD)	ΔH^0_{BF3}	D_S	D_H	B_i
THF	20.0	287	145		17	34.0	0.7
Toluene	0.1	58	54	135.9			0.5
Triethylamine	61.0	650					0.2
Trimethyl phosphate	23.0			84.8	23		
Water	18.0	156			17	24.7	1.0
Reference[a]	*a*	*b*	*u*	*c*	*f*	*f*	*h*

Solvent	β	μ	E^*_B	C^*_B	D_π	$\Delta \upsilon_{CI}$
1,2-Dichloroethane	0.10	0.03			−1.220	
1,2-Dimethoxyethane	0.41		1.86	1.29	0.704	
1,2-Ethanediol	0.52					17.0
1,4-Dioxane	0.37				0.590	
1-Butanol	0.84	0.18				
1-Hexanol	0.84					
1-Octanol	0.77					
1-Pentanol	0.86					
1-Propanol	0.90	0.16				
2-Butanol	0.80					
2-Butanone			1.67	1.24	0.177	
2-Methyl-1-propanol	0.84					
2-Propanol	0.84					
Acetic acid	0.45					18.0
Acetone	0.43	0.03	1.74	1.26	0.261	17.0
Acetonitrile	0.40	0.35	1.64	0.71	−0.440	10.0
Acetophenone			1.72	1.15		
Aniline	0.50					
Anisole						12.5

Benzene	0.10		0.70	0.45	0.000	
Benzonitrile	0.37	0.34	1.65	0.75	−0.398	7.5
Benzyl alcohol	0.56					
Bromobenzene	0.06				−1.030	0.0
Carbon disulfide	0.07					
Carbon tetrachloride	0.10					5.5
Chlorobenzene	0.07				−0.903	1.0
Chloroform	0.10				−1.560	
Cyclohexanol	0.84					19.0
Cyclohexanone	0.53		1.45	2.14	0.243	32.0
Dichloromethane	0.10				−1.300	20.5
Diethyl ether	0.47		1.80	1.63	−0.298	
Diisopropylether			1.95	1.66	−0.982	
Dimethyl sulfoxide	0.76	0.22	2.40	1.47		
di-*n*-Butylether	0.46		1.89	1.67	−0.372	2.0
Ethanol	0.75	0.08	1.84	1.09		
Ethyl acetate	0.45		1.62	0.98	0.289	
Fluorobenzene	0.07				−0.818	
Formamide	0.48	0.09				
HMPA		0.29				
Methanol	0.66	0.02	1.79	0.65		
Methyl acetate			1.63	0.95	0.196	
N,*N*-Dimethylacetamide	0.76	0.17	2.35	1.31		30.0
N,*N*-Dimethylformamide	0.69	0.11	2.19	1.31		
n-Heptane	0.00					
n-Hexane	0.00					11.0
Nitrobenzene	0.30	0.23			−0.583	
Nitromethane	0.06	0.03			−0.724	57.5
N-Methylacetamide	0.80					
N-Methylformamide	0.80	0.12				48.0
o-Dichlorobenzene	0.03					

TABLE 2.16 (*Continued*)

Solvent	β	μ	E_B^*	C_B^*	D_π	Δv_{CI}
Piperidine			1.44	4.93		91.0
Propylene carbonate	0.40	−0.09				
Pyridine	0.64	0.64	1.78	3.54		
Quinoline			2.28	2.89		
tert-Butanol	0.93		1.92	1.22		25.0
Tetramethylene sulfone		0.00				11.5
THF	0.55		1.64	2.18	0.639	
Toluene	0.11				0.394	88.0
Triethylamine	0.71		1.32	5.73		
Trimethyl phosphate	0.77	−0.03				26.0
Water	0.47	0.00	2.28	0.10		
Reference	*i*	*p*	*k*	*m*	*m*	*o*

[a]References correspond to those given in Table 2.15.

$$\text{C}_4\text{H}_3\text{S-SO}_2\text{Cl} + 2\ \text{R-C}_6\text{H}_4\text{-NH}_2 \longrightarrow \text{C}_4\text{H}_3\text{S-SO}_2\text{-NH-C}_6\text{H}_4\text{-R} + \text{R-C}_6\text{H}_4\text{-NH}_3^+\text{Cl}^-$$

The following LSER equations were reported:

$$\log k = -5.93 + 3.51\pi^* + 1.80\alpha$$
$$R^2 = 0.992 \qquad s = 0.10 \qquad n = 8$$

for the reaction with the unsubstituted aniline (R = H),

$$\log k = -5.27 + 3.42\pi^* + 1.52\alpha$$
$$R^2 = 0.992 \qquad s = 0.09 \qquad n = 8$$

for the reaction with the 4-methylaniline (R = CH_3), and

$$\log k = -6.96 + 3.29\pi^* + 2.30\alpha$$
$$R^2 = 0.990 \qquad s = 0.12 \qquad n = 8$$

for the reaction with the 4-chloraniline (R = Cl). Notably, the dipolarity/polarizability term is nearly constant whereas the contribution of the solvent acidity term increases with decreasing basicity of the aniline derivative. This has been rationalized in terms of "push–pull" mechanism where the lower the nucleophilicity of the attacking group the greater the importance of electrophilic assistance of solvent by hydrogen bonding to the leaving group. Attempts to find the correlations for this reaction using Koppel–Palm [274] and Krygowski–Fawcett [357] models have failed [388].

Nevertheless, the latter equations have been successfully employed for the correlation of the solvent effect on other reaction rates or molecular properties in solution. For instance, the LFER analysis of rate constants for the reaction of diazodiphenylmethane with benzoic acid according to Eq. (2.16) has lead to the following equation [389]:

$$\log k = -3.190 + (4.463 \pm 0.346)Y + (12.330 \pm 1.544)P$$
$$+ (0.210 \pm 0.020)E + (0.0181 \pm 0.0008)B$$
$$R = 0.958 \qquad s = 0.176 \qquad n = 43$$

TABLE 2.17 Numerical Values of Some Common Empirical Acidity or Electrophilicity Scales for Description of Specific Solvent Effects[a]

Solvent Name	AN	$A(^{14}N)$	A_j	E	α	α_2^H	E'_A	C'_A
1,2-Dichloroethane	16.7	1.5655	0.3	3.0	0.00			
1,2-Dimethoxyethane	10.2	1.5424	0.2		0.00			
1,2-Ethanediol		1.6364	0.8		0.90		1.63	1.49
1,3,5-Trimethylbenzene				0.8				
1,4-Dioxane	10.3	1.5452	0.2	4.2	0.00	0.00		
1-Butanol	36.8	1.6018	0.6	10.3	0.84	0.33	1.05	0.74
1-Hexanol				9.6	0.80		1.14	0.7
1-Octanol				10.1	0.77		0.89	0.87
1-Pentanol				9.7	0.84		1.27	0.68
1-Propanol	37.3		0.6	10.6	0.84	0.33	1.17	0.68
2-Butanol				7.4	0.69	0.32		
2-Butanone			0.2	2.0				
2-Methyl-1-propanol				7.4	0.79			
2-Propanol	33.5	1.5973	0.6	8.7	0.76	0.32	1.19	0.69
Acetic acid	52.9	1.6420	0.9		0.12	0.58	3.39	0.91
Acetone	12.5	1.5527	0.3	2.1	0.08	0.04		
Acetonitrile	18.9	1.5666	0.4	5.2	0.19	0.09		
Acetophenone			0.2	0.7				
Aniline			0.4		0.26	0.26		
Anisole			0.2	1.4				
Benzene	8.2	1.5404	0.2	2.1	0.00	0.00		
Benzonitrile	15.5		0.3	0.0	0.00	0.00		
Benzyl alcohol	36.8			10.3		0.38	1.19	0.55
Bromobenzene		1.5479	0.2	0.0	0.00			
Carbon disulfide		1.5289	0.1		0.00			
Carbon tetrachloride		1.5331	0.1	0.0	0.00			
Chlorobenzene		1.5472	0.2	0.0	0.00			
Chloroform	23.1	1.5863	0.4	3.3	0.20	0.20	1.56	0.44
Cyclohexane			0.0		0.00			

Cyclohexanol				7.4	0.66		1.23	0.5
Cyclohexanone			0.3	0.5	0.00			
Dichloromethane	20.4	1.5752	0.3	2.7	0.13	0.13	0.86	0.11
Diethyl ether		1.5334	0.1		0.00			
Diisopropylether				0.0				
Dimethyl sulfoxide	19.3	1.5692	0.3	3.2	0.00			
di-*n*-Butylether			0.1	0.0	0.00			
Ethanol	37.1	1.6030	0.7	11.6	0.86	0.33	1.15	0.67
Ethyl acetate	9.3		0.2	1.6	0.00	0.00		
Fluorobenzene				0.0	0.00			
Formamide	39.8		0.7		0.71		2.00	0.32
HMPA	10.6		0.0					
Methanol	41.3	1.6210	0.8	14.9	0.98	0.37	1.27	0.74
Methyl acetate	10.7			2.1				
N,N-Dimethylacetamide	13.6		0.3	2.4	0.00			
N,N-Dimethylformamide	16.0	1.5635	0.3	2.6	0.00			
n-Heptane	0.0		0.0		0.00			
n-Hexane	0.0	1.5134	0.0		0.00			
Nitrobenzene	14.8		0.3	0.0	0.00			
Nitromethane	20.5	1.5759	0.4	5.2	0.22	0.12		
N-Methylacetamide					0.47	0.38		
N-Methylformamide	32.1				0.62	0.38	1.25	0.15
o-Dichlorobenzene				0.0	0.00			
Propylene carbonate	18.3				0.00			
Pyridine	14.2	1.5608	0.2		0.00	0.00		
tert-Butanol	27.1	1.5860	0.5	5.2	0.42	0.33	1.14	0.66
Tetramethylene sulfone								
THF	8.0		0.2	0.0	0.00			
Toluene		1.5347	0.1	1.3	0.00			
Triethylamine	1.4		0.1		0.00			
Trimethyl phosphate	16.3				0.00			
Water	54.8	1.7175	1.0		1.17	0.35	1.35	0.78
References[a]	*r*	*p*	*h*	*u*	*v*	*j*	*m*	*m*

[a]References correspond to those given in Table 2.15.

for a set of 43 aprotic solvents. However, it was pointed out that a three-parameter LFER equation employing the solvatochromic polarity parameter E_T instead of Y and E scales had almost the same statistical quality [389]:

$$\log k = -5.379 + (7.659 \pm 1.707)P + (0.1351 \pm 0.0101)E_T + (0.0189 \pm 0.0009)B$$

$$R^2 = 0.941 \qquad s = 0.210 \qquad n = 43$$

This result lead to the questioning of the validity of the subtraction of polarity and polarizability contributions from the original E_T scale [Eq (2.72)] as leading to the increase of the number of correlation parameters [389]. Still, in the case of other reactions, the separate parameters may be needed. For instance, a comprehensive treatment of data on rate constants for the same reaction of diazodiphenylmethane with benzoic acid in 44 aprotic solvents, 22 alcohols, and 2-methoxyethanol has resulted in the following four-parameter equation [390]:

$$\log k = -2.850 + (4.434 \pm 0.425)Y + (11.27 \pm 1.50)P + (0.1068 \pm 0.0072)E + (0.0180 \pm 0.00096)B$$

$$R^2 = 0.912 \qquad s = 0.224 \qquad n = 67$$

where all terms are statistically significant.

The Krygowski–Fawcett equation (2.69) has been applied successfully for the description of solvent effects on the NMR chemical shifts, enthalpies of ionic solvation, and the standard potentials of reduction of organic molecules [357]. For example, the $^{23}Na^+$ NMR chemical shifts in the $NaClO_4$ and $NaBPh_4$ solutions of different concentrations in different solvents have been correlated just with the donor number, DN. In 0.05 M $NaClO_4$ solutions, the $\delta(^{23}Na^+)$ were described by the following equation:

$$\delta(^{23}\mathrm{Na}^+) = 19.7 - (0.54 \pm 0.05)DN$$

$$R = 0.968 \qquad n = 8$$

The solvent activity coefficient for the Ag^+ ion in different solvents have been presented as follows [357]:

$$\ln {}^0\gamma^S_{\mathrm{Ag}^+} = -13.7 + (0.31 \pm 0.01)E_T - (0.25 \pm 0.07)\mathrm{DN}$$

$$R^2 = 0.774 \qquad n = 9$$

and the enthalpy of solvation of the K^+ ion by the following equation:

$$\Delta H^S_{\mathrm{K}^+} = -19.2 + (0.38 \pm 0.04)E_T - (0.24 \pm 0.04)\mathrm{DN}$$

$$R^2 = 0.955 \qquad n = 8$$

In the last two equations, the coefficient in front of the E_T term is positive and the coefficient in front of the *DN* term negative. Consequently, as expected, the increase of the acidity of the solvent destabilizes the cationic solute whereas the increase of the basicity of solvent leads to the stabilization of this solute.

The Grünwald–Winstein approach has been further developed by inclusion of an original solvent nucleophilicity scale, N_T [391]. The respective extended equation has been applied for the description of the solvent effects on the rates of solvolysis of several *N,N*-disubstituted carbamoyl chlorides [392–395].

The empirical solvent effect approach has been applied for the description of the free energies of solvation of compounds in different media. For example, the free energies of transfer of the tetramethyl- and tetraethylammonium chloride, bromide and iodide ion pairs, and dissociated ions from methanol up to 17 widely varying solvents were well correlated by the Kamlet–Taft equation (2.17) [396]. The respective ΔG_t^0 for tetramethyl iodide ion pairs were presented by the following LSER equation:

$$\Delta G_t^0 = (10.9 \pm 0.2) - (15.6 \pm 0.4)\pi^* - (6.2 \pm 0.3)\alpha + (2.2 \pm 0.1)10^{-2}\delta_H^2$$

$$R^2 = 0.994 \qquad s = 0.34 \qquad n = 18$$

and the ΔG_t^0 for the individual ions of this salt are as follows:

$$\Delta G_t^0 = (28.2 \pm 1.3) - (42.0 \pm 2.4)\pi^* - (16.4 \pm 1.3)\alpha + (6.3 \pm 0.6)10^{-2}\delta_H^2$$

$$R^2 = 0.976 \qquad s = 1.1 \qquad n = 14$$

The comparison of last two equations reveals that the individual terms of the solute–solvent interaction are substantially diminished by ion pairing. The cation–anion complexation reduces naturally the terms accounting for the polar interactions and the acid–base interaction between the ions and the solvent. The reduction of the cavity term has been ascribed to the change of the effective volume of ions and to the solvent reorganization due to the ion pair formation.

The LFER solvent effect parameters have been also practical for the study of various complex formation processes in the gas phase. For instance, the log K_c of the complex formation between the hydrogen bond donors (alcohols, pyrrole, and HF) and the hydrogen bond acceptors (alcohols, amines, ketones, ethers, aldehydes, and nitriles) in the gas phase correlate with Abraham's solvent hydrogen-bonding parameters [Eq. (2.74)] as [385]:

$$\log K_c = -(0.86 \pm 0.27) + (8.85 \pm 0.42)\alpha_2^H\beta_2^H$$

$$R^2 = 0.994 \qquad s = 0.18 \qquad n = 8$$

This result confirms the applicability of parameters derived from the data in the condensed media as well as in the gas phase. Furthermore, the association of the methyl ammonium cation, $CH_3NH_3^+$, with various hydrogen bond acceptors in the gas phase has been described with the following two-parameter LSER equation [387]:

$$\log K_c = (4.1 \pm 1.1) + (6.9 \pm 1.6)\pi^* + (16.3 \pm 4.2)\beta_2^H$$
$$R^2 = 0.975 \qquad s = 1.0 \qquad n = 9$$

where π^* denotes the Kalmet–Taft dipolarity/polarizability parameter and β_2^H is Abraham's hydrogen-bonding basicity.

Various solvent effect scales have been employed for the description of the solute partition between the liquid and the micelles of a surfactant. The partition of 132 compounds between water and sodium dodecyl sulfate at 298 K has been correlated through the following equation [397]:

$$\log K_x = (1.201 \pm 0.058) + (0.542 \pm 0.057)R_2 - (0.400 \pm 0.071)\pi_2^H$$
$$- (0.133 \pm 0.060) \sum \alpha_2^H - (1.580 \pm 0.082) \sum \beta_2^o$$
$$+ (2.793 \pm 0.073)V_x$$
$$R^2 = 0.970 \qquad s = 0.171 \qquad F = 817 \qquad n = 132$$

where R_2 is the excess molar refraction, π_2^H is the dipolarity/polarizability, $\Sigma\ \alpha_2^H$ and $\Sigma\ \beta_2^H$ are the hydrogen bond acidity and basicity, respectively, and V_X is the McGowan characteristic volume of the molecule [398]. The partition of 46 compounds between water and hexadecylpyridinium chloride micelles has been described by an analogous equation as [399]:

$$\log K_x = (-0.437 \pm 0.198) + (0.973 \pm 0.150)R_2 - (0.736 \pm 0.107)\pi_2^H$$
$$+ (0.769 \pm 0.088) \sum \alpha_2^H - (2.840 + 0.166) \sum \beta_2^o$$
$$+ (3.386 \pm 0.210)V_x$$
$$R^2 = 0.948 \qquad s = 0.147 \qquad F = 146 \qquad n = 46$$

According to the last two equations, the main factors that influence the partition are the solute hydrogen-bonding basicity, which reduces the transfer into the surfactant micelles, and the solute volume, which increases this transfer.

Analogous equations have been employed for the description of the solubility of gases and vapors in a given solvent. The solubility of 408 gaseous compounds in water at 298 K has been correlated by the following equation [400]:

$$\log L^w = (-0.994 \pm 0.031) + (0.577 \pm 0.032)R_2 + (2.549 \pm 0.037)\pi_2^H$$
$$+ (3.813 \pm 0.040) \sum \alpha_2^H + (4.841 \pm 0.040) \sum \beta_2^o$$
$$- (0.869 \pm 0.031)V_x$$
$$R^2 = 0.995 \qquad s = 0.151 \qquad F = 16810 \qquad n = 408$$

where L^w is the Ostwald solubility coefficient defined as the ratio of the solute concentrations in a given solvent (water) and in the gas phase, respectively:

$$L^w = \frac{C_s}{C_g} \tag{2.78}$$

The solubility of gases and vapors in methanol at 298 K was described by the following LSER equation:

$$\begin{aligned}\log L^{\mathrm{MeOH}} &= (-0.004 \pm 0.026) - (0.215 \pm 0.061)R_2 + (1.173 \pm 0.077)\pi_2^H \\ &\quad + (3.701 \pm 0.088)\sum \alpha_2^H + (1.432 \pm 0.084)\sum \beta_2^o \\ &\quad + (0.769 \pm 0.011)\log L^{16} \\ &\quad R^2 = 0.990 \qquad s = 0.13 \qquad F = 3681 \qquad n = 93\end{aligned}$$

where L^{16} is the Ostwald solubility coefficient on hexadecane at 298 K. The different relative contributions of different terms in the LSER representation of data in water and methanol have been discussed proceeding from the solvational characteristics of these two solvents [401].

The solvent effect parameters have been used for the treatment of solvent effects on various spectroscopic data. The electronic spectral transition maxima of 3-nitroaniline in different solvents have been correlated using the following LSER equation [402]:

$$\begin{aligned}v(i)_{\max} &= 38.96 - 1.741\pi^* - 2.700\beta \\ R &= 0.982 \qquad s = 0.15 \qquad n = 33\end{aligned}$$

where π^* is the Kamlet–Taft dipolarity/polarizability and β is the basicity of the solvent. Notably, both the specific and nonspecific solvation contribute to the solvatochromic shift of the $p \rightarrow \pi^*$ transition energy of this compound.

The C=O, C≡N, and NO_2 asymmetric stretch frequencies in the infrared spectra of several compounds have been correlated with the solvent donor number, acceptor number, solvent polarity Y, and solvent polarizability P [337]. A similar approach was applied for the description of the solvent-induced frequency shifts in the IR spectrum of dimethyl sulfoxide and deuterated dimethyl sulfoxide (DMSO-d_6) [403]. The Kamlet–Taft equation involving the dipolarity/polarizability and solvent basicity parameters has been applied for the LSER description of the solvent effect on IR maxima of a wide selection of compounds [404–408]. The same equation has been used to correlate the solvent effects on various chemical shifts in the NMR spectra [369,405,409–413] and the ESR hyperfine splitting constants [346,414–416].

Although not extensively, the empirical solvent effect parameters have found some utilization in correlating the biological activity of compounds. As an example, the narcotic activity of aqueous compounds toward the tadpole [4] was correlated with the Abraham–Taft parameters as [417]:

$$\log\left(\frac{1}{C_{\mathrm{nar}}}\right) = (0.579 \pm 0.085) - (0.824 \pm 0.083)R_2 - (0.334 \pm 0.111)\pi_2^H$$
$$- (2.871 \pm 0.134)\sum \beta_2^o + (3.097 \pm 0.113)V_X$$
$$R^2 = 0.947 \qquad s = 0.246 \qquad F = 351 \qquad n = 84$$

where C_{nar} is the minimum concentration needed for the narcosis. It has been concluded from this equation that the solute hydrogen-bonding basicity reduces significantly the narcotic activity whereas the solute size expressed by the excess molar refraction and the solute volume enhances this activity substantially. A similar equation has been developed for the prediction of nasal pungency thresholds in humans [418,419].

2.3.3 Hydrophobicity Scales

The concept of hydrophobicity has been one of the most useful and productive for the description of the solvent effects and intermolecular interactions in condensed media, particularly as related to many QSAR biomedical applications. The term *hydrophobic effect* was probably first introduced by Frank and Evans [420] who postulated the formation of "iceberg"-type structured regions of water near the nonpolar solute molecules. It is essential to distinguish between the hydrophobic effect and hydrophobic interactions in solution. The former has been commonly regarded as the relative insolubility of certain nonpolar solutes in water compared to their higher solubility in nonaqueous solvents. The hydrophobic interaction is usually defined as the specific interaction between two or more solute molecules due to their desolvation. The free energy of the solute–solvent system diminishes as a result of this interaction and thus it is energetically advantageous for apolar molecules or apolar groups to aggregate in the solution with expulsion of water molecules from their hydration shells. It had been widely accepted that the hydrophobic interaction is driven by the entropy change due to the restructuring of the solvent. However, the computer modeling of the interaction of *n*-alkanes in aqueous solutions has revealed the importance of the enthalpic term in the interaction between the methylene groups. Thus the question of the origin of the hydrophobic interaction is still debatable, and it may depend on the molecular structure of the system studied [421–426].

The hydrophobic effect is related to the free energy of transfer of a solute between two different solvents or two different phases. Abraham [427] has pointed out that already about a century ago that the first studies were reported relating the biological activity of a series of compounds to the water–oil partition coefficient of these compounds [2,3,428]. The relationships were derived between the activity (usually defined as the reciprocal of the active concentration of the compound, $1/C$) and the water–oil partition coefficient, defined as:

$$P = \frac{C_{\text{org}}}{C_{\text{water}}} \tag{2.79}$$

where C_{org} and C_{water} are the equilibrium concentrations of a compound in the organic phase (oil) and in water, respectively.

The most popular quantitative scale to measure the hydrophobicity of compounds is the logarithm of the partition coefficient between 1-octanol and water, introduced by Hansch and Leo as:

$$\log P = \log \left(\frac{C_{\text{1-octanol}}}{C_{\text{water}}} \right) \tag{2.80}$$

which is well known as the log P parameter. The experimental values of log P are known for a large number of compounds. For example, the 1979 compilation of log P by Hansch and Leo [429] included almost 15,000 individual values of this descriptor. A small selection of log P values for a representative set of compounds is given in Table 2.18. A variety of programs are also available for the theoretical estimation of log P values from the structure of compounds [MacLogP 2.0, Biobyte Corp., Claremont, CA 91711; PCModels 4.6, Daylight Chemical Information Systems, Inc., Mission Viejo, CA 92691; AcculogP, Microsimulations, Inc., Mahwah, NJ 07430; SYBYL Molecular Modelling Software Version 6.2, Tripos Inc., St. Louis, MO 63144; ACD/logP, Advanced Chemistry Development, Inc., Toronto, Canada M5H 2L3; CLIP, Section de Pharmacie, Université de Lausanne, CH-1015, Lausanne-Dorigny, Switzerland; ChemPlus, MD Kemi AB, 30595 Halmstad, Sweden].

The most widespread methods for the calculation of log P are the group contribution schemes, first suggested in the 1970s [429a]. The Rekker scheme was proposed in the following form:

$$\log P = \sum a_n f_n + \sum b_m F_m \tag{2.81}$$

where a_n is the number of fragments with a given additive contribution f_n, and b_m is the number of occurrences of a certain correction factor F_m, related to a specific interaction of the solute molecule with adjacent water molecules. These corrections arise in situations where two polar groups are separated by only one or two aliphatic carbon atoms. The CLOGP methodology [430–434] for the calculation of log P follows the scheme given by Eq. (2.81). However, the methods to account for steric, electronic, and hydrogen-bonding interactions are quite different from those used in Rekker's original approach. These interactions include the aliphatic chain flexibility and branching effects, polar fragment interactions in aliphatic and aromatic systems, ortho-effects in aromatic rings, intramolecular hydrogen bonding, and other effects. Similar schemes

TABLE 2.18 Experimental log *P* Values for Selection of Organic Compounds [63a]

Compound	log *P*	Compound	log *P*
Methane	1.09	Methylamine	−0.57
n-Butane	2.89	Dimethylamine	−0.38
n-Octane	5.18	Propylamine	0.48
2,2-Dimethyl propane	3.11	Trimethylamine	0.16
Cyclopentane	3.00	Pyrrolidine	0.46
1-Butene	2.40	Diethylamine	0.58
Ethyne	0.37	Dipropylamine	1.67
1-Hexyne	2.73	Pyridine	0.65
Benzene	2.13	2-Ethylpyrazine	0.69
m-Xylene	3.20	2,5-Dimethylpyridine	1.68
Naphthalene	3.30	Ethanonitrile	−0.34
Methanol	−0.50	Benzonitrile	1.56
Ethanol	−0.31	Nitroethane	0.18
1-Propanol	0.25	2-Nitropropane	0.80
2-Propanol	0.05	2-Methyl-1-nitrobenzene	2.42
1-Butanol	0.88	(*E*)-*N*-Methylacetamide	−1.05
t-Butanol	0.35	Morpholine	−0.86
1,2-Ethanediol	−1.36	*N*-Methylmorpholine	−0.33
p-Cresol	2.00	Hydrazine	−2.07
Tetrahydrofuran	0.46	Thiophenol	2.52
1,4-Dioxane	−0.27	Dimethyl disulfide	1.77
Methyl propyl ether	1.21	Fluoromethane	0.51
1,2-Dimethoxyethane	−0.21	1,1-Difluoroethane	0.75
Anisole	2.11	Fluorobenzene	2.27
Butanal	0.88	Dichloromethane	1.25
Benzaldehyde	1.48	1,1,2-Trichloroethane	1.89
Butanone	0.29	2-Chloropropane	1.90
3-Pentanone	0.99	2-Chloropropene	2.00
Methyl phenyl ketone	1.58	Trichloroethene	2.61
2-Octanone	2.37	Chlorobenzene	2.89
Ethanoic acid	−0.17	*p*-Dichlorobenzene	3.44
Pentanoic acid	1.39	Bromomethane	1.19
Methyl methanoate	0.03	2-Bromopropane	2.14
Methyl butanoate	1.29	Bromobenzene	2.99
Butyl ethanoate	1.78	Diiodomethane	2.30
2-Propen-1-ol	0.17	Iodobenzene	3.25
2-Methoxyethanol	−0.77	1,1,1-Trifluoropropan-2-ol	0.71
m-Hydroxybenzaldehyde	1.38	*p*-Bromophenol	2.59

have been developed for the prediction of log P using the atomic contribution, and the respective interaction terms including the proximity effects, differences in the hybridization of atoms, intramolecular hydrogen bonds, ring structures, and amphoteric properties [435–444].

Another approach associates the hydrophobicity with the solvent-accessible surface of a compound. Various schemes have been developed to quantify this effect by the computation of the respective contributions into log P [445–450]. Furthermore, the algorithms have been proposed to calculate log P values for limited sets of compounds using the correlations with topological, topographical, and geometrical descriptors of a molecule [451–456]. Bodor et al. have suggested a correlation with a number of molecular descriptors including indicator variables for alkanes: molecular weight, number of carbon atoms, molecular surface area together with its square, "ovality" of the molecule together with its square and the fourth power, the sum of absolute values of the charges on oxygen and nitrogen atoms, the square root of the sum of squared charges on oxygen atoms together with the square and the fourth power of that number, the last parameters for nitrogen, and the calculated dipole moment [457,458]. The best correlation described the log P values for 302 solutes with the standard deviation of 0.305. Also, the log P has been modeled using the molecular electrostatic potentials from the ab initio and semiempirical quantum chemical calculations [459–461] and other quantum chemical molecular descriptors [462–465].

Charton [466] has suggested a complex intermolecular force equation (IMF) to model quantitatively log P and other transport and chromatographic properties of compounds. The IMF equation has been originally proposed in the following 10-parameter form [467,468]:

$$Q_X = L\sigma_{lX} + D\sigma_{dX} + R\sigma_{eX} + A\alpha_X + H_1 n_{HX} + H_2 n_{nX} + I_{IX} + B_{DX} n_{DX} + B_{AX} n_{AX} + S\Psi_X + B^0 \tag{2.82}$$

where Q_X is the quantity to be correlated (log P); σ_{lX} denotes the localized electrical effect parameter, identical to the σ_I and σ_F constants discussed above (cf. Section 2.2); σ_{dX} is the intrinsic delocalized electrical effect parameter and σ_{lX} is the electronic demand sensitivity electrical effect parameter. In Eq. (2.82), α_X denotes the polarizability, calculated by

$$\alpha_X = \frac{\mathrm{MR}_X - \mathrm{MR}_H}{100} \tag{2.83}$$

where MR_X and MR_H are the group molar refractivities of a given group X and of the hydrogen atom ($\mathrm{MR}_H = 0.0103$), respectively. In Eq. (2.82), n_H and n_n are the hydrogen-bonding parameters; n_H is the number of the OH and NH bonds in group X while n_n is equal to the total number of lone electron pairs on O or N atoms in group X. The parameter i_X in Eq. (2.82) reflects the effect of charged or ionized groups. Its value is taken equal to one when side chain X is charged by zero otherwise. The n_D and n_A are the charge-transfer parameters; n_D is equal to one when group X can act as an electron donor (zero otherwise), and n_A is equal to one in the case when group X acts as an electron acceptor (zero otherwise). The steric effect of X is reflected by the Ψ_X parameter in Eq. (2.82). Notably, even this complex equation had to be modified to preduct successfully the log P values for structurally variable compounds.

Alternatively, the Kamlet–Taft and Abraham equations have also been employed for the prediction of log P values of compounds [469–471]. The following LSER equation has been developed using the data for 174 compounds [472]:

$$\log P = (0.01 \pm 0.06) + (0.58 \pm 0.11)R_2 - (0.80 \pm 0.11)\pi_2^H - (1.47 \pm 0.11)\sum \alpha_2^H - (4.92 \pm 0.12)\sum \beta_2^o + (4.17 \pm 0.08)V_x$$

$$R^2 = 0.99 \qquad s = 0.15 \qquad F = 5841 \qquad n = 174$$

where R_2 is the excess molar refraction, π_2^H is the dipolarity/polarizability, $\Sigma\ \alpha_2^H$ and $\Sigma\ \beta_2^H$ are the hydrogen bond acidity and basicity, respectively, and V_x is the McGowan characteristic volume of the molecule.

The hydrophobicity constants of compounds have been also estimated using the comparative molecular field analysis (CoMFA) [473–479] and three-dimensional representation of molecular structure [480–482].

A disadvantage of the log P scale is that the related hydrophobic effect or hydrophobicity is a complex phenomenon involving a multitude of solute–solvent interactions in two different media. Therefore, the interpretation of the QSAR correlations with this parameter may obscure the different physical solute–solvent interactions that give different relative contributions to the log P values of different compounds. It has been also speculated that the log P parameter may not be appropriate for the analysis of the biological activity of gases and vapors as the solvation is missing in the gaseous phase [427].

Menger and Venkataram [483] have suggested a microscopic hydrophobicity (MH) substituent parameter as based on the experimental data on aggregation equilibrium. The basic hydrolysis of 4-nitrophenyl laurate is hindered due to the aggregation of the bulky ester in aqueous solutions. The addition of alkyl- or arylammonium salts $RNMe_3Hal$ to the solution destroys the aggregates and deshields the ester group for the hydrolysis. The more hydrophobic the alkyl- or aryl-group R of the ammonium salt, the greater its disaggregation power and the greater the enhancement of the reaction rate, taken as the measure of the hydrophobicity of the substituent R.

A quantitative measure for the solvophobic effect of compounds has been introduced by Abraham et al. [484]. It is based on the linear correlation of the standard molar Gibbs energies of transfer of nonpolar solutes (argon, alkanes, and alkanelike compounds) from water to other solvents with empirical M parameters. A scale of the solvophobic power SP has been defined according to the following equation:

$$SP = 1 - \frac{M(\text{solvent})}{M(n\text{-hexadecane})} \tag{2.84}$$

where the SP values of water and n-hexadecane are conventionally taken as one and zero, respectively. The SP values provide a simple quantitative measure of the solvophobicity of solvents.

The log P parameter refers to the individual chemical compound as a whole. Fujita et al. [485,486] have also defined a hydrophobicity parameter, π_X, for organic and inorganic substituents according to the following equation:

$$\pi_X = \log P^{C_6H_5X} - \log P^{C_6H_6} \tag{2.85}$$

where $P^{C_6H_5X}$ and $P^{C_6H_6}$ are the partition coefficients between 1-octanol and water for the substituted benzene, C_6H_5X, and benzene itself, respectively. The positive π_X values correspond to the substituents that favor the organic phase relative to the hydrogen atom as a substituent. The substituents with the negative value of π_X cause the preferential partitioning into the water as compared to the hydrogen substituent. The π_X values for a wide selection of substituents are given in Table 2.19.

The log P descriptor and the π scale of hydrophobicity have been most widely used for the correlation of biological activity of compounds, both in the case of in vitro and in vivo data [63a,487–491]. In many cases, the emergence of hydrophobicity in LFER equations for various biological activities is related to the importance of distribution of the substrate between the aqueous and lipid phases and to the penetration of cell membranes and other barriers. For instance, the rate of penetration of barbiturates through silastic membrane has been described by the following equation [492]:

$$\log k_{rel} = -(1.04 \pm 0.29) + (0.91 \pm 0.15)\log P$$
$$R^2 = 0.956 \qquad s = 0.161 \qquad n = 11$$

The log P descriptor has been also involved, together with the molecular weight of a compound (MW), in the LFER description of the drug penetration through the rat brain capillaries [493]:

$$\log Pen = -(1.84 \pm 1.3) + (0.50 \pm 0.01)\log P - (1.43 \pm 0.58)\log \text{MW}$$
$$R^2 = 0.859 \qquad s = 0.461 \qquad n = 23$$

The nonspecific toxicity has been associated with the ability of compounds to reach the target centers in biological systems. This has been confirmed by the

TABLE 2.19 Selection of π_X Values for Substituents [63a]

Substituent	π_X	Substituent	π_X
H	0.00	$OCONH_2$	−1.05
CH_3	0.56	OCH_2COOH	−0.87
CH_2CH_3	1.02	OCH_2CONH_2	−1.37
$CH_2CH_2CH_3$	1.55	$OCH_2CON(CH_3)_2$	−1.36
$CH(CH_3)_2$	1.53	$OCH_2CONHC_6H_5$	0.60
$CH_2CH_2CH_2CH_3$	2.13	$OCH_2CON(CH_3)(C_6H_5)$	0.12
$C(CH_3)_3$	1.98	OSO_2CH_3	−0.88
$CH{=}CH_2$	0.82	$OSO_2C_6H_5$	0.93
$C{\equiv}CH$	0.40	$OP(O)(OC_6H_5)_2$	2.46
$CH_2CH{=}CH_2$	1.10	OCH_2CH_2OH	−0.97
trans-$CH{=}CHCH_3$	1.22	$OCH_2CH_2OC_6H_5$	1.68
Cyclopropyl	1.14	$NHCH_3$	−0.47
C_6H_5	1.96	$N(CH_3)_2$	0.18
$CH_2C_6H_5$	2.01	$N(CH_3)_3^+$	−5.96
$CH{=}CHC_6H_5$	2.68	NHC_2H_5	0.08
$C{\equiv}CC_6H_5$	2.65	$N(C_2H_5)_2$	1.18
F	0.14	NHC_4H_9	1.45
Cl	0.71	NHC_6H_5	1.37
Br	0.86	$NHCH_2C_6H_5$	1.00
I	1.12	$N(CH_3)CH_2C_6H_5$	2.09
OH	−0.67	$N(C_6H_5)_2$	3.62
O^-	−3.87	NHCHO	−0.98
NH_2	−1.23	$NHCOCH_3$	−0.97
NO	−0.12	$NHCOC_2H_5$	−0.52
NO_2	−0.28	$NHCOCH(CH_3)_2$	−0.18
NH(OH)	−1.34	$NHCOC_6H_5$	0.49
$NHNH_2$	−0.88	$NHCOCH_2OH$	−1.42
NCS	1.15	$NHCOCH_2Cl$	−0.50
SCN	0.41	$NHCOCF_3$	0.08
SF5	1.23	$NHCONH_2$	−1.30
SO_2F	0.05	$NHCON(CH_3)_2$	−1.15
SO_3H	−2.68	$NHCON(CH_3)(OCH_3)$	−0.84
SO_3^-	−4.76	$NHCOOC_2H_5$	0.17
$SO_2(NH_2)$	−1.82	$NHCOCH_2OC_6H_5$	0.60
$B(OH)_2$	−0.55	$NHCSCH_3$	−0.42
$P(O)(OH)_2$	−1.59	$NHCSNH_2$	−1.40
$As(O)(OH)_2$	−2.13	$NHCS(NHC_2H_5)$	−0.71
OCH_3	−0.02	NHCN	−0.26
OCH_2CH_3	0.38	$NHSO_2CH_3$	−1.18
$OCH_2CH_2CH_3$	1.05	$NHSO_2C_6H_5$	0.45
OC_6H_5	2.08	$N(CHO)(CH_2C_6H_5)$	0.49
$OCH_2C_6H_5$	1.66	$N(CH_3)COCH_2OC_6H_5$	0.12
$OCHF_2$	0.31	$N(COCH_3)SO_2CH_3$	−1.84
OCF_3	1.04	$N(SO_2CH_3)_2$	−1.51
$OCOCH_3$	−0.64	CHO	−0.65
$OCOC_6H_5$	1.46	$COCH_3$	−0.55

TABLE 2.19 (*Continued*)

Substituent	π_X	Substituent	π_X
COC_6H_5	1.05	CH_2CH_2COOH	−0.29
COOH	−0.32	$CH_2CH_2CONH_2$	−1.22
COO^-	−4.36	$CH_2CH(NH_3^+)COO^-$	−3.56
$COOCH_3$	−0.01	$(CH_2)_3N(CH_3)_2$	0.60
$COOC_2H_5$	0.51	$(CH_2)_3N(CH_3)_3^+$	−4.15
$COOC_6H_5$	1.46	CH=CHCHO	−0.23
$COOCH_2C_6H_5$	1.84	$CH{=}CHCOCH_3$	−0.06
$CONH_2$	−1.49	$CH{=}CHCOC_6H_5$	0.95
$CONHCH_3$	−1.27	CH=CHCOOH	0.00
$CON(CH_3)_2$	−1.51	$CH{=}CHCOOC_2H_5$	0.86
$CONHC_6H_5$	0.49	CH=CHCN	−0.17
CONHOH	−1.87	*trans*-$CH{=}CHNO_2$	0.11
$CONHNH_2$	−1.90	$CH{=}C(CN)_2$	0.05
$CONHNHCONH_2$	−2.63	$CH{=}NC_6H_5$	−0.29
$COCF_3$	0.02	*trans*-CH=NOH	−0.38
$COCOC_6H_5$	1.25	$CH{=}NOCH_3$	0.40
CN	−0.57	$CH{=}NNHCOC_6H_5$	0.43
CH_2Cl	0.17	$CH{=}NNHCONH_2$	−0.86
CH_2Br	0.79	$CH{=}NNCSNH_2$	−0.27
CF_3	0.88	$CH{=}NNHCONHNH_2$	−1.32
CH_2OH	−1.03	$N{=}NC_6H_5$	1.69
CH_2OCH_3	−0.78	SCH_3	0.61
$C(OCH_3)_3$	0.14	SC_2H_5	1.07
$CH_2OC_6H_5$	1.66	SC_6H_5	2.32
$CH_2NH_3^+$	−4.09	SCF_3	1.44
$CH_2N(CH_3)_2$	−0.15	$SCOCH_3$	0.10
$CH_2N(CH_3)C_6H_5$	2.09	$SOCH_3$	−1.58
CH_2COOH	−0.72	SOC_6H_5	−0.07
CH_2COOCH_3	−0.30	SO_2CH_3	−1.63
CH_2OCOCH_3	−0.17	$SO_2C_6H_5$	0.27
CH_2CONH_2	−1.68	SO_2CF3	0.55
CH_2CN	−0.57	$SO_2OCH_2CH_3$	−0.90
CH_2NO_2	−0.38	$SO_2N(CH_3)_2$	−0.78
CH_2SCN	−0.14	$SO_2NHC_6H_5$	0.45
$CH_2NHC_6H_5$	1.00	$P(CH_3)_2$	0.44
$CH_2NHCOCH_2OH$	−1.58	$P(O)(C_6H_5)_2$	0.70
$CH_2N(CHO)(C_6H_5)$	0.49	$P(O)(OCH_3)_2$	−1.18
$CH_2Si(CH_3)_3$	2.00	$Si(CH_3)_3$	2.59
$CH_2CH_2C_6H_5$	2.66	$SeCH_3$	0.74
CH_2CH_2Cl	0.82	2-Pyridyl	0.50
CH_2CH_2Br	0.96	4-Pyridyl	0.46
CH_2CH_2OH	−0.77		

respective LFER correlations of toxic activity with hydrophobicity parameters [494]. For example, the inhibition of a frog heart by miscellaneous chemicals has been described by the following one-parameter equation [495]:

$$\log \frac{1}{I_{100}} = (0.14 \pm 0.12) + (0.91 \pm 0.07)\log P$$

$$R^2 = 0.951 \qquad s = 0.103 \qquad n = 34$$

where I_{100} is the concentration creating 100% inhibition. The minimum lethal dose of alcohols and ketones to *Salmonella typhosa* is also well correlated with log P as [496]:

$$\log \frac{1}{C} = -(0.43 \pm 0.06) + (0.94 \pm 0.04)\log P$$

$$R^2 = 0.996 \qquad s = 0.067 \qquad n = 12$$

Notably, the slopes of the last two relationships are close to one. Consequently, 1-octanol has been an excellent choice as the model solvent to describe the distribution of compounds between water and biomembranes.

The hydrophobicity of compounds has been also used in LFER correlations of enzymatic activity. For instance, the biomolecular rat constants k_{II} of enzymatic hydrolysis of esters by chymotrypsin have been presented by the following equation [497,498]:

$$\log k_{II} = (1.54 \pm 0.10) + (1.96 \pm 0.13)\log P$$

$$R^2 = 0.986 \qquad s = 0.043 \qquad n = 5$$

The biomolecular rate constants of enzyme-catalyzed hydrolysis of *N*-acetyl-L-aminoacid esters

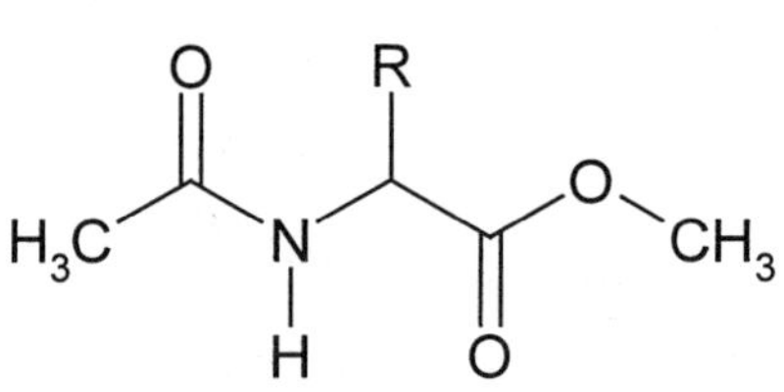

by thermitase and protease from *Thermoactinomyces vulgaris PA II-4a* have been described by the following equation [498]:

$$\log k_{II} = (3.54 \pm 0.06) + (1.16 \pm 0.08)\log P$$

$$R^2 = 0.953 \qquad s = 0.163 \qquad n = 12$$

The larger than one slopes of these relationships suggest the specific hydrophobic interaction between the substrate and the active center of the enzyme.

The bacteriostatic activity of substituted *trans*-3-benzoylacrylic acids

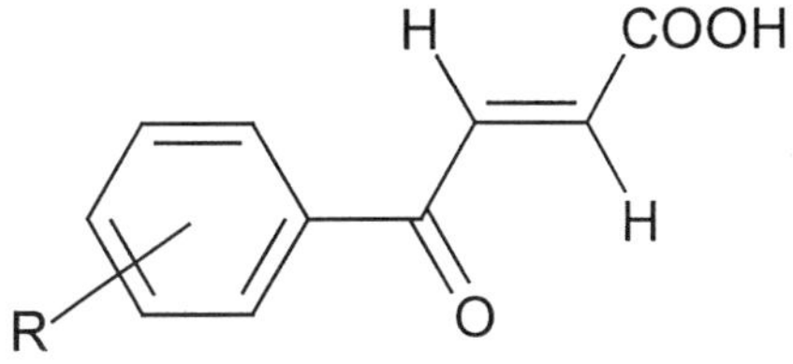

measured by the minimum inhibitory concentration, C, versus *M. tuberculosis Inh. H37 Rv* has been correlated with Hansch–Fujita hydrophobicity constants π for the substituents in phenyl ring as [499]:

$$\log\left(\frac{1}{C}\right) = -0.530 + 0.962\pi$$

$$R^2 = 0.903 \qquad s = 0.423 \qquad n = 19$$

The LFER correlation for the bacteriostatic activity *Staphylococcus aureus 209* with π involves also a quadratic term on that parameter [499]:

$$\log\left(\frac{1}{C}\right) = -0.294 + 1.424\pi - 0.166\pi^2$$

$$R^2 = 0.933 \qquad s = 0.294 \qquad n = 19$$

It has been pointed out that the presence of a quadratic term in a correlation equation permits to find an optimum (minimum or maximum) value of the descriptor. This feature may be useful in the QSAR molecular design of new biologically active compounds. A similar quadratic dependence on log P was obtained in the description of the activity of triazenoimidazolecarboxamide derivatives

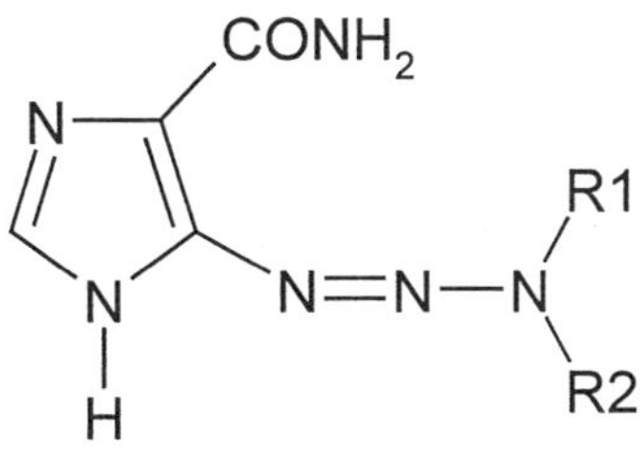

against L1210 leukemia as:

$$\log\left(\frac{1}{C}\right) = (3.454 \pm 0.18) + (0.586 \pm 0.24)\log P - (0.276 \pm 0.19)(\log P)^2$$

$$R^2 = 0.863 \qquad s = 0.146 \qquad n = 10$$

In both of these cases, the square term has an opposite sign to the term involving natural hydrophobicity scale. This observation points to the different nature of the interactions as described by the linear and quadratic terms, respectively.

Depending on the selection of compounds, the correlation with log P does not necessarily imply that the partitioning between two phases determines the biological activity. It has been pointed out that for congeneric series of compounds (e.g., for alcohols) the log P is perfectly collinear with the size of the variable radical [500]. However, the importance of hydrophobicity is certainly confirmed if similar correlations exist for sets of compounds with variable congeneric structure.

REFERENCES

1. S. Borman, *Chem. Eng. News* **68,** 20 (1990).
2. H.H. Meyer, *Arch. Exp. Pathol. Pharmakol.* **42,** 109 (1899).
3. H.H. Meyer, *Arch. Exp. Pathol. Pharmakol.* **46,** 338 (1901).
4. C.E. Overton, *Studien über die Narkose*. Fischer, Jena, 1901.
5. P. Müller, *Pure Appl. Chem.* **66,** 1077 (1994).
6. L.P. Hammett, *Chem. Rev.* **17,** 125 (1935).
7. L.P. Hammett, *Physical Organic Chemistry*. McGraw-Hill, New York, 1940.
8. Y. Tsuno, T. Ybata, and Y. Yukawa, *Bull. Chem. Soc. Jpn.* **32,** 960 (1959).
9. Y. Tsuno and Y. Yukawa, *Bull. Chem. Soc. Jpn.* **32,** 965 (1959).
10. Y. Tsuno and Y. Yukawa, *Bull. Chem. Soc. Jpn.* **32,** 971 (1959).
11. S. Ehrenson, R.T.C. Brownlee, and R.W. Taft, *Prog. Phys. Org. Chem.* **10,** 1 (1973).
12. R.W. Taft, Jr. and R.D. Topsom, *Prog. Phys. Org. Chem.* **16,** 1 (1987).
13. R.W. Taft, *J. Am. Chem. Soc.* **74,** 2729 (1952).
14. R.W. Taft, *J. Am. Chem. Soc.* **74,** 3120 (1952).
15. R.W. Taft, *J. Am. Chem. Soc.* **75,** 4321 (1953).
16. R.W. Taft, *J. Am. Chem. Soc.* **75,** 4538 (1953).
17. W.A. Pavelich and R.W. Taft, *J. Am. Chem. Soc.* **79,** 4935 (1957).
18. C.J.F. Böttcher and P. Bordewijk, *Theory of Electric Polarization,* 2nd ed., Vol. II. Elsevier, Amsterdam, 1978.
19. C. Reichardt, *Solvents and Solvent Effects in Chemistry,* 2nd ed. VCH, Weinheim, 1990.
20. M. Berthelot and L. Péan de Saint-Gilles, *Ann. Chim. Phys.* [3] **65,** 385 (1862).
21. M. Berthelot and L. Péan de Saint-Gilles, *Ann. Chim. Phys.* [3] **66,** 5 (1862).

22. L. Claisen, *Justus Liebigs Ann. Chem.* **291,** 25 (1896).
23. L. Knorr, *Justus Liebigs Ann. Chem.* **293,** 70 (1896).
24. W. Wislicenus, *Justus Liebigs Ann. Chem.* **291,** 147 (1896).
25. N. Menshutkin, *Z. Phys. Chem.* **5,** 589 (1890).
26. J.-L. Abboud, R. Notario, J. Bertrán, and M. Solá, *Prog. Phys. Org. Chem.* **19,** 1 (1993).
27. E. Grünwald and S. Winstein, *J. Am. Chem. Soc* **70,** 846 (1948).
28. V. Gutmann, *Coord. Chem. Rev.* **2,** 239 (1967).
29. V. Gutmann, *Coord. Chem. Rev.* **18,** 225 (1976).
30. I.A. Koppel and A.I. Paju, *Org. React.* **11,** 121 (1974).
31. I.A. Koppel and V.A. Palm, *Org. React.* **8,** 296 (1971).
32. I.A. Koppel and V.A. Palm, in *Linear Free Energy Relationships* (N.B. Chapman and J. Shorter, eds.), Chapter 5. Plenum, London, 1972.
33. M.J. Kamlet and R.W. Taft, *J. Chem. Soc, Perkin Trans. 2,* p. 337 (1979).
34. J.-L.M. Abboud, M.J. Kamlet, and R.W. Taft, *Prog. Phys. Org. Chem.* **13,** 485 (1981).
35. M.J. Kamlet, J.-L.M. Abboud, M.H. Abraham, and R.W. Taft, *J. Org. Chem.* **48,** 2877 (1983).
36. M.J. Kamlet and R.W. Taft, *Acta Chem. Scand., Ser. B* **39,** 611 (1985).
37. M.J. Kamlet and R.W. Taft, *Acta Chem. Scand., Ser. B* **40,** 619 (1986).
38. A.R. Katritzky and R.D. Topsom, *J. Chem. Educ.* **48,** 427 (1971).
39. L.M. Stock, *J. Chem. Educ.* **49,** 400 (1972).
40. R.D. Topsom, *Prog. Phys. Org. Chem.* **12,** 1 (1976).
41. W.F. Reynolds and G.K. Hamer, *J. Am. Chem. Soc.* **98,** 7296 (1976).
42. A.J. Hoefnagel and M.A. Hoefnagel, *J. Org. Chem.* **43,** 4720 (1978).
43. K. Bowden and E.J. Grubbs, *Prog. Phys. Org. Chem.* **19,** 183 (1993).
44. K. Bowden and E.J. Grubbs, *Chem. Soc. Rev.* **25,** 171 (1996).
45. J.G. Kirkwood and F.H. Westheimer, *J. Chem. Phys.* **6,** 506 (1938).
46. J.D. Roberts and W.T. Moreland, *J. Am. Chem. Soc.* **75,** 2167 (1953).
47. M.J.S. Dewar and P.J. Grisdall, *J. Am. Chem. Soc.* **84,** 2547 (1960).
48. F.H. Westheimer, *Tetrahedron* **51,** 3 (1995).
49. K.F. Reid, *Properties and Reactions of Bonds in Organic Molecules*. Longmans, Green, London, 1968.
50. M. Charton, *J. Am. Chem. Soc.* **97,** 3691 (1975).
51. M. Charton, *J. Am. Chem. Soc.* **99,** 5687 (1977).
52. M. Charton, *Prog. Phys. Org. Chem.* **13,** 119 (1981).
53. P.J. Taylor and A.R. Wait, *J. Chem. Soc. Perkin Trans. 2,* p. 1765 (1986).
54. R.W. Taft and I.C. Lewis, *J. Am. Chem. Soc.* **80,** 2436 (1958).
55. R.W. Taft, S. Ehrenson, I.C. Lewis, and R.E. Glick, *J. Am. Chem. Soc.* **81,** 5352 (1959).
56. Y. Yukawa, Y. Tsuno, and M. Sawada, *Bull. Soc. Chem. Jpn.* **45,** 1198 (1972).
57. M. Sjöström and S. Wold, *Chem. Scr.* **9,** 200 (1976).
58. J.D. Roberts and W.T. Moreland, *J. Am. Chem. Soc.* **75,** 2167 (1953).

59. R.W. Taft and C.A. Grob, *J. Am. Chem. Soc.* **96,** 1236 (1974).
60. C.A. Grob and M.G. Schlageter, *Helv. Chim. Acta* **59,** 264 (1976).
61. C.A. Grob, *Helv. Chim. Acta* **68,** 882 (1985).
62. C. Hansch, A. Leo, and R.W. Taft, *Chem. Rev.* **91,** 165 (1991).
63. V.A. Palm, ed., *Tables of Rate and Equilibrium Constants of Heterolytic Organic Reactions,* Vol. 5 (II). VINITI, Moscow, 1979.
63a. C. Hansch, A. Leo, and D. Hoekman, *Exploring QSAR. Hydrophobic, Electronic and Steriuc Constants*, ACS Professional Reference Book. American Chemical Society, Washington, DC, 1995.
64. J. Shorter, in *Advances in Linear Free Energy Relationships* (N.B. Chapman and J. Shorter, eds.), p. 71, Plenum, London, 1972.
65. C.-R. Kramer, *Z. Phys. Chem. (Leipzig)* **267,** 277 (1986).
66. C.-R. Kramer, *Z. Phys. Chem. (Leipzig)* **267,** 1077 (1986).
67. H.H. Jaffé, *Chem. Rev.* **53,** 191 (1953).
68. K. Kalfus, J. Kroupa, M. Vecera, and O. Exner, *Collect. Czech. Chem. Commun.* **40,** 3009 (1975).
69. O. Exner, *Correlation Analysis of Chemical Data.* Plenum, New York, 1988.
70. C. Hansch, A. Leo, S.H. Unger, K.H. Kim, D. Nikaitani, and E.J. Lein, *J. Med. Chem.* **16,** 1207 (1973).
71. V.A. Palm, *Foundations of the Quantitative Theory of Organic Reactions* (in Russian). Khimiya, Leningrad, 1967.
72. I.A. Koppel, M.M. Karelson, and V.A. Palm, *Org. React.* **10,** 497 (1973).
73. I.A. Koppel, M.M. Karelson, and V.A. Palm, *Org React.* **11,** 101 (1974).
74. M.M. Karelson, *Org. React.* **11,** 679 (1975).
75. M.M. Karelson, Ph.D. Thesis, Tartu State University, Tartu, 1975.
76. C.D. Ritchie and E.S. Lewis, *J. Am. Chem. Soc.* **84,** 591 (1962).
77. H.R. Timotheus and A.I. Talvik, *Org. React.* **3,** 158 (1966).
78. L.P. Hammett, *Physical Organic Chemistry.* McGraw-Hill, New York, 1970.
79. I.A. Koppel and M.M. Karelson, *Org. React.* **11,** 985 (1975).
80. R.D. Topsom, *Prog. Phys. Org. Chem.* **16,** 125 (1987).
81. S. Marriott, W.F. Reynolds, R.W. Taft, and R.D. Topsom, *J. Org. Chem.* **49,** 959 (1984).
82. S. Marriott and R.D. Topsom, *J. Am. Chem. Soc.* **106,** 7 (1984).
83. W.J. Hehre, C.-F. Pau, A.D. Headley, R.W. Taft, and R.D. Topsom, *J. Am. Chem. Soc.* **108,** 1711 (1986).
84. P. Čarsky, P. Nauš, and O. Exner, *J. Phys. Org. Chem.* **11,** 485 (1998).
85. A.R. Cherkasov, V.I. Galkin, and R.A. Cherkasov, *J. Phys. Org. Chem.* **11,** 437 (1998).
86. G. Briegleb and A. Bieber, *Z. Elektrochem.* **55,** 250 (1951).
87. A. Fisher, B.R. Mann, and J. Vaughan, *J. Chem. Soc.*, p. 1093 (1961).
88. O. Exner, *Collect. Czech. Chem. Commun.* **31,** 65 (1966).
89. O. Exner and J. Lakomy, *Collect. Czech. Chem. Commun.* **35,** 1371 (1970).
90. I.M. Kolthoff and M.K. Chantooni, Jr., *J. Am. Chem. Soc.* **93,** 3843 (1971).

91. O. Exner and K. Kalfus, *Collect. Czech. Chem. Commun.* **41,** 569 (1976).
92. F.G. Bordwell and G.D. Cooper, *J. Am. Chem. Soc.* **74,** 1058 (1952).
93. A.I. Biggs and R.A. Robinson, *J. Chem. Soc.*, p. 388 (1961).
94. G. Schwarzenbach and E. Rudin, *Helv. Chim. Acta* **22,** 360 (1939).
95. I. Leito, T. Rodima, I.A. Koppel, R. Schwesinger, and V.M. Vlasov, *J. Org. Chem.* **62,** 8479 (1997).
96. R.W. Taft, *Prog. Phys. Org. Chem.* **14,** 247 (1983).
97. J.H. Smith and F.M. Menger, *J. Org. Chem.* **34,** 77 (1969).
98. A.D. Campbell, S.Y. Chooi, L. Deady, and R.A. Shanks, *Aust. J. Chem.* **23,** 203 (1970).
99. C. Hansch, *Acc. Chem. Res.* **26,** 147 (1993).
100. J.M. Blaney, C. Hansch, C. Silipo, and A. Vittoria, *Chem. Rev.* **84,** 333 (1984).
101. C. Hansch, D. Hoekman, and H. Gao, *Chem. Rev.* **96,** 1045 (1996).
102. C. Hansch and A. Leo, *Exploring QSAR. Fundamentals and Applications in Chemistry and Biology,* ACS Prof. Ref. Book. American Chemical Society, Washington, DC, 1995.
103. C. Hansch, D. Hoekman, A. Leo, L. Zhang, and P. Li, *Toxicol. Lett.* **79,** 45 (1995).
104. C. Hansch, B.R. Telzer, and L. Zhang, *Crit. Rev. Toxicol.* **25,** 67 (1995).
105. B.D. Palmer, W.R. Wilson, S.M. Pullen, and W.A. Denny, *J. Med. Chem.* **33,** 112 (1990).
106. T.A. Gourdie, K.K. Valu, G.L. Gravatt, J. Boritzki, B.C. Baguley, L.P. Wakelin, S.M. Pullen, and W.A. Denny, *J. Med. Chem.* **33,** 1177 (1990).
107. C. Hansch, B.R. Telzer, and L. Zhang, *Crit. Rev. Toxicol.* **25,** 67 (1995).
108. S.P. Gupta, *Chem. Rev.* **94,** 1507 (1994).
109. W.A. Remers and R.T. Dorr, in *Alkaloids: Chemical and Biological Perspectives* (S.W. Pelletier, ed.), Vol. 6, pp. 1–74. Wiley-Interscience, New York, 1988.
110. S.M. Sami, B.S. Iyengar, S.E. Tarnow, W.A. Remers, and W.T. Bradner, *J. Med. Chem.* **27,** 701 (1984).
111. K.R. Kunz, B.S. Iyengar, R.T. Dorr, D.S. Alberts, and W.A. Remers, *J. Med. Chem.* **34,** 2281 (1991).
112. I. Moriguchi, K. Komatsu, and Y. Matsushita, *Anal. Chim. Acta* **133,** 625 (1981).
113. I. Moriguchi, S. Hirono, Q. Liu, Y. Matsushita, and T. Nakagawa, *Chem. Pharm. Bull.* **38,** 3373 (1990).
114. Q. Liu, S. Hirono, and I. Moriguchi, *Quan. Struct.–Act. Relat.* **11,** 318 (1992).
115. W.J. Dunn, III, M.J. Greenberg, and S.S. Callejas, *J. Med. Chem.* **19,** 1299 (1976).
116. G.L. Tong, M. Cory, W.W. Lee, D.W. Henry, and G. Zbinden, *J. Med. Chem.* **21,** 732 (1978).
117. S.I. Fink, A. Leo, M. Yamakawa, C. Hansch, and F.R. Quinn, *Farmaco, Ed. Sci.* **35,** 965 (1980).
118. E.J. Lien and G.L. Tong, *Cancer Chemother. Rep., Part 1* **57,** 251 (1973).
119. R.W. Taft, E. Price, I.R. Fox, I.C. Lewis, K.K. Andersen, and G.T. Davis, *J. Am. Chem. Soc.* **85,** 709 (1963).

120. W. Adcock, A.N. Abeywickrema, V.S. Iyer, and G.B. Kok, *Magn. Reson. Chem.* **24,** 213 (1986).
121. W. Adcock and A.N. Abeywickrema, *J. Org. Chem.* **47,** 2957 (1982).
122. W. Adcock and A.N. Abeywickrema, *Tetrahedron Lett.* **22,** 1135 (1981).
123. W. Adcock, A.N. Abeywickrema, and G.B. Kok, *J. Org. Chem.* **49,** 1387 (1984).
124. W. Adcock, A.N. Abeywickrema, and G.B. Kok, *Tetrahedron Lett.* **23,** 365 (1982).
125. W. Adcock, G.B. Kok, A.N. Abeywickrema, W. Kitching, G.M. Drew, H.A. Olszowy, and I. Schott, *J. Am. Chem. Soc.* **105,** 290 (1983).
126. W. Adcock and G.B. Kok, *J. Org. Chem.* **52,** 356 (1987).
127. R.T.C. Brownlee, A.R. Katritzky, and R.D. Topsom, *J. Am. Chem. Soc.* **87,** 3620 (1965).
128. R.T.C. Brownlee, A.R. Katritzky, and R.D. Topsom, *J. Am. Chem. Soc.* **88,** 1413 (1966).
129. A.R. Katritzky, *Chem. Ind. (London)*, p. 601 (1967).
130. A.R. Katritzky, R.D. Topsom, M.V. Sinnott, and T.T. Tidwell, *J. Am. Chem. Soc.* **91,** 628 (1969).
131. A.R. Katritzky, T.T. Tidwell, C.R. Palmer, F.J. Swinbourne, and R.D. Topsom, *J. Am. Chem. Soc.* **91,** 636 (1969).
132. A.R. Katritzky, J.M. Angelelli, R.F. Pinzelli, and R.D. Topsom, *Tetrahedron* **28,** 2037 (1972).
133. A.R. Katritzky, L. Deady, R.A. Shanks, and R.D. Topsom, *Spectrochim. Acta, Part A* **29A,** 115 (1973).
134. A.R. Katritzky, R.T.C. Brownlee, D.G. Cameron, R.D. Topsom, and A.F. Pozharskii, *J. Chem. Soc., Perkin Trans. 2,* p. 247 (1974).
135. A.R. Katritzky, N.C. Cutress, T.B. Grindley, and R.D. Topsom, *J. Chem. Soc., Perkin Trans. 2,* p. 263 (1974).
136. N.A. Puttnam, *J. Chem. Soc.* **5,** 100 (1960).
137. H.H. Freedman, *J. Am. Chem. Soc.* **82,** 2454 (1960).
138. H.C. Brown and Y. Okamoto, *J. Am. Chem. Soc.* **80,** 2454 (1960).
139. D.G. O'Sullivan and P.W. Sadler, *J. Chem. Soc.*, p. 4144 (1957).
140. M.R. Mander and H.W. Thompson, *Trans. Faraday Soc.*, **53,** 1402 (1957).
141. J.P. Jessen and H.W. Thompson, *Spectrochim. Acta* **13,** 217 (1958).
142. F. Gerson and E. Heilbronner, *Helv. Chim. Acta* **42,** 1877 (1959).
143. P. Noble, G. Borkhardt, and W.L. Reed, *Chem. Rev.* **64,** 51 (1963).
144. J.E. Dubois and A. Bienvenke, *C.R. Hebd. Seances Acad. Sci.* **256,** 5351 (1963).
145. W.D. Closson and P. Haug, *J. Am. Chem. Soc.* **86,** 2384 (1964).
146. R.W. Taft, *J. Am. Chem. Soc.* **75,** 4231 (1953).
147. C.N.R. Rao, W.H. Wahl, and E.J. Williams, *Can. J. Chem.* **35,** 1575 (1957).
148. A.R. Katritzky and P. Simmons, *J. Chem. Soc.*, p. 2051 (1959).
149. O. Exner, *Collect. Czech. Chem. Commun.* **25,** 642 (1960).
150. A.H. Sharpe and S. Walker, *J. Chem. Soc.*, p. 4522 (1961).
151. M. Gomel and H. Lumbroso, *Bull. Soc. Chim.*, p. 1196 (1962).

152. J.H. Richards and S. Walker, *Tetrahedron* **20,** 841 (1964).
153. I.V. Martynov and V.I. Anufriev, *Dokl. Akad. Nauk SSSR* **283,** 377 (1985).
154. A. Streitwieser, *Prog. Phys. Org. Chem.* **1,** 6 (1964).
155. G.F. Crable and G.L. Kearns, *J. Phys. Chem.* **66,** 436 (1962).
156. A.G. Harrison, P. Kebarle, and F.P. Lossing, *J. Am. Chem. Soc.* **83,** 777 (1961).
157. R.F. Pottie and F.P. Lossing, *J. Am. Chem. Soc.* **85,** 269 (1963).
158. J.M.S. Tait, T.M. Shannon, and A.G. Harrison, *J. Am. Chem. Soc.* **84,** 4 (1962).
159. G.W. Dillow and P. Kebarle, *J. Am. Chem. Soc.* **111,** 5592 (1989).
160. Y.P. Kitaev and G.K. Budnikov, *Usp. Khim.* **31,** 670 (1962).
161. L. Pauling and G.W. Wheland, *J. Chem. Phys.* **1,** 362 (1933).
162. G.W. Wheland, *Resonance in Organic Chemistry*. Wiley, New York, 1955.
163. C.G. Swain and E.C. Lupton, *J. Am. Chem. Soc.* **90,** 4328 (1968).
164. C.G. Swain, S.H. Unger, N.R. Rosenquist, and M.S. Swain, *J. Am. Chem. Soc.* **105,** 492 (1983).
165. C.G. Swain, *J. Org. Chem.* **49,** 2005 (1984).
166. W.F. Reynolds and R.D. Topsom, *J. Org. Chem.* **49,** 1989 (1984).
167. A.J. Hoefnagel, W. Oosterbeek, and B.M. Wepster, *J. Org. Chem.* **49,** 1993 (1984).
168. M. Charton, *J. Org. Chem.* **49,** 1997 (1984).
169. L.P. Hammett, *J. Am. Chem. Soc.* **59,** 96 (1937).
170. A. Fisher, G.J. Leary, R.D. Topsom, and J. Vaughan, *J. Chem. Soc. B,* p. 782 (1966).
171. Y. Okamoto and H.C. Brown, *J. Org. Chem,* **22,** 485 (1957).
172. L.M. Stock and H.C. Brown, *Adv. Phys. Org. Chem.* **1,** 35 (1963).
173. V. Kvasnička, Š. Sklenák, and J. Pospichal, *J. Am. Chem. Soc.* **115,** 1495 (1993).
174. H.H. Jaffé, *J. Chem. Phys.* **20,** 279 (1952).
175. H.H. Jaffé, *J. Chem. Phys.* **20,** 778 (1952).
176. H.H. Jaffé, *J. Chem. Phys.* **20,** 1554 (1952).
177. F.L. Sixma, *J. Recl.* **72,** 673 (1953).
178. R.D. Gilliom, J.-P. Beck, and W.P. Purcell, *J. Comput. Chem.* **6,** 437 (1985).
179. S. Böhm and J. Kuthan, *Int. J. Quantum Chem.* **16,** 21 (1984).
180. T. Sotomatsu, Y. Morita, and T. Fujita, *J. Comput. Chem.* **10,** 94 (1989).
181. J. Niwa, *Bull. Chem. Soc. Jpn.* **62,** 226 (1989).
182. K. Bowden and M.J. Price, *J. Chem. Soc. B,* p. 1784 (1971).
183. M. Fujio, M. Goto, M. Mishima, and Y. Tsuno, *Bull. Chem. Soc. Jpn.* **63,** 1121 (1990).
184. J. Toullec, M. El-Alaoui, and P. Kleffert, *J. Org. Chem.* **48,** 4808 (1983).
185. J. Toullec and M. El-Alaoui, *J. Org. Chem.* **50,** 4928 (1985).
186. M.R. Bridge, D.H. Davies, A. Maccoll, R.A. Ross, B. Stephenson, and O. Banjoko, *J. Chem. Soc. B,* p. 805 (1968).
187. W.A. Darlington, R.D. Partes, and K.W. Ratts, *Toxicol. Appl. Pharmacol.* **18,** 542 (1971).
188. C. Hansch, D. Hoekman, and H. Gao, *Chem. Rev.* **96,** 1045 (1996).

189. R.L. Metcalf and T.R. Fukuto, *J. Econ. Entomol.* **53,** 340 (1962).
190. C. Hansch and R. Kerley, *J. Med. Chem.* **13,** 957 (1970).
191. C.K. Hancock, E.A. Meyers, and B.J. Jager, *J. Am. Chem. Soc.* **83,** 4211 (1961).
192. J.J. Blanksma and H.H. Schreinemachers, *Recl. Trav. Chim. Pays-Bas* **52,** 428 (1933).
193. C. Hansch and E. Coats, *J. Pharm. Sci.* **59,** 731 (1970).
194. C. Hansch and A. Leo, *Correlation Analysis in Chemistry and Biology.* Wiley-Interscience, New York, 1979.
195. J.K. Seydel, K.-J. Shaper, F. Wempe, and H.P. Cordes, *J. Med. Chem.* **19,** 483 (1976).
196. E.A. Coats, K.J. Shah, S.R. Milstein, C.S. Genther, D.M. Nene, J. Roesner, J. Schmidt, M. Pleiss, E. Wagner, and J.K. Baker, *J. Med. Chem.* **25,** 57 (1982).
197. M. Karelson and G.H.F. Diercksen, in *Problem Solving in Computational Molecular Science,* pp. 215–248. Kluwer Acad. Publ., Dordrecht, The Netherlands, 1997.
198. J. Tomasi and M. Persico, *Chem. Rev.* **94,** 2027 (1994).
199. D.J. Giesen, C.J. Cramer, and D.G. Truhlar, *J. Phys. Chem.* **98,** 4141 (1994).
200. M. Karelson, *Adv. Quantum Chem.* **28,** 141 (1997).
201. C. Reichardt, *Angew. Chem.* **91,** 119 (1979).
202. E.M. Kosower, *J. Am. Chem. Soc.* **80,** 3253 (1958).
203. E.M. Kosower, J.A. Skorcz, W.M. Schwarz, and J.W. Patton, *J. Am. Chem. Soc.* **82,** 2188 (1960).
204. E.M. Kosower, D. Hofmann, and K. Wallenfels, *J. Am. Chem. Soc.* **84,** 2755 (1962).
205. E.M. Kosower, *An Introduction to Physical Organic Chemistry.* Wiley, New York, 1968.
206. K. Dimroth, C. Reichardt, T. Siepmann, and F. Bohlmann, *Justus Liebigs Ann. Chem.* **661,** 1 (1963).
207. C. Reichardt and K. Dimroth, *Chem. Forsch.* **11,** 1 (1968).
208. K. Dimroth and C. Reichardt, *Justus Liebigs Ann. Chem.* **93,** 727 (1969).
209. C. Reichardt, *Chem. Rev.* **94,** 2319 (1994).
210. P. Suppan and N. Ghoneim, *Solvatochromism.* The Royal Society of Chemistry Information Services, Cambridge, UK, 1997.
211. N.G. Bakhshiev, ed., *Solvatochromism: Problems and Methods* (in Russian), Izd. Leningr. Univ., Leningrad, 1989.
212. C. Reichardt and E. Harbusch-Görnert, *Liebigs Ann. Chem.*, p. 721 (1983).
213. V. Bekárek and J. Juřina, *Collect. Czech. Chem. Commun.* **47,** 1060 (1982).
214. C. Laurence, P. Nicolet, and C. Reichardt, *Bull. Soc. Chim. Fr.*, p. 125 (1987).
215. C. Laurence, P. Nicolet, M. Lucon, and C. Reichardt, *Bull. Soc. Chim. Fr.*, p. 1001 (1987).
216. C. Laurence, P. Nicolet, M. Lucon, T. Dalati, and C. Reichardt, *J. Chem. Soc., Perkin Trans. 2,* p. 873 (1989).
217. C. Reichardt, M. Eschner, and G. Schäfer, *Liebigs Ann. Chem.*, p. 57 (1990).

218. S. Spange, M. Lauterbach, K. Kyra, and C. Reichardt, *Liebigs Ann. Chem.*, p. 323 (1991).
219. S. Spange and D. Keutel, *Liebigs Ann. Chem.*, p. 423 (1992).
220. K. Dimroth and C. Reichardt, *Z. Anal. Chem.* **215,** 344 (1966).
221. T.G. Beaumont and K.M.C. Davis, *J. Chem. Soc. B,* p. 1010 (1968).
222. W. Koehler, P. Froelich, and R. Radeglia, *Z. Phys. Chem. (Leipzig)* **242,** 220 (1969).
223. K. Tamura, Y. Ogo, and T. Imoto, *Bull. Chem. Soc. Jpn.* **46,** 2988 (1973).
224. Z.B. Maksimović, C. Reichardt, and A. Spirić, *Z. Anal. Chem.* **270,** 100 (1974).
225. E.M. Kosower, H. Dodiuk, K. Tanizawa, M. Ottolenghi, and N. Orbach, *J. Am. Chem. Soc.* **97,** 2167 (1975).
226. S. Balakrishnan and A.J. Easteal, *Aust. J. Chem.* **34,** 933 (1981).
227. S. Balakrishnan and A.J. Easteal, *Aust. J. Chem.* **34,** 943 (1981).
228. M. de Vijlder, *Bull. Soc. Chim. Belg.* **91,** 947 (1982).
229. V. Bekárek and T. Nevěčná, *Collect. Czech. Chem. Commun.* **50,** 1928 (1985).
230. T.M. Krygowski, P.K. Wrona, U. Zielkowska, and C. Reichardt, *Tetrahedron* **41,** 4519 (1985).
231. H. Langhals, *Tetrahedron Lett.* **27,** 339 (1986).
232. J.G. Dawber, J. Ward, and R.A. Williams, *J. Chem. Soc., Faraday Trans.*, **84,** 713 (1988).
233. J.G. Dawber, *J. Chem. Soc., Faraday Trans.* **86,** 287 (1990).
234. R.I. Zalewski, I. Adamczewska, and C. Reichardt, *J. Chem. Res., Miniprint,* p. 2157 (1990).
235. R.A. Ransdell and C.C. Wamser, *J. Phys. Chem.* **96,** 10572 (1992).
236. R.D. Skwierczynski and K.A. Connors, *J. Chem. Soc., Perkin Trans. 2,* p. 467 (1994).
237. P.K. Kipkemboi and A. Easteal, *Aust. J. Chem.* **47,** 1771 (1994).
238. R.S. Drago, M.S. Hirsch, D.C. Ferris, and C.W. Chronister, *J. Chem. Soc., Perkin Trans. 2,* p. 219 (1994).
239. W. Schroth H.-D. Schädler, and J. Andersch, *Z. Chem.* **29,** 56 (1989).
240. W. Schroth, H.-D. Schädler, and J. Andersch, *Z. Chem.* **29,** 129 (1989).
241. S.K. Poole, P.H. Shetty, and C.F. Poole, *Anal. Chim. Acta* **218,** 241 (1989).
242. W.B. Harrod and N.J. Pienta, *J. Phys. Org. Chem.* **3,** 534 (1990).
243. K.A. Zachariasse, N. Van Phuc, and B. Kozankiewicz, *J. Phys. Chem.* **85,** 2676 (1981).
244. P. Plieniger and H. Baumgärtel, *Ber. Bunsenges. Phys. Chem.* **86,** 161 (1982).
245. C.J. Drummond, F. Grieser, and T.W. Healy, *Chem. Phys. Lett.* **140,** 493 (1987).
246. G.G. Warr and D.F. Evans, *Langmuir* **4,** 217 (1987).
247. M.A. Kessler and O.S. Wolfbeis, *Chem. Phys. Lett.* **50,** 51 (1989).
248. R. Varadaraj, J. Bock, N. Brons, and S. Pace, *J. Phys. Chem.* **97,** 12991 (1993).
249. R.A. Hobson, F. Grieser, and T.W. Healy, *J. Phys. Chem.* **98,** 274 (1994).
250. L.G.S. Brooker, A.C. Craig, D.W. Heseltine, P.W. Jenkins, and L.L. Lincoln, *J. Am. Chem. Soc.* **87,** 2443 (1965).

251. D. Walther, *J. Prakt. Chem.* **316,** 604 (1974).
252. D.M. Manuta and A.J. Lees, *Inorg. Chem.* **26,** 3212 (1986).
253. W. Kaim, B. Olbrich-Deussner, and T. Roth, *Organometallics* **10,** 410 (1991).
254. W. Walter and O. Bauer, *Liebigs Ann. Chem.*, p. 407 (1977).
255. A. Janowski, I. Turowska-Tyrk, and P.K. Wrona, *J. Chem. Soc., Perkin Trans. 2,* p. 821 (1985).
256. J.-E. Dubois and A. Bienvenüe, *Tetrahedron Lett.*, p. 1809 (1966).
257. M.J. Kamlet, J.-L. Abboud, and R.W. Taft, *J. Am. Chem. Soc.* **99,** 6027 (1977).
258. M.J. Kamlet, J.-L. Abboud, and R.W. Taft, *J. Am. Chem. Soc.* **99,** 8325 (1977).
259. C. Laurence, P. Nicolet, M.T. Dalati, J.-L.M. Abboud, and R. Notario, *J. Phys. Chem.* **98,** 5807 (1994).
260. E. Buncel and S. Rajagopal, *J. Org. Chem.* **54,** 798 (1989).
261. S. Dähne, F. Shob, K.-D. Nolte, and R. Radeglia, *Ukr. Khim. Zh.* **41,** 1170 (1975).
262. I.A. Zhmyreva, V.V. Zelinskii, V.P. Kolobkov, and N.D. Kranitskaya, *Dokl. Akad. Nauk SSSR, Ser. Khim.* **129,** 1089 (1959).
263. F. Armand, H. Sakuragi, and K. Tokumaru, *J. Chem. Soc., Faraday Trans.* **89,** 1021 (1993).
264. D.J. Raber, R.C. Bingham, J.M. Harris, J.L. Fry, and P. von R. Schleyer, *J. Am. Chem. Soc.* **92,** 5977 (1970).
265. T.W. Bentley and P. von R. Schleyer, *Adv. Phys. Org. Chem.* **14,** 1 (1977).
266. M.H. Abraham, R.W. Taft, and M.J. Kamlet, *J. Org. Chem.* **46,** 3053 (1981).
267. M.H. Abraham, R.M. Doherty, M.J. Kamlet, J.M. Harris, and R.W. Taft, *J. Chem. Soc., Perkin Trans. 2,* p. 913 (1987).
268. S.G. Smith, A.H. Fainberg, and S. Winstein, *J. Am. Chem. Soc.* **83,** 618 (1961).
269. Y. Drougard and D. Decroocq, *Bull. Soc. Chim. Fr.*, p. 2972 (1969).
270. M. Gielen and J. Nasielski, *J. Organomet. Chem.* **1,** 173 (1963).
271. M. Gielen and J. Nasielski, *J. Organomet. Chem.* **7,** 273 (1967).
272. J.A. Berson, Z. Hamlet, and W.A. Mueller, *J. Am. Chem. Soc.* **84,** 297 (1962).
273. F.W. Fowler, A.R. Katritzky, and R.J.D. Rutherford, *J. Chem. Soc. B,* p. 460 (1971).
274. I.A. Koppel and V.A. Palm, in *Advances in Linear Free Energy Relationships* (N.B. Chapman and J. Shorter, eds.), Chapter 5. Plenum, London, 1972.
275. N.S. Bayliss and E.G. McRae, *J. Phys. Chem.* **58,** 1002 (1954).
276. E.G. McRae, *J. Phys. Chem.* **61,** 562 (1957).
277. N. Mataga, Y. Kaifu, and M. Koizumi, *Bull. Chem. Soc. Jpn.* **29,** 456 (1956).
278. T. Abe, J.-L.M. Abboud, F. Belio, E. Bosch, J.I. Garcia, J.A. Mayoral, R. Notario, J. Ortega, and M. Rosés, *J. Phys. Org. Chem.* **11,** 193 (1998).
279. A.R. Katritzky, M. Karelson, and A.P. Wells, *J. Org. Chem.* **61,** 1619 (1996).
280. U. Maran, T.A. Pakkanen, and M. Karelson, *J. Chem. Soc., Perkin Trans. 2,* p. 2445 (1994).
281. U. Maran, M. Karelson, and A.R. Katritzky, *Int. J. Quantum Chem.* **23S,** 41 (1996).
282. J. Leis, G.P. Schiemenz, and M. Karelson, *ACH-Models Chem.* **135,** 157 (1998).

283. J. Leis, U. Maran, G.P. Schiemenz, and M. Karelson, *ACH Models Chem.* **135,** 173 (1998).
284. M. Karelson and J. Leis, *Proc. Estonian Acad. Sci. Chem.* **48,** 3 (1999).
285. J. Hylton McCreery, R.E. Christoffersen, and G.G. Hall, *J. Am. Chem. Soc.* **98,** 7191 (1976).
286. J.J. Urban, C.J. Cramer, and G.R. Famini, *J. Am. Chem. Soc.* **114,** 8226 (1992).
287. M. Vásquez, G. Nemethy, and H.A. Scheraga, *Chem. Rev.* **94,** 2183 (1994).
288. M.R. Pincus, *Biopolymers* **32,** 347 (1992).
289. J. Kostrowicki and H.A. Scheraga, *J. Phys. Chem.* **96,** 7442 (1992).
290. C.J. Cramer and D.G. Truhlar, in *Quantitative Treatments of Solute/Solvent Interactions* (P. Politzer and J.S. Murray, eds.). pp. 9–54. Elsevier, Amsterdam, 1994.
291. K.B. Wiberg and M.W. Wong, *J. Phys. Chem.* **96,** 668 (1992).
292. F.J. Luque and M. Orozco, *J. Org. Chem.* **58,** 6397 (1993).
293. N.A. Burton, S.C.-L. Chiu, M.M. Davidson, D.V.S. Green, I.H. Hillier, J.J.W. McDouall, and M.A. Vincent, *J. Chem. Soc., Faraday Trans.* **89,** 2631 (1993).
294. K.B. Wiberg, P.R. Rablen, D.J. Rush, and T.A. Keith, *J. Am. Chem. Soc.* **117,** 4261 (1995).
295. M.M. Karelson, T. Tamm, A.R. Katritzky, M. Szafran, and M.C. Zerner, *Int. J. Quantum Chem.* **37,** 1 (1990).
296. A.R. Katritzky, M.C. Zerner, and M.M. Karelson, *J. Am. Chem. Soc.* **108,** 7213 (1986).
297. M. Karelson, in *Problem Solving in Computational Molecular Science* (S. Wilson and G.H.F. Dierckson, eds.), pp. 353–387. Kluwer Acad. Publ., Dordrecht, The Netherlands, 1997.
298. A.T. Balaban, *Rev. Roum. Chim.* **16,** 725 (1971).
299. A.T. Balaban and R. Istratoiu, *Tetrahedron Lett.*, p. 1879 (1973).
300. R.W. Baldock, P. Hudson, A.R. Katritzky, and F. Soti, *J. Chem. Soc., Perkin Trans. 2,* p. 1422 (1974).
301. F.G. Bordwell and T.Y. Lynch, *J.* Am. *Chem. Soc.* **111,** 7558 (1989).
302. H.G. Viehe, R. Merenyi, L. Stella, and Z. Janousek, *Angew. Chem., Int. Ed. Engl.* **18,** 917 (1979).
303. H.G. Viehe, Z. Janousek, R. Merenyi, and L. Stella, *Acc. Chem. Res.* **18,** 148 (1985).
304. M.J. Kamlet, M.E. Jones, J.-L.M. Abboud, and R.W. Taft, *J. Chem. Soc., Perkin Trans. 2,* p. 342 (1979).
305. M.J. Kamlet and R.W. Taft, *J. Chem. Soc., Perkin Trans. 2,* p. 349 (1979).
306. R.W. Taft and J.S. Murray, in *Quantitative Treatments of Solute/Solvent Interactions* (P. Politzer and J.S. Murray, eds.), pp. 55–82. Elsevier, Amsterdam, 1994.
307. L.Y. Wilson and G.R. Famini, *J. Med. Chem.* **34,** 1668 (1991).
308. G.R. Famini, R.J. Kasel, J.W. King, and L.Y. Wilson, *Quant. Struct.-Act. Relat.* **10,** 344 (1991).
309. G.R. Famini, W.P. Ashman, A.P. Mickiewicz, and L.Y. Wilson, *Quant. Struct.-Act. Relat.* **11,** 162 (1992).

310. G.R. Famini, C.E. Penski, and L.Y. Wilson, *J. Phys. Org. Chem.* **5,** 395 (1992).

311. G.R. Famini, B.C. Marquez, and L.Y. Wilson, *J. Chem. Soc., Perkin Trans. 2,* p. 773 (1993).

312. G.R. Famini and L.Y. Wilson, in *Quantitative Treatments of Solute/Solvent Interactions* (P. Politzer and J.S. Murray, eds.), pp. 213–241. Elsevier, Amsterdam, 1994.

313. M.H. Abraham, *Chem. Soc. Rev.* **22,** 73 (1993).

314. M.H. Abraham, in *Quantitative Treatments of Solute/Solvent Interactions* (P. Politzer and J.S. Murray, eds.), pp. 83–134. Elsevier, Amsterdam, 1994.

315. M.H. Abraham, P.L. Grellier, and R.A. McGill, *J. Chem. Soc., Perkin Trans. 2,* p. 797 (1987).

316. M.H. Abraham and J.C. McGowan, *Chromatographia* **23,** 213 (1987).

317. J.-C. Dutoit, *J. Chromatogr.* **555,** 191 (1991).

317a. M.H. Abraham, G.S. Whiting, R.M. Doherty, and M.J. Shuely, *J. Chromatogr.* **587,** 213 (1991); M.H. Abraham, *ibid.* **644,** 95 (1993).

318. M.H. Abraham, in *Quantitative Treatments of Solute/Solvent Interactions* (P. Politzer and J.S. Murray, eds.), pp. 83–134. Elsevier, Amsterdam, 1994.

319. D. Svozil, J.G. Ševčík, and V. Kvasnička, *J. Chem. Inf. Comput. Sci.* **37,** 338 (1997).

320. R.S. Drago, *J. Chem. Soc., Perkin Trans. 2,* p. 1827 (1992).

321. R.S. Drago, M.S. Hirsch, D.C. Ferris, and C.W. Chronister, *J. Chem. Soc., Perkin Trans. 2,* p. 219 (1994).

322. S. Joerg, R.S. Drago, and J. Adams, *J. Chem. Soc., Perkin Trans. 2,* p. 2431 (1997).

323. V. Gutmann, *Chimia* **23,** 285 (1969).

324. V. Gutmann, *Electrochim. Acta* **21,** 661 (1976).

325. G. Olofson and I. Olofson, *J. Am. Chem. Soc.* **95,** 7231 (1973).

326. U. Mayer and V. Gutmann, *Struct. Bonding (Berlin)* **12,** 113 (1972).

327. V. Gutmann, *Coordination Chemistry in Non-Aqueous Solutions.* Springer, Wien, 1968.

328. V. Gutmann, *Pure Appl. Chem.* **27,** 72 (1971).

329. R. Schmid, *Rev. Inorg. Chem.* **1,** 117 (1979).

330. R.S. Drago, *Coord. Chem. Rev.* **33,** 251 (1980).

331. R.W. Taft, N.J. Pienta, M.J. Kamlet, and E.M. Arnett, *J. Org. Chem.* **46,** 661 (1981).

332. Y. Marcus, *J. Solut. Chem.* **13,** 599 (1984).

333. P.-C. Maria and J.-F. Gal, *J. Phys. Chem.* **89,** 1296 (1985).

334. P.-C. Maria, J.-F. Gal, J. de Franceschi, and E. Fargin, *J. Am. Chem. Soc.* **109,** 483 (1987).

335. M. Berthelot, J.-F. Gal, M. Helbert, C. Laurence, and P.-C. Maria, *J. Chim. Phys.* **82,** 427 (1987).

335a. C. Laurence, M. Berthelot, J.-Y. Le Questel, and M.J. El Ghomari, *J. Chem. Soc., Perkin Trans. 2,* p. 2075 (1995).

336. M. Berthelot, C. Laurence, M. Safar, and F. Besseau, *J. Chem. Soc., Perkin Trans. 2,* p. 283 (1998).
337. W.R. Fawcett, in *Quantitative Treatments of Solute/Solvent Interactions* (P. Politzer and J.S. Murray, eds.), pp. 183–212. Elsevier, Amsterdam, 1994.
338. I. Persson, M. Sandström, and P.L. Goggin, *Inorg. Chim. Acta* **129,** 183 (1987).
339. M. Sandström, I. Persson, and P. Persson, *Acta Chem. Scand.* **44,** 653 (1990).
340. I. Persson, *Pure Appl. Chem.* **58,** 1153 (1986).
341. U. Mayer, V. Gutmann, and W. Gerger, *Monatsh. Chem.* **106,** 1235 (1975).
342. U. Mayer, V. Gutmann, and W. Gerger, *Monatsh. Chem.* **108,** 489 (1977).
343. U. Mayer, *Pure Appl. Chem.* **51,** 169 (1979).
344. V. Gutmann and G. Resch, in *Ions and Molecules in Solution* (N. Tanaki, H. Ohtaki, and R. Tamamushi, eds.). Elsevier, Amsterdam, 1982, p. 203.
345. B.R. Knauer and J.J. Napier, *J. Am. Chem. Soc.* **98,** 4395 (1976); O.W. Kolling, *Anal. Chem.* **49,** 591 (1977).
346. O.W. Kolling, *Anal. Chem.* **55,** 143 (1983).
347. M.C.R. Symons and A.S. Pena-Nu˜nez, *J. Chem. Soc., Faraday Trans. 1* **81,** 2421 (1985).
348. Z. Kecki, B. Lyczkowski, and W. Kolodziejski, *J. Solut. Chem.* **15,** 413 (1986).
349. A.H. Reddoch and S. Konishi, *J. Chem. Phys.* **70,** 2121 (1979).
350. W.R. Fawcett, *J. Phys. Chem.* **97,** 9540 (1993).
351. R.S. Drago and B.B. Wayland, *J. Am. Chem. Soc.* **87,** 3571 (1965).
352. R.S. Drago, *Struct. Bonding (Berlin)* **15,** 73 (1973).
353. J. George and R.S. Drago, *Inorg. Chem.* **35,** 239 (1996).
354. C.G. Swain, M.S. Swain, A.L. Powell, and J. Alumni, *J. Am. Chem. Soc.* **105,** 502 (1983).
355. R.W. Taft, J.M. Abboud, and M.J. Kamlet, *J. Org. Chem.* **49,** 2001 (1984).
356. C.G. Swain, *J. Org. Chem.* **49,** 2005 (1984).
357. T.M. Krygowski and W.R. Fawcett, *J. Am. Chem. Soc.* **97,** 2143 (1975).
358. T.M. Krygowski and W.R. Fawcett, *Aust. J. Chem.* **28,** 2115 (1975).
359. T.M. Krygowski and W.R. Fawcett, *Can. J. Chem.* **54,** 3283 (1975).
360. W.G. Jackson, G.A. Lawrence, P.A. Lay, and A.M. Sargeson, *Aust. J. Chem.* **35,** 1561 (1982).
361. U. Mayer, *Monatsh. Chem.* **109,** 421 (1978).
362. U. Mayer, *Monatsh. Chem.* **109,** 775 (1978).
363. M.J. Kamlet and R.W. Taft, *J. Am. Chem. Soc.* **98,** 2886 (1976).
364. K. Dimroth, C. Reichardt, T. Sepmann, and F. Bohlmann, *Justus Liebigs Ann. Chem.* **661,** 1 (1963).
365. S. Brownstein, *Can. J. Chem.* **80,** 3253 (1960).
366. J. Burgess, *Spectrochim. Acta, Part A* **26A,** 1957 (1970).
367. M.H. Abraham, *J. Chem. Soc., Perkin Trans. 2,* p. 1343 (1972).
368. I.A. Koppel and V.A. Palm, *Org. React.* **6,** 504 (1969).
369. M.J. Kamlet and R.W. Taft, *J. Am. Chem. Soc.* **98,** 377 (1976)
370. M.J. Kamlet and R.W. Taft, *J. Am. Chem. Soc* **98,** 3233 (1976).

371. M.H. Abraham, P.L. Grellier, D.V. Prior, R.W. Taft, J.J. Morris, P.J. Taylor, C. Lauren, M. Berthelot, R.M. Doherty, M.J. Kamlet, J.-L.M. Abboud, K. Sraidi, and J. Guiheneuf, *J. Am. Chem. Soc.* **10,** 8534 (1988).
372. P.C. Maria, J.-F. Gal, J. de Franceschia, and E. Fargin, *J. Am. Chem. Soc* **109,** 483 (1987).
373. M.H. Abraham, P.L. Grellier, D.V. Prior, P.P. Duce, J.J. Morris, and P.J. Taylor, *J. Chem. Soc., Perkin Trans. 2,* p. 699 (1989).
374. M.H. Abraham, P.L. Grellier, D.V. Prior, J.J. Morris, and P.J. Taylor, *J. Chem. Soc., Perkin Trans. 2,* p. 521 (1990).
375. C. Laurence, M. Berthelot, M. Helbert, K. Sraidi, *J. Phys. Chem.* **93,** 3799 (1989).
376. J.-Y. Le Questel, C. Laurence, A. Lachkar, M. Helbert, and M. Berthelot, *J. Chem. Soc., Perkin Trans. 2,* p. 2091 (1992).
377. E.D. Raczyinska, C. Laurence, and M. Berthelot, *Can. J. Chem.* **70,** 2203 (1992).
378. C. Laurence, M. Helbert, and A. Lachkar, *Can. J. Chem.* **71,** 254 (1993).
379. M. Berthelot, M. Helbert, C. Laurence, and J.-Y. Le Questel, *J. Phys. Org. Chem.* **6,** 302 (1993).
380. C. Laurence, M. Berthelot, M. Lucon, and D.G. Morris, *J. Chem. Soc., Perkin Trans. 2,* p. 491 (1994).
381. Y. Marcus, *J. Phys. Chem.* **91,** 4422 (1987).
382. R.C. Dougherty, *Tetrahedron Lett.*, p. 385 (1975).
383. V. Bekárek, *J. Chem. Soc., Perkin Trans. 2,* p. 1293 (1983).
384. M.J. Kamlet and R.W. Taft, *Acta Chem. Scand., Ser. B* **B39,** 611 (1985).
385. J.-L.M. Abboud, R. Notario, and V. Botella, in *Quantitative Treatments of Solute/Solvent Interactions* (P. Politzer and J.S. Murray, eds.), pp. 135–182. Elsevier, Amsterdam, 1994.
386. M.H. Abraham, R.M. Doherty, M.J. Kamlet, J.M. Harris, and R.W. Taft, *J. Chem. Soc., Perkin Trans. 2,* p. 913 (1987).
387. M.J. Kamlet and R.W. Taft, *J. Chem. Soc., Perkin Trans. 2,* p. 349 (1979).
388. A. Arcoria, V. Librando, E. Maccarone, G. Musumarra, and G.A. Tomaselli, *Tetrahedron* **33,** 105 (1977).
389. A.G. Burden, N.B. Chapman, H.F. Duggue, and J. Shorter, *J. Chem. Soc., Perkin Trans. 2,* p. 296 (1978).
390. M.H. Aslam, G. Collier, and J. Shorter, *J. Chem. Soc., Perkin Trans. 2,* p. 1572 (1981).
391. D.N. Kevill, in *Advances in Quantitative Structure-Property Relationships* (M. Charton, ed.), Vol. 1, pp. 81–115. JAI Press, Greenwich, CT.
392. M.J. D'Souza, D.N. Kevill, T.W. Bentley, and A.C. Devaney, *J. Org. Chem.* **60,** 1632 (1995).
393. D.N. Kevill, A.J. Oldfield, and M.J. D'Souza, *J. Chem. Res., Synop.*, p. 122 (1996).
394. D.N. Kevill, A.J. Casamassa, and M.J. D'Souza, *J. Chem. Res., Synop.*, p. 472 (1996).
395. D.N. Kevill, M.W. Bond, and M.J. D'Souza, *J. Phys. Org. Chem.* **11,** 273 (1998).

396. R.W. Taft, M.H. Abraham, R.M. Doherty, and M.J. Kamlet, *J. Am. Chem. Soc.* **107,** 3105 (1985).
397. M.H. Abraham, H.S. Chadha, J.P. Dixon, C. Rafols, and C. Treiner, *J. Chem. Soc., Perkin Trans. 2,* p. 887 (1995).
398. M.H. Abraham and J.C. McGowan, *Chromatographia* **23,** 243 (1987).
399. M.H. Abraham, H.S. Chadha, J.P. Dixon, C. Rafols, and C. Treiner, *J. Chem. Soc., Perkin Trans. 2,* p. 19 (1997).
400. M.H. Abraham, J. Andonian-Haftvan, G.S. Whiting, A. Leo, and R.W. Taft, *J. Chem. Soc., Perkin Trans. 2,* p. 1777 (1994).
401. M.H. Abraham, G.S. Whiting, P.W. Carr, and H. Ouyang, *J. Chem. Soc., Perkin Trans. 2,* p. 1385 (1998).
402. M.J. Kamlet, M.E. Jones, R.W. Taft, and J.-L. Abboud, *J. Chem. Soc., Perkin Trans. 2,* p. 342 (1979).
403. W.R. Fawcett and A.W. Kloss, *J. Phys. Chem.* **100,** 2019 (1996).
404. M.J. Kamlet and R.W. Taft, *J. Chem. Soc., Perkin Trans. 2,* p. 337 (1979).
405. M.J. Kamlet, A. Solomonovici, and R.W. Taft, *J. Am. Chem. Soc.* **101,** 3734 (1979).
406. R.W. Taft, T. Gramstad, and M.J. Kamlet, *J. Org. Chem.* **47,** 4557 (1982).
407. S.Y. Khorshev, V.L. Tsvetkova, and A.N. Egorochkin, *J. Organomet. Chem.* **264,** 169 (1984).
408. T. Yokoyama, R.W. Taft, and M.J. Kamlet, *J. Chem. Soc., Perkin Trans. 2,* p. 875 (1987).
409. O.W. Kolling, *Anal. Chem.* **51,** 1324 (1979).
410. O.W. Kolling, *Anal. Chem.* **51,** 1327 (1979).
411. C.W. Fong and H.G. Grant, *Org. Magn. Reson.* **14,** 147 (1980).
412. R.W. Taft and M.J. Kamlet, *Org. Magn. Reson.* **14,** 485 (1980).
413. O. Kolling, *Anal. Chem.* **56,** 430 (1984).
414. B.R. Knauer and J.J. Napier, *J. Am. Chem. Soc.* **98,** 4395 (1976).
415. R.W. Taft, J.-L. Abboud, and M.J. Kamlet, *J. Am. Chem. Soc.* **103,** 1080 (1981).
416. V. Bekárek, *J. Chem. Soc., Perkin Trans. 2,* p. 1293 (1983).
417. M.H. Abraham and C. Rafols, *J. Chem. Soc., Perkin Trans. 2,* p. 1843 (1995).
418. M.H. Abraham, R. Kumarsingh, J.E. Cometto-Muñiz, and W.S. Cain, *Arch. Toxicol.* **72,** 227 (1998).
419. M.H. Abraham, R. Kumarsingh, J.E. Cometto-Muñiz, W.S. Cain, M. Rosés, E. Bosch and M.L. Diaz, *J. Chem. Soc., Perkin Trans. 2,* p. 2405 (1998).
420. H.S. Frank and M.W. Evans, *J. Phys. Chem.* **13,** 507 (1945).
421. C. Tanford, *The Hydrophobic Effect; Formation of Micelles and Biological Membranes.* Wiley-Interscience, New York 1973.
422. H. Hildebrand, *Proc. Natl. Acad. Sci. U.S.A.* **76,** 194 (1979).
423. A. Ben-Naim, *Hydrophobic Interactions.* Plenum, New York, 1980.
424. A. Hvidt, *Acta Chem. Scand., Ser. A* **A37,** 99 (1983).
425. L.R. Pratt, *Annu. Rev. Phys. Chem.* **36,** 433 (1985).
426. W. Glokzij and J.B.F.N. Engberts, *Angew. Chem.* **105,** 1610 (1993).

427. M.H. Abraham, in *Quantitative Treatments of Solute/Solvent Interactions* (P. Politzer and J.S. Murray, eds.), pp. 83–134. Elsevier, Amsterdam, 1994.
428. K.H. Meyer and H. Gottlieb-Billroth, *Hoppe-Seyler's Z. Physiol. Chem.* **112,** 55 (1920).
429. C. Hansch and A. Leo, *Substituent Constants for Correlation Analysis in Chemistry and Biology*. Wiley-Interscience, New York, 1979.
429a. G. Nys and R. Rekker, *Chim. Ther.* **8,** 521 (1973); R. Rekker, *The Hydrophobic Fragmental Constant*. Elsevier, New York, 1977; R. Rekker and H. de Kort, *J. Med. Chem.* **14,** 47 (1979); J. Chou and P. Jurs, *J. Chem. Inf. Comput. Sci.* **19,** 172 (1979).
430. A. Leo, in *Comprehensive Medicinal Chemistry* (C. Hansch, ed.), Vol. 4, p. 315. Pergamon, Oxford, 1990.
431. A. Leo, *Methods Enzymol.* **202,** 544 (1991).
432. A. Leo, *Chem. Rev.* **93,** 1281 (1993).
433. A. Leo, in *Classical and Three-Dimensional QSAR in Agrochemistry* (C. Hansch and T. Fujita, eds.), p. 62. American Chemical Society, Washington, DC, 1995.
434. C. Hansch and A. Leo, *Exploring QSAR,* Chapter 5. American Chemical Society, Washington, DC, 1995.
435. P. Broto, G. Moreau, and C. Vandycke, *Eur. J Med. Chem.* **19,** 71 (1984).
436. A.K. Ghose and G.M. Crippen, *J. Med. Chem.* **28,** 333 (1985).
437. A.K. Ghose and G.M. Crippen, *J. Comput. Chem.* **7,** 565 (1986).
438. A.K. Ghose and G.M. Crippen, *J. Chem. Inf. Comput. Sci.* **1,** 21 (1987).
439. A.K. Ghose, A. Pritchett, and G.M. Crippen, *J. Comput. Chem.* **9,** 80 (1988).
440. V. Viswanadhan, V. Ghose, G. Revankar, and R. Robins, *J. Chem. Inf. Comput. Sci.* **29,** 163 (1989).
441. T. Suzuki and Y. Kudo, *J. Comput.-Aided Mol. Des.* **4,** 155 (1990).
442. G. Klopman and S. Wang, *J. Comput. Chem.* **12,** 1025 (1991).
443. I. Moriguchi, S. Hirono, Q. Liu, I. Nakagome, and Y. Matsushita, *Chem. Pharm. Bull.* **40,** 127 (1992).
444. T. Convard, J.P. Dubost, H. Le Solleu, and E. Kummer, *Quant. Struct.-Act. Relat.* **13,** 34 (1994).
445. K. Iwase, K. Komatsu, S. Hirono, S. Nakagawa, and I. Moriguchi, *Chem. Pharm. Bull.* **33,** 2114 (1985).
446. W. Dunn, III, S. Grigoras, and M. Koehler, *J. Med. Chem.* **30,** 1121 (1987).
447. M. Koehler, S. Grigoras, and W. Dunn, III, *Quant. Struct.-Act. Relat.* **7,** 150 (1988).
448. N.G.J. Richards, P.B. Williams, and M. Tute, *Int. J. Quantum Chem., Quantum Biol. Symp.* **18,** 299 (1991).
449. N.G.J. Richards, P.B. Williams, and M. Tute, *Int. J. Quantum Chem.* **44,** 219 (1992).
450. P. Pixner, W. Heiden, H. Mers, G. Moeckel, A. Möller, and J. Brickmann, *J. Chem. Inf. Comput. Sci.* **34,** 1309 (1994).
451. M.L. Gargas, P.G. Seybold, and M.E. Andersen, *Toxicol. Lett.* **43,** 235 (1988).
452. G.J. Niemi, S.C. Basak, G.D. Veith, and G. Grunwald, *Environ. Toxicol. Chem.* **11,** 893 (1992).

453. Y. Yao, L. Xu, Y. Yang, and X. Yuan, *J. Chem. Inf. Comput. Sci.* **33,** 590 (1993).
454. P. Gaillard, P.A. Carrupt, B. Testa, and A. Boudon, *J. Comput.-Aided Mol. Des.* **8,** 83 (1994).
455. C. Waller, *Quant. Struct.-Act. Relat.* **13,** 172 (1994).
456. S.C. Basak, B.D. Gute, and G.D. Grunwald, *J. Chem. Inf. Comput. Sci.* **36,** 1054 (1996).
457. N. Bodor, Z. Gabanyi, and C.-K. Wong, *J. Am. Chem. Soc.* **111,** 3783 (1989).
458. N. Bodor and M.-J. Huang, *J. Pharm. Sci.* **81,** 272 (1992).
459. Y. Sasaki, H. Kubodera, T. Matuszaki, and H. Umeyama, *J. Pharmacobio-Dyn.* **14,** 207 (1991).
460. K.H. Kim and D.H. Kim, *Bioorg. Med. Chem.* **3,** 1389 (1995).
461. A. Breindl, B. Beck, T. Clark, and R.C. Glen, *J. Mol. Model* **3,** 142 (1997).
462. G. Klopman and L.D. Iroff, *J. Comput. Chem.* **2,** 157 (1981).
463. G. Klopman, K. Namboodiri, and M. Schochet, *J. Comput. Chem.* **6,** 28 (1985).
464. K. Kasai, H. Umeyama, and A. Tomonaga, *Bull. Chem. Soc. Jpn.* **61,** 2701 (1988).
465. T. Katagi, M. Miyakado, C. Takayama, and S. Tanaka, in *Classical and Three-Dimensional QSAR in Agrochemistry* (C. Hansch and T. Fujita, eds.), pp. 48–61. American Chemical Society, Washington, DC, 1995.
466. M. Charton, in *Classical and Three-Dimensional QSAR in Agrochemistry* (C. Hansch and T. Fujita, eds.), p. 75. American Chemical Society, Washington, DC, 1995.
467. M. Charton, *Top. Curr. Chem.* **114,** 107 (1983).
468. M. Charton, *Prog. Phys. Org. Chem.* **18,** 163 (1990).
469. M. Kamlet, R.M. Doherty, M.H. Abraham, Y. Marcus, and R.W. Taft, *J. Phys. Chem.* **93,** 5244 (1988).
470. M.H. Abraham, G.S. Whiting, Y. Alarie, J.J. Morris, J.J. Taylor, R.M. Doherty, R.W. Taft, and G.D. Nielson, *Quant. Struct.-Act. Relat.* **9,** 6 (1990).
471. M.H. Abraham, *Chem. Soc. Rev.* **22,** 73 (1993).
472. A. Pagliara, G. Caron, G. Lisa, W. Fan, P. Gaillard, P.-A. Carrupt, B. Testa, and M.H. Abraham, *J. Chem. Soc., Perkin Trans. 2,* p. 2639 (1997).
473. K.H. Kim, *Med. Chem. Res,* **1,** 259 (1991).
474. K.H. Kim, *Quant. Struct.-Act. Relat.* **11,** 309 (1992).
475. K.H. Kim, G. Greco, E. Novellino, C. Silipo, and A. Vittoria, *J. Comput.-Aided Mol. Des.* **7,** 263 (1993).
476. K.H. Kim, *J. Comput.-Aided Mol. Des.* **7,** 71 (1993).
477. K.H. Kim, *Quant. Struct.-Act. Relat.* **12,** 232 (1993).
478. A.K. Debnath, C. Hansch, K.H. Kim, and Y.C. Martin, *J. Med. Chem.* **36,** 1007 (1993).
479. K.H. Kim, *J. Comput.-Aided Mol. Des.* **9,** 308 (1995).
480. A. Hopfinger and R. Battershell, *J. Med. Chem.* **19,** 569 (1976).
481. M. Koehler, S. Grigoras, and W. Dunn, III, *Quant. Struct.-Act. Relat.* **7,** 150 (1988).

482. G.E. Kellogg, S.F. Semus, and D.J. Abraham, *J. Comput.-Aided Mol. Des.* **5,** 545 (1991).
483. F.M. Menger and U.V. Venkataram, *J. Am. Chem. Soc.* **108,** 2980 (1986).
484. M.H. Abraham, P.L. Grellier, and R.A. McGill, *J. Chem. Soc., Perkin Trans. 2,* p. 339 (1988).
485. T. Fujita, J. Iwasa, and C. Hansch, *J. Am. Chem. Soc.* **86,** 5175 (1964).
486. T. Fujita, *Prog. Phys. Org. Chem.* **14,** 75 (1985).
487. J.M. Blaney, C. Hansch, C. Silipo, and A. Vittoria, *Chem. Rev.* **84,** 333 (1983).
488. S.P. Gupta, *Chem. Rev.* **87,** 1183 (1987).
489. S.P. Gupta, *Chem. Rev.* **89,** 1765 (1989).
490. C. Hansch and T. Fujita, eds., *Classical and Three-Dimensional QSAR in Agrochemistry.* American Chemical Society, Washington, DC, 1995.
491. K.J. Rossifer, *Chem. Rev.* **96,** 3201 (1996).
492. W.J. Dunn, III and C. Hansch, *Chem.-Biol. Interact.* **9,** 75 (1974).
493. C. Hansch, J.P. Björkroth, and A. Leo, *J. Pharm. Sci.* **76,** 663 (1987).
494. C. Hansch and A. Leo, *Exploring QSAR,* Chapter 6. American Chemical Society, Washington, DC, 1995.
495. C. Hansch and W.J. Dunn, III, *J. Pharm. Sci.* **61,** 1 (1972).
496. C. Hansch, D. Kim, A.J. Leo, E. Novellino, C. Silipo, and A. Vittoria, *CRC Crit. Rev. Toxicol.* **19,** 185 (1989).
497. J.B. Jones, T. Kunitake, C. Niemann, and G.E. Hein, *J. Am. Chem. Soc.* **87,** 1777 (1965).
498. M.M. Peips, P.F. Sikk, and A.A. Aaviksaar, *Org. React.* **23,** 261 (1986).
499. K. Bowden and M.P. Henry, *Adv. Chem. Ser.* **114,** 130 (1973).
500. C. Hansch and A. Leo, *Exploring QSAR,* p. 180. American Chemical Society, Washington, DC, 1995.

3

THEORETICAL MOLECULAR DESCRIPTORS

3.1 CLASSIFICATION OF THEORETICAL MOLECULAR DESCRIPTORS

The molecular descriptors described in Chapter 2 are all obtained from some experimentally measured data, either thermodynamic, kinetic, spectroscopic, or other chemical or physical data. Each experimental method has, however, its limits as related to the molecular structure, physical state, and chemical stability of the compound. Consequently, depending on the characteristics of a compound, the determination of empirical molecular descriptors may be difficult, expensive in terms of cost and time, or simply impossible for a given compound. As an alternative, the theoretical descriptors derived solely from the chemical structure and the quantum mechanically calculated wave function of a molecule can, in principle, be obtained for any individual molecular structure. Furthermore, the empirical molecular descriptors often reflect complex and multiple physical interactions. Hence, the interpretation of the respective quantitative structure–activity/property relationships (QSAR/QSPR) can be difficult and ambiguous. The theoretical descriptors have, in most cases, explicit mathematical definition based on fundamental physical equations of molecular matter or consistent physical models. Therefore, the QSAR/QSPR equations based on theoretical descriptors embody distinct physical meaning related to clearly defined physical molecular models and model processes. These features of the theoretical molecular descriptors make them particularly attractive and advantageous in the case of molecular design of novel compounds and materials, even those not yet synthesized in the laboratory.

Theoretical molecular descriptors can be conventionally divided into separate groups based on their physical origin or methods of calculation. In our presentation, the following classes of theoretical descriptors are distinguished:

- *Constitutional*
- *Geometrical*
- *Topological*
- *Electrostatic or charge distribution related*
- *Quantum chemical or molecular orbital (MO) related*
- *Solvational*
- *Thermodynamic*
- *Combined*

This classification is somewhat subjective; however, it is useful in finding appropriate descriptors for a given molecular system or phenomenon. The conceptual elimination of irrelevant theoretical descriptors from the possibly very large assortment available for different molecular systems or phenomena may substantially simplify the subsequent derivation of the QSAR/QSPR equations.

3.2 CONSTITUTIONAL AND GEOMETRICAL DESCRIPTORS

The simple constitutional descriptors reflect only the chemical composition of the compound, without any reference to the geometry or electronic structure of the molecule. Typical representatives of constitutional descriptors include:

Total number of atoms in the molecule

Absolute and relative numbers of atoms of certain chemical identity (C, H, O, S, N, F, etc.) in the molecule

Absolute and relative numbers of certain chemical groups and functionalities in the molecule

Total number of bonds in the molecule

Absolute and relative numbers of single, double, triple, aromatic, or other bonds in the molecule

Total number of rings, number of rings divided by the number of atoms

Total and relative number of phenyl or other aromatic rings

Molecular weight and average atomic weight

The constitutional descriptors are especially attractive because of their extreme simplicity, both conceptually and from the point of view of computation. Historically, one of the first approaches of this type was suggested by Free and Wilson [1]. According to their method, the multiple regression treatment of a property P is carried out using the following equation:

$$P = P_0 + \sum_k c_k I_k \tag{3.1}$$

where I_k is the number of certain structural features k in the molecule, c_k is the least-squares regression coefficient, and P_0 is the standard value of the property corresponding to the compound without accountable structural features. It has been pointed out that the applicability of Eq. (3.1) depends on the validity of the assumption about the additivity of substituent effects in the series of molecules [2].

The indicator or "dummy" variables quantifying the presence or absence of certain chemical groups or structural fragments in the molecule have been widely used in QSAR/QSPR work, often in combination with some other molecular descriptors [3–10]. For instance, the following QSAR equation was developed to correlate the anesthetic pressure p of a wide variety of anesthetics [11]:

$$\log\left(\frac{1}{p}\right) = -(2.11 \pm 0.39) + (1.17 \pm 0.25)\log P + (1.88 \pm 0.33)I$$

$$R^2 = 0.897 \qquad s = 0.438 \qquad n = 30$$

The anesthetic pressure p is the partial pressure of the compound necessary to inhibit an animal (e.g., mouse) from standing. In the last equation, indicator variable I accounts for the presence of polar hydrogen atoms in the molecule, with $I = 1$ if such an atom is present and $I = 0$ otherwise. The polar hydrogen was defined as attached to a carbon linked to an electronegative element or group. The large contribution by this term indicates the importance of locally polar groups in determining the anesthetic activity.

The antitumor activity of colchicine analogs

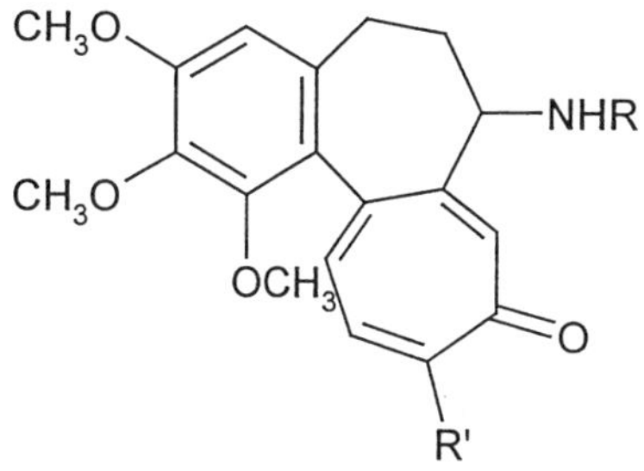

against P388 leukemia in mice has been correlated with log P and indicator variable I according to the following quadratic equation [12]:

$$\log\left(\frac{1}{C}\right) = (4.11 \pm 0.42) + (0.70 \pm 0.22)\log P - (0.30 \pm 0.08)(\log P)^2 + (2.16 \pm 0.44)I$$

$$R^2 = 0.869 \qquad s = 0.412 \qquad n = 24$$

Indicator variable I has the value 1 if R = —OCH_3 and $I = 0$ otherwise. The toxicity of the same series of compounds has been correlated with a very similar equation [13]:

$$\log\left(\frac{1}{C}\right) = (3.54 \pm 0.41) + (0.58 \pm 0.22)\log P - (0.24 \pm 0.08)(\log P)^2 + (1.71 \pm 0.43)I$$

$$R^2 = 0.817 \qquad s = 0.401 \qquad n = 24$$

It was concluded from the similarity of the two last equations that the colchicine derivatives are not very selective as potential antitumor agents because the antitumor activity increases in parallel with the toxicity of compounds [14]. In both cases, the methoxy-substituted compounds consist of the group with enhanced activity.

The antitumor activity of a very large set of anilinoacridines

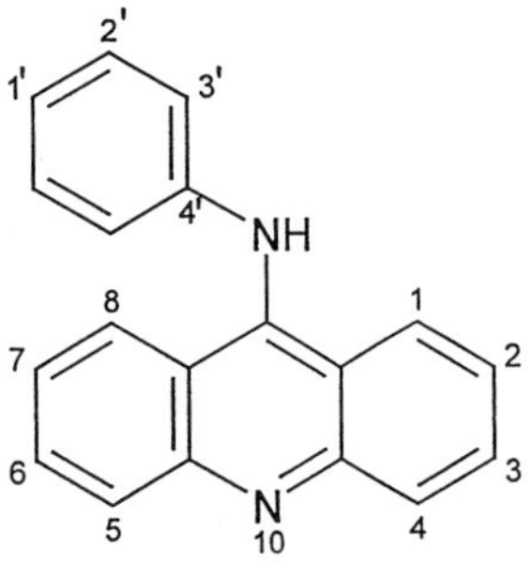

against L1210 leukemia in mice has been correlated by the following 13-parameter equation [15]:

$$\log\left(\frac{1}{C}\right) = (3.73 \pm 0.07) - (0.14 \pm 0.03)\sum \pi - (0.01 \pm 0.006) \cdot\left(\sum \pi\right)^2 - (1.08 \pm 0.09)\sum \sigma - (1.25 \pm 0.37)R_{BS} - (0.32 \pm 0.16)\mathrm{MR}_2 + (1.04 \pm (0.13)\mathrm{MR}_3 - (0.25 \pm 0.05)(\mathrm{MR}_3)^2 - (1.68 \pm 0.21)E_{s,3'} - (1.60 \pm 0.22)(E_{s,3'})^2 + (0.78 \pm 0.13)I_{NO_2} + (0.70 \pm 0.32)I_{DAT} + (0.50 \pm (0.18)I_{BS}$$

$$R^2 = 0.771 \qquad s = 0.323 \qquad n = 509$$

where $\Sigma\,\pi$ and $\Sigma\,\sigma$ denote the sum of hydrophobicity π constants and Hammett σ constants for all substituents, respectively, MR_k is the molecular refraction and $E_{s,k}$ is the steric constant for the substituent in kth position. Indicator variables I_{NO_2}, I_{DAT}, and I_{BS} have the value 1 each for 3-NO_2 group, 3,3-trialkyltriazene, and 1′-$NHSO_2C_6H_5$ group, respectively. The parameter R_{BS} denotes the Swain–Lupton resonance constant for a substituent in the 4-position of the last group. The last equation illustrates the complexity of biological activity that is revealed particularly distinctly in the case of large and structurally variable sets of active compounds.

However, note that although the constitutional descriptors typify simple QSAR/QSPR approaches, their interpretation in terms of individual physical or chemical interactions may be frequently cumbersome. For instance, any physical interaction that can be quantified with the number atoms or bonds of certain type in the molecule may be responsible for the rise of the respective QSPR/QSAR linear equation. Unfortunately, Occam's razor principle exludes the reverse interpretation. Thus, the existence of a linear relationship between the experimental property and the number of specific atoms or bonds of the molecular system studied is not sufficient to prove that only a single physical mechanism is bound to this property. Consequently, additional (independent) information about the possible leading molecular interaction is required to discriminate between different interaction mechanisms correlated linearly with the number of atoms or bonds in the system.

The geometrical descriptors represent more advanced structural molecular descriptors, derived from the three-dimensional coordinates of the atomic nuclei and the atomic masses and/or atomic radii in the molecule. The descriptors depending only on the atomic coordinates and masses are the principal moments of inertia of the molecule. In the rigid rotator approximation, the principal moments of inertia of a single molecule, I_A, I_B, and I_C, are calculated as:

$$I_A = \sum_i m_i r_{ix}^2$$
$$I_B = \sum_i m_i r_{iy}^2 \qquad (3.2)$$
$$I_C = \sum_i m_i r_{iz}^2$$

where m_i are the atomic masses and r_{ix}, r_{iy}, and r_{iz} denote the distance of the ith atomic nucleus from the main rotational axes, x, y, and z, of the molecule [16]. The moments of inertia characterize the mass distribution in the molecule and the susceptibility of the molecule to different rotational transitions. The formulas (3.2) can also be applied to rotationally flexible fragments or substituents in the molecule.

By orientation of the molecule in the space along the axes of inertia (X coordinate defined along the main axis of inertia, Y coordinate defined along the next longest, and Z coordinate along the shortest axis of inertia), the areas of the shadows S_1, S_2, and S_3 of the molecule can be calculated and employed as molecular descriptors. Additionally, the projections of shadows on the respective XY, YZ, and XZ planes can be found [17]. These shadow indices can be further normalized as the ratios $S_1/(X_{max}Y_{max})$, $S_2/(Y_{max}Z_{max})$, and $S_3/(X_{max}Z_{max})$, where X_{max}, Y_{max}, and Z_{max} are the maximum dimensions of the molecule along the corresponding axes. The shadow areas are usually calculated by applying a two-dimensional square grid on the molecular projection and by the subsequent summation of the areas of squares overlapped with the projection. The natural and normalized shadow indices reflect the size (natural shadow indices) and geometrical shape (normalized shadow indices) of the molecule.

The molecular surface area is one of the most popular geometrical descriptors of a compound. It has been closely related through various physical and quantum mechanical models to the intermolecular dispersion energy and the free energy of cavity formation in condensed media [18–23]. Different algorithms have been invented to calculate the molecular surface area and the related solvent-accessible surface area. These algorithms can be divided into two groups, employing numerical integration and analytical formulas, respectively.

For instance, according to one possible numerical algorithm, the molecule can be divided into a number of slices along the main principal axis of inertia of the molecule (X axis) with step ΔX (Fig. 3.1). For each slice, a set of circles corresponding to intersections of the van der Waals spheres and cutting plane is generated. Finally, the lengths of nonoccluded arcs are calculated for each circle using step dl, and the surface area is calculated as the sum of the lengths of nonoccluded arcs multiplied by the slice thickness ΔX:

$$S_M = \sum_i l_i \, \Delta X \tag{3.3}$$

Another important molecular surface-related descriptor is the solvent-accessible surface area of the molecule. Connolly has developed analytical equations for the calculation of this area proceeding from the model of a spherical solvent molecule rolling on the molecular van der Waals surface (Fig. 3.2) [24]. The total solvent-accessible surface area is calculated from the pieces corresponding to the convex areas (A_+), saddle areas (A_s), and concave areas (A_-), respectively.

$$S_{\mathrm{SA}} = A_+ + A_s + A_- \tag{3.4}$$

The convex area is simply the area defined by the center of the rolling solvent

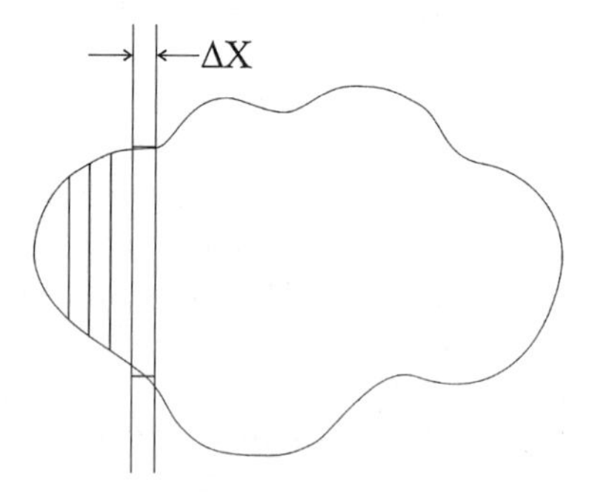

Figure 3.1 Numerical integration of the surface area of a molecule.

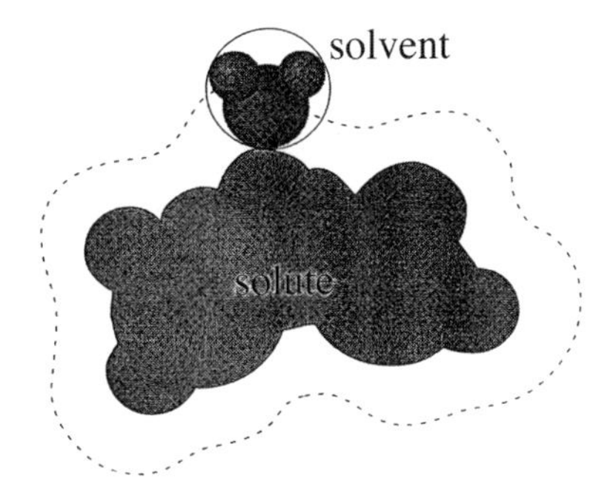

Figure 3.2 Connolly's model of rolling solvent spheres on the solute surface.

sphere when in contact with the van der Waals surface of one atom in the molecule. The saddle area and concave area correspond to the areas defined by the center of the rolling solvent sphere when in contact with the van der Waals surface of two atoms and three or more atoms in the molecule, respectively. Numerous other algorithms and methods are available to calculate the molecular surface area and the closely related solvent-accessible surface area [25–51].

The molecular volume and the solvent-excluded volume of the molecule are also widely used as geometric molecular descriptors. Various algorithms have been proposed to calculate these useful molecular characteristics [52–71]. The numerical calculation can be based on a Cartesian grid of cubes in the parallelepiped box with the dimensions X_{max}, Y_{max}, and Z_{max} containing the molecule. The total volume of a molecule is defined then as a sum of the individual volumes of the cubes overlapped with atomic spheres A (Fig. 3.3) as:

$$V_M = \sum_A \sum_{i \in A} \Delta V_i \tag{3.5}$$

The factorized molecular volume can be calculated as the ratio of the molecular volume and the volume of the box defined by X_{max}, Y_{max}, and Z_{max}, the maximum dimensions of the molecule along the principal axes of inertia.

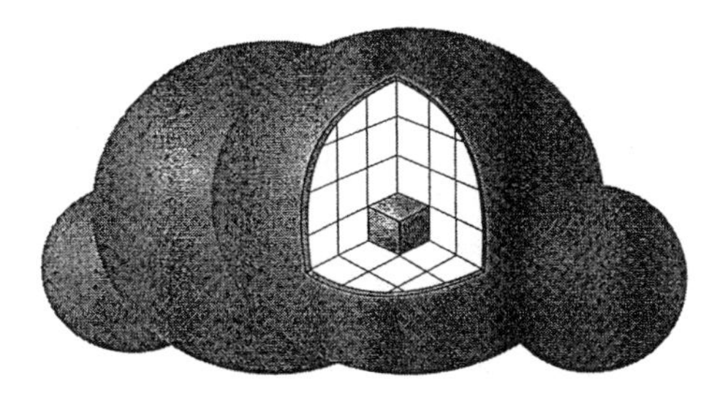

Figure 3.3 Calculation of total volume of a molecule using numerical integration over the volumes of cubes overlapped with atomic spheres.

An analytical formulation to calculate the solvent-excluded volume of a molecule proceeds from the division of this volume into two parts [72]. The first corresponds to the intrinsic van der Waals volume of the molecule. The second involves the interstitial volume consisting of packing defects between atoms that are too small to admit a solvent molecule sphere of a given radius. The total solvent-excluded molecular volume is calculated as a sum of following terms:

$$V_{\mathrm{mol(SA)}} = V_p + \sum V_+ + \sum V_s + \sum V_- + V_{\mathrm{ac}} + V_{\mathrm{nc}} \qquad (3.6)$$

corresponding to the volume of internal polyhedron (V_p), the pieces of volume between the center of an atom and the convex face of the solvent-accessible surface ($\Sigma\ V_+$), the volume corresponding to the saddle ($\Sigma\ V_s$), and concave pieces ($\Sigma\ V_-$), respectively, The last two terms account for the possible cusp pieces appearing in the case of separated atomic spheres with the limited access of solvent molecule between them. One of these terms has been presented by an analytical formula (V_{ac}), whereas the other, corresponding to cusps involving three or more concave pieces, has to be computed numerically [72].

The solvent-accessible surface area and molecular volume have been used extensively to identify binding regions on protein surfaces in structure-based drug design [73–76]. In this approach, the possible binding pockets on a protein surface are identified according to the complementarity of the protein and ligand surfaces, respectively [77–80]. The molecular and solvent-accessible surface areas have been also applied for the prediction of the hydrophobicity of compounds [81–83]. An empirical relationship has been established between the nuclear magnetic resonance (NMR) measured rotational correlation time and solvent-accessible surface area of compounds [84].

The van der Waals volume of molecules or substituents, V_w, has been used in the QSAR description of various biological activities of compounds. Thus, the in vitro inhibition of human lymphoblastic leukemia cells by adriamycin analogs

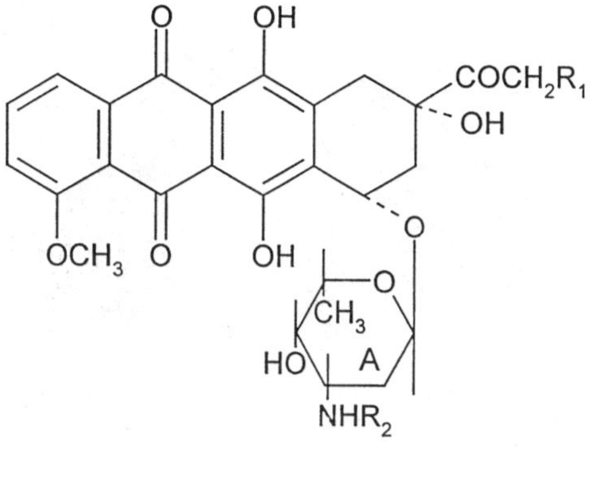

O OH COCH$_2$R$_1$ OH OCH$_3$ O OH O O CH$_3$ HO A NHR$_2$

given by inhibitory concentration IC_{50} has been expressed by the following regression equation [85]:

$$\log\left(\frac{1}{IC_{50}}\right) = 7.621 - (1.632 \pm 0.369)V_w(NHR_2) - (1.115 \pm 0.251)I_1$$
$$- (0.632 \pm 0.293)I_2$$
$$R^2 = 0.865 \qquad s = 0.218 \qquad n = 22 \qquad F = 38.32$$

The last equation includes two indicator variables. The first (I_1) was given a value 0 if R_1 = —H, —OH, or —OR and a value of 1 if R_1 = —SR or —SeR. The second indicator variable (I_2) had the value of 0 if the ring A is a natural six-membered glycoside ring and the value of 1 otherwise. The anticancer activity of hydroxyguanidine derivatives against cultured L1210 cells and the antiviral activity against the *Rous sarcoma* virus given by inhibitory dose ID_{50} have been correlated with the van der Waals volume of compounds alone as

$$\log\left(\frac{1}{ID_{50}}\right) = 3.92 + 0.62V_W$$
$$R^2 = 0.83 \qquad s = 0.16 \qquad n = 9$$

and

$$\log\left(\frac{1}{ID_{50}}\right) = 3.77 + 0.72V_W$$
$$R^2 = 0.67 \qquad s = 0.27 \qquad n = 9$$

respectively. Because of an approximate linear relationship between the dispersion energy and the size of structurally similar compounds, the last equations suggest that the dispersion interaction may be predominant in drug–receptor interactions of this kind. Similar correlations have been developed for the description of enzymatic reactions. The inhibitory constants K_i of the reaction of some adenosine monophosphate (AMP) and inosine monophosphate (IMP) derivatives with inosine monophosphate dehydrogenase were found well correlated with the van der Waals volume of compounds according to the following quadratic equation [14]:

$$\log\left(\frac{1}{K_i}\right) = 3.924 + (2.123 \pm 0.767)V_W - (3.720 \pm 0.993)V_w^2$$
$$- (0.136 \pm 0.074)I$$
$$R^2 = 0.823 \qquad s = 0.132 \qquad n = 13$$

The indicator variable I in the last equation distinguishes between two series of compounds ($I = 0$ for AMP analogs and $I = 1$ for IMP analogs).

An interesting group of molecular geometrical descriptors has been defined as gravitational indexes [86]. These indices are complex geometrical descriptors reflecting the molecular shape and the mass distribution in the molecule. Formally, they are calculated according to Newton's gravitational law, either for all pairs of atoms (G_p) or for all bonded pairs of atoms (G_b) as:

$$G_p = \sum_{i<j}^{Na} \frac{m_i m_j}{r_{ij}^2} \tag{3.7}$$

$$G_b = \sum_{i<j}^{Nb} \frac{m_i m_j}{r_{ij}^2} \tag{3.8}$$

where m_i and m_j are the atomic masses of atoms i and j, respectively, r_{ij} is the interatomic distance, N_a is the number of atoms, and N_b is the number of bonds in the molecule. The gravitational energy descriptors can be defined by substituting the square of the interatomic distances in Eqs. (3.7) and (3.8) with the distances themselves:

$$E_g = \sum_{i<j}^{Na \text{ or } Nb} \frac{m_i m_j}{r_{ij}} \tag{3.9}$$

Of course, the emergence of gravitational indices in the QSAR/QSPR equations does not originate from the interatomic gravitational interaction. The latter is at least 36 orders of magnitude weaker than the electrostatic interaction between the atomic nuclei in the molecule and thus negligible in determining any molecular property. However, apparently a formally similar mathematical equation may be used to quantify effectively the mass- and surface-distribution-dependent molecular properties.

For instance, a quantitative structure–property relationship treatment of the normal boiling points for a structurally widely variable set of 298 organic compounds resulted in the following highly significant two-parameter correlation [87]:

$$T_b = -(170.7 \pm 7.5) + (65.88 \pm 0.86)\sqrt[3]{G_b} + (18470 \pm 540)\text{HDCA(2)}$$

$$R^2 = 0.9544 \qquad s = 16.51 \qquad n = 298 \qquad F = 3126.5$$

where HDCA(2) denotes a hydrogen bonding donor characteristic charged area of the molecule. Interestingly, the best correlation was obtained not with the natural gravitational index for all bonded atoms in the molecule but with the cube root of this descriptor. Bearing in mind the possible nonlinear dependence of experimental properties of compounds on the molecular descriptors, this is not unexpected. For a subset of compounds containing only the carbon and

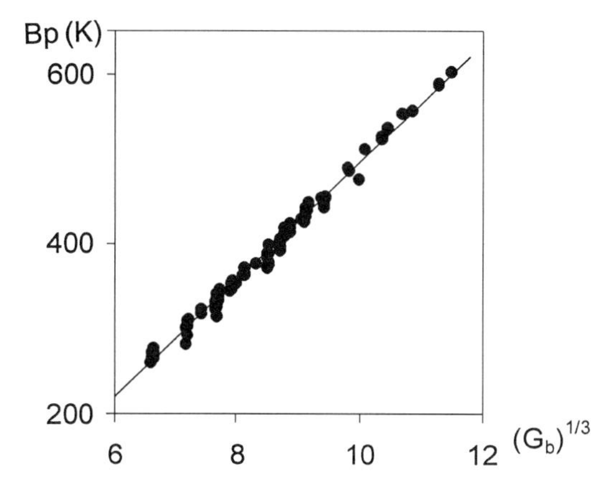

Figure 3.4 Correlation of boiling points of hydrocarbons with the cube root of gravitational index.

hydrogen atoms, the cube root of the gravitational index $\sqrt[3]{G_b}$ alone gives an excellent correlation. It was also shown that the boiling points of aliphatic, nonaliphatic, cyclic, acyclic, and aromatic hydrocarbons are predicted with a standard error of less than 8 K using the following excellent linear relationship (Fig. 3.4) [88]:

$$T_b = -190.7 + 68.6\sqrt[3]{G_b}$$

$$R^2 = 0.9908 \qquad s = 7.6 \qquad n = 113 \qquad F = 11965$$

The gravitational indices seem to quantify effectively the bulk cohesiveness of the compound arising mainly from the dispersion and hydrophobic interactions. Later, it was demonstrated that these descriptors or their root derivatives contribute to the main term in the QSPR correlations of other physical properties of liquids. Thus, the critical temperatures of hydrocarbons have been described by the following linear relationship [89]:

$$T_c = -(100 \pm 17) + (79.7 \pm 2.1)\sqrt[3]{G_b}$$

$$R^2 = 0.9526 \qquad s = 18.9 \qquad n = 76 \qquad F = 1506$$

whereas the best two-parameter correlation of the gas solubilities for the set of 95 hydrocarbons in aqueous solutions (Fig. 3.5) involved the gravitational index for the bonded atoms and a topological index, the complementary information content of the zeroth order ^{0}CIC, (cf. Eq. (3.95) [90]):

$$\log L^w = -(1.37 \pm 0.06) + (0.0067 \pm 0.0001)G_b - (0.050 \pm 0.001)^0\text{CIC}$$

$$R^2 = 0.9765 \qquad s = 0.20 \qquad n = 95 \qquad F = 1988$$

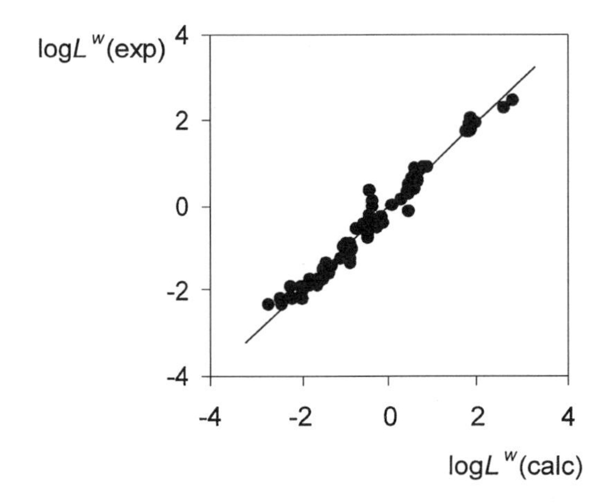

Figure 3.5 Correlation between the QSPR calculated and experimental logarithms of solubilities of hydrocarbons in water.

In the last case, the most significant correlation involves the gravitational index in the natural form.

Verloop and co-workers have developed so-called Sterimol parameters to characterize the geometry of substituent groups in the molecule [91–95]. By using the standard bond lengths, bond angles, and reasonably defined conformational angles, six Sterimol descriptors can be computed for a given substituent. The length parameter, L, is calculated as the length of the substituent along the axis defined by the bond connecting the substituent with the parent structural unit of the molecule. Four width parameters, B_1, B_2, B_3, and B_4, were introduced to characterize the distribution of atoms in the substituent with respect of the connecting bond. Parameter B_5 corresponds to the maximum width of the substituent. Notably, a poor correlation was found between the Taft steric constants E_s and the Sterimol parameters. By definition, the E_s constants are related to the steric interactions accompanied with the change of hybridization of the connecting carbon atom of a substituent. Apparently, these interactions are not directly correlated to the geometric dimensions of the substituent. However, a better correlation has been established between the molecular refraction MR and Sterimol parameters as [96]:

$$0.1\ \mathrm{MR} = -(2.70 \pm 0.47) + (0.38 \pm 0.08)L + (1.08 \pm 0.22)B_1 + (0.19 \pm 0.11)B_5$$

$$R^2 = 0.834 \qquad s = 0.388 \qquad n = 101$$

The Sterimol parameters have been used mostly for the correlation of biological activities. For example, a QSAR description of the activity of a set of homologous 1,4-dihydropyridine calcium (Ca^{2+}) channel blockers involved, apart from the Hansch–Leo lipophilicity term (π) and the Hammett electronic parameter (σ), three Sterimol parameters [97].

Arteca has reviewed the molecular shape descriptors [98] that also characterize the geometry and the mass distribution in a molecule. These descriptors were grouped according to the dimensionality or as being the relative and the absolute shape descriptors. According to this classification, the zeroth-order shape descriptors are presented by a single number (e.g., the molecular volume), the one-dimensional descriptors such as the radial distribution functions or knot polynomials have the vector representation, and the two-dimensional descriptors have the matrix representation. A relative shape descriptor is a *d*-dimensional function associated with a pair of molecules and measuring their similarity. The absolute shape descriptors are calculated by a certain mathematical algorithm, and they must be, as a rule, invariant with respect of the translation or rotation of the molecule. The definition of relative descriptors may not be unique and depends on the mutual configuration of the two molecules compared. For instance, the comparison of two molecules at the configuration, corresponding to the minimum root-mean-square deviations of the positions of the pairs of analogous atoms, may not be relevant for the comparison of their activity [99,100].

The simplest molecular shape descriptor is the span, R. This descriptor has been defined as the radius of the smallest sphere, fixed at the center of mass, which completely encloses the distribution of points, for example, atomic nuclei in the molecule [101]. The distribution of n nuclei in the space can be described by the instantaneous radius of gyration, R_G, defined as [102]:

$$R_G^2 = \frac{1}{M_T} \sum_{i=1}^{n} m_i r_i^2 \tag{3.10}$$

where m_i is the mass of the ith nucleus, r_i are the center-of-mass coordinates in the system, and M_T is the total nuclear mass. The degree by which a molecule deviates from a spherical form can be described by molecular eccentricity defined in terms of the principal moments of inertia of the molecule, λ_i, as follows:

$$\varepsilon = \frac{\{(\max[\lambda_i])^2 - (\min[\lambda_i])^2\}^{1/2}}{\max[\lambda_i]} \tag{3.11}$$

This parameter takes values between zero and one, with zero corresponding to molecules of spherical top symmetry having three degenerate principal moments of inertia. The asphericity of the molecule, Ω, is defined as [103,104]:

$$\Omega = \frac{1}{2} \left\{ \sum_{i=1}^{2} \sum_{j=i+1}^{3} (\lambda_i - \lambda_j)^2 \right\} \left\{ \sum_{i=1}^{3} \lambda_i \right\}^{-2} \tag{3.12}$$

and measures also the deviation of the molecule from the spheroidal form.

Todeschini and co-workers have invented the weighted holistic invariant molecular (WHIM) descriptors that consist of another perspective class of molecular geometric parameters [105–107]. These descriptors are built in a way that they capture the details of molecular size, shape, and atomic distribution with respect to invariant reference frame. The derivation of the WHIM descriptors proceeds with the principal component analysis of the weighted covariance matrix of atomic coordinates, **S**. The elements of this matrix are defined as:

$$s_{jk} = \frac{\sum_{i=1}^{n} w_i(q_{ij} - \bar{q}_j)(q_{ij} - \bar{q}_k)}{\sum_{i=1}^{n} w_i} \tag{3.13}$$

where n is the number of atoms in the molecule, w_i is the weight of the ith atom, q_{ij} is the jth Cartesian coordinate of atom i ($j = 1, 2, 3$) and $\bar{q}_j$ is the average of the jth coordinate. Different weighting schemes can be used to calculate the factors w_i. In the simplest case, all $w_i = 1$. In addition, atomic masses, van der Waals volumes, Mulliken electronegativities, atomic polarizabilities, and electrotopological indices of Kier and Hall [108] have been suggested for this purpose [107]. In principle, any atomic property may be considered as suitable for the definition of these descriptors, especially considering their further application for the development of particular QSPR/QSAR equations. Depending on the weighting scheme, different covariance matrices and different principal axes of the molecule are obtained. The WHIM descriptors are divided into two classes, the directional and the nondirectional descriptors.

The directional WHIM descriptors are calculated individually for each of three principal components. These include the eigenvalues λ_1, λ_2, and λ_3 of the respective covariance matrix and the eigenvalue proportions υ_1, υ_2, and υ_3, calculated as:

$$\upsilon_m = \frac{\lambda_m}{\sum_{m=1}^{3} \lambda_m} \qquad m = 1, 2, 3 \tag{3.14}$$

The WHIM descriptors called symmetries, γ_1, γ_2, and γ_3, are defined by the following equations:

$$\gamma_m = \frac{1}{1 + \gamma'_m}, \qquad m = 1, 2, 3 \tag{3.15}$$

and

$$\gamma'_m = -\left[\frac{n_s}{n} \log_2 \frac{n_s}{n} + n_a \left(\frac{1}{n} \log_2 \frac{1}{n}\right)\right] \tag{3.16}$$

In Eq. (3.16), n_s, n_a, and n denote the number of central symmetric atoms along the mth component, the number of nonsymmetric atoms, and the total number of atoms in the molecule, respectively. The last group of directional WHIM descriptors consists of the inverse of the kurtosis κ_1, κ_2, and κ_3, calculated from the fourth-order moments of the principal component analysis (PCA) scores $\mathbf{t}_m$ as:

$$\eta_m = (\kappa_m)^{-1} = \frac{\lambda_m^2 n}{\sum_i t_{im}^4} \qquad m = 1, 2, 3 \tag{3.17}$$

This group of descriptors can be related to the density of atomic distribution, and it has been called the emptiness. The greater η_m values correspond to the greater projected unfilled space in the molecule.

The nondirectional WHIM descriptors are derived from the above descriptors, and they are expressed by a single number for the whole molecule. The descriptors related to the total molecular size have been defined using different formulations as:

$$T = \lambda_1 + \lambda_2 + \lambda_3 \tag{3.18}$$

$$A = \lambda_1\lambda_2 + \lambda_1\lambda_3 + \lambda_2\lambda_3 \tag{3.19}$$

$$V = \prod_{m=1}^{3} (1 + \lambda_m) - 1 = T + A + \lambda_1\lambda_2\lambda_3 \tag{3.20}$$

with the same notations as above. The molecular shape is defined by the following equation:

$$K = \frac{3 \sum_m \left| \frac{\lambda_m}{\sum_m \lambda_m} - \frac{1}{3} \right|}{4} \tag{3.21}$$

with $0 \leq K \leq 1$ and $K = 0$ for the ideally spherical molecule. The larger value of K corresponds to larger eccentricity of the molecule. The total molecular symmetry is defined by the following expression:

$$G = (\gamma_1\gamma_2\gamma_3)^{1/3} \tag{3.22}$$

and the total molecular density as:

$$D = \eta_1 + \eta_2 + \eta_3 \tag{3.23}$$

The last two descriptors quantify the symmetry and the atomic distribution in the molecule, respectively. A large number of modified WHIM descriptors can be developed using different weighting factors in the definition of covariance matrix **S** (3.13).

The WHIM descriptors have demonstrated a good performance in the description of a variety of molecular properties, particularly in comparison with traditional topological descriptors. The properties investigated include melting points, boiling points, flash points, solubility, hydrophobicity, bioconcentration factor, and other properties of haloaromatic and polycyclic aromatic compounds [109,110]. In addition, successful correlations have been developed for the description of toxicities of various classes of compounds [111]. The performance of WHIM descriptors as well as of the classical molecular connectivity indices in QSAR correlations has been questioned in terms of both predictivity and interpretability [112,113]. Nevertheless, they encode precisely defined information about the geometrical structure of the molecule.

The molecular distance approach [114–116] has been also employed for the development of QSAR equations, for example, in the development of QSARs for macromolecule–ligand binding [117]. In the molecular shape and QSAR analysis of a family of substituted dichlorodiphenyl aromatase inhibitors, the best one- (linear) and two- (quadratic) parameter correlations were obtained with the distance between the nitrogen atom in the heterocycle and the α-carbon atom of the side chain at this heterocycle, D [118]. For instance, in the case of heterosubstituted (dichlorodiphenyl) methanes, the following equation gave the best representation of experimental data:

$$-\log(\mathrm{EC}_{50}) = -21.48 + (16.82 \pm 2.72)D - (2.41 \pm 0.41)D^2$$

$$R^2 = 0.83 \qquad s = 0.51 \qquad R^2_{cv} = 0.44 \qquad F = 28.7$$

The distance between the two active atomic sites in the substrate can thus be considered as a useful geometrical descriptor. The presence of the quadratic term in the last equation permits to predict the distance D, corresponding to the maximum activity (3.6 ± 0.1 Å). Consequently, this type of equation can be used, in principle, for the molecular design of more effective inhibitors.

The biological activity of a set of 54 antioccidial triazine analogs has been analyzed in terms of molecular shape analysis (MSA) to establish a QSAR model [119]. The procedure included the identificiation of the active conformation of each compound and the alignment of molecules by placing the N—N—C fragment of the triazine ring of the molecules compared on top of each other. The common overlap steric volume, V_{ov}, between each compound u in the set and the reference compound, v, was then calculated as:

$$V_{\mathrm{ov}} = V_u \cap V_v \tag{3.24}$$

where V_u and V_v are the volumes of isolated molecules. The other molecular shape similarity descriptor, the nonoverlapping volume V_{non}, was calculated as:

$$V_{\text{non}} = V_{uv} - V_v \tag{3.25}$$

where V_{uv} is the composite steric volume of the aligned pair of molecules u and v. Satisfactory correlations of antioccidial activities were obtained using these geometrical descriptors together with the semiempirical quantum-chemical CNDO/2 calculated nontriazine dipole moment contributions to the total dipole moment of the molecule.

Randić has introduced a scheme that gives a numerical characterization for a molecular shape and shapes in general [120]. In his scheme, a sequence that represents molecular shape was derived from the powers of interatomic distances for all atoms in the given atomic periphery. For instance, a geometrical descriptor ${}^1R$ was defined as:

$${}^1R = \sum_{i=1}^{n} \frac{R_i}{n} \tag{3.26}$$

where the R_i are the row (or column) sums of the geometrical distance matrix $\mathbf{D} = \{d_{ij}\}$ of a molecule, and n is the number of atoms in this molecule. Analogously, the next descriptor in this series

$${}^2R = \sum_{i=1}^{n} \frac{{}^2R_i}{n} \tag{3.27}$$

involves the row sums ${}^2R_i$ of the matrix built up from the squares of the interatomic geometrical distances, d_{ij}^2. Such descriptors can be used for building a power series

$$S = n + {}^1Rx + \frac{{}^2R}{2!}x^2 + \frac{{}^3R}{3!}x^3 + \frac{{}^4R}{4!}x^4 + \ldots \tag{3.28}$$

that itself can be a molecular shape descriptor (with $x = 1$, e.g., and ${}^3R$, ${}^4R$, . . . involving the higher powers of geometrical distances). The use of factorials in the denominators of the individual terms ensures the convergence of this sequence.

The molecular similarity is one of the fundamental terms in computer-aided molecular design [121,122]. However, as discussed in the Introduction to this book, there is no "absolute" measure of similarity as it depends on the property used to compare a given set of molecules. In principle, any molecular property P can be used for this purpose. A number of molecular similarity evaluation methods have been presented [123–136]. When the characteristics of molecular geometry such as the Cartesian coordinates of atomic nuclei, molecular surfaces, and volumes are applied to calculate the similarity, the resepctive similarity measures can be considered as geometrical molecular descriptors. A widely accepted family of similarity indices is based on the measurement of the Euclidean distance between molecular properties. In the case of geometrical properties, these descriptors characterize the similarity of the three-dimensional structure of molecules. One possible descriptor of this type is the root-mean-squared difference (RMSD) between the properties of two molecules A and B [137]:

$$\mathrm{RMSD}_{AB} = \left[\sum_{i=1}^{n} \left\{ \frac{(P_{A,i} - P_{B,i})^2}{n} \right\} \right]^{1/2} \tag{3.29}$$

where the summation is carried out over all n properties used, for example, the equivalent coordinates of atomic nuclei in two molecules. Two similar descriptors are the sum of squared errors [138] and the mean-squared difference [139] of properties.

A more general similarity measure is the Carbó similarity index defined as [140,141]:

$$R_{AB} = \frac{\int P_A P_B \, dv}{\left(\int P_A^2 \, dv \right)^{1/2} \left(\int P_B^2 \, dv \right)^{1/2}} \tag{3.30}$$

where a structural property P is compared for two molecules, A and B (P_A and P_B, respectively). The numerator of the ratio measures property overlap, while the denominator normalizes the similarity result. The quantum chemically derived electron density was orginally applied by Carbó as the structural property P [141]. The technique has since been extended to cover electrostatic potentials, electric fields, and other spatial properties of molecules and to increase the sensitivity of the measure to property magnitude [142–145]. Notably, it had been suggested to weight the Carbó index over the possible energetically close conformations of molecules using the respective Boltzmann factor [146]:

$$C_{AB} = \left[\frac{\int P_A P_B \, dv}{\left(\int P_A^2 \, dv \right)^{1/2} \left(\int P_B^2 \, dv \right)^{1/2}} \right] e^{-(c\Delta E/RT)} \tag{3.31}$$

where ΔE is the difference of energies of a given conformation and the global conformational minimum, respectively, and c is some user-defined weighting factor. Notably, this modified index is a temperature-dependent descriptor involving the absolute temperature T in its definition.

Meyer and Richards [147] have suggested another modification of the Carbó index to permit the evaluation of the molecular shape similarity using the grid technique. The molecules are surrounded by a rectilinear grid, and the property is evaluated at each intersection. In the case of the shape as the property, every grid point is testeed to see whether it falls inside the van der Waals surface of the molecules compared. The following equation measures the shape similarity:

$$S_{AB} = \frac{B}{(T_A T_B)^{1/2}} \tag{3.32}$$

where B is the number of grid points falling inside of both molecules compared, and T_A and T_B are the total number of grid points falling inside of each molecule. Grid-based shape and electrostatic potential similarity evaluations are rather time-consuming processes since very fine grids (about 0.2 Å) are required to obtain satisfactory results. However, a method for much more rapid screening of grid data has been developed recently [148].

The so-called linear similarity index [149] has been also introduced as based on the grid data:

$$l_{AB} = 1 - \frac{|P_A - P_B|}{\max(|P_A|, |P_B|)} \tag{3.33}$$

In this formula, $\max(|P_A|, |P_B|)$ equals the larger property magnitude between P_A and P_B at the grid point where the similarity is calculated. Another modification of the Carbó index is the Tanimoto coefficient [150], which has been used for the measurement of the shape similarity of molecule A with some shape query B [151]:

$$R_{AB} = \frac{\int P_A P_B \, dv}{\int w_1(P_A - P_B)\, dv + \int w_2(P_B - P_A)\, dv + \int P_A P_B \, dv} \tag{3.34}$$

The query B may be the shape of another molecule or it may be derived for a group of molecules. In Eq. (3.34), the difference $P_A - P_B$ measures the volume of structure A not overlapping with query B, and the difference $P_B - P_A$ is the query volume not overlapping with structure A. The coefficients w_1 and w_2 are some user-defined weights. The three-dimensional similarity of molecules has been described also by the Spearman rank-correlation coefficient [152–156] defined as:

$$S_{AB} = 1 \frac{6 \sum_{i=1}^{n} d_i^2}{n^3 - n} \tag{3.35}$$

where d_i is the property ranking at equivalent points i of two molecular structures and n is the total number of points over which the property is measured.

The work carried out using Gaussian functions to speed up the evaluation of electrostatic potential similarity [157,158] has been extended by development of the procedure to speed up the evaluation of the shape similarity index using Gaussian functions [121,159]. The respective procedure proceeds from the fitting of Gaussian functions to the STO-3G atomic orbital derived electron density functions of different atomic types. The Gaussian functions are then used to determine the Carbó index analytically. Similarity results produced by this procedure are evaluated rapidly (2 to 3 orders of magnitude faster than the grid results), with the same precision as by using the grid technique.

A list of typical geometrical descriptors is given in Table 3.1. The numerical values of some common descriptors for a set of structurally variable compounds are given in Table 3.2.

3.3 TOPOLOGICAL DESCRIPTORS

The classical chemical formula gives information not only about the atomic constitution of a compound but also about the presence and character of chemical bonds by which the atoms are connected to each other. The information about the constitute atoms and the connecting bonds between them can be formalized using special mathematical objects called graphs. The foundations of graph theory have been applied in a variety of areas of chemical research such as chemical informatics, quantum chemistry, and organic retrosynthesis [160–167]. The graph-theoretic models of molecular structure have also been extensively used in the development of quantitative structure–property/activity relationships [108,168–173]. The respective topological descriptors, also called topological indices, have been conventionally divided into several groups. Some describe the atomic connectivity in the molecule whereas others are based on the information theory. Before discussing the application of these descriptors for the development of QSAR/QSPR, we will proceed with an outline of the basic terms used in the definition of graph-theoretic molecular descriptors [174,175].

In mathematics, a graph G is defined as an ordered array consisting of two sets of objects, V and R,

$$G = [V, R] \tag{3.36}$$

where V is a finite nonempty set and R is a binary relation defined on V [175]. The elements of V are called *vertices* and the elements of R are called *edges*. The latter are often also denoted as E. The vertices can be visualized as mathematical points, connected by lines (edges). In chemistry, the graph representation of a molecule involves the atoms as the vertices and the chemical bonds between them as the edges. Two vertices, v_i and v_j, are called adjacent if they are connected by an edge $x = (v_i, v_j)$. Two vertices, v_0 and v_n, are joined by a *walk* of the form $v_0, x_1, v_1, x_2, \ldots, v_n$. The *length of a walk* is defined by the number of edges of which it consists. A *path* is an open walk in which all

TABLE 3.1 Typical Geometrical Descriptors

Notation	Description	Reference
I_A, I_B, I_C	Principal moments of inertia of a molecule	*a*
S_1, S_2, S_3	Molecular shadow areas projected on the planes defined by the principal axes of inertia of a molecule	*b*
S_M	Molecular surface area	*c*
S_{SA}	Solvent-accessible molecular surface area	*d*
V_M	Molecular volume	*e*
$V_{mol(SA)}$	Solvent-excluded volume of a molecule	*f*
G_p, G_b	Gravitation index over all atomic pairs or all chemical bonds in a molecule	*g*
E_p, E_b	Gravitation energy over all atomic pairs or all chemical bonds in a molecule	*h*
L	Sterimol length parameter	*i*
$B_1, \ldots, B_5$	Sterimol width parameters	*i*
R	Molecular span	*j*
R_G	Instantaneous radius of gyration of a molecule	*k*
ε	Eccentricity of a molecule	*k*
Ω	Asphericity of a molecule	*l*
λ_1, λ_2, λ_3	WHIM eigenvalues of the weighted covariance matrix of coordinates	*m*
υ_1, υ_2, υ_3	WHIM eigenvalue proportions	*m*
γ_1, γ_2, γ_3	WHIM symmetries	*m*
κ_1, κ_2, κ_3	WHIM inverse kurtosis	*m*
T, A, V	WHIM total molecular size descriptors	*m*
K	WHIM molecular shape descriptor	*m*
G	WHIM total molecular symmetry	*m*
D	WHIM total molecular density	*m*
V_{ov}	Common overlap steric volume	*n*
${}^iR$	Randić's molecular shape descriptors	*o*

[a] *Handbook of Chemistry and Physics*, R. C. Weast (ed.) p. F-112. CRC Press, Cleveland, OH, 1974.
[b] R.H. Rohrbaugh and P.C. Jurs, *Anal. Chim. Acta* **199,** 99 (1987).
[c] B. Lee and F.M. Richards, *J. Mol. Biol.* **55,** 379 (1971).
[d] M.L. Connolly, *J. Appl. Crystallogr.* **16,** 548 (1983).
[e] F.M. Richards, *Annu. Rev. Biophys. Bioeng.* **6,** 151 (1977).
[f] M.L. Connolly, *J. Am. Chem. Soc.* **107,** 1118 (1985).
[g] A.R. Katritzky, V.S. Lobanov, and M. Karelson, *CODESSA User's Manual*. University of Florida, Gainesville, 1994.
[h] This book, Chapter 3.
[i] A. Verloop, *The STERIMOL Approach to Drug Design*. Dekker, New York, 1987.
[j] M.V. Volkenstein, *Configurational Statistics of Polymeric Chains*. Wiley-Interscience, New York, 1963.
[k] C. Tanford, *Physical Chemistry of Macromolecules*. Wiley, New York, 1961.
[l] J. Rudnick and G. Gaspari, *Science* **237,** 384 (1986).
[m] R. Todeschini and P. Gramatica, in *3D QSAR in Drug Design* (H. Kubinyi, G. Folkers, and Y.C. Martin, eds.), Vol. 2, pp. 355–380. Kluwer/ESCOM, Dordrecht, The Netherlands, 1998.
[n] K.-B. Rhyu, H.C. Patel, and A.J. Hopfinger, *J. Chem. Inf. Comput. Sci.* **35,** 771 (1995).
[o] M. Randić, *J. Chem. Inf. Comput. Sci.* **35,** 373 (1995).

TABLE 3.2 Numerical Values of Selection of Constitutional and Geometrical Descriptors for Set of Structurally Variable Compounds

Compound	n_C	n_H	n_{bonds}	I_A	I_B	I_C	V_M	S_M	G_b
Water	0	2	2	26.2244	14.7054	9.4220	19.05	28.56	34.8
Ammonia	0	3	3	10.2566	10.2352	6.3920	22.40	32.56	42.2
Trichloromethane	1	1	4	0.1102	0.1100	0.0572	69.81	81.56	427.3
Tribromomethane	1	1	4	0.0417	0.0417	0.0212	82.58	93.96	801.5
Formaldehyde	1	2	3	9.4103	1.2417	1.0970	30.74	48.80	145.9
Dibromomethane	1	2	4	0.8925	0.0415	0.0400	64.47	76.52	550.6
Dichloromethane	1	2	4	1.1003	0.1116	0.1034	55.88	69.36	300.4
Chloromethane	1	3	4	4.9989	0.4568	0.4567	42.21	57.56	168.5
Bromomethane	1	3	4	5.0391	0.3265	0.3265	46.44	61.32	282.8
Iodomethane	1	3	4	5.0776	0.2693	0.2693	51.65	63.56	391.9
Nitromethane	1	3	6	0.4374	0.3373	0.1980	56.28	73.88	413.0
Methane	1	4	4	5.0761	5.0741	5.0727	28.32	37.88	39.2
Methanol	1	4	5	4.0435	0.8350	0.8034	36.91	50.24	142.3
Methylamine	1	5	6	3.3461	0.7810	0.7472	39.93	53.96	138.3
Tetrafluoromethane	1	0	4	0.1809	0.1804	0.1803	49.33	58.88	495.2
Tetrachloromethane	1	0	4	0.0576	0.0575	0.0575	83.77	93.08	549.6
Pentachloroethane	2	1	7	0.0389	0.0324	0.0251	113.82	121.68	760.3
Trichloroethene	2	1	5	0.1313	0.0512	0.0368	79.76	97.92	537.4
1,1,1,2-Tetrachloroethane	2	2	7	0.0627	0.0368	0.0347	100.32	112.92	631.7
1,1-Dichloroethene	2	2	5	1.7674	0.0518	0.0503	66.33	85.84	399.0
trans-1,2-Dichloroethene	2	2	5	1.7813	0.0516	0.0502	66.29	86.20	398.8
cis-1,2-Dichloroethene	2	2	5	0.4040	0.0827	0.0686	66.05	83.80	400.6
1,1,2,2-Tetrachloroethane	2	2	7	0.0559	0.0447	0.0254	99.88	111.20	635.3
2,2,2-Trifluoroethanol	2	3	8	0.1746	0.0905	0.0888	69.69	83.76	557.0
Acetonitrile	2	3	5	5.0435	0.3082	0.3081	47.99	72.68	222.4
1,1,1-Trichloroethane	2	3	7	0.0791	0.0789	0.0569	86.61	98.52	501.2
1,1,2-Trichloroethane	2	3	7	0.1179	0.0491	0.0360	86.81	99.00	507.5
Ethene	2	4	5	4.8921	0.9972	0.8284	39.39	59.96	122.3

1,1-Dichloroethane	2	4	7	0.2145	0.1080	0.0762	72.68	85.24	377.1
Acetaldehyde	2	4	6	1.8272	0.3371	0.3014	47.33	62.40	229.5
Acetic acid	2	4	7	0.3891	0.3009	0.1755	55.35	71.36	340.3
1,2-Dibromoethane	2	4	7	0.9761	0.0194	0.0192	81.78	97.80	618.9
1,2-Dichloroethane	2	4	7	0.9809	0.0505	0.0490	72.74	89.16	379.0
Methyl formate	2	4	7	1.4511	0.1589	0.1476	56.56	74.72	362.1
Chloroethane	2	5	7	1.0637	0.1843	0.1675	59.34	73.04	249.1
Bromoethane	2	5	7	1.0357	0.1255	0.1171	63.21	77.52	369.6
Iodoethane	2	5	7	1.0118	0.1038	0.0978	68.90	82.84	468.6
Nitroethane	2	5	9	0.3511	0.1332	0.1054	72.86	92.76	493.9
Ethane	2	6	7	2.5533	0.6818	0.6818	45.11	59.72	122.4
Ethanol	2	6	8	1.1355	0.3164	0.2744	53.50	67.72	223.5
Ethanethiol	2	6	8	0.9405	0.1934	0.1712	63.32	77.12	252.6
Ethylamine	2	7	9	1.1166	0.2929	0.2588	57.00	71.80	219.1
Dimethylamine	2	7	9	1.1577	0.3117	0.2763	57.47	73.44	233.7
Tetrachloroethene	2	0	5	0.0580	0.0465	0.0258	93.32	112.36	674.8
Propyne	3	4	6	5.0221	0.2928	0.2928	53.19	80.00	211.0
Methoxyflurane	3	4	11	0.0663	0.0365	0.0274	109.37	125.00	811.0
1-Chloro-2-propene	3	5	8	0.4603	0.1271	0.1016	70.09	86.84	334.1
Propanonitrile	3	5	8	0.9311	0.1579	0.1427	64.65	86.96	303.9
Ethyl formate	3	6	10	0.5110	0.0951	0.0875	73.47	92.32	443.5
Methyl acetate	3	6	10	0.3387	0.1367	0.1013	73.10	89.88	446.8
1,3-Dichloropropane	3	6	10	0.1690	0.0575	0.0458	89.88	100.88	459.6
Cyclopropane	3	6	9	0.6675	0.6667	0.4168	54.23	67.20	251.5
Propene	3	6	8	1.5150	0.3159	0.2756	56.42	72.40	206.5
Propanone	3	6	9	0.3306	0.2900	0.1644	64.33	81.48	312.4
Propanoic acid	3	6	10	0.3343	0.1257	0.0949	72.59	90.36	422.2
2-Propenol	3	6	9	0.8782	0.1442	0.1403	64.56	81.44	308.5
1,2-Dichloropropane	3	6	10	0.2292	0.0490	0.0423	89.62	104.56	457.9
Propionaldehyde	3	6	9	0.5619	0.1920	0.1544	64.43	80.12	312.2
1-Bromopropane	3	7	10	0.3781	0.0753	0.0673	80.44	92.84	451.2

TABLE 3.2 (*Continued*)

Compound	n_C	n_H	n_{bonds}	I_A	I_B	I_C	V_M	S_M	G_b
2-Bromopropane	3	7	10	0.2764	0.0950	0.0758	80.40	95.80	445.2
1-Nitropropane	3	7	12	0.3285	0.0646	0.0558	89.78	111.32	576.4
1-Chloropropane	3	7	10	0.4109	0.1097	0.0958	76.01	88.68	331.4
2-Chloropropane	3	7	10	0.2745	0.1520	0.1078	75.97	91.40	328.6
1-Iodopropane	3	7	10	0.3821	0.0597	0.0549	85.77	96.52	550.3
N,*N*-Dimethylformamide	3	7	11	0.2956	0.1392	0.0983	77.75	98.16	443.2
2-Nitropropane	3	7	12	0.1775	0.0970	0.0829	89.62	106.00	573.2
Propane	3	8	10	0.9707	0.2868	0.2545	61.61	78.12	204.5
Methyl ethyl ether	3	8	11	0.9060	0.1389	0.1298	70.99	90.24	330.7
1-Propanol	3	8	11	0.4971	0.1664	0.1452	70.48	83.76	306.4
2-Propanol	3	8	11	0.2880	0.2706	0.1603	70.40	85.44	305.1
2-Methoxyethanol	3	8	12	0.2825	0.1213	0.0952	79.14	91.24	432.7
n-Propanethiol	3	8	11	0.3975	0.1130	0.0988	79.98	90.96	335.1
Trimethylamine	3	9	12	0.2904	0.2869	0.1630	74.35	91.08	327.3
Propylamine	3	9	12	0.4431	0.1739	0.1431	73.84	86.52	301.4
Thiophene	4	4	9	0.2716	0.1851	0.1101	82.00	102.28	538.5
1-Butyne	4	6	9	0.9162	0.1548	0.1397	70.17	98.76	292.4
(*E*)-2-Butenal	4	6	10	0.6432	0.0863	0.0773	75.05	98.52	398.2
1-Cyanopropane	4	7	11	0.8072	0.0760	0.0725	81.45	106.92	386.7
Isobutyraldehyde	4	8	12	0.2347	0.1412	0.1090	81.45	96.60	393.2
Tetrahydrofuran	4	8	13	0.2353	0.2291	0.1278	76.78	91.00	451.5
Ethyl acetate	4	8	13	0.2434	0.0777	0.0687	89.86	106.76	528.6
Methyl propanoate	4	8	13	0.2517	0.0735	0.0643	89.93	107.36	528.6
1,4-Dichlorobutane	4	8	13	0.1479	0.0308	0.0267	106.57	120.88	542.3
1-Butene	4	8	11	0.7291	0.1425	0.1400	73.26	90.84	288.7
Butanone	4	8	12	0.3081	0.1222	0.0922	81.30	100.88	394.5
Isopropyl formate	4	8	13	0.1714	0.0944	0.0896	90.00	105.76	528.4
Butanoic acid	4	8	13	0.1920	0.0780	0.0723	89.17	102.48	504.7

Propyl formate	4	8	13	0.2235	0.0762	0.0620	89.88	105.80	525.3
Butyraldehyde	4	8	12	0.3038	0.1121	0.0923	81.18	97.04	393.6
1,4-Dioxane	4	8	14	0.1664	0.1557	0.0902	85.91	97.12	580.0
1-Bromobutane	4	9	13	0.1957	0.0518	0.0436	97.27	107.56	532.7
1-Chlorobutane	4	9	13	0.2197	0.0775	0.0631	92.72	102.28	412.9
1-Nitrobutane	4	9	15	0.2850	0.0368	0.0335	106.38	131.40	658.6
2-Bromo-2-methylpropane	4	9	13	0.1547	0.0670	0.0668	97.27	108.00	520.6
1-Bromo-2-methylpropane	4	9	13	0.1915	0.0554	0.0518	97.10	110.40	532.2
2-Chloro-2-methylpropane	4	9	13	0.1542	0.1006	0.1005	92.80	103.32	407.7
2-Chlorobutane	4	9	13	0.2250	0.0778	0.0625	92.77	107.40	410.0
1-Iodobutane	4	9	13	0.2038	0.0412	0.0363	102.76	109.32	632.0
Morpholine	4	9	15	0.1626	0.1508	0.0871	89.07	100.28	562.0
n-Butane	4	10	13	0.4437	0.1634	0.1399	78.81	92.96	286.6
Diethyl ether	4	10	14	0.4458	0.0860	0.0788	88.39	105.28	412.6
2-Methylpropane	4	10	13	0.2644	0.2614	0.1534	79.00	93.44	285.5
2-Methyl-1-propanol	4	10	14	0.2555	0.1186	0.0900	87.29	101.72	387.8
2-Butanol	4	10	14	0.2632	0.1174	0.0908	87.33	101.52	387.9
2-Methyl-2-propanol	4	10	14	0.1570	0.1559	0.1519	87.37	101.24	384.8
Diethyl sulfide	4	10	14	0.4856	0.0604	0.0560	96.65	116.96	470.6
Diethyl disulfide	4	10	15	0.1940	0.0338	0.0317	116.19	132.48	696.8
2-Ethoxyethanol	4	10	15	0.2350	0.0631	0.0547	96.34	110.04	514.4
1-Butanol	4	10	14	0.2466	0.1199	0.0913	87.25	97.24	387.8
n-Butanethiol	4	10	14	0.2152	0.0803	0.0649	96.55	108.08	416.2
Diethylamine	4	11	15	0.4027	0.0815	0.0751	91.30	107.28	395.5
Butylamine	4	11	15	0.2412	0.1172	0.0892	90.36	102.40	383.4
2-Chloropyridine	5	4	11	0.1925	0.0538	0.0421	95.94	124.80	662.3
3-Chloropyridine	5	4	11	0.1921	0.0534	0.0418	95.94	124.52	667.2
Pyridine	5	5	11	0.1990	0.1913	0.0975	82.79	105.32	529.1
2-Methylthiophene	5	6	12	0.1792	0.1069	0.0678	98.65	122.08	623.2
2-Methylpyrazine	5	6	13	0.1895	0.0890	0.0613	95.67	122.04	635.5
Cyclopentene	5	8	13	0.2432	0.2419	0.1307	79.48	92.56	408.9

TABLE 3.2 (*Continued*)

Compound	n_C	n_H	n_{bonds}	I_A	I_B	I_C	V_M	S_M	G_b
2-Methyl-1,3-butadiene	5	8	12	0.2887	0.1399	0.0961	84.18	106.88	373.4
Cyclopentanone	5	8	14	0.2211	0.1105	0.0782	87.46	106.40	517.3
1,4-Pentadiene	5	8	12	0.3113	0.1225	0.0966	84.33	100.12	372.2
1-Pentyne	5	8	12	0.3183	0.1157	0.0914	86.99	106.24	374.8
1-Cyanobutane	5	9	14	0.4986	0.0455	0.0431	98.03	123.52	469.6
Cyclopentane	5	10	15	0.2168	0.2165	0.1213	85.03	96.04	408.0
1-Pentene	5	10	14	0.2829	0.1131	0.0919	89.95	103.16	371.0
Ethyl propanoate	5	10	16	0.1472	0.0537	0.0423	107.13	126.80	610.2
Isobutyl formate	5	10	16	0.2067	0.0473	0.0414	106.63	123.16	608.2
2-Methyl-2-butene	5	10	14	0.2573	0.1286	0.0944	90.03	109.72	372.1
3-Methyl-1-butene	5	10	14	0.2282	0.1354	0.1062	89.96	105.96	369.6
Isopropyl acetate	5	10	16	0.1527	0.0595	0.0583	106.59	124.08	609.9
Pentanal	5	10	15	0.4962	0.0422	0.0413	97.93	119.32	476.4
Methyl butanoate	5	10	16	0.1543	0.0524	0.0500	106.99	117.32	610.0
Pentanoic acid	5	10	16	0.2636	0.0352	0.0319	106.15	128.84	586.9
3-Methylbutanoic acid	5	10	16	0.1645	0.0505	0.0457	106.00	124.16	585.9
Cyclopentanol	5	10	16	0.1988	0.1055	0.0780	93.85	101.40	509.1
3-Methyl-2-butanone	5	10	15	0.1532	0.0882	0.0840	97.91	115.04	475.4
3-Pentanone	5	10	15	0.2079	0.0746	0.0616	98.33	116.28	476.9
Propyl acetate	5	10	16	0.1581	0.0587	0.0501	106.48	119.84	609.3
2-pentanone	5	10	15	0.1754	0.0810	0.0702	98.05	108.96	476.4
2-Methyltetrahydrofuran	5	10	16	0.1902	0.1134	0.0859	94.71	102.48	531.9
Tetrahydropyran	5	10	16	0.1595	0.1456	0.0892	93.18	97.84	536.4
1-Chloropentane	5	11	16	0.2002	0.0392	0.0350	109.58	124.76	494.7
1-Nitropentane	5	11	18	0.2381	0.0225	0.0220	123.48	146.96	739.8
1-Bromopentane	5	11	16	0.1984	0.0256	0.0239	113.86	129.20	615.3
1-Iodopentane	5	11	16	0.2023	0.0199	0.0189	119.06	133.24	714.6
N-Methylmorpholine	5	11	18	0.1575	0.0769	0.0561	106.21	123.20	656.8

n-Pentane	5	12	16	0.5531	0.0671	0.0636	95.03	117.72	368.8
2-Methyl-1-butanol	5	12	17	0.1378	0.0946	0.0774	103.66	112.24	470.2
2-Methyl-2-butanol	5	12	17	0.1486	0.0869	0.0862	103.94	116.84	467.3
3-Methyl-1-butanol	5	12	17	0.2029	0.0713	0.0579	104.00	116.12	469.3
1-Pentanol	5	12	17	0.2154	0.0590	0.0504	104.37	117.72	470.3
2-Pentanol	5	12	17	0.2087	0.0683	0.0575	104.21	117.20	468.6
3-Pentanol	5	12	17	0.1153	0.1075	0.0619	103.76	115.20	469.1
Methyl *tert*-butyl ether	5	12	17	0.1477	0.0919	0.0918	104.42	118.44	492.0
2,2-Dimethylpropane	5	12	16	0.1508	0.1498	0.1491	95.61	105.48	365.6
2-Propoxyethanol	5	12	18	0.2964	0.0301	0.0290	113.75	131.08	596.1
2-Methylbutane	5	12	16	0.2064	0.1287	0.1033	95.54	109.44	367.6
n-Pentylamine	5	13	18	0.2063	0.0590	0.0509	107.62	120.68	465.0
1,2,3,4-Tetrachlorobenzene	6	2	12	0.0387	0.0183	0.0124	139.27	172.00	1054.8
1,2,3,5-Tetrachlorobenzene	6	2	12	0.0283	0.0195	0.0115	139.30	172.60	1054.8
1,2,4,5-Tetrachlorobenzene	6	2	12	0.0386	0.0157	0.0111	139.32	172.92	1055.3
1,2,3-Trichlorobenzene	6	3	12	0.0504	0.0282	0.0181	126.22	155.80	916.2
1,3,5-Trichlorobenzene	6	3	12	0.0287	0.0286	0.0143	126.44	157.72	915.7
1,2,4-Trichlorobenzene	6	3	12	0.0617	0.0200	0.0151	126.27	159.40	916.9
1,2-Dichlorobenzene	6	4	12	0.0642	0.0480	0.0275	113.34	143.56	777.8
1,3-Dichlorobenzene	6	4	12	0.0947	0.0287	0.0200	113.30	143.44	778.2
1,4-Dichlorobenzene	6	4	12	0.1870	0.0224	0.0200	113.31	149.00	777.9
3-Cyanopyridine	6	4	12	0.1927	0.0519	0.0409	101.46	141.76	713.1
4-Cyanopyridine	6	4	12	0.1989	0.0509	0.0405	101.57	142.50	712.3
Bromobenzene	6	5	12	0.1894	0.0330	0.0281	105.45	132.72	767.8
Chlorobenzene	6	5	12	0.1886	0.0528	0.0412	100.14	128.64	641.2
4-Fluorophenol	6	5	13	0.1839	0.0481	0.0381	100.12	135.20	723.9
2-Chlorophenol	6	5	13	0.0964	0.0518	0.0337	108.00	138.64	748.0
3-Chlorophenol	6	5	13	0.1147	0.0398	0.0295	108.11	137.80	748.5
4-Chlorophenol	6	5	13	0.1845	0.0327	0.0278	107.95	142.28	748.2
4-Bromophenol	6	5	13	0.1851	0.0212	0.0190	113.28	146.24	875.2
2-Iodophenol	6	5	13	0.0946	0.0250	0.0198	118.80	145.76	974.0

TABLE 3.2 (*Continued*)

Compound	n_C	n_H	n_{bonds}	I_A	I_B	I_C	V_M	S_M	G_b
2-Nitrophenol	6	5	15	0.0795	0.0413	0.0272	131.25	162.48	986.1
3-Nitrophenol	6	5	15	0.0912	0.0323	0.0238	122.22	157.36	986.6
4-Nitrophenol	6	5	15	0.1323	0.0269	0.0224	132.70	173.12	986.4
3-Formalpyridine	6	5	13	0.1782	0.0525	0.0405	101.61	131.36	721.1
4-Formalpyridine	6	5	13	0.1819	0.0518	0.0403	101.42	132.40	719.9
2-Fluorophenol	6	5	13	0.1076	0.0740	0.0438	100.05	130.04	725.2
Fluorobenzene	6	5	12	0.1881	0.0839	0.0580	92.31	121.32	617.2
Iodobenzene	6	5	12	0.1896	0.0258	0.0227	110.82	137.20	867.8
Nitrobenzene	6	5	14	0.1347	0.0418	0.0319	114.68	145.68	877.6
Benzene	6	6	12	0.1899	0.1895	0.0948	87.05	109.32	505.2
Phenol	6	6	13	0.1856	0.0875	0.0595	94.83	123.04	611.9
4-Chloroaniline	6	6	14	0.1821	0.0327	0.0278	111.75	147.04	741.6
2-Nitroaniline	6	6	16	0.0783	0.0409	0.0268	125.58	154.72	979.9
3-Nitroaniline	6	6	16	0.0909	0.0322	0.0238	126.35	159.52	980.7
4-Nitroaniline	6	6	16	0.1318	0.0269	0.0223	125.38	164.56	981.5
3-Chloroaniline	6	6	14	0.1137	0.0399	0.0295	111.95	140.40	742.1
2-Chloroaniline	6	6	14	0.0962	0.0517	0.0337	111.57	141.96	742.2
Thiophenol	6	6	13	0.1860	0.0543	0.0420	103.59	132.76	646.4
2-Methylpyridine	6	7	14	0.1866	0.0874	0.0602	100.24	126.20	611.5
3-Methylpyridine	6	7	14	0.1861	0.0868	0.0599	100.20	128.64	613.5
4-Methylpyridine	6	7	14	0.1917	0.0849	0.0595	100.27	127.04	613.3
N-Acetylpyrrolidine	6	7	15	0.1587	0.0541	0.0409	108.87	139.12	729.9
2-Ethylpyrazine	6	8	16	0.1628	0.0516	0.0398	112.53	142.68	716.9
Cyclohexene	6	10	16	0.1594	0.1532	0.0861	95.72	106.40	496.4
Cyclohexanone	6	10	17	0.1409	0.0833	0.0598	103.36	113.52	600.6
1,5-Hexadiene	6	10	15	0.1976	0.0767	0.0609	101.16	113.20	454.6
2,3-Dimethyl-1,3-butadiene	6	10	15	0.1633	0.1124	0.0683	100.27	123.80	456.3
trans-2-Hexenal	6	10	16	0.1796	0.0401	0.0345	109.24	127.84	561.4

1-Hexyne	6	10	15	0.1616	0.0944	0.0657	103.21	119.16	456.8
Cyclohexane	6	12	18	0.1454	0.1413	0.0819	101.28	108.04	492.7
Methylcyclopentane	6	12	18	0.1890	0.1040	0.0782	101.87	108.40	489.6
2-Methyl-1-pentene	6	12	17	0.2128	0.0746	0.0603	106.33	122.12	453.6
2-Hexanone	6	12	18	0.1712	0.0453	0.0399	114.64	129.80	558.4
Hexanoic acid	6	12	19	0.1321	0.0297	0.0274	123.09	135.88	669.1
Isobutyl acetate	6	12	19	0.1284	0.0355	0.0337	123.50	143.24	691.4
n-Propyl propanoate	6	12	19	0.1178	0.0344	0.0292	123.44	140.28	691.4
1-Hexene	6	12	17	0.2982	0.0514	0.0476	107.13	127.96	452.8
Isoamyl formate	6	12	19	0.1262	0.0374	0.0357	123.71	129.84	691.4
Ethyl butanoate	6	12	19	0.1215	0.0351	0.0322	124.05	134.72	691.5
Methyl pentanoate	6	12	19	0.1262	0.0372	0.0347	123.55	129.60	692.0
Hexanal	6	12	18	0.1348	0.0461	0.0370	115.15	127.64	558.3
4-Methyl-2-pentanone	6	12	18	0.1433	0.0564	0.0535	114.48	132.48	558.4
Butyl acetate	6	12	19	0.1289	0.0361	0.0317	123.48	135.64	692.1
Cyclohexylamine	6	13	20	0.1197	0.0858	0.0652	112.83	110.12	587.5
n-Butylacetamide	6	13	20	0.1137	0.0327	0.0310	128.09	142.32	674.6
1-Bromohexane	6	13	19	0.1071	0.0201	0.0177	130.49	142.72	697.3
1-Chlorohexane	6	13	19	0.1250	0.0283	0.0244	126.49	139.24	576.7
1-Iodohexane	6	13	19	0.1184	0.0148	0.0137	136.32	149.76	797.1
N-Methylpiperidine	6	13	20	0.1433	0.0772	0.0561	114.13	132.52	613.6
2,2-Dimethylbutane	6	14	19	0.1453	0.0854	0.0852	111.88	124.52	447.7
2,3-Dimethylbutane	6	14	19	0.1442	0.0877	0.0767	112.02	127.48	448.7
n-Hexane	6	14	19	0.1962	0.0626	0.0519	112.09	130.32	450.3
2-Methylpentane	6	14	19	0.2002	0.0692	0.0575	112.38	125.12	449.3
3-Methylpentane	6	14	19	0.1341	0.0924	0.0748	112.43	116.72	449.7
Diisopropyl ether	6	14	20	0.1379	0.0552	0.0432	121.90	136.44	573.2
Di-*n*-propyl ether	6	14	20	0.4360	0.0270	0.0262	121.68	147.84	576.4
1-Hexanol	6	14	20	0.4209	0.0249	0.0242	120.94	146.36	553.1
Di-*n*-propyl sulfide	6	14	20	0.3036	0.0235	0.0224	130.65	154.88	635.0
Diisopropyl sulfide	6	14	20	0.1231	0.0406	0.0399	130.19	142.28	630.3

TABLE 3.2 ***(Continued)***

Compound	n_C	n_H	n_{bonds}	I_A	I_B	I_C	V_M	S_M	G_b
4-Methyl-2-pentanol	6	14	20	0.1318	0.0578	0.0463	120.63	132.08	549.8
2-Methyl-3-pentanol	6	14	20	0.1145	0.0640	0.0455	120.43	133.40	549.7
3-Hexanol	6	14	20	0.1093	0.0532	0.0392	121.23	137.00	550.9
2-Butoxyethanol	6	14	21	0.1025	0.0310	0.0252	129.76	139.00	677.9
2-Methyl-2-pentanol	6	14	20	0.1340	0.0549	0.0530	120.77	129.08	549.0
Di-*n*-Propylamine	6	15	21	0.2693	0.0295	0.0280	125.43	145.44	559.9
Hexylamine	6	15	21	0.1450	0.0387	0.0330	124.61	134.48	547.7
Diisopropylamine	6	15	21	0.1273	0.0565	0.0459	124.97	134.60	552.8
Triethylamine	6	15	21	0.1036	0.0645	0.0456	125.31	140.52	569.8
Benzotrifluoride	7	5	15	0.0946	0.0303	0.0261	120.54	148.44	918.1
3-Cyanophenol	7	5	14	0.1149	0.0398	0.0296	113.70	152.12	795.9
4-Cyanophenol	7	5	14	0.1854	0.0330	0.0280	113.41	159.52	797.5
Benzonitrile	7	5	13	0.1884	0.0516	0.0405	105.66	145.72	688.9
Benzaldehyde	7	6	14	0.1742	0.0520	0.0400	106.03	136.36	695.9
3-Hydroxybenzaldehyde	7	6	15	0.1207	0.0379	0.0289	113.60	146.16	804.0
4-Hydroxybenzaldehyde	7	6	15	0.1671	0.0332	0.0277	113.91	149.36	803.9
2-Chlorotoluene	7	7	15	0.0957	0.0514	0.0337	117.62	140.80	724.5
4-Bromotoluene	7	7	15	0.1821	0.0213	0.0191	122.16	155.52	850.6
2-Nitrotoluene	7	7	17	0.0778	0.0412	0.0271	130.83	158.16	960.9
3-Nitrotoluene	7	7	17	0.0904	0.0320	0.0237	131.43	162.44	960.6
Benzamide	7	7	16	0.1265	0.0399	0.0313	118.70	146.52	799.3
4-Chloro-3-methylphenol	7	7	16	0.0933	0.0316	0.0237	125.42	155.64	831.4
3-Acetylpyridine	7	7	16	0.1234	0.0402	0.0319	118.94	147.72	803.3
4-Acetylpyridine	7	7	16	0.1259	0.0399	0.0313	118.65	149.68	802.9
Toluene	7	8	15	0.1827	0.0854	0.0588	104.33	132.28	589.0
Benzyl alcohol	7	8	16	0.1634	0.0502	0.0387	112.57	141.64	690.0
o-Cresol	7	8	16	0.1082	0.0737	0.0442	111.98	141.68	696.1
p-Cresol	7	8	16	0.1800	0.0492	0.0389	112.32	146.00	696.4

1,3,5-Cycloheptatriene	7	8	15	0.1239	0.1226	0.0668	99.85	117.68	587.8
Methyl phenyl ether	7	8	16	0.1646	0.0526	0.0402	113.15	143.28	717.1
2-Methoxyphenol	7	8	17	0.0846	0.0520	0.0324	121.21	152.56	825.0
Phenyl methyl sulfide	7	8	16	0.1354	0.0384	0.0340	122.39	146.36	779.0
3-Methoxyphenol	7	8	17	0.1201	0.0375	0.0287	121.16	152.64	825.6
o-Toluidine	7	9	17	0.1062	0.0736	0.0439	115.33	143.20	690.0
p-Toluidine	7	9	17	0.1789	0.0490	0.0388	115.51	147.68	690.5
2-Methoxyaniline	7	9	18	0.0828	0.0520	0.0322	124.33	153.28	817.2
3-Methoxyaniline	7	9	18	0.1185	0.0376	0.0287	124.47	156.24	818.4
4-Methoxyaniline	7	9	18	0.1575	0.0332	0.0276	124.45	160.12	818.1
2,3-Dimethylpyridine	7	9	17	0.1045	0.0749	0.0444	116.43	143.00	694.4
2,4-Dimethylpyridine	7	9	17	0.1228	0.0599	0.0409	116.82	144.04	695.5
2,5-Dimethylpyridine	7	9	17	0.1793	0.0497	0.0395	116.53	150.64	695.2
2,6-Dimethylpyridine	7	9	17	0.1192	0.0617	0.0413	116.83	144.96	693.0
3,4-Dimethylpyridine	7	9	17	0.1075	0.0729	0.0442	116.56	145.04	697.3
3,5-Dimethylpyridine	7	9	17	0.1214	0.0601	0.0408	116.44	143.52	696.9
N-Methylaniline	7	9	17	0.1606	0.0508	0.0399	116.57	145.68	700.4
2-Ethylpyridine	7	9	17	0.1629	0.0508	0.0394	116.88	145.88	693.7
3-Ethylpyridine	7	9	17	0.1594	0.0508	0.0395	116.65	146.24	695.7
4-Ethylpyridine	7	9	17	0.1648	0.0499	0.0389	116.83	145.68	695.3
1-Methylcyclohexene	7	12	19	0.1500	0.0773	0.0551	112.41	132.40	580.8
1-Heptyne	7	12	18	0.1278	0.0540	0.0408	120.27	134.84	540.0
1-Heptene	7	14	20	0.1386	0.0443	0.0363	123.50	136.80	534.5
Heptanal	7	14	21	0.2209	0.0215	0.0205	131.78	153.72	641.0
n-Propyl butanoate	7	14	22	0.1043	0.0234	0.0218	140.62	155.60	773.8
Ethyl pentanoate	7	14	22	0.1173	0.0245	0.0231	140.20	150.76	772.9
Methyl hexanoate	7	14	22	0.1211	0.0250	0.0239	140.67	146.28	774.0
Cycloheptanol	7	14	22	0.0988	0.0591	0.0415	126.33	133.48	674.2
Methylcyclohexane	7	14	21	0.1170	0.0866	0.0659	118.11	113.52	573.9
2,4-Dimethyl-3-pentanone	7	14	21	0.0885	0.0494	0.0442	131.68	140.96	638.8
Isoamyl acetate	7	14	22	0.1252	0.0222	0.0214	140.34	158.96	773.4

TABLE 3.2 (*Continued*)

Compound	n_C	n_H	n_{bonds}	I_A	I_B	I_C	V_M	S_M	G_b
4-Heptanone	7	14	21	0.0991	0.0371	0.0304	131.64	145.72	641.4
trans-2-Heptene	7	14	20	0.1565	0.0392	0.0337	123.78	137.08	536.9
2-Heptanone	7	14	21	0.1115	0.0317	0.0295	131.54	146.60	641.5
1-Bromoheptane	7	15	22	0.0805	0.0144	0.0127	147.52	156.36	779.5
1-Chloroheptane	7	15	22	0.0872	0.0207	0.0177	143.21	159.84	658.6
1-Iodoheptane	7	15	22	0.0815	0.0105	9.5903	153.12	166.16	879.4
2,3-Dimethylpentane	7	16	22	0.1219	0.0613	0.0523	128.90	133.44	530.5
n-Heptane	7	16	22	0.1250	0.0441	0.0357	129.25	136.04	532.4
2-Methylhexane	7	16	22	0.1867	0.0386	0.0344	129.38	142.88	531.3
3-Methylhexane	7	16	22	0.1109	0.0576	0.0468	128.97	140.32	531.8
1-Heptanol	7	16	23	0.0966	0.0287	0.0236	137.81	149.12	635.3
2,4-Dimethylpentane	7	16	22	0.1288	0.0545	0.0485	129.04	142.08	530.5
3,3-Dimethylpentane	7	16	22	0.1078	0.0637	0.0560	128.60	138.92	529.7
2,2-Dimethylpentane	7	16	22	0.1256	0.0581	0.0553	128.75	137.68	528.9
n-Heptylamine	7	17	24	0.3634	0.0169	0.0165	140.58	169.72	630.9
Styrene	8	8	16	0.1688	0.0524	0.0406	115.28	146.24	672.3
Acetophenone	8	8	17	0.1218	0.0405	0.0314	122.95	154.52	778.9
4-Methylbenzaldehyde	8	8	17	0.1655	0.0329	0.0276	123.02	159.00	779.9
Methyl benzoate	8	8	18	0.1186	0.0276	0.0225	132.04	166.64	911.4
Phenyl acetate	8	8	18	0.1224	0.0271	0.0240	132.42	162.88	916.4
Ethylbenzene	8	10	18	0.1590	0.0501	0.0387	121.08	150.36	670.9
m-Xylene	8	10	18	0.1197	0.0596	0.0404	120.96	149.56	672.6
p-Xylene	8	10	18	0.1769	0.0488	0.0388	120.55	153.12	673.1
o-Xylene	8	10	18	0.1043	0.0731	0.0437	120.46	148.88	672.6
Ethyl phenyl ether	8	10	19	0.1177	0.0375	0.0320	130.51	153.28	798.1
2,3-Dimethylphenol	8	10	19	0.0751	0.0590	0.0335	128.88	157.12	778.9
2,5-Dimethylphenol	8	10	19	0.1009	0.0436	0.0308	128.90	158.28	779.5
2,6-Dimethylphenol	8	10	19	0.0766	0.0582	0.0335	128.31	157.00	779.0

3,4-Dimethylphenol	8	10	19	0.0979	0.0440	0.0307	128.59	157.84	778.7
3,5-Dimethylphenol	8	10	19	0.0594	0.0593	0.0300	128.72	159.88	779.1
4-Ethylphenol	8	10	19	0.1520	0.0322	0.0268	129.28	163.32	777.2
2-Phenylethanol	8	10	19	0.1482	0.0288	0.0257	129.55	159.68	772.3
2,4-Dimethylphenol	8	10	19	0.0995	0.0440	0.0309	128.97	159.08	780.2
3-Ethylphenol	8	10	19	0.1163	0.0361	0.0281	128.62	158.24	778.1
2,6-Dimethylaniline	8	11	20	0.0754	0.0584	0.0334	131.58	160.84	773.4
2-Isobutylpyrazine	8	12	22	0.1040	0.0248	0.0233	145.98	165.60	879.3
trans-2-Octenal	8	14	22	0.0766	0.0215	0.0176	142.76	157.44	726.8
1-Octyne	8	14	21	0.1373	0.0282	0.0244	137.55	166.32	620.4
n-Propylcyclopentane	8	16	24	0.1718	0.0327	0.0293	135.87	154.04	653.7
trans-1,4-Dimethylcyclohexane	8	16	24	0.1070	0.0529	0.0454	134.24	139.08	653.6
Octanal	8	16	24	0.2451	0.0134	0.0130	148.57	177.03	723.7
n-Pentyl propanoate	8	16	25	0.1393	0.0142	0.0136	157.23	178.99	856.5
Ethyl hexanoate	8	16	25	0.0876	0.0199	0.0190	157.20	163.24	855.6
Isobutyl isotuanoate	8	16	25	0.0936	0.0203	0.0180	156.99	178.43	854.7
cis-1,2-Dimethylcyclohexane	8	16	24	0.0745	0.0721	0.0529	135.09	141.52	654.9
Hexyl acetate	8	16	25	0.0825	0.0167	0.0151	157.05	176.03	857.4
2-Octanone	8	16	24	0.0698	0.0304	0.0250	148.40	161.60	723.4
1-Octene	8	16	23	0.1030	0.0306	0.0250	140.68	157.64	615.4
1-Bromooctane	8	17	25	0.0482	0.0126	0.0102	164.28	170.08	860.3
n-Octane	8	18	25	0.3474	0.0170	0.0166	146.42	175.47	615.5
2,2,4-Trimethylpentane	8	18	25	0.0947	0.0464	0.0418	145.43	153.28	610.4
2,3,4-Trimethylpentane	8	18	25	0.0862	0.0473	0.0434	145.05	153.68	611.1
Di-*n*-butyl ether	8	18	26	0.3031	0.0128	0.0125	154.77	185.87	740.8
3-Methylheptane	8	18	25	0.1821	0.0240	0.0225	145.88	168.36	614.5
1-Octanol	8	18	26	0.0672	0.0222	0.0177	154.84	165.52	716.6
Di-*n*-butylamine	8	19	27	0.3087	0.0123	0.0121	158.48	191.51	724.4
Octylamine	8	19	27	0.0656	0.0225	0.0175	157.69	169.92	712.0
Quinoline	9	7	18	0.1042	0.0420	0.0299	128.26	167.52	910.0

TABLE 3.2 ***(Continued)***

Compound	n_C	n_H	n_{bonds}	I_A	I_B	I_C	V_M	S_M	G_b
α-Methylstyrene	9	10	19	0.1195	0.0406	0.0313	131.74	161.96	755.5
Ethyl benzoate	9	10	21	0.0944	0.0196	0.0168	148.86	182.15	993.4
4-Methylacetophenone	9	10	20	0.1189	0.0259	0.0225	139.78	173.59	862.6
Indane	9	10	20	0.1177	0.0500	0.0358	127.28	152.48	793.7
1,2,3-Trimethylbenzene	9	12	21	0.0747	0.0579	0.0332	136.64	164.92	755.3
1,2,4-Trimethylbenzene	9	12	21	0.0972	0.0437	0.0307	136.71	165.40	756.6
1,3,5-Trimethylbenzene	9	12	21	0.0593	0.0584	0.0300	137.19	162.36	756.4
Isopropylbenzene	9	12	21	0.1123	0.0368	0.0335	137.80	163.84	751.7
2-Ethyltoluene	9	12	21	0.0809	0.0494	0.0312	137.37	162.68	754.1
4-Ethyltoluene	9	12	21	0.1499	0.0322	0.0270	137.53	171.20	754.9
4-*n*-Propylphenol	9	12	22	0.1032	0.0244	0.0228	146.00	172.04	859.3
Propylbenzene	9	12	21	0.1537	0.0290	0.0248	137.81	170.28	752.9
3-Phenyl-1-propanol	9	12	22	0.1395	0.0187	0.0167	146.32	180.03	854.7
Nonanal	9	18	27	0.2144	9.6959	9.4134	165.27	196.99	805.6
5-Nonanone	9	18	27	0.0848	0.0158	0.0144	165.61	182.27	805.5
2-Nonanone	9	18	27	0.0647	0.0194	0.0167	165.29	185.59	805.3
1-Nonene	9	18	26	0.0701	0.0225	0.0180	157.56	171.32	699.4
n-Nonane	9	20	28	0.0576	0.0290	0.0205	162.92	172.44	696.8
2,2,5-Trimethylhexane	9	20	28	0.0965	0.0249	0.0234	162.79	179.99	691.6
1-Nonanol	9	20	29	0.0539	0.0162	0.0130	171.58	183.71	799.2
Naphthalene	10	8	19	0.1038	0.0412	0.0295	132.34	175.91	886.1
1-Naphthol	10	8	20	0.0645	0.0376	0.0238	140.45	175.55	994.0

2-Naphthol	10	8	20	0.0948	0.0275	0.0213	140.57	179.55	993.7
1-Naphthylamine	10	9	21	0.0645	0.0374	0.0237	143.30	178.11	988.3
2-Naphthylamine	10	9	21	0.0936	0.0277	0.0214	143.58	184.07	987.9
n-Butylbenzene	10	14	24	0.1359	0.0190	0.0169	154.82	189.71	835.3
tert-Butylbenzene	10	14	24	0.0857	0.0322	0.0276	153.84	176.27	830.9
4-Isopropyltoluene	10	14	24	0.1090	0.0247	0.0232	154.43	185.75	835.4
4-*tert*-Butylphenol	10	14	25	0.0849	0.0217	0.0194	161.84	189.11	936.6
Isobutylbenzene	10	14	24	0.1076	0.0245	0.0221	154.56	180.43	833.9
sec-Butylbenzene	10	14	24	0.0990	0.0256	0.0228	154.98	181.35	833.3
n-Pentylcyclopentane	10	20	30	0.1484	0.0139	0.0132	169.95	193.55	816.2
2-Decanone	10	20	30	0.0484	0.0137	0.0115	182.18	194.75	887.9
n-Decane	10	22	31	0.0505	0.0176	0.0137	180.46	183.87	776.7
Decyl alcohol	10	22	32	0.0307	0.0188	0.0123	187.87	191.47	878.9
1-Methylnaphthalene	11	10	22	0.0637	0.0376	0.0238	150.77	183.71	970.1
n-Pentylbenzene	11	16	27	0.1219	0.0138	0.0130	171.53	201.95	918.3
2-Undecanone	11	22	33	0.0361	0.0126	0.0100	198.95	209.99	969.8
Biphenyl	12	10	23	0.0946	0.0184	0.0162	162.31	203.51	1054.8
Acenaphthene	12	10	24	0.0474	0.0397	0.0218	157.32	193.99	1092.8
2,3-Dimethylnaphthalene	12	12	25	0.0720	0.0226	0.0173	167.09	210.87	1055.1
2,6-Dimethylnaphthalene	12	12	25	0.0622	0.0246	0.0178	167.06	203.63	1054.1
1,3-Dimethylnaphthalene	12	12	25	0.0529	0.0274	0.0182	166.64	205.07	1054.7
1,4-Dimethylnaphthalene	12	12	25	0.0401	0.0372	0.0194	166.40	208.23	1054.4
1-Ethylnaphthalene	12	12	25	0.0465	0.0308	0.0193	167.38	194.63	1052.3
m-Hexylbenzene	12	18	30	0.1040	0.0104	9.9461	188.41	222.63	1000.1
Fluorene	13	10	25	0.0720	0.0197	0.0155	169.45	211.07	1179.2

vertices are distinct. Therefore, a walk is any sequence of adjacent graph edges; in walking from v_i to v_j, it is allowed to go back and forth and to visit vertices repeatedly. In contrast, a path is a sequence of adjacent graph edges without repetition.

The *distance* $d_{ij} = d(v_i, v_j)$ is the length of the shortest path connecting vertices v_i and v_j. All distances d_{ij} in a molecular graph compose the *distance matrix* $\boldsymbol{D}$. Another matrix, called the *adjacency matrix* $\boldsymbol{A}$, is defined as consisting of elements $a_{ij} = 1$ if the vertices v_i and v_j are adjacent and $a_{ij} = 0$ otherwise. For the diagonal elements of $\boldsymbol{A}$, $a_{ii} = 0$, too. The *degree* δ_i of the vertex v_i in graph G is equal to the number of edges (lines) adjacent to this vertex. In a hydrocarbon skeleton, the primary (terminal) carbon atom has the vertex degree $\delta = 1$, for the secondary, tertiary, and quaternary carbon atom, the vertex degree is 2, 3, and 4, respectively. A different definition of the vertex degree, often denoted as D_i, is obtained by summing over rows or columns the elements of the adjacency matrix $\boldsymbol{A}$. The *radius* ρ of a graph is given by the smallest of the maximum path lengths, called *eccentricities*, between any two vertices in G, that is,

$$\rho = \min\{\max_{u,v \in V} d(u, v)\} \tag{3.37}$$

The *center* of a graph is defined as the set of all vertices whose eccentricity equals to ρ. A graph G is *connected* if every pair of its vertices is connected by a path. A graph G is a *multigraph* if it contains more than one edge between at least one pair of adjacent vertices, otherwise G is a linear graph. In the following, the chemical structure, hydrogen-suppressed multigraph and linear graph of para-cyanophenol are presented.

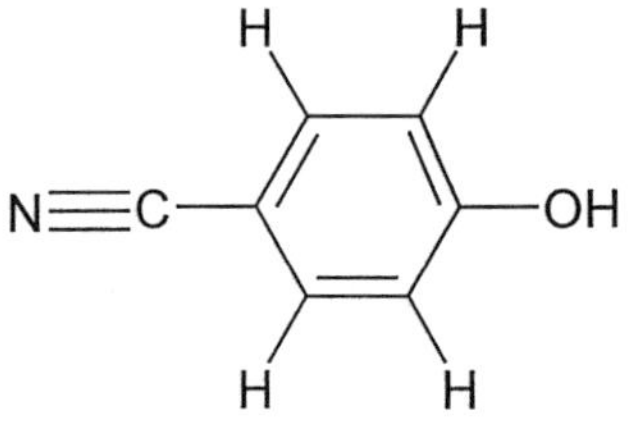

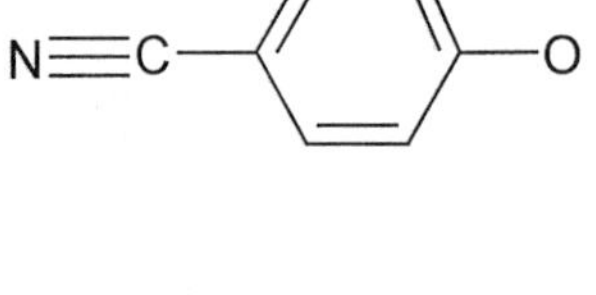

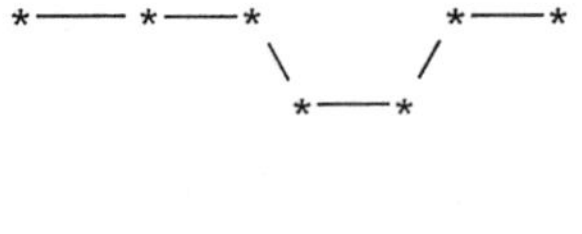

```
              *——*
             /    \
*——*——*          *——*
             \    /
              *——*
```

For a vertex v in G, the *first-order neighborhood* $\Gamma^1(v)$ is defined as a subset of vertices u that are directly connected to this vertex, that is, $d(v, u) = 1$. If ρ is the radius of a graph G, the neighborhoods of ith order up to ρ can be constructed for each vertex in the graph. By extrapolation of this approach to the path of length 0, the neighborhood of the zeroth order could be defined as well.

The topological indices are defined as graph invariants, that is, they should be independent on the numbering of vertices. However, the opposite does not hold in general. Due to the degeneracy of topological indices, the same value of the index may correspond to different chemical structures. The topological indices represent usually the local vertex invariants, that is, they are the numbers associated with vertices symbolizing atoms in such a way as to be independent of arbitrary vertex labeling. Wiener introduced one of the first and simplest topological indices for molecules in 1947 [176] in terms of the distance matrix **D**. The Wiener index W was defined as:

$$W = \frac{1}{2} \sum_{(i,j)}^{N_{SA}} d_{ij} \tag{3.38}$$

where the distances d_{ij} correspond to the number of bonds in the shortest path connecting the pair of atoms i and j, and N_{SA} is the number of atoms (or nonhydrogen atoms) in the molecule. The Wiener index characterizes the "compactness" of a molecule, being larger for extended chains and smaller for branched compounds. According to Balaban [177], the Wiener index belongs to the first-generation topological descriptors, which are integer numbers obtained by operations on integer number local vertex invariants. The operations involve one vertex at a time. A generalization of the Wiener index W, the hyper-Wiener index R, has been introduced by Randić as follows [178,179]:

$$R = \frac{1}{2} \sum_{i} \sum_{j \neq i} K_{ij} \tag{3.39}$$

where the K_{ij} are the matrix elements obtained by the following two-step algorithm. First, two nonconnected subgraphs are obtained by removing the path between vertices i and j. K_{ij} is then calculated as the product of the number of vertices in both subgraphs. A "product" definition of the Wiener index has been extended also for cyclic structures [180–183] as:

$$W_c = \sum_{e} \sum_{k<j} \frac{k_{ij}^c}{k_{ij}} \tag{3.40}$$

where k_{ij}^c denotes the number of all paths of minimal length that connect vertices i and j and involve edge e, and k_{ij} is the number of all paths of minimal length between i and j. For monocyclic compounds, simple formulas can be applied for the calculation of this index [184]:

$$W_c = \frac{n^2}{8} \quad \text{if } n \text{ is an even number}$$
$$W_c = \frac{(n^2 - 1)^2}{8} \quad \text{if } n \text{ is an odd number} \tag{3.41}$$

The hyper-Wiener index R has been developed also for dendrimers [184a].

A number of other first-generation topological indices have been developed by different research groups. For example, the Zagreb group index M_1 is defined as [185]:

$$M_1 = \sum_i D_i \tag{3.42}$$

where the summation is performed over all rows D_i (or columns) of the adjacency matrix $\boldsymbol{A}$. The Gordon–Scantlebury index denoted by N_2 [186] has been defined according to the following equation:

$$N_2 = \sum_i \frac{D_i(D_i - 1)}{2} \tag{3.43}$$

where the summation is again carried out over all rows (or columns) of matrix $\boldsymbol{A}$. The topological Hosoya index denoted as Z [187] is based on counting the nonadjacent bonds in a molecular graph:

$$Z = \sum_{k=0}^{m} p(G, k) \tag{3.44}$$

where $p(G, 0)$ is by definition 1, $p(G, 1)$ corresponds to the number of bonds in the molecular graph, and $p(G, k)$ denotes the maximum number of combinations for given k bonds in the graph. It has been applied for the QSPR description of the thermodynamic properties of alkanes [188] and for the prediction of the gas chromatographic retention indices [189,190]. Recently, an analogous topological index was introduced as [191]:

$$Z_i = \sum_{k=0}^{m} Z_{ik} \tag{3.45}$$

where i is the chosen path length (number of atoms included), k is the number of combined paths with length i, and m is the possible number of k. Whereas the combinations of the Z_i indices have a very low degree of degeneracy, it was found that these indices are less suitable as the descriptors in QSPR models.

The quadratic index Q has been proposed also using local vertex invariants as [192]:

$$Q = \frac{\sum_i (i^2 - 2i)D_i + 2}{2} \tag{3.46}$$

with the same notations as above. Furthermore, the first-generation topological indices include the centric indices denoted by B and C and the Merrifield-Simmon σ index [193]. The B index is defined as [192,194]:

$$B = \sum_i \Delta_i^2 \tag{3.47}$$

where Δ_i represents the number of vertices eliminated in the ith step of the stepwise elimination of first-degree vertices from the molecular graph. In general, the first-generation topological indices are characterized by high degeneracy, that is, the same value of the index may correspond to several different chemical structures.

The second-generation topological indices have been defined as the real numbers obtained from local vertex invariants by sophisticated operations involving more than one vertex at a time. A typical second-generation index is the Randić molecular connectivity [195] defined as the following sum over all pairs of edges, i and j, in the molecule:

$$\chi = \sum_{\text{edges } ij} (D_i D_j)^{-1/2} \tag{3.48}$$

where D_i and D_j are the edge degrees. The latter have been also called atom connectivities and denoted δ_i. The generic formula for the calculation of the Randić indices of different orders has been presented as [170]:

$$^m\chi = \sum_{\text{path}} (D_i D_j \cdots D_k)^{-1/2} \tag{3.49a}$$

for continuous paths of certain length, $m > 1$. The length of a continuous path between two vertices has been defined as the number of nonrepeating edges between them. It has been suggested that the substitution of the exponential $-\frac{1}{2}$ by $-\frac{1}{3}$ in the formulation of the Randić index

$$\chi = \sum_{\text{edges } ij} (D_i D_j)^{-1/3} \tag{3.49b}$$

where the summation is performed over all pairs of edges, i and j, in the molecule, reduces substantially the degeneracy of this topological index [196].

Two other second-generation topological indices have been defined by Balaban as the mean-square distance between all vertices in the molecular graph [197]:

$$D^{(2)} = \frac{1}{n(n-1)}\left(\sum_{i,j} d_{ij}^2\right)^{-1/2} \tag{3.50}$$

and the average distance sum connectivity, J, given by the following formula:

$$J = \frac{q}{\mu + 1} \sum_{\text{edges } ij} (S_i S_j)^{-1/2} \tag{3.51}$$

where q denotes the number of edges in the molecular graph, $\mu = (q - n + 1)$ is the so-called cyclomatic number of the graph and S_i are distance sums calculated as the sums over the rows or columns of the topological distance matrix $\boldsymbol{D}$ [198–200]. The topological indices derived by applying information theory equations to topological distance or adjacency matrix elements have been also classified as the second-generation indices [201]. The same applies to the cluster, path-cluster, and cycle types of connectivity indices calculated by the method of Kier and Hall [170].

The Schultz matrix topological indices [202,203] have been defined as:

$$\text{MTI} = \sum_{i=1}^{N} e_i \tag{3.52}$$

where e_i are the elements of the ith row in the matrix built from the sum of the adjacency and the distance matrices as:

$$\boldsymbol{E} = \boldsymbol{\nu}(\boldsymbol{A} + \boldsymbol{D}) \tag{3.53}$$

where $\boldsymbol{\nu}$ is the $1 \times n$ valency matrix of a tree. An alternative Schultz index is defined as:

$$\text{MTI}' = \sum_{i=1}^{N} e_i' \tag{3.54}$$

where the elements e_i' are the elements of the ith row of the following matrix:

$$\boldsymbol{E}' = \boldsymbol{\nu}\boldsymbol{D} \tag{3.55}$$

Following a similar line, the determinant of adjacency-plus-distance matrix has been introduced as a topological index (TI) to characterize the alkanes [204,205]:

$$\mathrm{TI} = \det|\boldsymbol{A} + \boldsymbol{D}| \tag{3.56}$$

Also, it has been suggested [206] to apply the permanents of the topological distance matrix or the adjacency matrix as the molecular descriptors. The permanent of any square ($n \times n$) matrix $\boldsymbol{M}$ is determined as:

$$\mathrm{per}(\boldsymbol{M}) = \sum_{\sigma} a_{1\sigma(1)} a_{2\sigma(2)} \ldots a_{n\sigma(n)} \tag{3.57}$$

where the summation extends over all possible pairs of vertices. Thus, the permanent of $\boldsymbol{M}$ is the sum of all possible diagonal products of this matrix. By definition, the permanent differs from the determinant just by the omission of the factor $(-1)^m$ in front of each term of the latter sum, where m is the number of permutations needed to obtain the necessary diagonal product. In addition, the hafnians and pfaffians of the distance and adjacency matrices have been proposed as the topological descriptors of molecules [207]. The Schultz indices have been extended to various specific molecular structures including cycloalkanes [208], unsaturated and aromatic hydrocarbons [209], heteronuclear systems [210], homonuclear and heteronuclear stereoisomeric organic compounds [211], and conformational isomers [212].

The two Mohar indices [213] have been applied in a QSPR study of octane isomers [214]. The definition of them has been based on the Laplacian spectrum of a molecular graph G and on the Laplacian x_2 eigenvalue of G, respectively. The Laplacian matrix of a simple graph G is defined as the following difference matrix:

$$\boldsymbol{L} = \boldsymbol{V} - \boldsymbol{A} \tag{3.58}$$

where $\boldsymbol{A}$ is the adjacency matrix and $\boldsymbol{V}$ is the degree or valency matrix. The latter is a diagonal matrix with the elements

$$(\boldsymbol{V})_{ii} = D(i) = \sum_{j} (\boldsymbol{A})_{ij} \tag{3.59}$$

where $D(i)$ is defined as the degree (valency) of the vertex i. The Mohar indices are thus defined as:

$$(\mathrm{TI})_1 = 2N \log\left(\frac{M}{N}\right) \sum_{i=2}^{N} \left(\frac{1}{x_i}\right) \tag{3.60}$$

$$(\mathrm{TI})_2 = \frac{4}{N x_2} \tag{3.61}$$

where N is the number of vertices (atoms) and M is the number of edges (bonds) in a molecular graph. The x_i denote the eigenvalues of the respective Laplacian. It has been shown, however, that the Mohar indices are closely related to the Wiener index, W [214]. In fact, for all isomers of a certain alkane, $(TI)_1$ is simply proportional to the Wiener index, for example, for octane isomers

$$(\mathrm{TI})_1 = -0.11598W \tag{3.62}$$

The correlation of various physical properties (boiling point, molar volume, heats of vaporization, critical pressures, critical temperatures, etc.) of alkanes with the Mohar indices has, however, revealed their inferiority in comparison with other topological indices [214].

The plain connectivity indices oversimplify the chemical bonding in the molecule because they do not account for the change in bond orders, bond lengths and angles, and do not differentiate between atom types in the molecule. Therefore, the simple graph theory has been extended for molecules to account for these changing features by introduction of the appropriately weighted graphs. Whereas the original adjacency and distance matrices of molecular graphs contain the integer numbers, the presence of multiple bonds in the molecule can be accounted for by the introduction of rational numbers into these matrices. These numbers can be defined as the inverse of the respective chemical bond orders. For instance, the topological distance of a double bond has been defined as $\frac{1}{2}$, by a triple bond as $\frac{1}{3}$, and by an aromatic bond as $\frac{2}{3}$, respectively [177]. The corresponding graph invariants S_i are thus no longer just integer numbers but also rational numbers.

The presence of heteroatoms (noncarbon and nonhydrogen atoms) in the molecule can be accounted for by multiplying the corresponding local vertex invariant or all entries in the respective row or column of the distance matrix with a characteristic factor for this atom. The ratio $6/Z_i$ has been proposed for such a factor, where Z_i is the atomic number of the atom, having a natural value 1 for the carbon atom [215]. It has been advocated that other characteristics such as the covalent radii of atoms or electronegativities of elements are better suited as weighting factors in the QSPR description of physico-chemical properties of compounds [200,216]. Coefficients reflecting the variation of these atomic characteristics have been devised and applied in the calculation of the average distance sum connectivity descriptor, J [200]. It has also been proposed to use the so-called chemical distances as multipliers of distance sums. The chemical distances have been defined [217] as $b^{-1/4}$, where b is the bond distance relative to the length of the C—C single bond.

The structural differentiation of vertices can also be based on the topological properties of the molecular graph. A topological matrix, $\boldsymbol{R}$, has been constructed for this purpose from the reciprocal distances [218,219] with the elements defined as:

$$(\boldsymbol{R})_{ij} = (\boldsymbol{R})_{ji} = (\boldsymbol{D})_{ij}^{-1} \tag{3.63}$$

where $\boldsymbol{D}$ is the topological distance matrix. The sum of elements over the rows or columns of matrix $\boldsymbol{R}$ give local vertex invariants that are reciprocal distance sums, $r_i = \Sigma_j (\boldsymbol{R})_{ij}$. The summation of all r_i in the upper triangle of the reciprocal distance matrix gives a global topological index called the Harare number. These numbers are somewhat less degenerate than the Wiener indices but otherwise very similar to the latter.

Klein and Randić [220] introduced another type of topological distance based on the Kirchoff laws for electrical circuits. By identifying each edge of the molecular graph with an Ohm resistor, the resistance between the vertices i and j is called the resistance distance. The resistance distances for acyclic graphs are identical with the topological distances. However, for cyclic graphs, the resistance distances emanating from one vertex may be summed up. The sum of all such local vertex invariants, called the Kirchoff number Kf, has been proposed as a much less degenerate topological index for cyclic structures. Additional third-generation topological indices have been proposed using the ratio of distance sums and vertex degrees as the less degenerate local vertex invariants [216]:

$$t_i = \frac{s_i}{v_i} \tag{3.64}$$

A modified topological index J_t can be defined as:

$$J_t = \frac{q}{\mu + 1} \sum_{\text{edges } ij} (t_i t_j)^{-1/2} \tag{3.65}$$

where q is the number of edges and μ is the cyclomatic number of the molecular graph.

When the distance vectors in the molecular graph are different but the distance sum is the same, the degeneracy of topological indices may be diminished by applying locally the information theory according to the Shannon formula. The respective information-theoretic topological indices have been proposed as based on four types of the local vertex invariants [177]. By multiplying each of the distances for a given vertex, i, by a factor g_j characterizing the other vertex (atom), j, new distance sums have been obtained as:

$$S_i = \sum_j j g_j \tag{3.66}$$

The local information and the mean local information on the magnitude of distances were defined as follows, respectively:

$$v_i = S_i \log_2 S_i - u_i \tag{3.67}$$

$$u_i = -\sum_j \frac{jg_j}{S_i} \log_2 \frac{j}{S_i} \tag{3.68}$$

The extended local information and the mean extended local information on the magnitude of distances have the following appearance, respectively:

$$x_i = S_i \log_2 S_i - y_i \tag{3.69}$$

$$y_i = \sum_j jg_j \log_2 j \tag{3.70}$$

All distance vectors used in the above formulas may include data about the presence of multiple bonds and/or heteroatoms. Thus, low degeneracy topological indices can be obtained by applying the definition of the *J* index (3.51) and the four above-listed definitions for the connectivity. The respective topological indices, denoted as *U, V, X,* and *Y*, respectively, have been analyzed from the point of view of suitability for the differentiation of chemical structures. It was found that the intramolecular orderings by *V* and *X* are practically identical, but *Y* and especially *U* lead to different orderings of alkanes.

The exponential distance sum connectivities have been introduced as the topological indices obtained by the summation of real-number local vertex invariants, $\log c_i$. These invariants were defined as [221]:

$$\log c_i = \left(\sum_{k=1}^{k_{i,m}} k \sum_{k=1}^{k_{i,m}} G_{i,k} \right) \log G_i \tag{3.71}$$

where

$$\log G_i = \left(\prod_{j=1}^{v_i} g_j \right) \log g_j \tag{3.72}$$

and

$$g_i = v_i^{-1/2} \left(\sum_{j=1}^{v_i} v_j \right)^{-1} \tag{3.73}$$

Alternatively,

$$g_i = g_i' = v_i^{-1/2} \left(\sum_{j=1}^{v_i} v_j \prod_{j=1}^{v_i} v_j \right)^{-1} \tag{3.74}$$

In the last formulas (3.71–3.74), v_i is the degree of vertex *i*, and the sum in

Eq. (3.71) extends to all immediate neighbors of vertex i. The sum in Eq. (3.73) involves the topological distances k from vertex i to all other vertices, including the maximum distance $k_{i,m}$ and $G_{i,k}$ denotes the G_i values for each vertex situated at distance k from vertex i. The resulting vertex invariants have a very low degeneracy and by summation yield topological indices that have a good correlating ability with respect of physical properties of congeneric compounds. Notably, the fragment descriptors have been also proposed as calculated from the above-defined local vertex invariants for the dimers built up from two identical fragments (substituents) [216].

Another approach to the definition of the third-generation topological indices is based on the combination of the distance matrix $\boldsymbol{D}$ (or the adjacency matrix $\boldsymbol{A}$) with some prespecified $1 \times n$ two-column matrices [222]. One of the columns in these matrices contains the diagonal elements of the original matrix, whereas the other includes some atomic data. The local vertex invariants are obtained as the solutions of the respectively formed linear equations. Different atomic characteristics have been tested in the column vectors including the atomic number, electronegativity, covalent radius, vertex degree D_i, distance sum S_i, and other atomic properties. Another definition of the low degeneracy topological indices is based on the eigenvalues of distance or adjacency matrices, modified by accounting for the chemical specificity of the molecular graph vertices (presence of heteroatoms, multiple bonding). It has been shown that the lowest eigenvalue of the distance matrix may serve as a useful topological index for alkanes [223]. By analogy, the lowest eigenvalue of the modified distance matrix would differentiate between the members of functional derivatives of organic compounds.

The valence connectivity indices ${}^m\chi^v$ were suggested by Kier and Hall [224–226] to account for the presence of heteroatoms and the hybridization of atoms in the molecule. The definition of those descriptors proceeds from the atomic valence connectivity for the ith atom in the molecular skeleton, calculated by the following formula:

$$\delta_i^v = \frac{Z_i^v - H_i}{Z_i - Z_i^v - 1} \tag{3.75}$$

where Z_i is the total number of electrons in the ith atom, Z_i^v is the number of valence electrons, and H_i is the number of hydrogen atoms directly attached to the ith nonhydrogen atom. Different orders of valence connectivity indices relate to atoms (zeroth order, $m = 0$), one bond paths (first order, $m = 1$), two bond fragments (second order, $m = 2$), three contiguous bond fragments (third order, $m = 3$), and higher order paths, clusters, path-clusters, or chain fragments, respectively. The definition of descriptor ${}^m\chi^v$ is given as:

$${}^m\chi^v = \sum_{i=1}^{N_s} \prod_{k=1}^{m+1} \left(\frac{1}{\delta_k^v}\right)^{1/2} \tag{3.76}$$

where the summation is carried out over all subgraphs of a given type of order

m in the entire molecule. It is possible to remove that part of each ${}^m\chi^v$ index that encodes only the σ electrons and leave only that part that encodes the π and the lone-pair electrons. The resulting $\Delta^m\chi^v$ indices are obtained by subtraction of the simple connectivity index ${}^m\chi$ (3.49) from the corresponding valence index of the same order [227,228]. This index encodes the intra- and intermolecular interactions involving only the more active π and lone-pair electrons.

The Kier shape index (of different order) depends on the number of skeletal atoms, the molecular branching, and a special parameter α_i that is calculated as the ratio of the atomic radius (r_i) and the radius of the carbon atom in the sp^3 hybridization state (r_C) [229–234]:

$$^1\kappa = (N_{SA} + \alpha)(N_{SA} + \alpha - 1)^2(^1P + \alpha)^2 \quad (3.77)$$

$$^2\kappa = (N_{SA} + \alpha - 1)(N_{SA} + \alpha - 2)^2(^2P + \alpha)^2 \quad (3.78)$$

$$^3\kappa = (N_{SA} + \alpha - 1)(N_{SA} + \alpha - 3)^2(^3P + \alpha)^2 \quad \text{if } N_{SA} \text{ is odd} \quad (3.79)$$

$$^3\kappa = (N_{SA} + \alpha - 3)(N_{SA} + \alpha - 2)^2(^3P + \alpha)^2 \quad \text{if } N_{SA} \text{ is even} \quad (3.80)$$

where ${}^nP$ is the number of paths of length n in the molecular skeleton, and α is the sum of the α_i parameters for all skeletal atoms minus 1. From the indices of molecular shape ${}^1\kappa$ and ${}^2\kappa$, the Kier flexibility index is calculated as [233]:

$$\Phi = \frac{(^1\kappa^2\kappa)}{N_{SA}} \quad (3.81)$$

It has been pointed out that the first-order shape index ${}^1\kappa$ encodes the number of atoms in the molecule and the relative degree of cyclicity. The second-order index ${}^2\kappa$ describes the branching and spatial density of the molecule.

A local flexibility index LS and a global flexibility index GS have been developed to carry out the search of molecular properties in CAS registry files for closely related molecular structures [235]. The LS index between atoms i and j in a molecule was defined as:

$$\text{LS}_{ij} = \text{SPN} - (\text{NRB} + 0.75\text{BX} + 0.50\text{BY})/2 \quad (3.82)$$

with the following notations: SPN is the number of centers in the shortest path between atoms i and j, NRB is the number of nonrotable bonds, BX is the number of branching atoms with four attached nonhydrogen atoms, and BY is the number of branching atoms with three attached atoms. The global flexibility index GS was calculated as:

$$\mathrm{GS} = \frac{2}{N(N-1)} \sum_{i<j} \mathrm{LS}_{ij} \tag{3.83}$$

where the summation is carried out for all pairs of atoms in the molecule. An algorithm has been suggested [236] to calculate the molecular flexibility within the theory of fuzzy logic [237].

A class of real-number topological indices has been developed proceeding from the regressive distance sums. Those sums are defined as the real-number local vertex invariants as:

$$r_i = \sum_{k=0}^{\mathrm{diam}} 10^{-nk} r_{ik} \tag{3.84}$$

where *diam* is the diameter of the molecular graph and n denotes the number of digits for the maximum r_{ik} value in the graph. The elements of matrix $\boldsymbol{R}$ are

$$r_{ik} = \sum_{j,d_{ij}=k} D_j \tag{3.85}$$

built up from the columns of distance sums. The first column, corresponding to $k = 0$, involves the distance sums ($D_{i0} = \Sigma_j d_{ij}$, with d_{ij} the distance matrix elements) for each vertex. The ith column adds up all distance sums D_j of vertices j belonging to a shell at a distance $d_{ij} = k$, around the ith vertex. Four additional local vertex invariants have been defined [238] to simplify the handling of the $\boldsymbol{R}$ matrix. These invariants are given by the following equations:

$$r_i^* = \left[\sum_{k=0}^{\mathrm{diam}} 10^{-nk} r_{ik} \right]^{-1} = r_i^{-1} \tag{3.86}$$

$$rc_i = \left[\sum_{k=0}^{\mathrm{diam}} (r_{ik})^{k/d_{\mathrm{spec}}} \right]^{-1} \tag{3.87}$$

$$rx_i = \left[\frac{r_i}{dg_i} - m_i \right]^{-1} w_i \tag{3.88}$$

$$rj_i = \sum_{(i,j)} [r_i/(w_i c_i) r_j/(w_j c_j)]^{-1/2} \tag{3.89}$$

In the last equations, d_{spec} denotes a specified distance value, usually larger than the longest path in the graph; w_i is the weighting factor to account for the heteroatoms; m_i is a local parameter for multiple bonds; and c_i and f_i refer to the connectivity around the vertex i;

TABLE 3.3 Topological and Information-Theoretic Molecular Descriptors

Notation	Name	Reference
W	Wiener index	*a*
R	Hyper-Wiener index	*b*
W_c	Wiener index for cyclic structures	*c*
M_1	Zagreb group index	*d*
N_2	Gordon–Scantlebury index	*e*
Z	Hosoya index	*f*
Q	Balaban quadratic index	*g*
B, C	Balaban centric indices	*g*
σ	Merrifield–Simmon index	*h*
$^m\chi$	Randić molecular connectivity index of *m*th order	*i*
$D^{(2)}$	Mean-square distance between all molecular graph vertices	*j*
J, J_t	Balaban average distance sum connectivity index	*j*
MTI, MTI′	Schultz indices	*k*
TI	Determinant of the adjacency-plus-distance matrix	*l*
per($\boldsymbol{M}$)	Permanent of the distance or adjacency matrix	*m*
$(\mathrm{TI})_1$, $(\mathrm{TI})_2$	Mohar indices	*n*
Kf	Kirchoff number	*o*
$\log c_i$, $\log G_i$	Exponential distance sum connectivities	*p*
$^m\chi^v$	Kier–Hall valence connectivity indices	*q*
$^m\kappa$	Kier shape indices	*r*
Φ	Kier flexibility index	*s*
LS, GS	Local and global flexibility indices	*t*
R, RC, RX, RJ	Regressive distance sum indices	*u*
U, V, X, Y	Local information indices	*v*
^{k}IC	Mean information content index of *k*th order	*w*
^{k}SIC	Structural information content index of *k*th order	*x*
^{k}CIC	Complementary information content index of *k*th order	*x*
^{k}BIC	Bonding information content index of *k*th order	*x*
I_D^W	Information-theoretic index on graph distance	*y*
S_i, S_i^b, δ_i^v	Electrotopological state indices	*z*
T_1^E, T_2^E	Topological electronic indices	*aa*
$\delta(q_i)$	Electron charge density connectivity	*bb*
$\Omega(q)$	3D valence connectivity index	*bb*
G_k, J_k	Topological charge indices	*cc*
3-W	3D Wiener index	*dd*
3-J	3D Balaban index	*dd*
3-TI	3D topographical index	*dd*
ε	Edge adjacency index	*ee*

[a] H. Wiener, *J. Am. Chem. Soc.* **69,** 17 (1947)
[b] M. Randić, *Chem. Phys. Lett.* **211,** 478 (1993).
[c] I. Lukovits and W. Linert, *J. Chem. Inf. Comput. Sci.* **34,** 899 (1994).
[d] I. Gutman, B. Ruscič, N. Trinajstić, and C.F. Wilcox, Jr., *J. Chem. Phys.* **62,** 3399 (1975).
[e] M. Gordon and G.R. Scantlebury, *Trans. Faraday Soc.* **60,** 605 (1964).
[f] H. Hosoya, *Bull. Chem. Soc. Jpn.* **44,** 2332 (1971); A. Hermann and P. Zinn, *J. Chem. Inf. Comput. Sci.* **35,** 551 (1995).

$$m_i = f_i[r_{i0}/10 + r_{i1}/100] \tag{3.90}$$

$$f_i = \sum_j (c_{ij} - 1) \tag{3.91}$$

$$c_i = 1 + f_i \tag{3.92}$$

where c_{ij} is the conventional bond order, that is, 1, 2, 3, and 1.5 for single, double, triple, and aromatic bonds, respectively. Summation of above-defined indices over all vertices in molecular graphs provides the global topological indices, denoted by the corresponding capital letters (*R, RC, RX, RJ*) [239–241].

Pogliani has successfully introduced so-called supramolecular connectivity indices [242–244] and molecular connectivity terms [245,246] to model different properties of amino acids, purines and pyrimidines, alkanes, and inorganic salts. To account for the van der Waals effects and hydrogen bonding within the topological approach, special molecular connectivity terms were introduced that make use of empirical dielectric constant of the medium. In addition, three ad hoc parameters, indirectly connected with the dielectric properties of the medium, were also used [247]. The combination of the topology

---

[g] A.T. Balaban, *Theor. Chim. Acta* **53,** 355 (1979).
[h] R.E. Merrifield and H.E. Simmons, *Topological Methods in Chemistry*. Wiley, New York, 1989.
[i] M. Randić, *J. Am. Chem. Soc.* **97,** 6609 (1975); L.B. Kier and L.H. Hall, *Molecular Connectivity in Structure-Activity Analysis*. Wiley, New York, 1986.
[j] A.T. Balaban, *Pure Appl. Chem.* **55,** 199 (1983); *J. Chem. Inf. Comput. Sci.* **34,** 398 (1994).
[k] H.P. Schultz, *J. Chem. Inf. Comput. Sci.* **29,** 227 (1989).
[l] H.P. Schultz, E.B. Schultz, and T.P. Schultz, *J. Chem. Inf. Comput. Sci.* **30,** 27 (1990).
[m] H.P. Schultz, E.B. Schultz, and T.P. Schultz, *J. Chem. Inf. Comput. Sci.* **32,** 69 (1992).
[n] B. Mohar, in *MATH/CHEM/COMP 1988* (A. Graovac, ed.), pp. 1–8. Elsevier, Amsterdam, 1989.
[o] D.J. Klein and M. Randić, *J. Math. Chem.* **12,** 85 (1993).
[p] A.T. Balaban and C. Catana, *J. Comput. Chem.* **14,** 155 (1993).
[q] L.B. Kier and L.H. Hall, *Eur. J. Med. Chem.* **12,** 307 (1977).
[r] L.B. Kier, *Quant. Struct.-Act. Relat.* **4,** 109 (1985).
[s] L.B. Kier, in *Computational Chemical Graph Theory* (D.H. Rouvray, ed.). Nova Science Publishers, New York, 1990, pp. 329ff.
[t] W. Fisanick, K. Criss, and A. Rusink, III, *Tetrahedron Comput. Methodol.* **3,** 635 (1990).
[u] M.V. Diudea, *J. Chem. Inf. Comput. Sci.* **34,** 1064 (1994).
[v] A.T. Balaban, *J. Chem. Inf. Comput. Sci.* **32,** 23 (1992).
[w] L.B. Kier, *J. Pharm. Sci.* **69,** 807 (1980).
[x] S.C. Basak, D.K. Harriss, and V.R. Magnuson, *J. Pharm. Sci.* **73,** 429 (1984).
[y] D. Bonchev and N. Trinajstić, *J. Chem. Phys.* **67,** 4517 (1977).
[z] L.B. Kier and L.H. Hall, *Pharm. Res.* **7,** 801 (1990); L.H. Hall and G.E. Kellogg, *Molconn-Z: Software Package for Molecular Topology Analysis, User's Guide, Version, 3.15*. eduSoft, LC, Ashland, VA, 1997.
[aa] K. Osmialowski, J. Halkiewicz, and R. Kaliszan, *J. Chromatogr.* **63,** 361 (1986).
[bb] E. Estrada, *J. Chem. Inf. Comput. Sci.* **35,** 708 (1995).
[cc] J. Gálvez, R. Garcia, M.T. Salabert, and R. Soler, *J. Chem. Inf. Comput. Sci.* **34,** 520 (1994).
[dd] Z. Mihalić, S. Nikolić, and N. Trinajstić, *J. Chem. Inf. Comput. Sci.* **32,** 28 (1992).
[ee] E. Estrada, *J. Chem. Inf. Comput. Sci.* **35,** 31 (1995).

with empirical properties is thus another possibility to obtain highly nondegenerate molecular descriptors.

A list of common topological descriptors is given in Table 3.3. The numerical values of some topological indices for a set of structurally variable compounds are given in Table 3.4.

Another group of molecular descriptors based on the constitution and topology of molecules are the molecular complexity indices. These descriptors are based on information theory. For instance, the mean information content index and its derivatives of different order are defined on the basis of the Shannon information theory as follows [248]:

$$^{k}\mathrm{IC} = -\sum_{i=1}^{k} \frac{n_i}{n} \log_2 \frac{n_i}{n} \tag{3.93}$$

where n_i is a number of atoms in the ith class and n is a total number of atoms in the molecule. The division of atoms into different classes depends on the size of coordination sphere taken into account near a given atom. This leads to the indices of different order k. The overall information content (IC) is equal to mean information content multiplied by the total number of atoms. Other information content indices (SIC, structural IC; CIC, complementary IC; BIC, bonding IC) are defined as [249]:

$$^{k}\mathrm{SIC} = {}^{k}\mathrm{IC}/\log_2 n \tag{3.94}$$

$$^{k}\mathrm{CIC} = \log_2 n - {}^{k}\mathrm{IC} \tag{3.95}$$

$$^{k}\mathrm{BIC} = {}^{k}\mathrm{IC}/\log_2 q \tag{3.96}$$

where q is a number of edges in the structural graph of the molecule. The information-theoretic index on graph distance, I_D^W, is calculated as [250]:

$$I_D^W = W \log_2 W - \sum_h g_h h \log_2 h \tag{3.97}$$

where g_h is the number of nonordered pairs of vertices whose distance is h. The mean information index, $\bar{I}_D^W$, is calculated as the ratio of I_D^W and the Wiener index, W [251]. The values of some information-theoretic descriptors for a set of structurally variable compounds are presented in Table 3.5.

Kier and Hall [252–256] have invented electrotopological state indices that couple the information about the topological environment of atoms and the atom–atom electronic interactions in the molecule. The electrotopological state index (*E*-state index) for a given atom i is calculated as:

$$S_i = I_i + \Delta I_i \tag{3.98}$$

where I_i is the intrinsic state of an atom, defined as the ratio of the Kier–Hall electronegativity [252] and the number of skeletal bonds for that atom:

TABLE 3.4 Numerical Data on Some Representative Topological Descriptors for Set of Structurally Variable Compounds

Compound	W	$^1\chi$	$^2\chi$	$^3\chi$	$^1\chi^v$	$^1\kappa$	Φ
Water	0	0.000	0.000	0.000	0.000	—	0.000
Ammonia	0	0.000	0.000	0.000	0.000	—	0.000
Tetrafluoromethane	16	2.000	3.000	0.000	0.756	4.720	0.794
Tetrachloromethane	16	2.000	3.000	0.000	2.268	6.160	2.146
Trichloromethane	9	1.732	1.732	0.000	1.964	4.870	2.591
Tribromomethane	9	1.732	1.732	0.000	3.402	5.440	3.625
Formaldehyde	1	1.000	0.000	0.000	0.289	1.670	0.000
Dibromomethane	4	1.414	0.707	0.000	2.778	3.960	3.907
Dichloromethane	4	1.414	0.707	0.000	1.604	3.580	3.079
Chloromethane	1	1.000	0.000	0.000	1.134	2.290	0.000
Bromomethane	1	1.000	0.000	0.000	1.964	2.480	0.000
Iodomethane	1	1.000	0.000	0.000	2.536	2.730	0.000
Nitromethane	9	1.732	1.732	0.000	0.812	3.400	0.694
Methane	0	0.000	0.000	0.000	0.000	—	0.000
Methanol	1	1.000	0.000	0.000	0.447	1.960	0.000
Methylamine	1	1.000	0.000	0.000	0.577	1.960	0.000
Tetrachloroethene	29	2.643	2.488	1.333	2.518	6.900	3.422
Pentachloroethane	42	2.943	3.521	1.732	3.299	8.450	3.426
Trichloroethene	18	2.270	1.802	0.817	2.077	5.610	3.172
1,1,1,2-Tetrachloroethane	28	2.561	2.914	1.061	2.856	7.160	2.939
1,1-Dichloroethene	10	1.914	1.000	0.500	1.643	4.320	3.586
trans-1,2-Dichloroethene	10	1.914	1.000	0.500	1.643	4.320	3.586
cis-1,2-Dichloroethene	10	1.914	1.000	0.500	1.643	4.320	3.586
1,1,2,2-Tetrachloroethane	29	2.643	2.488	1.333	2.952	7.160	3.818
2,2,2-Trifluoroethanol	28	2.561	2.914	1.061	1.237	5.750	1.405
Acetonitrile	4	1.414	0.707	0.000	0.724	2.490	1.237
1,1,1-Trichloroethane	16	2.000	3.000	0.000	2.201	5.870	1.814
1,1,2-Trichloroethane	18	2.270	1.802	0.817	2.519	5.870	3.610
Ethene	1	1.000	0.000	0.000	0.500	1.740	0.000
1,1-Dichloroethane	9	1.732	1.732	0.000	1.887	4.580	2.129
Acetaldehyde	4	1.414	0.707	0.000	0.813	2.670	1.486
Acetic acid	9	1.732	1.732	0.000	0.928	3.630	0.917
1,2-Dibromoethane	10	1.914	1.000	0.500	3.278	4.960	4.910
1,2-Dichloroethane	10	1.914	1.000	0.500	2.104	4.580	4.099
Methyl formate	10	1.914	1.000	0.500	0.880	3.630	2.387
Chloroethane	4	1.414	0.707	0.000	1.509	3.290	2.511
Bromoethane	4	1.414	0.707	0.000	2.096	3.480	2.877
Iodoethane	4	1.414	0.707	0.000	2.500	3.730	3.394
Nitroethane	18	2.270	1.802	0.817	1.389	4.400	1.491
Ethane	1	1.000	0.000	0.000	1.000	2.000	0.000
Ethanol	4	1.414	0.707	0.000	1.023	2.960	1.934
Ethanethiol	4	1.414	0.707	0.000	1.656	3.350	2.624
Ethylamine	4	1.414	0.707	0.000	1.115	2.960	1.934
Dimethylamine	4	1.414	0.707	0.000	1.000	2.960	1.934
Propyne	4	1.414	0.707	0.000	0.789	2.560	1.331

TABLE 3.4 (*Continued*)

Compound	W	$^1\chi$	$^2\chi$	$^3\chi$	$^1\chi^v$	$^1\kappa$	Φ
Methoxyflurane	62	3.504	3.497	2.474	2.588	8.400	2.943
1-Chloro-2-propene	10	1.914	1.000	0.500	1.618	4.030	3.053
Propanonitrile	10	1.914	1.000	0.500	1.284	3.490	2.173
Ethyl formate	20	2.414	1.354	0.707	1.467	4.630	3.361
Methyl acetate	18	2.270	1.802	0.817	1.317	4.630	1.765
1,3-Dichloropropane	20	2.414	1.354	0.707	2.604	5.580	5.111
Cyclopropane	3	1.500	1.061	0.000	1.500	1.333	0.099
Propene	4	1.414	0.707	0.000	0.986	2.740	1.589
Propanone	9	1.732	1.732	0.000	1.204	3.670	0.958
Propanoic acid	18	2.270	1.802	0.817	1.488	4.630	1.765
2-Propenol	10	1.914	1.000	0.500	1.133	3.700	2.498
1,2-Dichloropropane	18	2.270	1.802	0.817	2.442	5.580	3.123
Propionaldehyde	10	1.914	1.000	0.500	1.351	3.670	2.450
1-Bromopropane	10	1.914	1.000	0.500	2.596	4.480	3.898
2-Bromopropane	9	1.732	1.732	0.000	2.289	4.480	1.979
1-Nitropropane	32	2.770	2.183	0.866	1.889	5.400	2.365
1-Chloropropane	10	1.914	1.000	0.500	2.009	4.290	3.529
2-Chloropropane	9	1.732	1.732	0.000	1.809	4.290	1.710
1-Iodopropane	10	1.914	1.000	0.500	3.000	4.730	4.411
N,N-Dimethylformamide	18	2.270	1.802	0.817	1.388	4.470	1.572
2-Nitropropane	29	2.643	2.488	1.333	1.778	5.400	1.570
Propane	4	1.414	0.707	0.000	1.414	3.000	2.000
Methyl ethyl ether	10	1.914	1.000	0.500	1.404	3.960	2.930
1-Propanol	10	1.914	1.000	0.500	1.523	3.960	2.930
2-Propanol	9	1.732	1.732	0.000	1.413	3.960	1.285
2-Methoxyethanol	20	2.414	1.354	0.707	1.513	4.920	3.857
n-Propanethiol	10	1.914	1.000	0.500	2.156	4.350	3.643
Trimethylamine	9	1.732	1.732	0.000	1.342	3.960	1.285
Propylamine	10	1.914	1.000	0.500	1.615	3.960	2.930
Thiophene	15	2.500	1.768	1.250	2.414	2.913	0.711
1-Butyne	10	1.914	1.000	0.500	1.349	3.560	2.278
(*E*)-2-Butenal	20	2.414	1.354	0.707	1.480	4.410	3.008
1-Cyanopropane	20	2.414	1.354	0.707	1.784	4.490	3.134
Isobutyraldehyde	18	2.270	1.802	0.817	1.724	4.670	1.814
Tetrahydrofuran	15	2.500	1.768	1.250	2.077	3.162	0.892
Ethyl acetate	32	2.770	2.183	0.866	1.904	5.630	2.671
Methyl propanoate	31	2.808	1.922	1.394	1.877	5.630	2.671
1,4-Dichlorobutane	35	2.914	1.707	0.957	3.104	6.580	6.119
1-Butene	10	1.914	1.000	0.500	1.524	3.740	2.562
Butanone	18	2.270	1.802	0.817	1.765	4.670	1.814
Isopropyl formate	32	2.770	2.183	0.866	1.862	5.630	2.671
Butanoic-acid	32	2.770	2.183	0.866	1.988	5.630	2.671
Propyl formate	35	2.914	1.707	0.957	1.967	5.630	4.345
Butyraldehyde	20	2.414	1.354	0.707	1.851	4.670	3.428
1,4-Dioxane	27	3.000	2.121	1.500	2.155	4.089	1.470
1-Bromobutane	20	2.414	1.354	0.707	3.096	5.480	4.910

TABLE 3.4 (*Continued*)

Compound	W	$^1\chi$	$^2\chi$	$^3\chi$	$^1\chi^v$	$^1\kappa$	Φ
1-Chlorobutane	20	2.414	1.354	0.707	2.509	5.290	4.539
1-Nitrobutane	52	3.270	2.536	1.135	2.389	6.400	3.278
2-Bromo-2-methylpropane	16	2.000	3.000	0.000	2.482	5.480	1.416
1-Bromo-2-methylpropane	18	2.270	1.802	0.817	2.952	5.480	2.963
2-Chloro-2-methylpropane	16	2.000	3.000	0.000	2.067	5.290	1.242
2-Chlorobutane	18	2.270	1.802	0.817	2.347	5.290	2.669
1-Iodobutane	20	2.414	1.354	0.707	3.500	5.730	5.421
Morpholine	27	3.000	2.121	1.500	2.285	4.089	1.470
n-Butane	10	1.914	1.000	0.500	1.914	4.000	3.000
Diethyl ether	20	2.414	1.354	0.707	1.992	4.960	3.928
2-Methylpropane	9	1.732	1.732	0.000	1.732	4.000	1.333
2-Methyl-1-propanol	18	2.270	1.802	0.817	1.879	4.960	2.195
2-Butanol	18	2.270	1.802	0.817	1.951	4.960	2.195
2-Methyl-2-propanol	16	2.000	3.000	0.000	1.724	4.960	0.969
Diethyl sulfide	20	2.414	1.354	0.707	3.146	5.350	4.655
Diethyl disulfide	35	2.914	1.707	0.957	4.646	6.700	6.365
2-Ethoxyethanol	35	2.914	1.707	0.957	2.101	5.920	4.854
1-Butanol	20	2.414	1.354	0.707	2.023	4.960	3.928
n-Butanethiol	20	2.414	1.354	0.707	2.656	5.350	4.655
Diethylamine	20	2.414	1.354	0.707	2.121	4.960	3.928
Butylamine	20	2.414	1.354	0.707	2.115	4.960	3.928
2-Chloropyridine	42	3.394	2.743	1.894	2.337	4.595	1.272
3-Chloropyridine	42	3.394	2.743	1.894	2.327	4.595	1.272
Pyridine	27	3.000	2.121	1.500	1.850	3.344	0.865
2-Methylthiophene	26	2.894	2.390	1.644	2.775	3.875	0.926
2-Methylpyrazine	42	3.394	2.743	1.894	2.120	4.245	1.023
Cyclopentene	15	2.500	1.768	1.250	2.150	2.951	0.738
2-Methyl-1,3-butadiene	18	2.270	1.802	0.817	1.551	4.480	1.584
Cyclopentanone	26	2.894	2.390	1.644	2.411	3.846	0.906
1,4-Pentadiene	20	2.414	1.354	0.707	1.633	4.480	3.118
1-Pentyne	20	2.414	1.354	0.707	1.849	4.560	3.247
1-Cyanobutane	35	2.914	1.707	0.957	2.284	5.490	4.108
Cyclopentane	15	2.500	1.768	1.250	2.500	3.200	0.922
1-Pentene	20	2.414	1.354	0.707	2.024	4.740	3.546
Ethyl propanoate	50	3.308	2.302	1.478	2.465	6.630	3.606
Isobutyl formate	52	3.270	2.536	1.135	2.323	6.630	3.606
2-Methyl-2-butene	18	2.270	1.802	0.817	1.914	4.740	1.903
3-Methyl-1-butene	18	2.270	1.802	0.817	1.896	4.740	1.903
Isopropyl acetate	48	3.126	3.023	0.943	2.299	6.630	2.601
Pentanal	35	2.914	1.707	0.957	2.351	5.670	4.413
Methyl butanoate	50	3.308	2.302	1.478	2.377	6.630	3.606
Pentanoic acid	52	3.270	2.536	1.135	2.488	6.630	3.606
3-Methylbutanoic acid	48	3.126	3.023	0.943	2.344	6.630	2.601
Cyclopentanol	26	2.894	2.390	1.644	2.575	4.128	1.105
3-Methyl-2-butanone	29	2.643	2.488	1.333	2.148	5.670	1.849
3-Pentanone	31	2.808	1.922	1.394	2.325	5.670	2.726

TABLE 3.4 (*Continued*)

Compound	W	${}^1\chi$	${}^2\chi$	${}^3\chi$	${}^1\chi^v$	${}^1\kappa$	Φ
Propyl acetate	52	3.270	2.536	1.135	2.404	6.630	3.606
2-Pentanone	32	2.770	2.183	0.866	2.265	5.670	2.726
2-Methyltetrahydrofuran	26	2.894	2.390	1.644	2.510	4.128	1.105
Tetrahydropyran	27	3.000	2.121	1.500	2.577	4.128	1.506
1-Chloropentane	35	2.914	1.707	0.957	3.009	6.290	5.546
1-Nitropentane	79	3.770	2.890	1.385	2.889	7.400	4.215
1-Bromopentane	35	2.914	1.707	0.957	3.596	6.480	5.918
1-Iodopentane	35	2.914	1.707	0.957	4.000	6.730	6.427
N-Methylmorpholine	42	3.394	2.743	1.894	2.657	5.065	1.653
n-Pentane	20	2.414	1.354	0.707	2.414	5.000	4.000
2-Methyl-1-butanol	31	2.808	1.922	1.394	2.417	5.960	3.141
2-Methyl-2-butanol	28	2.561	2.914	1.061	2.284	5.960	1.595
3-Methyl-1-butanol	32	2.770	2.183	0.866	2.379	5.960	3.141
1-Pentanol	35	2.914	1.707	0.957	2.523	5.960	4.927
2-Pentanol	32	2.770	2.183	0.866	2.451	5.960	3.141
3-Pentanol	31	2.808	1.922	1.394	2.489	5.960	3.141
Methyl *tert*-butyl ether	28	2.561	2.914	1.061	2.112	5.960	1.595
2,2-Dimethylpropane	16	2.000	3.000	0.000	2.000	5.000	1.000
2-Propoxyethanol	56	3.414	2.061	1.207	2.601	6.920	5.852
2-Methylbutane	18	2.270	1.802	0.817	2.270	5.000	2.250
n-Pentylamine	35	2.914	1.707	0.957	2.615	5.960	4.927
1,2,3,4-Tetrachlorobenzene	109	4.626	4.250	3.702	3.929	8.476	2.700
1,2,3,5-Tetrachlorobenzene	110	4.609	4.390	3.343	3.923	8.476	2.700
1,2,4,5-Tetrachlorobenzene	111	4.609	4.381	3.430	3.923	8.476	2.700
1,2,3-Trichlorobenzene	82	4.215	3.745	3.114	3.445	7.200	2.226
1,3,5-Trichlorobenzene	84	4.182	4.023	2.414	3.433	7.200	2.226
1,2,4-Trichlorobenzene	84	4.198	3.873	2.860	3.439	7.200	2.226
1,2-Dichlorobenzene	60	3.805	3.239	2.540	2.961	5.928	1.765
1,3-Dichlorobenzene	61	3.788	3.377	2.199	2.955	5.928	1.765
1,4-Dichlorobenzene	62	3.788	3.365	2.305	2.955	5.928	1.765
3-Cyanopyridine	64	3.932	2.912	2.302	2.234	4.791	1.246
4-Cyanopyridine	64	3.932	2.912	2.302	2.234	4.791	1.246
Bromobenzene	42	3.394	2.743	1.894	2.893	4.849	1.471
Chlorobenzene	42	3.394	2.743	1.894	2.478	4.664	1.324
4-Fluorophenol	62	3.788	3.365	2.305	2.234	5.251	1.262
2-Chlorophenol	60	3.805	3.239	2.540	2.618	5.604	1.512
3-Chlorophenol	61	3.788	3.377	2.199	2.612	5.604	1.512
4-Chlorophenol	62	3.788	3.365	2.305	2.612	5.604	1.512
4-Bromophenol	62	3.788	3.365	2.305	3.027	5.791	1.655
2-Iodophenol	60	3.805	3.239	2.540	3.319	6.036	1.854
2-Nitrophenol	114	4.715	4.170	3.034	2.640	6.539	1.533
3-Nitrophenol	117	4.698	4.276	2.920	2.634	6.697	1.639
4-Nitrophenol	120	4.698	4.264	3.003	2.634	6.539	1.533
3-Formalpyridine	64	3.932	2.912	2.302	2.285	4.967	1.373
4-Formalpyridine	64	3.932	2.912	2.302	2.285	4.967	1.373
2-Fluorophenol	60	3.805	3.239	2.540	2.240	5.251	1.262

TABLE 3.4 (*Continued*)

Compound	W	$^1\chi$	$^2\chi$	$^3\chi$	$^1\chi^v$	$^1\kappa$	Φ
Fluorobenzene	42	3.394	2.743	1.894	2.100	4.313	1.069
Iodobenzene	42	3.394	2.743	1.894	3.178	5.094	1.679
Nitrobenzene	88	4.305	3.642	2.593	2.499	5.751	1.444
Benzene	27	3.000	2.121	1.500	2.000	3.412	0.913
Phenol	42	3.394	2.743	1.894	2.134	4.342	1.089
4-Chloroaniline	62	3.788	3.365	2.305	2.677	5.447	1.398
2-Nitroaniline	114	4.715	4.170	3.034	2.705	6.539	1.533
3-Nitroaniline	117	4.698	4.276	2.920	2.699	6.539	1.533
4-Nitroaniline	120	4.698	4.264	3.003	2.699	6.539	1.533
3-Chloroaniline	61	3.788	3.377	2.199	2.677	5.447	1.398
2-Chloroaniline	60	3.805	3.239	2.540	2.683	5.447	1.398
Thiophenol	42	3.394	2.743	1.894	2.582	4.722	1.369
2-Methylpyridine	42	3.394	2.743	1.894	2.271	4.313	1.069
3-Methylpyridine	42	3.394	2.743	1.894	2.260	4.313	1.069
4-Methylpyridine	42	3.394	2.743	1.894	2.260	4.313	1.069
N-Acetylpyrrolidine	62	3.805	3.289	2.343	2.444	5.251	1.262
2-Ethylpyrazine	64	3.932	2.912	2.302	2.681	5.221	1.569
Cyclohexene	27	3.000	2.121	1.500	2.650	3.914	1.313
Cyclohexanone	42	3.394	2.743	1.894	2.911	4.820	1.447
1,5-Hexadiene	35	2.914	1.707	0.957	2.133	5.480	4.092
2,3-dimethyl-1,3-butadiene	29	2.643	2.488	1.333	1.957	5.480	1.650
trans-2-Hexenal	56	3.414	2.061	1.207	2.518	6.410	4.954
1-Hexyne	35	2.914	1.707	0.957	2.349	5.560	4.226
Cyclohexane	27	3.000	2.121	1.500	3.000	4.167	1.543
Methylcyclopentane	26	2.894	2.390	1.644	2.894	4.167	1.134
2-Methyl-1-pentene	32	2.770	2.183	0.866	2.414	5.740	2.823
2-Hexanone	52	3.270	2.536	1.135	2.765	6.670	3.665
Hexanoic acid	79	3.770	2.890	1.385	2.988	7.630	4.560
Isobutyl acetate	74	3.626	3.365	1.321	2.760	7.630	3.443
n-Propyl propanoate	76	3.808	2.656	1.747	2.965	7.630	4.560
1-Hexene	35	2.914	1.707	0.957	2.524	5.740	4.535
Isoamyl formate	79	3.770	2.890	1.385	2.823	7.630	4.560
Ethyl butanoate	75	3.808	2.683	1.563	2.965	7.630	4.560
Methyl pentanoate	76	3.808	2.656	1.747	2.877	7.630	4.560
Hexanal	56	3.414	2.061	1.207	2.851	6.670	5.403
4-Methyl-2-pentanone	48	3.126	3.023	0.943	2.621	6.670	2.649
Butyl acetate	79	3.770	2.890	1.385	2.904	7.630	4.560
Cyclohexylamine	42	3.394	2.743	1.894	3.150	5.104	1.687
n-Butylacetamide	79	3.770	2.890	1.385	3.015	7.470	4.318
1-Bromohexane	56	3.414	2.061	1.207	4.096	7.480	6.924
1-Chlorohexane	56	3.414	2.061	1.207	3.509	7.290	6.551
1-Iodohexane	56	3.414	2.061	1.207	4.500	7.730	7.432
N-Methylpiperidine	42	3.394	2.743	1.894	3.080	5.104	1.687
2,2-Dimethylbutane	28	2.561	2.914	1.061	2.561	6.000	1.633
2,3-Dimethylbutane	29	2.643	2.488	1.333	2.643	6.000	2.222
n-Hexane	35	2.914	1.707	0.957	2.914	6.000	5.000

TABLE 3.4 (*Continued*)

Compound	W	$^1\chi$	$^2\chi$	$^3\chi$	$^1\chi^v$	$^1\kappa$	Φ
2-Methylpentane	32	2.770	2.183	0.866	2.770	6.000	3.200
3-Methylpentane	31	2.808	1.922	1.394	2.808	6.000	3.200
Diisopropyl ether	48	3.126	3.023	0.943	2.781	6.960	3.010
Di-*n*-propyl ether	56	3.414	2.061	1.207	2.992	6.960	5.926
1-Hexanol	56	3.414	2.061	1.207	3.023	6.960	5.926
Di-*n*-propyl sulfide	56	3.414	2.061	1.207	4.146	7.350	6.668
Diisopropyl sulfide	48	3.126	3.023	0.943	3.724	7.350	3.533
4-Methyl-2-pentanol	48	3.126	3.023	0.943	2.807	6.960	3.010
2-Methyl-3-pentanol	46	3.181	2.630	1.782	2.862	6.960	3.010
3-Hexanol	50	3.308	2.302	1.478	2.989	6.960	4.104
2-Butoxyethanol	84	3.914	2.414	1.457	3.101	7.920	6.851
2-Methyl-2-pentanol	46	3.061	3.311	1.000	2.784	6.960	2.301
Di-*n*-Propylamine	56	3.414	2.061	1.207	3.121	6.960	5.926
Hexylamine	56	3.414	2.061	1.207	3.115	6.960	5.926
Diisopropylamine	48	3.126	3.023	0.943	2.887	6.960	3.010
Triethylamine	48	3.346	2.091	1.732	3.070	6.960	4.104
Benzotrifluoride	114	4.605	4.701	2.830	2.728	7.121	1.656
3-Cyanophenol	88	4.326	3.546	2.622	2.519	5.800	1.478
4-Cyanophenol	90	4.326	3.534	2.713	2.519	5.800	1.478
Benzonitrile	64	3.932	2.912	2.302	2.384	4.859	1.294
Benzaldehyde	64	3.932	2.912	2.302	2.435	5.035	1.424
3-Hydroxybenzaldehyde	88	4.326	3.546	2.622	2.569	5.977	1.605
4-Hydroxybenzaldehyde	90	4.326	3.534	2.713	2.569	5.977	1.605
2-Chlorotoluene	60	3.805	3.239	2.540	2.894	5.643	1.542
4-Bromotoluene	62	3.788	3.365	2.305	3.303	5.830	1.686
2-Nitrotoluene	114	4.715	4.170	3.034	2.916	6.736	1.666
3-Nitrotoluene	117	4.698	4.276	2.920	2.910	6.736	1.666
Benzamide	88	4.305	3.642	2.593	2.654	5.820	1.492
4-Chloro-3-methylphenol	84	4.198	3.873	2.860	3.029	6.588	1.740
3-Acetylpyridine	88	4.305	3.642	2.593	2.715	5.948	1.583
4-Acetylpyridine	88	4.305	3.642	2.593	2.715	5.948	1.583
Toluene	42	3.394	2.743	1.894	2.411	4.381	1.116
Benzyl alcohol	64	3.932	2.912	2.302	2.581	5.319	1.648
o-Cresol	60	3.805	3.239	2.540	2.551	5.319	1.308
p-Cresol	62	3.788	3.365	2.305	2.545	5.319	1.308
1,3,5-Cycloheptatriene	42	3.500	2.475	1.750	2.483	4.381	1.504
Methyl phenyl ether	64	3.932	2.912	2.302	2.523	5.319	1.648
2-Methoxyphenol	86	4.343	3.430	2.814	2.663	6.263	1.822
Phenyl methyl sulfide	64	3.932	2.912	2.302	3.748	5.702	1.978
3-Methoxyphenol	88	4.326	3.546	2.622	2.657	6.263	1.822
o-Toluidine	60	3.805	3.239	2.540	2.616	5.319	1.308
p-Toluidine	62	3.788	3.365	2.305	2.610	5.319	1.308
2-Methoxyaniline	86	4.343	3.430	2.814	2.728	6.263	1.822
3-Methoxyaniline	88	4.326	3.546	2.622	2.722	6.263	1.822
4-Methoxyaniline	90	4.326	3.534	2.713	2.722	6.263	1.822
2,3-Dimethylpyridine	60	3.805	3.239	2.540	2.687	5.290	1.288

TABLE 3.4 (*Continued*)

Compound	W	$^1\chi$	$^2\chi$	$^3\chi$	$^1\chi^v$	$^1\kappa$	Φ
2,4-Dimethylpyridine	61	3.788	3.377	2.199	2.681	5.290	1.288
2,5-Dimethylpyridine	62	3.788	3.365	2.305	2.681	5.290	1.288
2,6-Dimethylpyridine	61	3.788	3.377	2.199	2.691	5.290	1.288
3,4-Dimethylpyridine	60	3.805	3.239	2.540	2.677	5.290	1.288
3,5-Dimethylpyridine	61	3.788	3.377	2.199	2.671	5.290	1.288
N-Methylaniline	64	3.932	2.912	2.302	2.661	5.319	1.648
2-Ethylpyridine	64	3.932	2.912	2.302	2.831	5.290	1.624
3-Ethylpyridine	64	3.932	2.912	2.302	2.821	5.290	1.624
4-Ethylpyridine	64	3.932	2.912	2.302	2.821	5.290	1.624
1-Methylcyclohexene	42	3.394	2.743	1.894	3.051	4.888	1.503
1-Heptyne	56	3.414	2.061	1.207	2.849	6.560	5.211
1-Heptene	56	3.414	2.061	1.207	3.024	6.740	5.527
Heptanal	84	3.914	2.414	1.457	3.351	7.670	6.395
n-Propyl butanoate	108	4.308	3.036	1.832	3.465	8.630	5.524
Ethyl pentanoate	108	4.308	3.036	1.832	3.465	8.630	5.524
Methyl hexanoate	110	4.308	3.009	1.997	3.377	8.630	5.524
Cycloheptanol	61	3.894	3.097	2.144	3.575	6.086	2.343
Methylcyclohexane	42	3.394	2.743	1.894	3.394	5.143	1.722
2,4-Dimethyl-3-pentanone	65	3.553	3.347	2.103	3.091	7.670	2.735
Isoamyl acetate	108	4.126	3.719	1.563	3.260	8.630	4.318
4-Heptanone	75	3.808	2.683	1.563	3.325	7.670	4.621
trans-2-Heptene	56	3.414	2.061	1.207	3.026	6.740	5.527
2-Heptanone	79	3.770	2.890	1.385	3.265	7.670	4.621
1-Bromoheptane	84	3.914	2.414	1.457	4.596	8.480	7.929
1-Chloroheptane	84	3.914	2.414	1.457	4.009	8.290	7.554
1-Iodoheptane	84	3.914	2.414	1.457	5.000	8.730	8.435
2,3-Dimethylpentane	46	3.181	2.630	1.782	3.181	7.000	3.061
n-Heptane	56	3.414	2.061	1.207	3.414	7.000	6.000
2-Methylhexane	52	3.270	2.536	1.135	3.270	7.000	4.167
3-Methylhexane	50	3.308	2.302	1.478	3.308	7.000	4.167
1-Heptanol	84	3.914	2.414	1.457	3.523	7.960	6.925
2,4-Dimethylpentane	48	3.126	3.023	0.943	3.126	7.000	3.061
3,3-Dimethylpentane	44	3.121	2.871	1.914	3.121	7.000	2.344
2,2-Dimethylpentane	46	3.061	3.311	1.000	3.061	7.000	2.344
n-Heptylamine	84	3.914	2.414	1.457	3.615	7.960	6.925
Styrene	64	3.932	2.912	2.302	2.608	5.104	1.476
Acetophenone	88	4.305	3.642	2.593	2.865	6.017	1.634
4-Methylbenzaldehyde	90	4.326	3.534	2.713	2.846	6.017	1.634
Methyl benzoate	121	4.843	3.784	3.099	2.977	6.963	2.179
Phenyl acetate	126	4.788	4.134	2.530	3.023	6.963	2.179
Ethylbenzene	64	3.932	2.912	2.302	2.971	5.359	1.680
m-Xylene	61	3.788	3.377	2.199	2.821	5.359	1.335
p-Xylene	62	3.788	3.365	2.305	2.821	5.359	1.335
o-Xylene	60	3.805	3.239	2.540	2.827	5.359	1.335
Ethyl phenyl ether	94	4.432	3.293	2.422	3.111	6.302	2.279
2,3-Dimethylphenol	82	4.215	3.745	3.114	2.968	6.302	1.536

TABLE 3.4 (*Continued*)

Compound	W	$^1\chi$	$^2\chi$	$^3\chi$	$^1\chi^v$	$^1\kappa$	Φ
2,5-Dimethylphenol	84	4.198	3.873	2.860	2.962	6.302	1.536
2,6-Dimethylphenol	82	4.215	3.745	3.114	2.968	6.302	1.536
3,4-Dimethylphenol	84	4.198	3.873	2.860	2.962	6.302	1.536
3,5-Dimethylphenol	84	4.182	4.023	2.414	2.956	6.302	1.536
4-Ethylphenol	90	4.326	3.534	2.713	3.106	6.302	1.853
2-Phenylethanol	94	4.432	3.293	2.422	3.081	6.302	2.279
2,4-Dimethylphenol	84	4.198	3.873	2.860	2.962	6.302	1.536
3-Ethylphenol	88	4.326	3.546	2.622	3.106	6.302	1.853
2,6-Dimethylaniline	82	4.215	3.745	3.114	3.033	6.302	1.536
2-Isobutylpyrazine	126	4.788	4.134	2.530	3.537	7.190	2.372
trans-2-Octenal	120	4.414	2.768	1.707	3.518	8.410	6.924
1-Octyne	84	3.914	2.414	1.457	3.349	7.560	6.199
n-Propylcyclopentane	67	3.932	2.939	2.172	3.932	6.125	2.382
trans-1,4-Dimethylcyclohexane	62	3.788	3.365	2.305	3.788	6.125	1.929
Octanal	120	4.414	2.768	1.707	3.851	8.670	7.389
n-Pentyl propanoate	153	4.808	3.363	2.247	3.965	9.630	6.496
Ethyl hexanoate	150	4.808	3.390	2.082	3.965	9.630	6.496
Isobutyl isotuanoate	136	4.537	4.193	2.337	3.703	9.630	4.282
cis-1,2-Dimethylcyclohexane	60	3.805	3.239	2.540	3.805	6.125	1.929
Hexyl acetate	158	4.770	3.597	1.885	3.904	9.630	6.496
2-Octanone	114	4.270	3.243	1.635	3.765	8.670	5.588
1-Octene	84	3.914	2.414	1.457	3.524	7.740	6.521
1-Bromooctane	120	4.414	2.768	1.707	5.096	9.480	8.932
n-Octane	84	3.914	2.414	1.457	3.914	8.000	7.000
2,2,4-Trimethylpentane	66	3.417	4.159	1.021	3.417	8.000	2.520
2,3,4-Trimethylpentane	65	3.553	3.347	2.103	3.553	8.000	3.111
Di-*n*-butyl ether	120	4.414	2.768	1.707	3.992	8.960	7.925
3-Methylheptane	76	3.808	2.656	1.747	3.808	8.000	5.143
1-Octanol	120	4.414	2.768	1.707	4.023	8.960	7.925
Di-*n*-butylamine	120	4.414	2.768	1.707	4.121	8.960	7.925
Octylamine	120	4.414	2.768	1.707	4.115	8.960	7.925
Quinoline	109	4.966	4.089	3.466	3.265	5.418	1.139
α-Methylstyrene	88	4.305	3.642	2.593	3.014	6.086	1.685
Ethyl benzoate	164	5.343	4.164	3.199	3.565	7.952	2.807
4-Methylacetophenone	120	4.698	4.264	3.003	3.276	7.003	1.855
Indane	79	4.466	3.736	3.216	3.535	5.041	1.048
1,2,3-Trimethylbenzene	82	4.215	3.745	3.114	3.244	6.342	1.564
1,2,4-Trimethylbenzene	84	4.198	3.873	2.860	3.238	6.342	1.564
1,3,5-Trimethylbenzene	84	4.182	4.023	2.414	3.232	6.342	1.564
Isopropylbenzene	88	4.305	3.642	2.593	3.354	6.342	1.884
2-Ethyltoluene	86	4.343	3.430	2.814	3.388	6.342	1.884
4-Ethyltoluene	90	4.326	3.534	2.713	3.382	6.342	1.884
4-*n*-Propylphenol	127	4.826	3.915	2.832	3.606	7.289	2.459
Propylbenzene	94	4.432	3.293	2.422	3.471	6.342	2.315
3-Phenyl-1-propanol	133	4.932	3.646	2.691	3.581	7.289	2.966
Nonanal	165	4.914	3.121	1.957	4.351	9.670	8.384

TABLE 3.4 (*Continued*)

Compound	W	$^1\chi$	$^2\chi$	$^3\chi$	$^1\chi^v$	$^1\kappa$	Φ
5-Nonanone	149	4.808	3.390	2.101	4.325	9.670	6.561
2-Nonanone	158	4.770	3.597	1.885	4.265	9.670	6.561
1-Nonene	120	4.414	2.768	1.707	4.024	8.740	7.516
n-Nonane	120	4.414	2.768	1.707	4.414	9.000	8.000
2,2,5-Trimethylhexane	98	3.917	4.493	1.472	3.917	9.000	3.240
1-Nonanol	165	4.914	3.121	1.957	4.523	9.960	8.924
Naphthalene	109	4.966	4.089	3.466	3.405	5.482	1.175
1-Naphthol	140	5.377	4.617	3.933	3.545	6.375	1.370
2-Naphthol	144	5.360	4.723	3.802	3.539	6.375	1.370
1-Naphthylamine	140	5.377	4.617	3.933	3.610	6.375	1.370
2-Naphthylamine	144	5.360	4.723	3.802	3.604	6.375	1.370
n-Butylbenzene	133	4.932	3.646	2.691	3.971	7.329	3.007
tert-Butylbenzene	114	4.605	4.701	2.830	3.661	7.329	1.797
4-Isopropyltoluene	120	4.698	4.264	3.003	3.765	7.329	2.103
4-*tert*-Butylphenol	152	4.999	5.323	3.241	3.795	8.278	2.006
Isobutylbenzene	126	4.788	4.134	2.530	3.827	7.329	2.494
sec-Butylbenzene	121	4.843	3.784	3.099	3.892	7.329	2.494
n-Pentylcyclopentane	140	4.932	3.646	2.691	4.932	8.100	3.856
2-Decanone	212	5.270	3.950	2.135	4.765	10.670	7.540
n-Decane	165	4.914	3.121	1.957	4.914	10.000	9.000
Decyl alcohol	220	5.414	3.475	2.207	5.023	10.960	9.924
1-Methylnaphthalene	140	5.377	4.617	3.933	3.821	6.413	1.392
n-Pentylbenzene	182	5.432	4.000	2.941	4.471	8.318	3.742
2-Undecanone	277	5.770	4.304	2.385	5.265	11.670	8.523
Biphenyl	198	5.966	4.796	3.966	4.071	7.109	1.910
Acenaphthene	166	5.950	5.297	4.841	4.445	6.242	1.092
2,3-Dimethylnaphthalene	182	5.771	5.231	4.387	4.232	7.355	1.613
2,6-Dimethylnaphthalene	181	5.771	5.250	4.269	4.232	7.355	1.613
1,3-Dimethylnaphthalene	179	5.771	5.262	4.178	4.232	7.355	1.613
1,4-Dimethylnaphthalene	176	5.788	5.144	4.414	4.238	7.355	1.613
1-Ethylnaphthalene	182	5.915	4.808	4.248	4.382	7.355	1.826
n-Hexylbenzene	242	5.932	4.353	3.191	4.971	9.309	4.513
Fluorene	219	6.450	5.653	5.091	4.612	6.903	1.307
Reference[a]	*a*	*i*	*q*	*q*	*q*	*r*	*s*

[a]References correspond to those given in Table 3.3.

$$I_i = \frac{(2/N)^2 \delta_i^v + 1}{\delta_i} \tag{3.99}$$

The symbol N denotes here the principal quantum number for the valence shell of that atom, and δ_i^v and δ_i are the molecular connectivity indices calculated as:

TABLE 3.5 Numerical Data on Some Representative Information-Theoretic Descriptors for Set of Structurally Variable Compounds

Compound	^{0}IC	^{1}IC	^{2}IC	^{1}SIC	^{1}CIC	^{1}BIC
Water	2.755	2.755	2.755	1.738	2.000	2.755
Ammonia	3.245	3.245	3.245	1.623	4.755	2.047
Tetrafluoromethane	3.610	3.610	3.610	1.555	8.000	1.805
Tetrachloromethane	3.610	3.610	3.610	1.555	8.000	1.805
Trichloromethane	6.855	6.855	6.855	2.952	4.755	3.427
Tribromomethane	6.855	6.855	6.855	2.952	4.755	3.427
Formaldehyde	6.000	6.000	6.000	3.000	2.000	3.786
Dibromomethane	7.610	7.610	7.610	3.277	4.000	3.805
Dichloromethane	7.610	7.610	7.610	3.277	4.000	3.805
Chloromethane	6.855	6.855	6.855	2.952	4.755	3.427
Bromomethane	6.855	6.855	6.855	2.952	4.755	3.427
Iodomethane	6.855	6.855	6.855	2.952	4.755	3.427
Nitromethane	12.897	12.897	12.897	4.594	6.755	4.989
Methane	3.610	3.610	3.610	1.555	8.000	1.805
Methanol	7.510	10.755	10.755	4.161	4.755	4.632
Methylamine	8.042	12.897	12.897	4.594	6.755	4.989
Tetrachloroethene	5.510	5.510	5.510	2.132	10.000	2.373
Pentachloroethane	10.390	12.390	17.245	4.130	11.610	4.414
Trichloroethene	8.755	10.755	13.510	4.161	4.755	4.632
1,1,1,2-Tetrachloroethane	12.000	14.000	17.245	4.667	10.000	4.987
1,1-Dichloroethene	9.510	9.510	9.510	3.679	6.000	4.096
trans-1,2-Dichloroethene	9.510	9.510	9.510	3.679	6.000	4.096
cis-1,2-Dichloroethene	9.510	9.510	9.510	3.679	6.000	4.096
1,1,2,2-Tetrachloroethane	12.000	12.000	12.000	4.000	12.000	4.275
2,2,2-Trifluoroethanol	17.020	21.774	21.774	6.869	6.755	7.258
Acetonitrile	10.755	10.755	10.755	4.161	4.755	4.632
1,1,1-Trichloroethane	12.490	14.490	14.490	4.830	9.510	5.162
1,1,2-Trichloroethane	12.490	14.490	20.000	4.830	9.510	5.162
Ethene	5.510	5.510	5.510	2.132	10.000	2.373
1,1-Dichloroethane	12.000	14.000	17.245	4.667	10.000	4.987
Acetaldehyde	11.652	14.897	14.897	5.306	4.755	5.763
Acetic acid	16.000	19.245	19.245	6.415	4.755	6.855
1,2-Dibromoethane	12.000	12.000	12.000	4.000	12.000	4.275
1,2-Dichloroethane	12.000	12.000	12.000	4.000	12.000	4.275
Methyl formate	16.000	19.245	19.245	6.415	4.755	6.855
Chloroethane	10.390	12.390	17.245	4.130	11.610	4.414
Bromoethane	10.390	12.390	17.245	4.130	11.610	4.414
Iodoethane	10.390	12.390	17.245	4.130	11.610	4.414
Nitroethane	17.610	19.610	24.464	5.903	13.610	6.186
Ethane	6.490	6.490	6.490	2.163	17.510	2.312
Ethanol	11.020	16.920	21.774	5.338	11.610	5.640
Ethanethiol	11.020	16.920	21.774	5.338	11.610	5.640
Ethylamine	11.568	19.610	24.464	5.903	13.610	6.186
Dimethylamine	11.568	15.710	15.710	4.729	17.510	4.956
Propyne	9.652	14.897	14.897	5.306	4.755	5.763
Methoxyflurane	26.265	31.020	34.265	8.653	12.000	8.967

TABLE 3.5 ***(Continued)***

Compound	^{0}IC	^{1}IC	^{2}IC	^{1}SIC	^{1}CIC	^{1}BIC
1-Chloro-2-propene	14.920	21.774	24.529	6.869	6.755	7.258
Propanonitrile	14.920	16.920	21.774	5.338	11.610	5.640
Ethyl formate	20.544	26.444	31.299	7.644	11.610	7.961
Methyl acetate	20.544	22.544	28.544	6.517	15.510	6.786
1,3-Dichloropropane	15.789	18.544	24.054	5.360	19.510	5.582
Cyclopropane	8.265	8.265	8.265	2.607	20.265	2.607
Propene	11.020	19.020	21.774	6.000	9.510	6.340
Propanone	15.710	15.710	15.710	4.729	17.510	4.956
Propanoic acid	20.544	26.444	31.299	7.644	11.610	7.961
2-Propenol	15.710	26.464	29.219	7.967	6.755	8.349
1,2-Dichloropropane	15.789	20.544	31.299	5.939	17.510	6.184
Propionaldehyde	15.710	21.610	26.464	6.505	11.610	6.817
1-Bromopropane	13.647	18.402	29.299	5.319	19.652	5.540
2-Bromopropane	13.647	16.402	20.544	4.741	21.652	4.938
1-Nitropropane	21.699	26.454	37.351	7.149	21.652	7.379
1-Chloropropane	13.647	18.402	29.299	5.319	19.652	5.540
2-Chloropropane	13.647	16.402	20.544	4.741	21.652	4.938
1-Iodopropane	13.647	18.402	29.299	5.319	19.652	5.540
N,*N*-Dimethylformamide	21.368	25.510	25.510	7.116	17.510	7.374
2-Nitropropane	21.699	24.454	28.596	6.609	23.652	6.821
Propane	9.299	12.054	18.544	3.484	26.000	3.629
Methyl ethyl ether	14.265	19.020	31.510	5.305	24.000	5.498
1-Propanol	14.265	23.368	34.265	6.518	19.652	6.755
2-Propanol	14.265	21.368	25.510	5.961	21.652	6.177
2-Methoxyethanol	17.351	26.454	35.351	7.149	21.652	7.379
n-Propanethiol	14.265	23.368	34.265	6.518	19.652	6.755
Trimethylamine	14.822	14.822	14.822	4.005	33.284	4.134
Propylamine	14.822	26.454	37.351	7.149	21.652	7.379
Thiophene	12.529	16.529	20.529	5.214	12.000	5.214
1-Butyne	13.710	21.610	26.464	6.505	11.610	6.817
(*E*)-2-Butenal	17.789	28.544	33.299	8.251	9.510	8.593
1-Cyanopropane	18.613	23.368	34.265	6.518	19.652	6.755
Isobutyraldehyde	19.351	26.454	30.596	7.149	21.652	7.379
Tetrahydrofuran	16.106	20.106	28.106	5.433	28.000	5.433
Ethyl acetate	24.548	29.303	41.793	7.696	24.000	7.919
Methyl propanoate	24.548	29.303	41.793	7.696	24.000	7.919
1,4-Dichlorobutane	19.303	23.303	31.303	6.121	30.000	6.297
1-Butene	15.020	26.655	34.265	7.435	16.365	7.705
Butanone	19.351	24.106	36.596	6.514	24.000	6.724
Isopropyl formate	24.548	31.652	35.793	8.313	21.652	8.553
Butanoic acid	24.548	33.652	44.548	8.839	19.652	9.094
Propyl formate	24.548	33.652	44.548	8.839	19.652	9.094
Butyraldehyde	19.351	28.454	39.351	7.689	19.652	7.937
1,4-Dioxane	19.303	19.303	19.303	5.070	34.000	5.070
1-Bromobutane	16.774	22.774	38.548	5.982	30.529	6.154
1-Chlorobutane	16.774	22.774	38.548	5.982	30.529	6.154
1-Nitrobutane	25.471	31.471	47.245	7.868	32.529	8.055

TABLE 3.5 (***Continued***)

Compound	^0IC	^1IC	^2IC	^1SIC	^1CIC	^1BIC
2-Bromo-2-methylpropane	16.774	20.019	20.019	5.258	33.284	5.410
1-Bromo-2-methylpropane	16.774	22.774	33.793	5.982	30.529	6.154
2-Chloro-2-methylpropane	16.774	20.019	20.019	5.258	33.284	5.410
2-Chlorobutane	16.774	22.774	35.793	5.982	30.529	6.154
1-Iodobutane	16.774	22.774	38.548	5.982	30.529	6.154
Morpholine	22.074	30.603	38.603	7.833	28.000	7.833
n-Butane	12.084	16.084	25.793	4.224	37.219	4.346
Diethyl ether	17.384	21.384	31.094	5.473	37.219	5.617
2-Methylpropane	12.084	15.329	20.019	4.026	37.974	4.142
2-Methyl-1-propanol	17.384	28.074	39.094	7.186	30.529	7.374
2-Butanol	17.384	28.074	41.094	7.186	30.529	7.374
2-Methyl-2-propanol	17.384	25.319	25.319	6.481	33.284	6.650
Diethyl sulfide	17.384	21.384	31.094	5.473	37.219	5.617
Diethyl disulfide	20.781	24.781	34.490	6.195	39.219	6.343
2-Ethoxyethanol	20.781	30.716	43.735	7.679	33.284	7.862
1-Butanol	17.384	28.074	43.849	7.186	30.529	7.374
n-Butanethiol	17.384	28.074	43.849	7.186	30.529	7.374
Diethylamine	17.946	26.781	36.490	6.695	37.219	6.855
Butylamine	17.946	31.471	47.245	7.868	32.529	8.055
2-Chloropyridine	18.444	25.299	33.299	7.313	12.755	7.313
3-Chloropyridine	18.444	26.054	34.054	7.531	12.000	7.531
Pyridine	14.835	19.689	27.299	5.692	18.365	5.692
2-Methylthiophene	19.510	31.510	36.265	8.789	11.510	8.789
2-Methylpyrazine	22.596	31.841	36.596	8.605	16.265	8.605
Cyclopentene	17.351	26.596	32.106	7.187	21.510	7.187
2-Methyl-1,3-butadiene	16.106	29.741	35.351	8.037	18.365	8.296
Cyclopentanone	21.303	25.303	33.303	6.646	28.000	6.646
1,4-pentadiene	16.106	26.596	32.106	7.187	21.510	7.419
1-Pentyne	17.351	28.454	39.351	7.689	19.652	7.937
1-Cyanobutane	22.074	28.074	43.849	7.186	30.529	7.374
Cyclopentane	13.774	13.774	13.774	3.526	44.829	3.526
1-Pentene	18.629	34.197	47.849	8.753	24.406	8.982
Ethyl propanoate	28.268	34.268	49.977	8.384	35.219	8.567
Isobutyl formate	28.268	38.958	49.977	9.531	30.529	9.739
2-Methyl-2-butene	18.629	32.603	45.094	8.345	26.000	8.563
3-Methyl-1-butene	18.629	32.197	39.094	8.241	26.406	8.457
Isopropyl acetate	28.268	34.268	47.222	8.384	35.219	8.567
Pentanal	22.781	33.471	49.245	8.368	30.529	8.567
Methyl butanoate	28.268	36.268	55.977	8.873	33.219	9.067
Pentanoic acid	28.268	38.958	54.732	9.531	30.529	9.739
3-Methylbutanoic acid	28.268	38.958	49.977	9.531	30.529	9.739
Cyclopentanol	19.171	27.471	36.000	6.868	36.529	6.868
3-Methyl-2-butanone	22.781	28.781	41.735	7.195	35.219	7.367
3-Pentanone	22.781	26.781	36.490	6.695	37.219	6.855
Propyl acetate	28.268	36.268	55.977	8.873	33.219	9.067
2-Pentanone	22.781	30.781	50.490	7.695	33.219	7.879

TABLE 3.5 *(Continued)*

Compound	^0IC	^1IC	^2IC	^1SIC	^1CIC	^1BIC
2-Methytetrahydrofuran	19.171	28.781	49.245	7.195	35.219	7.195
Tetrahydropyran	19.171	24.026	36.490	6.007	39.974	6.007
1-Chloropentane	19.824	26.678	47.222	6.527	42.809	6.670
1-Nitropentane	29.047	35.902	56.446	8.452	44.809	8.610
1-Bromopentane	19.824	26.678	47.222	6.527	42.809	6.670
1-Iodopentane	19.824	26.678	47.222	6.527	42.809	6.670
N-Methylmorpholine	25.395	33.005	50.304	7.915	42.054	7.915
n-Pentane	14.858	19.712	34.467	4.823	49.774	4.928
2-Methyl-1-butanol	20.430	35.005	55.549	8.395	40.054	8.564
2-Methyl-2-butanol	20.430	32.250	42.529	7.734	42.809	7.890
3-Methyl-1-butanol	20.430	35.005	53.549	8.395	40.054	8.564
1-Pentanol	20.430	32.250	52.794	7.734	42.809	7.890
2-Pentanol	20.430	33.005	51.549	7.915	42.054	8.075
3-Pentanol	20.430	33.005	47.549	7.915	42.054	8.075
Methyl *tert*-butyl ether	20.430	27.284	37.020	6.543	47.774	6.675
2,2-Dimethylpropane	14.858	18.467	18.467	4.518	51.020	4.617
2-Propoxyethanol	24.081	37.902	58.446	8.923	42.809	9.089
2-Methylbutane	14.858	21.712	36.958	5.312	47.774	5.428
n-Pentylamine	20.995	35.902	56.446	8.452	44.809	8.610
1,2,3,4-Tetrachlorobenzene	17.510	23.020	27.020	6.421	20.000	6.421
1,2,3,5-Tetrachlorobenzene	17.510	23.020	29.020	6.421	20.000	6.421
1,2,4,5-Tetrachlorobenzene	17.510	23.020	23.020	6.421	20.000	6.421
1,2,3-Trichlorobenzene	18.000	24.000	29.510	6.695	19.020	6.695
1,3,5-Trichlorobenzene	18.000	24.000	24.000	6.695	19.020	6.695
1,2,4-Trichlorobenzene	18.000	24.000	29.510	6.695	19.020	6.695
1,2-Dichlorobenzene	17.510	23.020	27.020	6.421	20.000	6.421
1,3-Dichlorobenzene	17.510	23.020	29.020	6.421	20.000	6.421
1,4-Dichlorobenzene	17.510	23.020	23.020	6.421	20.000	6.421
3-Cyanopyridine	23.410	31.020	39.020	8.653	12.000	8.653
4-Cyanopyridine	23.410	31.020	35.020	8.653	12.000	8.653
Bromobenzene	15.900	19.800	24.655	5.523	23.219	5.523
Chlorobenzene	15.900	19.800	24.655	5.523	23.219	5.523
4-Fluorophenol	20.986	32.106	36.106	8.676	16.000	8.676
2-Chlorophenol	20.986	32.106	38.106	8.676	16.000	8.676
3-Chlorophenol	20.986	32.106	40.106	8.676	16.000	8.676
4-Chlorophenol	20.986	32.106	36.106	8.676	16.000	8.676
4-Bromophenol	20.986	32.106	36.106	8.676	16.000	8.676
2-Iodophenol	20.986	32.106	38.106	8.676	16.000	8.676
2-Nitrophenol	29.484	40.603	46.603	10.393	18.000	10.393
3-Nitrophenol	29.484	40.603	48.603	10.393	18.000	10.393
4-Nitrophenol	29.484	40.603	44.603	10.393	18.000	10.393
3-Formalpyridine	20.986	32.496	44.106	8.782	15.610	8.782
4-Formalpyridine	20.986	32.496	40.106	8.782	15.610	8.782
2-Fluorophenol	20.986	32.106	38.106	8.676	16.000	8.676
Fluorobenzene	15.900	19.800	24.655	5.523	23.219	5.523
Iodobenzene	15.900	19.800	24.655	5.523	23.219	5.523

TABLE 3.5 (*Continued*)

Compound	${}^{0}IC$	${}^{1}IC$	${}^{2}IC$	${}^{1}SIC$	${}^{1}CIC$	${}^{1}BIC$
Nitrobenzene	24.184	28.084	32.938	7.376	25.219	7.376
Benzene	12.000	12.000	12.000	3.347	31.020	3.347
Phenol	17.086	24.886	29.741	6.725	23.219	6.725
4-Chloroaniline	22.283	35.303	39.303	9.272	18.000	9.272
2-Nitroaniline	28.980	44.000	50.000	11.000	20.000	11.000
3-Nitroaniline	28.980	44.000	52.000	11.000	20.000	11.000
4-Nitroaniline	28.980	44.000	48.000	11.000	20.000	11.000
3-Chloroaniline	22.283	35.303	43.303	9.272	18.000	9.272
2-Chloroaniline	22.283	35.303	41.303	9.272	18.000	9.272
Thiophenol	17.086	24.886	29.741	6.725	23.219	6.725
2-Methylpyridine	22.042	35.793	43.793	9.401	17.510	9.401
3-Methylpyridine	22.042	36.548	44.548	9.599	16.755	9.599
4-Methylpyridine	22.042	36.548	40.548	9.599	16.755	9.599
N-Acetylpyrrolidine	27.342	41.849	45.849	10.712	16.755	10.712
2-Ethylpyrazine	28.000	40.881	50.490	10.220	23.119	10.220
Cyclohexene	20.781	32.000	40.000	8.000	32.000	8.000
Cyclohexanone	24.658	29.513	41.977	7.220	39.974	7.220
1,5-Hexadiene	20.781	34.490	40.000	8.623	29.510	8.828
2,3-Dimethyl-1,3-butadiene	20.781	34.490	34.490	8.623	29.510	8.828
trans-2-Hexenal	26.758	45.081	60.732	11.029	24.406	11.270
1-Hexyne	20.781	33.471	49.245	8.368	30.529	8.567
Cyclohexane	16.529	16.529	16.529	3.964	58.529	3.964
Methylcyclopentane	16.529	24.039	42.304	5.765	51.020	5.765
2-Methyl-1-pentene	22.039	39.839	59.549	9.554	35.219	9.747
2-Hexanone	26.081	35.691	61.201	8.402	45.020	8.559
Hexanoic acid	31.809	43.630	64.174	10.095	42.809	10.271
Isobutyl acetate	31.809	41.419	62.174	9.584	45.020	9.750
n-Propyl propanoate	31.809	41.419	64.929	9.584	45.020	9.750
1-Hexene	22.039	39.774	58.304	9.538	35.284	9.731
Isoamyl formate	31.809	46.385	64.929	10.732	40.054	10.919
Ethyl butanoate	31.809	41.419	64.929	9.584	45.020	9.750
Methyl pentanoate	31.809	41.419	66.929	9.584	45.020	9.750
Hexanal	26.081	37.902	58.446	8.923	42.809	9.089
4-Methyl-2-pentanone	26.081	35.691	56.446	8.402	45.020	8.559
Butyl acetate	31.809	41.419	66.929	9.584	45.020	9.750
Cyclohexylamine	22.823	34.775	44.464	8.046	51.663	8.046
n-Butylacetamide	32.523	47.219	72.729	10.750	45.020	10.926
1-Bromohexane	22.823	30.333	53.684	7.018	56.106	7.141
1-Chlorohexane	22.823	30.333	53.684	7.018	56.106	7.141
1-Iodohexane	22.823	30.333	53.684	7.018	56.106	7.141
N-Methylpiperidine	22.823	31.578	54.174	7.307	54.861	7.307
2,2-Dimethylbutane	17.626	25.136	36.664	5.816	61.303	5.917
2,3-Dimethylbutane	17.626	23.136	31.419	5.353	63.303	5.446
n-Hexane	17.626	23.136	40.929	5.353	63.303	5.446
2-Methylpentane	17.626	26.381	47.909	6.104	60.058	6.210
3-Methylpentane	17.626	26.381	45.909	6.104	60.058	6.210
Diisopropyl ether	23.426	29.936	37.219	6.588	63.303	6.695

TABLE 3.5 (*Continued*)

Compound	^{0}IC	^{1}IC	^{2}IC	^{1}SIC	^{1}CIC	^{1}BIC
Di-*n*-propyl ether	23.426	32.936	54.729	7.499	59.303	7.621
1-Hexanol	23.426	36.133	59.484	8.226	56.106	8.360
Di-*n*-propyl sulfide	23.426	32.936	54.729	7.499	59.303	7.621
Diisopropyl sulfide	23.426	28.936	37.219	6.588	63.303	6.695
4-Methyl-2-pentanol	23.426	39.378	59.709	8.965	52.861	9.111
2-Methyl-3-pentanol	23.426	39.378	59.709	8.965	52.861	9.111
3-Hexanol	23.426	37.378	59.219	8.510	54.861	8.649
2-Butoxyethanol	27.295	43.247	69.843	9.698	54.861	9.846
2-Methyl-2-pentanol	23.426	37.378	53.709	8.510	84.861	8.649
Di-*n*-Propylamine	23.994	38.805	60.598	8.702	59.303	8.835
Hexylamine	23.994	40.002	63.353	8.970	58.106	9.107
Diisopropylamine	23.994	34.805	43.088	7.805	63.303	7.924
Triethylamine	23.994	29.994	44.559	6.726	68.113	6.829
Benzotrifluoride	26.729	30.629	35.484	7.840	27.974	7.840
3-Cyanophenol	26.184	37.303	45.303	9.798	16.000	9.798
4-Cyanophenol	26.184	37.303	41.303	9.798	16.000	9.798
Benzonitrile	20.986	24.886	29.741	6.725	23.219	6.725
Benzaldehyde	18.142	26.184	34.938	6.877	27.119	6.877
3-Hydroxybenzaldehyde	23.442	38.994	50.603	9.981	19.610	9.981
4-Hydroxybenzaldehyde	23.442	38.994	46.603	9.981	19.610	9.981
2-Chlorotoluene	23.442	37.849	43.849	9.688	20.755	9.688
4-Bromotoluene	23.442	37.849	41.849	9.688	20.755	9.688
2-Nitrotoluene	32.326	46.732	52.732	11.433	22.755	11.433
3-Nitrotoluene	32.326	46.732	54.732	11.433	22.755	11.433
Benzamide	24.697	38.781	43.636	9.695	25.219	9.695
4-Chloro-3-methylphenol	28.839	49.735	54.490	12.434	14.265	12.434
3-Acetylpyridine	28.839	47.245	55.245	11.811	16.755	11.811
4-Acetylpyridine	28.839	47.245	51.245	11.811	16.755	11.811
Toluene	19.094	30.629	35.484	7.840	27.974	7.840
Benzyl alcohol	24.490	38.781	43.636	9.695	25.219	9.695
o-Cresol	24.490	43.245	49.245	10.811	20.755	10.811
p-Cresol	24.490	43.245	47.245	10.811	20.755	10.811
1,3,5-Cycloheptatriene	19.094	31.094	40.603	7.959	27.510	7.959
Methyl phenyl ether	24.490	36.026	40.881	9.007	27.974	9.007
2-Methoxyphenol	27.977	46.732	52.732	11.433	22.755	11.433
Phenyl methyl sulfide	24.490	36.026	40.881	9.007	27.974	9.007
3-Methoxyphenol	27.977	46.732	54.732	11.433	22.755	11.433
o-Toluidine	25.448	46.732	52.732	11.433	22.755	11.433
p-Toluidine	25.448	46.732	50.732	11.433	22.755	11.433
2-Methoxyaniline	31.020	52.304	58.304	12.543	22.755	12.543
3-Methoxyaniline	31.020	52.304	60.304	12.543	22.755	12.543
4-Methoxyaniline	31.020	52.304	56.304	12.543	22.755	12.543
2,3-Dimethylpyridine	27.348	45.222	51.977	11.064	24.265	11.064
2,4-Dimethylpyridine	27.348	45.222	51.977	11.064	24.265	11.064
2,5-Dimethylpyridine	27.348	45.222	51.977	11.064	24.265	11.064
2,6-Dimethylpyridine	27.348	40.467	43.222	9.900	29.020	9.900
3,4-Dimethylpyridine	27.348	43.222	49.977	10.574	26.265	10.574

TABLE 3.5 (*Continued*)

Compound	^{0}IC	^{1}IC	^{2}IC	^{1}SIC	^{1}CIC	^{1}BIC
3,5-Dimethylpyridine	27.348	43.222	45.977	10.574	26.265	10.574
N-Methylaniline	25.448	41.513	46.368	10.156	27.974	10.156
2-Ethylpyridine	27.348	45.122	57.977	11.039	24.365	11.039
3-Ethylpyridine	27.348	45.877	58.732	11.224	23.610	11.224
4-Ethylpyridine	27.348	45.877	54.732	11.224	23.610	11.224
1-Methylcyclohexene	24.081	38.657	57.956	9.100	42.054	9.100
1-Heptyne	24.081	37.902	58.446	8.923	42.809	9.089
1-Heptene	25.326	44.675	67.974	10.171	47.564	10.337
Heptanal	29.295	42.002	65.353	9.419	56.106	9.563
n-Propyl butanoate	35.229	46.739	74.532	10.332	57.303	10.481
Ethyl pentanoate	35.229	46.739	76.532	10.332	57.303	10.481
Methyl hexanoate	35.229	45.984	77.022	10.166	58.058	10.312
Cycloheptanol	25.153	34.492	45.088	7.735	63.616	7.735
Methylcyclohexane	19.284	27.326	47.510	6.221	64.913	6.221
2,4-Dimethyl-3-pentanone	29.295	34.805	43.088	7.805	63.303	7.924
Isoamyl acetate	35.229	48.739	77.777	10.775	55.303	10.929
4-Heptanone	29.295	38.805	60.598	8.702	59.303	8.835
trans-2-Heptene	25.326	43.219	70.729	9.840	49.020	10.000
2-Heptanone	29.295	40.050	71.088	8.981	58.058	9.118
1-Bromoheptane	25.787	33.829	59.313	7.478	70.213	7.586
1-Chloroheptane	25.787	33.829	59.313	7.478	70.213	7.586
1-Iodoheptane	25.787	33.829	59.313	7.478	70.213	7.586
2,3-Dimethylpentane	20.390	30.042	52.268	6.641	74.000	6.737
n-Heptane	20.390	26.432	46.558	5.843	77.610	5.927
2-Methylhexane	20.390	30.532	58.003	6.750	73.510	6.847
3-Methylhexane	20.390	30.532	58.003	6.750	73.510	6.847
1-Heptanol	26.388	39.826	65.310	8.686	70.213	8.804
2,4-Dimethylpentane	20.390	30.042	47.022	6.641	74.000	6.737
3,3-Dimethylpentane	20.390	30.042	47.022	6.641	74.000	6.737
2,2-Dimethylpentane	20.390	30.042	48.268	6.641	74.000	6.737
n-Heptylamine	26.958	43.883	69.367	9.450	72.213	9.571
Styrene	16.000	28.390	43.636	7.098	35.610	7.098
Acetophenone	25.835	41.513	46.368	10.156	27.974	10.156
4-Methylbenzaldehyde	25.835	45.122	52.732	11.039	24.365	11.039
Methyl benzoate	31.407	47.085	51.939	11.291	27.974	11.291
Phenyl acetate	31.407	47.085	51.939	11.291	27.974	11.291
Ethylbenzene	24.330	40.230	49.939	9.648	34.829	9.648
m-Xylene	24.330	39.549	45.549	9.484	35.510	9.484
p-Xylene	24.330	39.549	39.549	9.484	35.510	9.484
o-Xylene	24.330	39.549	43.549	9.484	35.510	9.484
Ethyl phenyl ether	29.982	45.882	55.591	10.801	34.829	10.801
2,3-Dimethylphenol	29.982	51.691	58.446	12.169	29.020	12.169
2,5-Dimethylphenol	29.982	51.691	56.446	12.169	29.020	12.169
2,6-Dimethylphenol	29.982	51.691	54.446	12.169	29.020	12.169
3,4-Dimethylphenol	29.982	51.691	56.446	12.169	29.020	12.169
3,5-Dimethylphenol	29.982	51.691	54.446	12.169	29.020	12.169

TABLE 3.5 (*Continued*)

Compound	${}^{0}IC$	${}^{1}IC$	${}^{2}IC$	${}^{1}SIC$	${}^{1}CIC$	${}^{1}BIC$
4-Ethylphenol	29.982	53.101	61.956	12.500	27.610	12.500
2-Phenylethanol	29.982	49.491	58.346	11.651	31.219	11.651
2,4-Dimethylphenol	29.982	51.691	58.446	12.169	29.020	12.169
3-Ethylphenol	29.982	53.101	65.956	12.500	27.610	12.500
2,6-Dimethylaniline	30.875	55.419	58.174	12.823	31.020	12.823
2-Isobutylpyrazine	37.088	56.068	71.843	12.573	42.039	12.573
trans-2-Octenal	34.374	56.478	81.777	12.485	47.564	12.665
1-Octyne	27.295	42.002	65.353	9.419	56.106	9.563
n-Propylcyclopentane	22.039	30.529	55.510	6.659	79.510	6.659
trans-1,4-Dimethylcyclohexane	22.039	34.039	56.529	7.424	76.000	7.424
Octanal	32.445	45.883	71.367	9.881	70.213	10.007
n-Pentyl propanoate	38.560	51.457	87.192	10.947	70.755	11.081
Ethyl hexanoate	38.560	51.457	87.192	10.947	70.755	11.081
Isobutyl isotuanoate	38.560	50.211	73.192	10.682	72.000	10.812
cis-1,2-Dimethylcyclohexane	22.039	34.039	60.529	7.424	76.000	7.424
Hexyl acetate	38.560	50.211	84.702	10.682	72.000	10.812
2-Octanone	32.445	44.096	78.587	9.496	72.000	9.618
1-Octene	28.529	49.179	75.284	10.726	60.861	10.872
1-Bromooctane	28.725	37.215	64.437	7.917	84.997	8.014
n-Octane	23.153	29.643	51.682	6.306	92.568	6.383
2,2,4-Trimethylpentane	23.153	35.543	54.853	7.562	86.668	7.654
2,3,4-Trimethylpentane	23.153	30.788	45.244	6.550	91.423	6.630
Di-*n*-butyl ether	29.323	41.323	72.872	8.691	87.059	8.791
3-Methylheptane	23.153	34.398	67.682	7.318	87.814	7.407
1-Octanol	29.323	43.385	70.608	9.124	84.997	9.230
Di-*n*-butylamine	29.895	47.547	79.096	9.891	87.059	10.000
Octylamine	29.895	47.609	74.832	9.903	86.997	10.013
Quinoline	21.306	34.326	49.977	8.398	35.161	8.232
α-Methylstyrene	23.491	44.695	55.591	10.522	36.016	10.522
Ethyl benzoate	37.368	57.410	67.119	13.071	34.829	13.071
4-Methylacetophenone	31.568	52.929	58.929	12.247	33.510	12.247
Indane	27.227	45.201	54.711	10.641	35.510	10.459
1,2,3-Trimethylbenzene	28.955	44.690	50.200	10.175	47.549	10.175
1,2,4-Trimethylbenzene	28.955	44.690	50.200	10.175	47.549	10.175
1,3,5-Trimethylbenzene	28.955	44.690	44.690	10.175	47.549	10.175
Isopropylbenzene	28.955	47.368	56.364	10.784	44.871	10.784
2-Ethyltoluene	28.955	50.239	68.729	11.438	42.000	11.438
4-Ethyltoluene	28.955	50.239	64.729	11.438	42.000	11.438
4-*n*-Propylphenol	34.823	62.456	77.353	14.005	35.652	14.005
Propylbenzene	28.955	49.368	65.119	11.240	42.871	11.240
3-Phenyl-1-propanol	34.823	59.378	73.743	13.315	38.729	13.315
Nonanal	35.547	49.609	76.832	10.320	84.997	10.433
5-Nonanone	35.547	47.547	79.096	9.891	87.059	10.000
2-Nonanone	35.547	47.938	85.122	9.972	86.668	10.082
1-Nonene	31.672	53.414	81.653	11.234	74.968	11.364
n-Nonane	25.914	32.791	56.459	6.750	108.090	6.821

TABLE 3.5 (*Continued*)

Compound	^{0}IC	^{1}IC	^{2}IC	^{1}SIC	^{1}CIC	^{1}BIC
2,2,5-Trimethylhexane	25.914	40.833	67.523	8.405	100.048	8.494
1-Nonanol	32.239	46.845	75.539	9.547	100.362	9.643
Naphthalene	17.839	25.059	33.059	6.009	50.000	5.899
1-Naphthol	23.491	39.408	51.549	9.277	41.303	9.118
2-Naphthol	23.491	39.408	52.304	9.277	41.303	9.118
1-Naphthylamine	24.690	43.136	55.277	9.981	43.303	9.821
2-Naphthylamine	24.690	43.136	56.032	9.981	43.303	9.821
n-Butylbenzene	33.226	56.291	76.920	12.277	53.749	12.277
tert-Butylbenzene	33.226	53.536	58.390	11.676	56.504	11.676
4-Isopropyltoluene	33.226	56.820	71.774	12.393	53.219	12.393
4-*tert*-Butylphenol	39.284	66.812	70.812	14.387	49.284	14.387
Isobutylbenzene	33.226	56.291	72.165	12.277	53.749	12.277
sec-Butylbenzene	33.226	56.291	74.165	12.277	53.749	12.277
n-Pentylcyclopentane	27.549	36.768	65.697	7.493	110.439	7.493
2-Decanone	38.612	51.632	91.051	10.422	101.948	10.522
n-Decane	28.673	35.893	60.980	7.179	124.108	7.245
Decyl alcohol	35.138	50.226	80.200	9.957	116.239	10.045
1-Methylnaphthalene	25.800	46.181	58.323	10.514	46.058	10.356
n-Pentylbenzene	37.263	62.354	87.753	13.114	66.028	13.114
2-Undecanone	41.647	55.215	96.551	10.853	117.759	10.946
Biphenyl	21.869	29.669	39.378	6.653	68.439	6.559
Acenaphthene	29.669	53.088	64.598	11.905	45.020	11.579
2,3-Dimethylnaphthalene	31.800	57.510	67.020	12.543	52.529	12.384
2,6-Dimethylnaphthalene	31.800	57.510	73.020	12.543	52.529	12.384
1,3-Dimethylnaphthalene	31.800	57.510	73.020	12.543	52.529	12.834
1,4-Dimethylnaphthalene	31.800	57.510	67.020	12.543	52.529	12.384
1-Ethylnaphthalene	31.800	57.127	74.123	12.460	52.913	12.302
n-Hexylbenzene	41.129	67.882	96.087	13.834	79.325	13.834
Fluorene	27.803	50.042	62.042	11.063	54.000	10.776
Reference[a]	*w*	*w*	*w*	*x*	*x*	*x*

[a]References correspond to those given in Table 3.3.

$$\delta_i^v = Z^v - h = \sigma + \pi + n - h \tag{3.100}$$

$$\delta = \sigma - h \tag{3.101}$$

In the last two formulas, σ, π, and n denote the number of electrons on the σ, π, and n orbitals, respectively; Z^v is total number of the valence electrons; and h is the number of hydrogen atoms at the given atom. The difference $(\delta_i^v - \delta_i)$ is equal to the number of π and n electrons. It has been shown that this difference is also proportional to the valence-state electronegativity [225]. The perturbation term in Eq. (3.98) is calculated as:

$$\Delta I_i = \sum_{j \neq i} \frac{I_i - I_j}{r_{ij}^2} \tag{3.102}$$

where r_{ij} is the topological distance between the atoms plus one ($r_{ij} = d_{ij} + 1$). The global molecular descriptors can be calculated as the sums of electrotopological state indices for given types of atoms or atoms of certain hybridization or connectivity. The above-discussed electrotopological state indices have also been defined for individual atoms in the molecule. However, an alternative definition is based on the electrotopological state value for chemical bonds [257]. The bond electrotopological state index is given as:

$$S_i^b = I_{ij} + \sum \frac{\Delta I_{ij}}{\bar{r}_{ij}^2} \tag{3.103}$$

where I_{ij} is the average of electrotopological state indices of two atoms involved in the given bond

$$I_{ij} = \frac{I_i + I_j}{2} \tag{3.104}$$

and $\bar{r}_{ij}$ is calculated as the average for the atoms in the two bonds.

Another type of topological electronic indices has been suggested according to the following formulas [258] for all pairs of atoms, T_1^E, and for all bonded pairs of atoms, T_2^E:

$$T_1^E = \sum_{(i<j)}^{N} \frac{|q_i - q_j|}{r_{ij}^2} \tag{3.105a}$$

$$T_2^E = \sum_{(i<j)}^{Nb} \frac{|q_i - q_j|}{r_{ij}^2} \tag{3.105b}$$

where q_i is a partial charge on the ith atom and r_{ij} is a distance between the ith and jth atoms. This descriptor encodes the electrostatic interactions in the molecule but not according to the Coulomb law. In particular, it overestimates the electrostatic attraction between the opposite charges as compared to the repulsion of similar charges in the molecule (the terms for numerically equal similar charges are zero). Therefore, it may be difficult to give a distinct physical meaning to the respective contribution into QSAR/QSPR.

The electron charge density connectivity $\delta(q_i)$ has been defined as [259]:

$$\delta(q_i) = q_i - h_i \tag{3.106}$$

where q_i is the quantum chemically calculated Mulliken partial charge on an atom, and h_I is the number of hydrogen atoms connected to that particular atom. The three-dimensional valence connectivity index is then calculated using the Randić-type invariant as:

$$\Omega(q) = \sum_k [\delta(q_i)\delta(q_j)]_k^{-1/2} \tag{3.107}$$

where the summation is carried out for all pairs of bonded atoms in the hydrogen-depleted molecular graph. An attempt has been made to develop topological charge indices using the topological information encoded in the molecule and the inverse square dependence of electrostatic interactions from distance [260]. For this purpose, the inverse square distance matrix $\boldsymbol{D}^*$ has been introduced, with the nondiagonal elements being the inverses of the respective elements of the ordinary topological distance matrix $d_{ij}^* = (d_{ij})^{-1}$. The diagonal elements of this matrix were postulated as zeroes. The charge terms have been defined as:

$$\text{CT}_{ij} = m_{ij} - m_{ji} \tag{3.108}$$

where m_{ij} are the elements of matrix $\boldsymbol{M} = \boldsymbol{A} \times \boldsymbol{D}^*$, where $\boldsymbol{A}$ is the adjacency matrix. The topological charge index of order k was derived as:

$$G_k = \sum_{i<j} |\text{CT}_{ij}|\delta(k, d_{ij}) \tag{3.109}$$

with the summation carried out over all pairs of vertices in the molecular graph. The δ denotes here the Kronecker symbol ($\delta = 1$ if $k = d_{ij}$, otherwise $\delta = 0$). Another topological charge index was defined as [260]:

$$J_k = \frac{G_k}{N - 1} \tag{3.110}$$

where N is the number of vertices in the molecular graph.

Finally, the topographical indices can be considered as combining the geometrical and topological information encoded in a molecule [261]. These indices are based on the geometric distance matrix $\boldsymbol{G}$ that comprises the shortest Cartesian distances l_{ij} between any two atomic centers, i and j, in the molecule

$$l_{ij} = \sqrt{(x_i - x_j)^2 + (y_i - y_j)^2 + (z_i - z_j)^2} \tag{3.111}$$

where x, y, and z are the Cartesian coordinates of the respective centers. The three-dimensional Wiener number, denoted as *3-W*, was thus defined as the half-sum of the elements of the geometric distance matrix:

$$3 = W = \frac{1}{2} \sum_{i=1}^{N} \sum_{j=1}^{N} l_{ij} \tag{3.112}$$

The three-dimensional Balaban index *3-J* has been calculated similarly to the original topological J index, with the difference in the definition of the distances between the vertices (atoms). In the case of $3\text{-}J$, the elements of the geometric distance matrix $\boldsymbol{G}$ were used in the calculation of the index. The definition of the three-dimensional Schultz indices resembles also formally the respective topological counterparts. However, the quantities e_I are now the elements of the ith row in the matrix built from the sum of the three-dimensional adjacency and the distance matrices. The elements of the three-dimensional adjacency matrix $\boldsymbol{3\text{-}A}$ are defined as:

$$(\boldsymbol{3\text{-}A})_{ij} = \begin{cases} l_{ij} & \text{if atom } i \text{ is bonded to atom } j \\ 0 & \text{otherwise} \end{cases} \tag{3.113}$$

where l_{ij} is the geometric distance between the two atoms, i and j, in the molecule. By analogy, the determinant of adjacency-plus-distance matrix can also be defined in the case of the geometrical matrices as:

$$\text{3-TI} = \det|\boldsymbol{3} - \boldsymbol{A} + \boldsymbol{G}| \tag{3.114}$$

Topographic matrices have been employed to incorporate the stereochemical features of molecules [262,263].

An edge adjacency graph-theoretic (topological) index has been defined as [264]:

$$\varepsilon = \sum_{l} [\delta(e_i)\delta(e_j)]_l^{-1/2} \tag{3.115}$$

where the sum is over all l adjacent edges in the molecular graph and $\delta(e_i)$ is the edge's degree:

$$\delta(e_k) = \sum_{i} g_{ik} = \sum_{j} g_{kj} \tag{3.116}$$

where the g_{ik} are the elements of the edge adjacency matrix $\boldsymbol{E}$. An analogous topographic index $\varepsilon(\rho)$has been calculated by the same formula (3.115). However, instead of topological edge degrees, the quantum chemically calculated bond orders ρ_{AB} between any two atoms A and B in the molecule were used for the edge adjacency [265,266].

The topological descriptors have been extensively used for the QSAR/QSPR description and prediction of different properties of compounds. Those include the boiling points and critical temperatures of compounds [267–270], viscosity [271], dipole moments [272], spectroscopic data [273–276], pK_a of acid–base equilibria [277–280], thermodynamic and thermochemical data [281–288], partition coefficients [289–292], chromatographic characteristics [293–300], toxicity [301–303], and other biological activities of compounds [304–324]. The large interest in the application of various topological descriptors in QSAR/

QSPR has been due to their success in many areas. For instance, Katritzky and Gordeeva [325] had carried out a comparison of the descriptive power of topological indices and other molecular descriptors. The latter involved various constitutional, electronic, geometrical, and combined descriptors. The properties studied included melting point, boiling point, refractive index, molar volume, and density of organic compounds. In addition, the correlations were developed to describe the anaesthetic and narcotic activity and the sweetness intensity of chemicals. In total 84 descriptors were used in this study, together with some normalized and square forms. Somewhat surprisingly, the best regression models of physico-chemical properties involving one to four parameters were mainly comprised of topological indices such as the Randić index, Wiener index, or molecular connectivity indices. For the correlation of biological activity, combinations of topological indices with geometrical descriptors resulted in the models of the best quality.

One of the most widely studied properties using the topological indices is the normal boiling point of compounds. The boiling points (bp) for the first 150 alkanes were successfully described by the following quadratic equation involving the topological distance index TI (3.56) [326]:

$$\text{bp} = (-135.07 \pm 5.90) + (35.30 \pm 1.49)\ln \text{TI} + (0.787 \pm 0.092)(\ln \text{TI})^2$$

$$R^2 = 0.9809 \qquad s = 5.93 \qquad n = 149 \qquad F = 3735$$

This model was a little inferior of the analogous correlation employing the Randić connectivity index, χ (3.48):

$$\text{bp} = (-172.93 \pm 4.93) + (104.14 \pm 3.02)\chi + (6.899 \pm 0.459)\chi^2$$

$$R^2 = 0.9833 \qquad s = 6.39 \qquad n = 150 \qquad F = 4318$$

An excellent correlation with $R^2 = 0.999$ was obtained for the boiling points of 74 normal and branched alkanes using a model that combined five different connectivity indices [327]. In this model, the first-order connectivity index ${}^1\chi$ was the most significant parameter, which already described the most of the variance ($R^2 = 0.969$). The data on the boiling points for a more limited set of 62 alkanes and cycloalkanes has been described using only one descriptor as [328]:

$$t_B = 46.6 \ln(R + 0.4) - 115.6$$

$$R = 0.982 \qquad s = 11.6 \qquad F = 783$$

where R denotes the hyper-Wiener index by Randić [329].

The data on boiling points of 41 alkanes were correlated by another highly significant single-parameter QSPR equation [206]:

$$\text{bp} = -103.706 + 80.384\{\log(\text{per}[\boldsymbol{D}])\}^{1/2}$$
$$R^2 = 0.993 \qquad F = 5514$$

where per[$\boldsymbol{D}$] denotes the permanent of the topological distance matrix (3.57). The data for methane was excluded in the last treatment. A similar equation was obtained for the boiling points of 16 linear alkenes as [209]:

$$\text{bp} = -163.40 + 103.70\{\log(\text{per}[\boldsymbol{D}])\}^{1/2}$$
$$R^2 = 0.981 \qquad s = 7.729 \qquad n = 16 \qquad F = 712$$

However, a better correlation for the alkene data has been developed using the product of the row sums of the topological distance matrix, PRS[$\boldsymbol{D}$] as the descriptor [206,209].

$$\text{bp} = -167.01 + 89.27\{\log(\text{PRS}[\boldsymbol{D}])\}^{1/2}$$
$$R^2 = 0.991 \qquad s = 5.296 \qquad n = 16 \qquad F = 1533$$

The electron charge density connectivity index $\Omega(q)$ (3.107) [259] correlates also with the boiling points of alkenes. For a set of 53 linear and branched alkenes, the following QSPR equation was developed [259]:

$$\text{bp} = -79.967 + 61.437\Omega(q)$$
$$R^2 = 0.9685 \qquad s = 5.62 \qquad n = 53 \qquad F = 1561$$

The addition of an indicator variable, the number of methyl groups bonded to the double bond of alkene, n_{Me}, has enabled a significant improvement of the last equation:

$$\text{bp} = -90.986 + 63.966\Omega(q) + 5.041n_{\text{Me}}$$
$$R^2 = 0.9910 \qquad s = 3.03 \qquad n = 53 \qquad F = 2743$$

Various Schultz indices have been applied for the description of the boiling points of aliphatic alcohols [210]. For a set of 26 alcohols, the best correlation was obtained with the MTI′ index (3.54) as:

$$\text{bp} = -79.967 + 61.437\Omega(q)$$
$$R^2 = 0.9685 \qquad s = 5.62 \qquad n = 53 \qquad F = 1561$$

A three-layer 19:5:1 neural network was trained to describe the boiling points of a much larger set of 298 structurally variable organic compounds [330]. These included aliphatic, alicyclic, and aromatic hydrocarbons, alcohols, amines, ethers, ketones, aldehydes, carboxylic acids, esters, nitriles, and halogen derivatives. The 19 input nodes corresponded to 19 atom-type electrotopological state indices [331]. The mean average error was 3.93 K for the whole set, 3.86 K for the training set, and 4.57 K for the test set of 30 compounds. Five atomic electrotopological state indices, corresponding to the fragments CH_3, CH_2, CH, C, and OH, respectively, had been applied earlier in the QSPR correlation of the boiling points of a structurally more restricted set of 245 alkanes and alkyl alcohols. The resulting QSPR linear regression had $R = 0.97$ and an average error of 8 K [331]:

$$\begin{aligned} \text{bp} = &-(43.95 \pm 3.53) + (8.21 \pm 0.30)\text{SsCH}_3 + (14.86 \pm 0.28)\text{SssCH}_2 \\ &+ (24.56 \pm 0.99)\text{SsssCH} + (43.76 \pm 2.58)\text{SssssCH} \\ &+ (11.63 \pm 0.22)\text{SsOH} \end{aligned}$$

$$R^2 = 0.941 \qquad s = 8.0 \qquad n = 245 \qquad F = 755$$

In the last equation, $SsCH_3$, $SssCH_2$, SsssCH, and SssssC denote the sum of E-state values for all methyl groups, methylene groups, CH-, and C-groups in the molecule, respectively, and SsOH is the value of the E-state for the hydroxyl group. Each term in this equation is highly significant, with the respective Student's t value spanning from 16.9 to 53.8.

A similar three-layer 19:5:1 neural network using 19 atom-type electrotopological state indices as the input has been employed to describe a set of critical temperatures of 165 structurally variable organic compounds [330]. The obtained mean average error was 4.52 K for the whole set, 4.39 K for the training set, and 5.59 K for the test set of 18 compounds.

The success of topological indices in the description of macroscopic bulk properties of compounds has been explained as a result of hidden relationships between the topological characteristics of a molecule and the intermolecular interactions in condensed media. Thus, in a factor analysis of six physical properties related to intermolecular interactions in the liquid state (including boiling point), two factors, namely "bulk" and "bulk-corrected cohesiveness" of a molecule, were found to determine the observed physical properties [332]. It has also been shown that Wiener and Randić molecular connectivity indices effectively represent the molecular van der Waals volume or the molecular bulk in the liquid state [333,334]. Cohesiveness can be described by a shape-dependent variable, such as a topological shape descriptor [335]. Although molecular connectivity indices also offer a suitable basis for attributing shape-dependent variance, this is often disguised by their bulk dependence [336]. The use of orthogonalized connectivity indices has been suggested by Randić as a possible solution to this difficulty [337].

The quantification of the intermolecular interaction effects using topological indices has been applied for the description of gas chromatographic retention characteristics of compounds [338–342]. Trinajstić and co-workers [343] used the natural and orthogonalized [344,345] path numbers, Randić connectivity indices, and Harare indices to develop a QSPR equation for the high-pressure liquid chromatography (HPLC) retention times of anthocyanidin malonyl-glucosides.

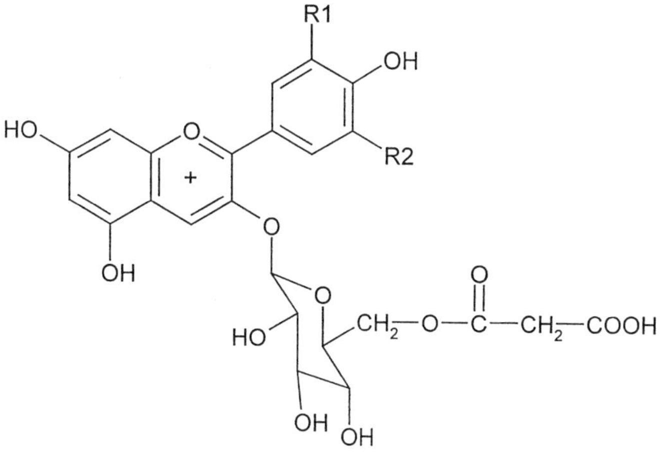

The best correlation was obtained with three sequential path numbers, p_1, p_2, and p_3, as:

$$t_R = -(6.800 \pm 1.483) + (12.503 \pm 0.344)p_1 - (8.777 \pm 0.297)p_2 + (0.630 \pm 0.077)p_3$$

$$R^2 = 0.9978 \qquad s = 0.29 \qquad n = 12 \qquad F = 4484$$

The path number of a given (lth) order is defined as the number of paths of this length (l) in the molecule. The correlation with the orthogonalized path numbers, ${}^1\Pi$, ${}^2\Pi$, and ${}^3\Pi$, had the same statistical characteristics.

$$t_R = -(37.643 \pm 1.128) + (1.436 \pm 0.029)^2\Pi - (7.181 \pm 0.223)^2\Pi + (0.630 \pm 0.077)^3\Pi$$

$$R^2 = 0.9978 \qquad s = 0.29 \qquad n = 12 \qquad F = 4484$$

However, the numerical values of the second and third descriptor and the regression coefficients were different because of the orthogonalization of scales.

The topological indices correlate with a number of physical properties of compounds. Thus, the edge adjacency index ε (3.115) has been used to describe the experimental molecular volume MV and molecular refraction MR of C_5—C_9 alkanes as [264]:

$$\mathrm{MV} = 57.8501 + 30.8559\varepsilon$$

$$R^2 = 0.9862 \qquad s = 2.032 \qquad n = 69 \qquad F = 4831$$

and

$$\mathrm{MR} = 7.6512 + 9.3742\varepsilon$$

$$R^2 = 0.9827 \qquad s = 0.698 \qquad n = 69 \qquad F = 3782$$

respectively. The use of an analogous topographical index $\varepsilon(\rho)$ has resulted in a slightly better correlation for the molecular refraction [346]:

$$\mathrm{MR} = 7.7314 + 9.2009\varepsilon(\rho)$$

$$R^2 = 0.9872 \qquad s = 0.598 \qquad n = 69 \qquad F = 5169$$

The good correlation between the $\varepsilon(\rho)$ index and the molecular refraction indicates that the former is a good measure of effective molecular volume in a condensed medium. It has been suggested [346] that this descriptor also contains information about polarizability and polarity of molecules in the pure liquid. As the respective electrostatic interactions are important in processes such as the partition of solutes between two phases, the $\varepsilon(\rho)$ index could be effective in the development of QSAR equations on biological activity.

The topological descriptors that reflect charge distribution of the molecule can also be effective in describing the related molecular properties. For example, the topological charge indices G_k (3.109) and J_k (3.110) have been employed to develop the following QSAR equation for the dipole moments of hydrocarbons [260]:

$$\mu(D) = -0.037 - 0.179G_1 - 0.134G_2 + 1.365G_4 + 0.789J_1 + 0.820J_2 + 2.421J_3$$

$$R^2 = 0.885 \qquad s = 0.107 \qquad n = 30 \qquad F = 28.0$$

This result had lead to a conclusion that the topological charge indices are indeed superior to the simple connectivity indices in describing the charge-distribution-dependent experimental properties of compounds.

The topological descriptors have also been useful for the correlation of spectroscopic and less common physical properties of compounds. For example, a single descriptor, the molecular walk count of the second order, mwc_2, has given a satisfactory correlation with the sum of the ^{13}C NMR shifts δ_C in a molecule for a set of 90 linear and branched alkanes [347].

$$\sum \delta_C = 16.86 + 5.98\mathrm{mwc}_2$$

$$R^2 = 0.9328 \qquad s = 27.46 \qquad n = 90 \qquad F = 1221$$

An interesting application of topological indices is the prediction of novel odorants [348]. For instance, the odor similarity, OS, of substituted benzaldehydes and nitrobenzenes has been estimated relative to the unsubstituted benzaldehyde, which is the standard for the smell of bitter almond [349]. The best correlation of this property was obtained with Kier connectivity indices as [350]:

$$OS = 13.6 - 8.08\,{}^{3}\chi^{v} + 2.19\,{}^{4}\chi^{v}_{pc}$$

$$R^2 = 0.926 \qquad s = 0.545 \qquad n = 26$$

where ${}^{3}\chi^{v}$ is the third-order valence molecular connectivity index and ${}^{4}\chi^{v}_{pc}$ is the fourth-order valence path cluster molecular connectivity index, respectively. Notably, this correlation was better than any correlation using the "physical" molecular descriptors such as log P, molar refractivity, energy of frontier molecular orbitals, or dipole moment of compounds. The Kier connectivity indices have also been used to predict successfully the smell of jasmine compounds and the odor quality of muguet (lily of the valley) aldehydes [351]. Using the predictions made from the latter relationship, the following new compound had been synthesized:

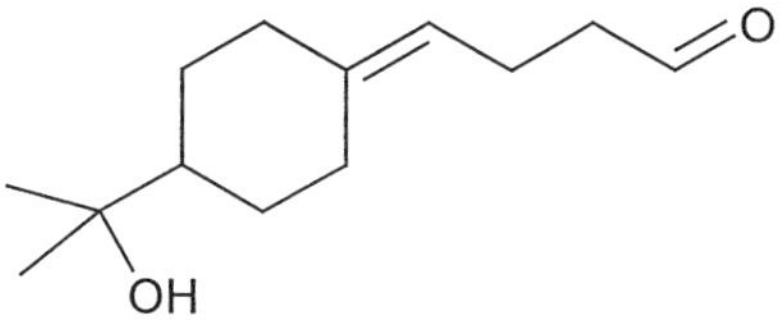

with excellent olfactive (perfume) properties.

The odor can be considered as a biological property of a compound. Numerous other biological activities have also been described using the QSAR equations involving the topological descriptors. The activity of 19 flavone derivatives

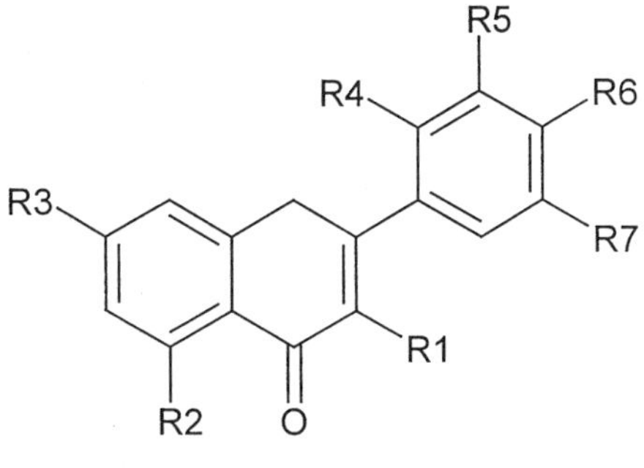

to inhibit cyclic adenosine monophosphate (cAMP) phosphodiesterase has been correlated with four topological indices as [352]:

$$\%\mathrm{INH} = (235.5 \pm 33.2) - (15.86 \pm 4.00)\,{}^{0}\chi^{v} - (7.856 \pm 1.482)\,{}^{5}\chi^{v} + (10.24 \pm 1.69)p_{10} + (0.606 \pm 0.146)H_2$$

$$R^2 = 0.905 \qquad s = 6.36 \qquad n = 19 \qquad F = 33.30$$

where %INH is the percent of the cAMP inhibition, p_{10} denotes the number of paths with length of 10 and H_2 is the Harare index [219] with squares of reciprocal distances. A notable improvement of the last correlation was obtained by employing additional electronic descriptors. Those included the absolute and the relative hardness of molecules [353], the energy of the lowest unoccupied orbital, the topological resonance energy per electron [354], and the sum of π charges in the chromone or phenyl moiety. The inhibition of cyclooxygenase by compounds of the following generic structure

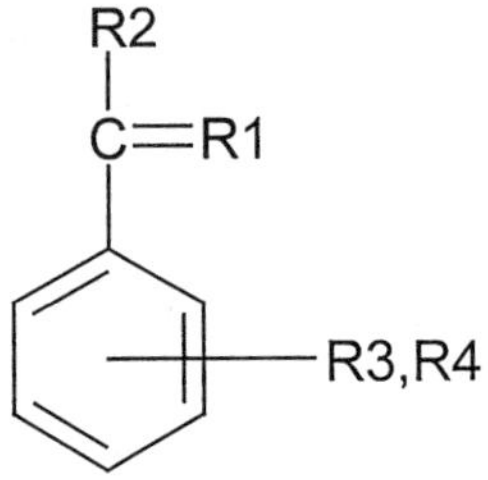

that is related to nonnarcotic analgesic action, was correlated with the Kier connectivity indices as [355]:

$$\log \mathrm{IC}_{50} = 3.056 - 2.64\,{}^{3}\chi_{p}^{v} + 1.8\,{}^{4}\chi_{p} - 10.28\,{}^{4}\chi_{c} + 1.24\,{}^{4}\chi_{pc}$$

$$R^2 = 0.715 \qquad s = 0.28 \qquad n = 171 \qquad F = 9.42$$

The relatively poor correlation witnesses the need to account for other types of molecular descriptors. Also, an unsuccessful attempt has been reported [356] to correlate the minimum inhibitory concentration (MIC) of 2-substituted isonicotine hydrazides

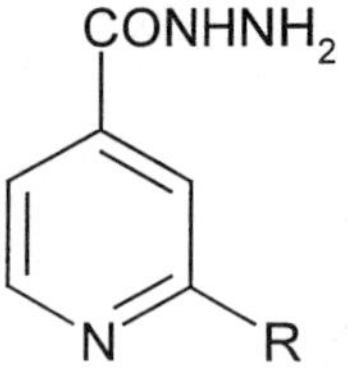

against *Mycobacterium tuberculosis* using linear and multiple linear regression with 14 different topological indices. The results were significantly improved when the principal component analysis [357] was carried out using the same 14 topological indices and the molecular weight (MW) of compound. The principal component derived from five parameters, Kier index of molecular shape $^3\kappa$ [230], information indices of neighborhood symmetry IC_0 and IC_1 [308], structural information content index SIC_0 [358], and the molecular weight MW described satisfactorily the 1/MIC data (R = 0.798). The square root transformation of the principal component scores led to a further improvement of the correlation (R = 0.891). The multiparameter QSPR on log(1/MIC) using the stepwise regressions technique resulted in the following best four-parameter relationship [359]:

$$\log(1/\mathrm{MIC}) = (-1.029 \pm 0.378)^1\kappa - (9.757 \pm 2.701)\mathrm{IC}_0 - (3.144 \pm 1.361)\mathrm{CIC}_0 = + (0.0148 \pm 0.0040)W + 29.372$$

$$R^2 = 0.9095 \qquad s = 0.2953 \qquad n = 17 \qquad F = 30$$

where $^1\kappa$ is the Kier index of molecular shape and W denotes the Wiener index. The topological descriptors have also been applied in the discriminant analysis and prediction of the biological activity of compounds [360].

The Wiener index [176,361,362] and the Randić connectivity index [195] have been applied for the prediction of the antiulcer activity of 128 analogs of 4-substituted-2-guanidino thiazole [363]. No correlation equations, however, were reported. Likewise, the Randić connectivity index has been used in the prediction of the antiinflammatory activity of 76 pyrazole carboxylic acid hydrazide analogs [364].

One of the most popular molecular descriptors used in the correlations of biological activity of compounds has been Hansch's water–octanol partition coefficient, log P. It has been attempted to predict the values of this empirical descriptor using the correlation with topological indices. For instance, the log P for a set of 137 organic compounds with weak hydrogen bonding ability (hydrocarbons, sulfides, and halogen derivatives) have been described using the following five-parameter equation [365]:

$$\log P = -3.127 - 1.644\mathrm{IC}_0 + 2.120\,^5\chi_c - 2.914\,^6\chi_{\mathrm{ch}} + 4.208\,^0\chi^\nu + 1.060\,^4\chi^\nu - 1.020\,^4\chi^\nu_{\mathrm{pc}}$$

$$R^2 = 0.97 \qquad s = 0.26 \qquad n = 137$$

where IC_0 denotes the information content of the zeroth order and $^n\chi^{(\nu)}_{(c,\mathrm{ch\ or\ pc})}$ are the valence (ν), cluster (c), chain or cycle (ch), and path cluster (pc) connectivity indices of the given order n, respectively. The correlation of the log P values for a set of 382 structurally more variable compounds involved also a hydrogen bonding descriptor, HB_1, and the number of paths with length 6 in the hydrogen-suppressed graph of the molecule, K_6 [365].

$$\log P = 1.76 - 5.28\mathrm{IC}_0 - 1.48\mathrm{CIC}_1 - 2.914\,{}^6\chi_{\mathrm{ch}} + 2.75\,{}^0\chi^{\nu} - 0.50\mathrm{HB}_1 + 0.41K_6$$

$$R^2 = 0.97 \qquad s = 0.26 \qquad n = 137$$

The CIC_1 denotes the first-order complementary information content index (3.95).

In conclusion, the topological and related information-theoretic and topographical molecular descriptors can be considered as convenient numerical scales for the correlation of properties of compounds. However, in most cases the physical interpretation of the resulting QSAR/QSPR equations is not easy because of the formal nature of those descriptors.

3.4 CHARGE-DISTRIBUTION-RELATED DESCRIPTORS

According to the contemporary theory of molecular structure, all chemical interactions are by nature either electrostatic (polar) or orbital (covalent). Obviously, the electrical charges in the molecule are the driving force for electrostatic interactions. Indeed, it has been long established that the local electron densities or charges determine the mechanism and rate of most chemical reactions and physico-chemical properties of compounds. Therefore, the charge-based descriptors have been widely employed as chemical reactivity indices or as measures of intermolecular interactions in QSAR/QSPR treatments. Many electrostatic descriptors have been derived from the electronic charge distribution in a molecule or from the electron densities on particular atoms. The electrostatic attraction and repulsion between the molecules or between different fragments of a molecule are often responsible for the interactions determining the chemical and physical properties of a compound in condensed media. Therefore, the appropriately defined electrostatic or charge-distribution-related molecular descriptors are directly related to the energy of the electrostatic interaction and applicable in the development of the respective QSAR/QSPR equations.

The charge distribution of the molecule can be described at various levels of theory. First, the empirical schemes based on the concept of electronegativity of atoms have been developed for the calculation of the atomic partial charges and the respective electrostatic field of the molecule. One of the first successful empirical schemes for the calculation of atomic partial charges was invented by Gasteiger and Marsili [366,367]. Their scheme is based on Sanderson's principle of equalization of atomic electronegativities in the molecule [368,369]. According to this principle, the charge redistribution during the bond formation leads to the equalization of electronegativities of all atoms in the molecule. However, the total equalization is unacceptable, as this implies that all atoms of the same chemical type will have the same electronegativity and, respectively, the same charge in the molecule. As pointed out in Gasteiger and

Marsili [367], all hydrogen atoms in acetic acid will thus receive the same charge whereas undoubtedly the acid-group hydrogen atom should have lower electron density than the hydrogen atoms of the methyl group. Therefore, Gasteiger and Marsili proceeded from the Mulliken definition electronegativity:

$$\chi_\nu = \frac{I_\nu + E_\nu}{2} \tag{3.117}$$

where I_ν and E_ν denote the ionization potential and electron affinity, respectively, and applied it to individual orbitals (ν) in atoms. A quadratic dependence of any given νth orbital electronegativity on the total charge Q_i on atom i was postulated then as:

$$\chi_{i\nu} = a_{i\nu} + b_{i\nu}Q_i + c_{i\nu}Q_i^2 \tag{3.118}$$

where the coefficients $a_{i\nu}$, $a_{i\nu}$, and $a_{i\nu}$ can be straightforwardly determined from the following equations:

$$a_{i\nu} = \frac{I_{i\nu}^0 + E_{i\nu}^0}{2} \tag{3.119}$$

$$b_{i\nu} = \frac{I_{i\nu}^0 + E_{i\nu}^+ - E_{i\nu}^0}{4} \tag{3.120}$$

$$b_{i\nu} = \frac{I_{i\nu}^+ - I_{i\nu}^0 + E_{i\nu}^+ - E_{i\nu}^0}{4} \tag{3.121}$$

In these equations, $I_{i\nu}^0$, $I_{i\nu}^+$, $E_{i\nu}^0$, and $E_{i\nu}^+$ correspond to the ionization potentials and electron affinities of the neutral atom (superscript 0) and the positive ion (superscript +), respectively. The atomic charge was then calculated iteratively according to the following equation:

$$q_i^{<\alpha>} = \left(\frac{1}{2}\right)^\alpha \sum_{\nu \in i} \left[\sum_{\mu \in j} \frac{\chi_{j\mu}^{<\alpha>} - \chi_{i\nu}^{<\alpha>}}{\chi_{i\nu}^+} + \sum_{\lambda \in k} \frac{\chi_{k\lambda}^{<\alpha>} - \chi_{i\nu}^{<\alpha>}}{\chi_{k\lambda}^+} \right] \tag{3.122}$$

where $q_i^{<\alpha>}$ is the contribution to the atomic charge obtained on αth iteration step. The first sum in square brackets applies for orbitals in all directly neighboring atoms j that are more electronegative than the given (ith) atom. The second summation is carried out over the orbitals belonging to the less electronegative atoms k directly bonded to the given atom. After each step, the total charge on ith atom was calculated using the contributions from each iterative step as:

$$Q_i = \sum_\alpha q_i^{<\alpha>} \tag{3.123}$$

and applied in Eq. (3.118) for the next step of iteration. Obviously, this procedure considers only the atoms that are directly bonded to a given atom. Furthermore, it has been pointed out that the geometry of a molecule is also not accounted for and therefore, in principle, this approach is purely topological by nature. Originally, the above-described method of calculation of atomic charges was developed only for molecules with σ and nonconjugated π bonds. However, it had been later extended also for the molecular systems with conjugated π bonds [370] and to small ring systems [371]. Notably, the calculated atomic partial charges correlated excellently ($R = 0.993$, $s = 0.18$ eV) with the experimentally measured electron spectroscopy for chemical analysis (ESCA) $1s$ core electron binding energies of the carbon atom in a series of simple organic compounds [370] and with the dipole moments of compounds [371]. In addition, a fair correlation was found between the pK_a of structurally variable compounds in aqueous solutions and the calculated partial charges on acidic hydrogen [372]. These results suggest that the empirically calculated charges are consistent with the electrostatic models of molecular phenomena.

Zefirov et al. [373,374] have proposed two schemes for the calculation of partial charges in the molecule based on the same Sanderson electronegativity equalization principle but using the representation of the molecular electronegativity as a geometric mean of atomic electronegativities:

$$\chi_i = \left(\chi_i^0 \prod_{k=1}^{n} S_{ik} \right)^{1/(n+1)} \tag{3.124}$$

where χ_i is the electronegativity of atom i to be found in the molecule, χ_i^0 is the electronegativity of the isolated atom, S_{ik} are some coefficients attributed to the neighboring atoms k in the molecule, and n is the overall number of atoms bonded directly to the given atom. The first scheme for the calculation of electronegativities in the molecule proceeds from the assumption that the coefficients S_{ik} are simply the electronegativities of other bonded atoms in the molecule, χ_k:

$$\chi_i = \left(\chi_i^0 \prod_{k=1}^{n} \chi_k \right)^{1/(n+1)} \tag{3.125}$$

It was shown that by taking the logarithm of the last equation

$$\log \chi_i = \frac{1}{n+1} \left(x_i^0 + \sum_{k=1}^{n} x_k \right) \tag{3.126}$$

a system of linear equations can be formed for any molecular graph and the unknown variables $x_k = \log \chi_k$ found as the solutions of this system of equations. Zefirov's second scheme for the calculation of atomic electronegativities in the molecule [373] is based on the equalization of the electronegativities of the edges of molecular graph. The atomic electronegativities obtained can be further directly applied for the calculation of partial charges.

Mullay [375,376] has proposed another empirical method of calculation for the atomic partial charges based on the orbital electronegativities. According to his scheme, the electronegativity of an orbital ν on atom i was expressed as:

$$\chi_{i\nu}(\delta_{i\nu}) = \chi_{i\nu}^{0}\left(1 + 0.5 \sum_{\substack{\mu \in i \\ \mu \neq \nu}} \delta_{i\mu} + 1.5\delta_{i\nu}\right) \tag{3.127}$$

where $\chi_{i\nu}^{0}$ is the electronegativity of the νth orbital in isolated neutral atom and $\delta_{i\mu}$ are the charges on the respective orbitals (μ). The summation is carried out over all other orbitals at the same atom. The total charge on ith atom is obtained by adding all orbital contributions:

$$\delta_i = \sum_{\mu \in i} \delta_{i\mu} \tag{3.128}$$

The atomic total partial charges were then derived by applying two constraints to each bond in the molecule. The first constraint requires that the electronegativities of both atomic orbitals participating in the bond formation should be equal. The second requirement implied the charge conservation during the bond formation. These two constraints were summarized by two simple equations for every bond k between atoms i and j as:

$$\chi_i(\delta_{ik}) = \chi_j(\delta_{jk}) \tag{3.129}$$

$$\delta_{ik} + \delta_{jk} = 1 \tag{3.130}$$

Therefore, for each bond in the molecule, two equations with two unknowns had been obtained that could be solved to obtain the values of δ_{ik} and the total atomic partial charges. Notably, whereas the numerical values are rather different, a good correlation may exist between the partial charges obtained using the different above-described methods for structurally similar compounds. Therefore, for the purpose of many QSAR applications, the choice of the method of calculation of charges is not crucial and depends mostly on user preference.

The atomic connectivity indices of Kier and Hall [170] have been found to be in good correlation with CNDO/2-calculated Mulliken atomic partial charges in alkanes [377,378]. Thus, it has been suggested to define the topological atomic charges, atomic valences, and the bond orders proceeding from the topological self-returning walks (SRWs) [379,380]. The SRWs are defined as the topological walks that begin and end at the same vertex (atom). It was shown that in the framework of the Hückel-type theory of molecular electronic structure, the molecular kth moment of energy, μ^k, defined as the sum of its Hamiltonian eigenvalues of kth order

$$\mu^k = \sum_i E_i^k = \mathrm{SRW}^k \beta^k \quad (3.131)$$

is related to the self-returning walk of the same order (β denotes the Hückel interaction integrals). It was also proven that the limit of relative atomic moments of energy

$$f_i = \lim_{k \gg 1} f_i^k = \lim_{k \gg 1} (\mu_i^k / \mu^k) = \lim_{k \gg 1} (\mathrm{SRW}_i^k / \mathrm{SRW}^k) \quad (3.132)$$

corresponds to atomic partial charge. The topological bond orders have been defined as the limit of the ratio of the number of self-returning walks beginning and ending on the edge ij, and the total number of self-returning walks of the same length in the molecule [381]:

$$p_{ij} = \lim_{k \gg 1} p_{ij}^k = \lim_{k \gg 1} [(\mathrm{SRW}_{ij}^k + \mathrm{SRW}_{ji}^k)/\mathrm{SRW}^k] \quad (3.133)$$

It should be emphasized that the applicability of these formulations has been proven only in the case of hydrocarbons and within the approximations of Hückel MO theory. Moreover, it was shown [381] that the atomic partial charges f_i correspond to the lowest occupied molecular orbital (LOMO) of the molecule that is usually chemically inactive and thus irrelevant in structure–activity correlations. For an overview, numerous other empirical schemes can be referred for the calculation of the atomic charges in the molecule [382–398].

The partial charges on atoms can also be obtained from quantum chemical calculations. Almost any standard quantum chemical program gives in output the Mulliken atomic partial charges. The definition of those charges is based on MO theory. In the framework of the linear combination of atomic orbitals (LCAO) MO theory, the total electronic charge of a molecule is given as [399]:

$$\rho_{\mathrm{el}} = |\Psi_{\mathrm{el}}|^2 = 2 \sum_{i=1}^{\mathrm{occ}} \sum_{k=1}^{n} \sum_{l=1}^{n} c_{ik} c_{il} S_{kl} = 2 \sum_{k=1}^{n} \sum_{l=1}^{n} \rho_{kl} S_{kl} = \sum_{k=1}^{n} \sum_{l=1}^{n} P_{kl} \quad (3.134)$$

where Ψ_{el} is the total many-electron wave function of the molecule, and c_{ik} and c_{il} are the expansion coefficients for an MO ϕ_i on a given set of atomic orbitals $\{\chi_k\}$:

$$\phi_i = \sum_{k=1}^{n} c_{ik} \chi_k \quad (3.135)$$

In Eqs. (3.134) and (3.135), n denotes the number of AOs in the basis set,

$$S_{kl} = \int_{(V)} \chi_k \chi_l \, dv \tag{3.136}$$

is the overlap integral for a pair of atomic orbitals, χ_k and χ_l, respectively, and P_{kl} are the elements of the population matrix. The Mulliken definition for atomic charge can be presented as [400]:

$$\delta_A = Z_A - \left(\sum_{k \in A} P_{kk} + \frac{1}{2} \sum_{l \neq k} P_{kl} + \frac{1}{2} \sum_{l \neq k} P_{lk} \right) = Z_A - \sum P_{kl} \tag{3.137}$$

where Z_A is the nuclear (core) charge of the atom, and the summation of Mulliken populations P_{kl} is made over all (valence) atomic orbitals at the given atom A. It is well known that the Mulliken definition of the atomic charge is arbitrary, and other definitions have been developed, though none of them correspond to any directly experimentally measurable quantity [401]. Moreover, the quantum chemical methods are mostly designed to optimize the total energy of the molecule. Various semiempirical methods are also parameterized to reproduce the heats of formation, ionization potentials, or geometric characteristics of the molecules. Therefore, the calculated atomic charges may not be as reliable. For these reasons, the values of atomic charges calculated by different quantum chemical methods are sometimes in poor agreement with each other. However, those numerical quantities are easy to obtain and they give at least a qualitative picture of the charge distribution in a molecule.

Nevertheless, different, more or less arbitrary, methods have been developed for the calculation of atomic charges using the analysis of the quantum mechanical wave function of the molecule. For instance, a molecule can be divided into atomic regions and each atomic charge calculated thereupon by the integration of the charge density in each such region [402–404]. A variety of approaches has been proposed for the partitioning of the molecule into atomic regions. One interesting partitioning algorithm is based on Shannon's information theory [405–407]. In a two-electron system, the probability that electron 1 is in space region V_A and electron 2 is simultaneously placed in region V_B (regions V_A and V_B are nonoverlapping), is given by the following equation:

$$p_1 = \int_{(V_A)} dv_1 \int_{(V_B)} dv_2 \Psi^2(\mathbf{x}_1, \mathbf{x}_2) \tag{3.138}$$

where $\Psi(\mathbf{x}_1, \mathbf{x}_2)$ is the total electronic wave function of the system. The probability that both electrons are in region V_A is presented as:

$$p_2 = \int_{(V_A)} dv_1 \, dv_2 \Psi^2(\mathbf{x}_1, \mathbf{x}_2) \tag{3.139}$$

whereas the probability that both electrons are in region V_B is

$$p_3 = \int_{(V_B)} dv_1\, dv_2 \Psi^2(\mathbf{x}_1, \mathbf{x}_2) \tag{3.140}$$

The maximum information about the distribution of two electrons can now be obtained by the partitioning of the whole physical space into two regions that correspond to the minimum of the respective information entropy:

$$I = \sum_{i=1}^{3} p_i \log_2 p_i \tag{3.141}$$

The corresponding regions in the atoms or molecules have been named the lodges. Other schemes have also been used for the separation of physical space into the atomic regions [408–411] from which the atomic charges can be obtained by the respective integration of the square of the molecular wave function.

The atomic partial charges can also be obtained by analyzing certain physical observables predicted from the wave function. For instance, the atomic charges can be determined by fitting the predicted interaction energies [412,413], dipole moments, or electrostatic potential values inside and around the molecule [414–432]. To cite some instances, Cramer, Truhlar, and co-workers have proposed two sets of parameters for the calculation of corrected partial charges in molecules based on semiempirical quantum chemical theory [433]. In both cases, the improved partial charge for a given atom k was obtained as:

$$q_k^{\mathrm{CM}} = f(q_k^{(0)}, A, B, \ldots) \tag{3.142}$$

where f is a semiempirical functional, $q_k^{(0)}$ are the charges obtained from the semiempirical wave function using Mulliken's method, and A, B, . . . are parameters adjusted to achieve the agreement with a set of physical observables. The first set of parameters (CM1A = charge model 1A) was developed using the AM1 (Austin Model 1) [434] semiempirical wave functions for 186 diverse molecules containing H, C, F, Si, S, Cl, Br, and I. The properties fitted included the experimental dipole moments for all compounds and the theoretical dipole moments calculated for nine compounds using ab initio MP2/cc-pVDZ theory [435]. In addition, partial charges in 25 ions were compared with the ab initio partial charges. The second set of parameters (CM1P) was based on initial Mulliken charges obtained using the semiempirical PM3 (Parametric Method 3) method [436,437]. Notably, only experimental dipole moments were used in respective fitting procedure.

On the basis of the atomic partial charges, a number of simple electrostatic descriptors can be calculated and applied in the development of QSAR/QSPR equations:

The minimum (most negative) and maximum (most positive) partial charges in the molecule ($q_{\min}$, $q_{\max}$)

The minimum and maximum partial charges for particular types of atoms (e.g., C, O, etc.)

A polarity parameter ($q_{\max} - q_{\min}$) and polarity parameter factorized by the division by a function of the distance r_{mx} between atoms bearing minimum and maximum partial charges as:

$$P_f = \frac{q_{\max} - q_{\min}}{F(r_{mx})} \tag{3.143}$$

The function $F(r_{mx})$ can be represented by the distance r_{mx} itself, the square of this distance, or by some other function. In the first case, the factorized polarity resembles the electrostatic interaction energy between these two charges whereas the square dependence on the distance corresponds to the electrostatic force between the charges. It may seem trivial, but still it is important to remember the dimensionality of descriptors while using it in the QSAR/QSPR correlation. In the case of energies, free energies, or properties proportional to them to be correlated, the electrostatic energy [$F(r_{mx}) = r_{mx}$] is the proper descriptor to use.

Atomic partial charges have been referred to as the static chemical reactivity indices [438]. According to this view, the calculated σ- and π-electron densities on a particular atom characterize the possible direction of the chemical transformations and are thus considered the directional reactivity indices. On the contrary, the overall electron densities and net charges on atoms are considered as nondirectional reactivity indices [439]. Various sums of absolute or squared values of partial charges have been used to describe intermolecular interactions, for example, solute–solvent interactions [440–442]. Other common charge-based descriptors are the averages of the absolute values of atomic partial charges [401,443].

Charged partial surface area (CPSA) descriptors have been invented by Jurs et al. [444,445] in terms of the surface area of the whole molecule or its fragments and in terms of the charge distribution in the molecule. These descriptors are expected to encode the features responsible for polar interactions between molecules. In most cases, the CPSA descriptors are calculated from the contributions related to the atomic partial charges and the molecular solvent-accessible surface area. The total molecular surface area (TMSA) can be calculated, in principle, by using any scheme described above in relation to the geometrical molecular descriptors. Originally, the grid calculation was used to estimate the TMSA. In the following, we will give the formal definitions of CPSA descriptors utilized in QSPR equations.

The partial positive surface area (PPSA1) is defined as the sum of the positively charged solvent-accessible atomic surface areas, S_a, in the molecule:

$$\mathrm{PPSA1} = \sum_{A} S_A \qquad A \in \{\delta_A > 0\} \tag{3.144}$$

The total charge weighted partial positively charged surface area (PPSA2) is calculated as:

$$\mathrm{PPSA2} = \sum_{A} q_A \cdot \sum_{A} S_A \tag{3.145}$$

where q_A is the atomic partial charge, calculated either from some empirical scheme or from the quantum chemical wave function, and S_A is the respective atomic solvent-accessible surface area. The atomic charge weighted partial positive surface area (PPSA3) is obtained by the summation of products of the individual atomic partial charges and the atomic solvent-accessible surface areas:

$$\mathrm{PPSA3} = \sum_{A} q_A \cdot S_A \tag{3.146}$$

Notably, the summation in each of the last three equations [(3.144)–(3.146)] is carried out over the atoms with positive partial charges in the molecule ($\delta_A > 0$). Similar equations can be developed for the description of the partial negative charge distribution in the molecule, by using the summation over the atoms with the partial negative charge. The respective descriptors are denoted as the partial negative surface area (PNSA1), the total charge weighted partial negative surface area (PNSA2), and the atomic charge weighted partial negative surface area (PNSA3). The descriptors defined as the differences between the CPSA descriptors for the positive partial charged atoms and the negative partial charged atoms in the molecule, respectively, have been introduced as a measure of the polarity of a compound:

$$\mathrm{DPSA1} = \mathrm{PPSA1} - \mathrm{PNSA1} \tag{3.147}$$

$$\mathrm{DPSA2} = \mathrm{PPSA2} - \mathrm{PNSA2} \tag{3.148}$$

$$\mathrm{DPSA3} = \mathrm{PPSA3} - \mathrm{PNSA3} \tag{3.149}$$

The fractional partial positive surface area (FPSA1) has been defined as the ratio of the positively charged solvent-accessible surface area of the molecule PPSA1 and TMSA

$$\mathrm{FPSA1} = \frac{\mathrm{PPSA1}}{\mathrm{TMSA}} \tag{3.150}$$

Analogous descriptors, normalized by the total molecular solvent-accessible surface area, TMSA, can be calculated using PPSA2 and PPSA3, respectively.

$$\mathrm{FPSA2} = \frac{\mathrm{PPSA2}}{\mathrm{TMSA}} \tag{3.151}$$

$$\mathrm{FPSA3} = \frac{\mathrm{PPSA3}}{\mathrm{TMSA}} \tag{3.152}$$

Another three fractional partial negative surface area descriptors have been defined as:

$$\mathrm{FNSA1} = \frac{\mathrm{PNSA1}}{\mathrm{TMSA}} \qquad \mathrm{FNSA2} = \frac{\mathrm{PNSA2}}{\mathrm{TMSA}} \qquad \mathrm{FNSA3} = \frac{\mathrm{PNSA3}}{\mathrm{TMSA}} \tag{3.153}$$

The surface weighted charged partial positive charged surface area related molecular descriptors (WPSA) are given by the following definitions:

$$\begin{aligned} \mathrm{WPSA1} &= \frac{\mathrm{PPSA1} \cdot \mathrm{TMSA}}{1000} \\ \mathrm{WPSA2} &= \frac{\mathrm{PPSA2} \cdot \mathrm{TMSA}}{1000} \\ \mathrm{WPSA3} &= \frac{\mathrm{PPSA3} \cdot \mathrm{TMSA}}{1000} \end{aligned} \tag{3.154}$$

and the surface weighted charged partial negative charged surface area related descriptors (WNSA) as:

$$\begin{aligned} \mathrm{WNSA1} &= \frac{\mathrm{PNSA1} \cdot \mathrm{TMSA}}{1000} \\ \mathrm{WNSA2} &= \frac{\mathrm{PNSA2} \cdot \mathrm{TMSA}}{1000} \\ \mathrm{WNSA3} &= \frac{\mathrm{PNSA3} \cdot \mathrm{TMSA}}{1000} \end{aligned} \tag{3.155}$$

The dispersion of the partial charges in the molecule has been characterized by the electrostatic descriptors defined as the relative positive charge (RPCG) and the relative negative charge (RNCG). These descriptors are calculated as the ratio of the maximum (by absolute value) atomic partial charge and the sum of the similar partial charges in the molecule:

$$\mathrm{RPCG} = \frac{\delta^{+}_{\max}}{\sum_{A} \delta_A} \qquad A \in \{\delta_A > 0\} \tag{3.156}$$

$$\mathrm{RNCG} = \frac{|\delta^{-}|_{\max}}{\sum_{B} \delta_B} \qquad B \in \{\delta_B < 0\} \tag{3.157}$$

A set of electrostatic descriptors has been defined to account for the possible hydrogen-bonding interactions between the molecules. Thus, a descriptor characterizing the hydrogen-bonding donor ability of the molecule, HDSA1, was introduced as the sum of the solvent-accessible areas of hydrogen atoms active as the possible hydrogen bonding acceptors:

$$\mathrm{HDSA1} = \sum_{D} s_D \qquad D \in H_{\text{H-donor}} \tag{3.158}$$

Usually, the hydrogen atoms directly connected with an electronegative atom in the molecule (e.g., O or N) should be accounted for as the possible hydrogen-bonding donors. It has been also suggested to modify this descriptor by the inclusion of active hydrogen atoms adjacent to the carbon atoms directly connected with strong electron-accepting substituents. For instance, the hydrogen atoms in the α position to carbonyl- and cyano-groups can also be possible hydrogen-bonding donors with, of course, much smaller activity reflected by smaller partial charges on them.

The area-weighted surface charge of a hydrogen-bonding donor atom in the molecule, HDSA2, is calculated as:

$$\mathrm{HDSA2} = \sum_{D} \frac{q_D \sqrt{s_D}}{\sqrt{S_{\text{tot}}}} \tag{3.159}$$

where q_D is the partial charge on hydrogen-bonding donor (H) atom, s_D denotes the surface area for this atom and S_{tot} is the total molecular surface area, calculated from the van der Waals' radii of the atoms within the approximation of overlapping spheres. Analogously, the solvent-accessible area of hydrogen-bonding acceptor atoms in the molecule, HASA1, can be defined as:

$$\mathrm{HASA1} = \sum_{A} s_A \qquad A \in H_{\text{H-acceptor}} \tag{3.160}$$

whereas the area-weighted surface charge of the hydrogen-bonding acceptor atoms in the molecule, HASA2, can be calculated as:

$$\mathrm{HASA2} = \sum_A \frac{q_A\sqrt{s_A}}{\sqrt{S_{\mathrm{tot}}}} \tag{3.161}$$

where q_A is the partial charge on the hydrogen-bonding donor (H) atom(s), s_A is the surface area for this atom, and S_{tot} is the total molecular surface area. The electronegative atoms with lone electron pairs are usually considered as the potential hydrogen-bonding acceptors in the molecule (O, N, etc.). Alternatively, the inclusion of a hydrogen atom into the list of the hydrogen-bonding donor atoms in the molecule or the inclusion of the other atoms into the list of hydrogen-bonding acceptors may be based on the comparison of their partial charges with some threshold charge. Thus, if the positive partial charge on a given hydrogen atom is larger than the threshold charge ($\delta_{\mathrm{H}} > \delta_{\mathrm{thr}}$), it will be considered as a hydrogen bond donor. Likewise, if the negative partial charge on any atom in the molecule is less than a given (negative) threshold ($\delta_A < \delta_{\mathrm{thr}}$), the atom will be listed as a hydrogen bond acceptor. The respective descriptors, calculated formally according to Eqs. (3.158) to (3.161), have been labeled as HDCA1, HDCA2, HACA1, and HACA2, respectively.

In single-component systems, it has been suggested to perform the summations in Eqs. (3.158) to (3.161) over all simultaneously possible hydrogen-bonding donor and acceptor pairs per solute molecule. In mixtures, the pairs in all components may be accounted for, depending on the property studied. The hydrogen-bonding related electrostatic descriptors could also be normalized by the total molecular surface area. The respective fractional descriptors are defined as:

$$\mathrm{FHASA1} = \frac{\mathrm{HASA1}}{\mathrm{TMSA}} \qquad \mathrm{FHASA2} = \frac{\mathrm{HASA2}}{\mathrm{TMSA}} \tag{3.162}$$

and

$$\mathrm{FHDSA1} = \frac{\mathrm{HDSA1}}{\mathrm{TMSA}} \qquad \mathrm{FHDSA2} = \frac{\mathrm{HDSA2}}{\mathrm{TMSA}} \tag{3.163}$$

The descriptors

$$\mathrm{HBSA1} = \mathrm{HDSA1} - \mathrm{HASA1} \tag{3.164}$$

$$\mathrm{HBSA2} = \mathrm{HDSA2} - \mathrm{HASA2} \tag{3.165}$$

are defined as the differences between the respective hydrogen-bonding donor and hydrogen-bonding acceptor descriptors.

Politzer et al. have invented a methodology to calculate original electrostatic molecular descriptors for the development of QSAR/QSPR based mostly on quantum-chemically computed charge distribution in the molecule [446,447].

This methodology, called the general interaction properties function (GIPF), involves six descriptors including the molecular surface area and five descriptors derived from the charge distribution in the molecule, which have been defined as follows. The average ionization energy, $\bar{I}(\mathbf{r})$ [448,449], has been expressed by the following equation:

$$\bar{I}(\mathbf{r}) = \frac{\sum_i \rho(\mathbf{r})|\varepsilon_i|}{\rho(\mathbf{r})} \tag{3.166}$$

where $\rho(\mathbf{r})$ is the electron density of the ith molecular orbital at the point $\mathbf{r}$ and ε_i is the orbital energy. The descriptor $\bar{I}(\mathbf{r})$ has been interpreted as the average energy needed to remove an electron from any given point in the space of the molecule. This explanation is based on the Koopmans' theorem that relates the orbital energies to the respective ionization potentials in the molecule [450]. It has been shown that the plot of this descriptor against the molecular surface gives a quantitative indicator of the reactivity toward electrophilic attack [451,452]. The minima ($\bar{I}_{S,\text{min}}$) on the surfaces of conjugate bases have been correlated with the aqueous pK_a values and with gas-phase protonation enthalpies of several classes of compounds [451,453]. Both properties can be viewed as reflecting the reactivity toward the proton as electrophile.

The next two descriptors, $V_{S,\text{min}}$ and $V_{S,\text{max}}$, were defined as the minimum and the maximum of the electrostatic potential at the molecular surface, respectively. The electrostatic potential at any point inside and around a molecule can be defined rigorously as:

$$V(\mathbf{r}) = \sum_A \frac{Z_A}{|\mathbf{R}_A - \mathbf{r}|} - \int \frac{\rho(\mathbf{r}')\,d\mathbf{r}'}{|\mathbf{r}' - \mathbf{r}|} \tag{3.167}$$

where Z_A is the charge on atomic nucleus A, located at $\mathbf{R}_A$, and $\rho(\mathbf{r})$ is the total electron density of the molecule. The two terms are of opposite sign and therefore the value of electrostatic potential can be both the negative and positive, depending on the position of the sample point in the molecule. The electrostatic potential $V(\mathbf{r})$ and the related descriptors have been suitable for the analysis of the behavior of molecules in electrophilic and nucleophilic processes [454–456]. It has been emphasized that the spatial maximum of electrostatic potential coincides with the positions of nuclei and reflects the magnitude of the nuclear charges rather than the relative reactivity toward nucleophiles [457]. On the other hand, it has been demonstrated that the potential maxima on the molecular surfaces ($V_{S,\text{max}}$) quantify well the nucleophilic interactions [458–460]. In the case of electrophilic attack, the electron distribution of the substrate molecule is determining the site of interaction. Therefore, both the spatial minima (V_{min}) and the maxima on the molecular surface ($V_{S,\text{max}}$) are applicable for the identification and ranking of the molecular sites reactive toward electrophiles [461,462].

The GIPF descriptor Π has been introduced as a measure of the local polarity in the molecule [463]. In fact, it represents the amount by which the electrostatic potential on the molecular surface deviates from its average value, $\bar{V}_S$:

$$\Pi = \frac{1}{A}\int_S |V(\mathbf{r}) - \bar{V}_S|\, dS \approx \frac{1}{n}\sum_{i=1}^{n} |V_i(\mathbf{r}) - \bar{V}_S| \tag{3.168}$$

where A is the molecular surface area and n is the number of points over which the numerical integration is performed. It was suggested that this descriptor would be particularly useful in the case of molecules that have the zero overall polarity (dipole moment zero) but still exhibit considerable internal charge redistribution. For most organic molecules, Π ranges between 10 and 80 kJ/mol, for the symmetric boron trifluoride with zero dipole moment, Π = 58.48 kJ/mol [447]. The Π descriptor has been shown to correlate well with various empirical polarity or polarizability scales and with the dielectric constant of the bulk compound [446].

Another GIPF descriptor, the total variance of the surface electrostatic potential, σ^2_{tot}, has been defined as follows [464,465]:

$$\sigma^2_{\text{tot}} = \sigma^2_{+} - \sigma^2_{-} = \frac{1}{m}\sum_{i=1}^{m}\left[V^{+}(\mathbf{r}_i) - \bar{V}^{+}_S\right]^2 + \frac{1}{n}\sum_{i=1}^{n}\left[V^{-}(\mathbf{r}_i) - \bar{V}^{-}_S\right]^2 \tag{3.169}$$

where $V^{+}(\mathbf{r}_i)$ and $V^{-}(\mathbf{r}_i)$ are the positive and negative values of the electrostatic potential $V(\mathbf{r})$ on the surface, and $\bar{V}^{+}_S$ and $\bar{V}^{-}_S$ denote the respective averages. This descriptor has been suggested as an indicator of molecule's ability to participate in noncovalent electrostatic interactions. For example, it has been found effective, in conjunction with the molecular surface area or volume, to correlate the solubility in supercritical carbon dioxide [465]. For some applications, it may be advantageous to use σ^2_{+} or σ^2_{-} instead of total variance of the electrostatic potential.

The last GIPF descriptor is the electrostatic "balance'' parameter, ν, defined as follows [466,467]:

$$\nu = \frac{\sigma^2_{+}\sigma^2_{-}}{[\sigma^2_{\text{tot}}]^2} \tag{3.170}$$

According to this equation, the ν parameter has a maximum value of 0.25 when σ^2_{+} and σ^2_{-} are equal. Consequently, the less the value of this parameter, the larger the difference between the negative and positive potentials on the molecular surface. For instance, ν = 0.246 has been calculated for benzene, and thus this compound is expected to behave similarly through its positively and negatively charged regions. The ν values for ethanol and dimethyl ether were obtained as 0.159 and 0.055, respectively [447]. This is in harmony with the general chemical knowledge about these compounds. Ethanol can act both as the hydrogen bond donor (positive area) and hydrogen bond acceptor (negative area) whereas dimethyl ether is able to participate only in the last interaction.

It has been argued that in the case of the use of comparative molecular field analysis (CoMFA, cf. Chapter 3.5) and related approaches for the development of quantitative relationships between the biological activity and continuous distribution of three-dimensional molecular properties, such as the molecular electrostatic potential, the results depend strongly on the adopted molecular alignment [107]. Various strategies have been developed to overcome this difficulty, but still the high number of independent variables (i.e., the number of grid points where the property is calculated) may create ambiguities in the extraction of the chemical information relevant to the quantitative structure–activity problem. To overcome these difficulties, a new methodology, based on grid–weighted holistic invariant molecular (G-WHIM) descriptors, has recently been developed [107,468,469]. The respective descriptors, calculated from the three-dimensional distribution of molecular scalar fields, are independent of molecule alignment and summarize all the information of the whole distribution in terms of dimension and shape indexes.

A series of G-WHIM descriptors involving the molecular electrostatic potential has been calculated according the following procedure. First, the molecule is embedded into the center of a three-dimensional grid that accommodates a predefined isopotential surface, so-called threshold surface. The molecular electrostatic potential can be then calculated at all grid points between the van der Waals and the threshold surface and the weighted covariance matrix of the grid points created by weighting each point with its potential value. The principal component analysis (PCA) on the weighted covariance matrix provides thus the directions of maximum property variance, and different G-WHIM descriptors can be calculated from the obtained eigenvalues and scores of the PCA.

Analogous to the WHIM descriptors considered above as the geometrical molecular descriptors, the G-WHIM descriptors include the directional and global descriptors of a molecule. The directional descriptors depend on the axis along which they are calculated. Those include the dimension factors defined as the eigenvalues λ_i, (i = 1, 2, 3) of the weighted covariance matrix. Those factors have been related to the extension of the electrostatic potential distribution along each principal axis. The shape factors υ_i have been defined as the eigenvalue proportions as:

$$\upsilon_i = \frac{\lambda_i}{\sum_{i=1}^{3} \lambda_i} \quad i = 1, 2, 3 \tag{3.171}$$

These descriptors relate to the shape of the distribution along each axis. Only two υ_i have been considered as their values are close to unity. The emptiness factors η_i (i = 1, 2, 3) have been calculated as the inverse of the curtosi, that is, of the function of the fourth-order moments similarly to Eq. (3.17). In this equation, only the G-WHIM weighted covariance matrix eigenvalues were used instead of λ_m. The emptiness factors represent the projection of nonoccupied space along each principal axis defined by this covariance matrix. The value $\eta_i = 1$ corresponds to a bimodal distribution, while $\eta_i = 0$ reflects a peak distribution of the potential along the axis.

The global G-WHIM descriptors have been defined for the whole distribution of the molecular electrostatic potential. Those involve the following five factors. A linear dimension descriptor, $\boldsymbol{T}$, has been defined as the sum of the eigenvalues of the PCA on the weighted covariance matrix of the grid points [cf. also Eq. (3.18)]. A quadratic dimension term, $\boldsymbol{A}$, was calculated as the sum of products between couples of eigenvalues λ_i [Eq. (3.19)]. The total volume, $\boldsymbol{V}$, has been defined as the sum of $\boldsymbol{T}$, $\boldsymbol{A}$ and the product of the three λ_i (3.20). The global shape factor, $\boldsymbol{K}$ (3.21) describes the global shape of the distribution ($\boldsymbol{K} = 0$ represents a spherical distribution). Finally, the global density factor, $\boldsymbol{D}$, has been calculated as the average value of three λ_i:

$$\boldsymbol{D} = \frac{\sum_{i=1}^{3} \lambda_i}{3} \tag{3.172}$$

The molecular electrostatic potential may acquire both the negative and positive values, and thus the above procedures have been applied separately to the negative and positive parts of the charge distribution. Therefore, two sets of G-WHIM descriptors can be obtained for any given molecule.

A list of electrostatic descriptors is presented in Table 3.6. The numerical data on some electrostatic descriptors, obtained from the AM1 quantum chemical partial charges and empirical atomic charges are given in Table 3.7. The values of a selection of charged partial surface area (CSPA) descriptors, obtained from the AM1 quantum chemical partial charges for structurally variable compounds are given in Table 3.8.

The electrostatic descriptors derived from either the empirical atomic charges or quantum chemical partial charges in the molecule have been widely used in the development of QSPR regressions for a large variety of physical and analytical properties of compounds. For example, the empirically calculated charges by the method of Abraham and co-workers [389,470] were employed in the development of atomic contribution QSPR model for the aqueous pK_a of organic oxyacids [471]. The atomic partial charges have been also used for the theoretical prediction of hydrophobicity constants, log P. For a set of 61 compounds, including hydrocarbons, alcohols, ethers, ketones, acids, esters, amines, nitriles, and amides, the following equation has been developed [442] that employs the sums of squared atomic charge densities calculated by the MINDO/3 procedure:

TABLE 3.6 Electrostatic Molecular Descriptors

Notation	Name	Reference
q_A	Atomic or group partial charges	*a*
μ	Dipole moment of the molecule	*a*
Q	Quadrupole moment of the molecule	*a*
α, β, γ	Polarizabilities of the molecule	*a*
P_f	Polarity parameter	*b*
PPSA1, PNSA1	Partial positive or negative surface area of a molecule	*c*
PPSA2, PNSA2	Total charge-weighted partial positively or negatively charged surface area	*c*
PPSA3, PNSA3	Atomic charge-weighted partial positive or negative surface area	*c*
DPSA1	DPSAI = PPSAI − PNSAI $I = 1,3$	*c*
FPSAI, FNSAI	Fractional partial positive or negative surface areas ($I = 1,3$)	*c*
WPSAI, WNSAI	Surface-weighted charged partial positively or negatively charged surface area ($I = 1,3$)	*c*
RPCG, RNCG	Relative positive or negative charge	*c*
HDSA1, HDCA1	Hydrogen bonding donor ability of a molecule	*c*
HDSA2, HDCA2	Area-weighted surface charge of a hydrogen bonding donor atom	*c*
HASA1, HACA1	Hydrogen bonding acceptor ability of a molecule	*c*
HASA2, HACA2	Area-weighted surface charge of a hydrogen bonding acceptor atom	*c*
FHASAI, FHDSAI	Fractional hydrogen bonding related descriptors	*c*
$\bar{I}(\mathbf{r})$	Average ionization energy of a molecule	*d*
$\bar{I}_{S,\min}$, $\bar{I}_{S,\max}$	Minimum and maximum values of the ionization energy	*e*
$V_{S,\min}$, $V_{S,\max}$	Minimum and maximum values of the electrostatic potential on the surface of the molecule	*e*
Π	Local polarity of a molecule	*f*
σ^2_{tot}	Total variance of the surface electrostatic potential	*g*
ν	Electrostatic balance parameter	*h*
υ_i	G-WHIM electrostatic potential shape factors ($i = 1,3$)	*i*
η_i	G-WHIM emptiness factors	*i*
$\boldsymbol{T}$	Linear dimension of molecular electrostatic potential	*i*
$\boldsymbol{A}$	Quadratic dimension of molecular electrostatic potential	*i*
D	Global density factor	*i*

[a] J.J.P. Stewart, *MOPAC Program Package 6.0 Manual*, QCPE, Indiana University, Bloomington, IN, No. 455, 1990.
[b] A.R. Katritzky, V.S. Lobanov, and M. Karelson, *CODESSA User's Manual*. University of Florida, Gainesville, 1994.
[c] D.T. Stanton and P.C. Jurs, *Anal. Chem.* **62,** 2323 (1990); D.T. Stanton, L.M. Egolf, P.C. Jurs, and M.G. Hicks, *J. Chem. Inf. Comput. Sci.* **32,** 306 (1992).
[d] P. Sjoberg, J.S. Murray, T. Brinck, and P. Politzer, *Can. J. Chem.* **68,** 1440 (1990); T. Brinck, J. S. Murray, and P. Politzer, *Int. J. Quantum Chem.* **48,** 73 (1993).
[e] T. Brinck, J.S. Murray, P. Politzer, and R.E. Carter, *J. Org. Chem.* **56,** 2934 (1991).
[f] T. Brinck, J.S. Murray, and P. Politzer, *Mol. Phys.* **76,** 609 (1992).
[g] P. Politzer, P. Lane, J.S. Murray, and T. Brinck, *J. Phys. Chem.* **96,** 7938 (1992).
[h] J.S. Murray, P. Lane, T. Brinck, and P. Politzer, *J. Phys. Chem.* **97,** 5144 (1993).
[i] R. Todeschini, G. Moro, R. Boggia, L. Bonati, U. Cosentino, M. Lasagni, and D. Pitea, *Chemom. Intell. Lab. Syst.* **36,** 65 (1997); R. Todeschini, P. Gramatica, in *3D QSAR in Drug Design* (H. Kubinyi, G. Folkers, and Y.C. Martin, eds.), Vol. 2, pp. 355–380, Kluwer/ESCOM, Dordrecht, The Netherlands, 1998.

$$\log P = 0.344 + 0.2078N_{\mathrm{H}} + 0.093N_{\mathrm{C}} - 2.119N_{\mathrm{N}} - 1.937N_{\mathrm{O}} - 1.389\sum q_{\mathrm{C}}^2 - 17.28\sum q_{\mathrm{N}}^2 + 0.7614\sum q_{\mathrm{O}}^2 + 2.844N_A + 0.910N_T + 1.709N_M$$

$$R^2 = 0.985 \qquad s = 0.15 \qquad n = 61$$

where N_{H}, N_{C}, N_{N}, and N_{O} are the counts of hydrogen, carbon, nitrogen, and oxygen atoms, q_{C}, q_{N}, and q_{O} are the corresponding atomic charges, and N_A, N_T, and N_M are the counts of carboxy-, cyano-, and amide groups in the molecule, respectively. It was argued that this method yields better results than the fragment approach and requires fewer parameters. Later, this approach was extended by involving more molecular descriptors including the calculated dipole moment, the sums of absolute values of atomic charges, and the charge dispersions [440,472]. The following equation was developed:

$$\log P = 27.273 - 1.167 \times 10^{-4}S^2 - 6.106 \times 10^{-2}S + 14.87O^2 - 43.67O + 0.9986I_{\mathrm{alkane}} + 9.57 \times 10^{-3}M_w - 0.13D - 4.929Q_{\mathrm{ON}} - 12.17Q_{\mathrm{N}}^4 + 26.81Q_{\mathrm{N}}^2 - 7.416Q_{\mathrm{N}} - 4.551Q_{\mathrm{O}}^4 + 17.92Q_{\mathrm{O}}^2 - 4.03Q_{\mathrm{O}}$$

$$R^2 = 0.9388 \qquad F = 115.1 \qquad n = 118$$

where S is the molecular surface, O is the ovality of the molecule, I_{alkane} is the indicator variable for alkanes, M_w is the molecular weight, D is the calculated dipole moment, Q_{ON} is the sum of absolute values of atomic charges on nitrogen and oxygen atoms, Q_{N} is the square root of the sum of the squared charges on nitrogen atoms, and Q_{O} is the square root of the sum of the squared charges on oxygen atoms. All quantum chemical parameters were calculated using the AM1 parameterization, and regression equations were extended to the larger set of partition coefficient data. The predictive power of the model was demonstrated by the accurate estimation of log P for complex molecules.

An approach for the estimation of physical properties of organic compounds (critical temperature, molar critical volume, and boiling point) has been based on the computation of the molecular surface interactions (MSI) [473]. This approach has employed MSI descriptors calculated from atomic surface areas and net atomic charges calculated using the extended Hückel method (EHT). For instance, the following correlation equation was developed for the boiling point:

$$\mathrm{bp} = 127.7 + 0.718A - 1.015A_- + 0.230A_+ + 8.800A_{\mathrm{HB}}$$

$$R^2 = 0.958 \qquad s = 14.1 \qquad n = 152= \qquad F = 745.1$$

where A is the total molecular surface area and A_- is the sum of the surface areas of negatively charged atoms multiplied by their corresponding scaled net

TABLE 3.7 Electrostatic Descriptors Obtained Using AM1 SCRF Partial Charges (A) and Zefirov's Empirical Atomic Charges (Z) for Structurally Variable Compounds

Compound	$q_{max}(Z)$	$q_{max}(A)$	$q_{min}(Z)$	$q_{min}(A)$	$\mu(A)$	α	$\gamma \times 10^{-3}$	T_f
Water	0.086	0.201	−0.172	−0.402	1.921	3.51	0.014	0.2586
Ammonia	0.042	0.132	−0.125	−0.396	1.931	5.79	0.515	0.1671
Tetrafluoromethane	0.251	0.577	−0.063	−0.145	0.014	11.89	0.394	0.3135
Tetrachloromethane	0.157	0.031	−0.039	−8.800	0.014	36.00	3.874	0.1962
Trichloromethane	0.108	0.161	−0.064	−0.043	1.184	27.83	1.263	0.1718
Tribromomethane	0.070	0.174	−0.045	−0.365	0.929	34.90	2.793	0.1144
Formaldehyde	0.046	0.147	−0.126	−0.313	2.598	11.30	0.535	0.1724
Dibromomethane	0.046	0.146	−0.063	−0.322	1.356	25.06	1.837	0.1083
Dichloromethane	0.059	0.132	−0.088	−0.102	1.552	20.36	0.842	0.1473
Chloromethane	0.034	0.102	−0.113	−0.170	1.670	14.17	1.132	0.1466
Bromomethane	0.028	0.111	−0.081	−0.279	1.612	16.46	0.038	0.1083
Iodomethane	0.015	0.111	−0.027	−0.378	1.477	19.25	3.038	0.0423
Nitromethane	0.038	0.511	−0.067	−0.377	4.558	23.77	0.236	0.1056
Methane	9.619	0.067	−0.039	−0.266	0.000	9.75	0.692	0.0481
Methanol	0.082	0.198	−0.182	−0.337	1.737	13.04	1.005	0.2632
Methylamine	0.042	0.140	−0.126	−0.349	1.644	15.91	1.795	0.1673
Tetrachloroethene	0.118	0.053	−0.059	−0.106	0.000	46.51	6.347	0.1766
Pentachloroethane	0.128	0.165	−0.070	−0.068	1.058	49.53	2.046	0.1981
Trichloroethene	0.099	0.176	−0.086	−0.151	0.831	38.72	2.439	0.1857
1,1,1,2-Tetrachloroethane	0.115	0.140	−0.093	−0.144	1.524	43.03	4.354	0.2079
1,1-Dichloroethene	0.051	0.166	−0.096	−0.155	0.009	31.93	1.267	0.1466
trans-1,2-Dichloroethene	0.051	0.165	−0.096	−0.154	0.012	31.94	2.801	0.1466
cis-1,2-Dichloroethene	0.051	0.165	−0.096	−0.157	1.617	30.92	1.540	0.1466
1,1,2,2-Tetrachloroethane	0.083	0.160	−0.076	−0.063	0.002	42.33	2.284	0.1590
2,2,2-Trifluoroethanol	0.179	0.415	−0.159	−0.310	3.793	23.93	3.240	0.3378
Acetonitrile	0.023	0.121	−0.064	−0.148	3.351	21.12	1.489	0.0870
1,1,1-Trichloroethane	0.102	0.114	−0.067	−0.231	1.802	37.62	0.435	0.1688

1,1,2-Trichloroethane	0.070	0.150	−0.099	−0.141	1.188	35.77	0.035	0.1688
Ethene	0.014	0.110	−0.029	−0.218	0.004	19.35	1.181	0.0433
1,1-Dichloroethane	0.057	0.134	−0.089	−0.231	1.960	30.33	0.067	0.1466
Acetaldehyde	0.045	0.190	−0.127	−0.319	2.915	20.38	3.848	0.1724
Acetic acid	0.096	0.310	−0.152	−0.376	2.091	23.55	0.968	0.2485
1,2-Dibromoethane	0.033	0.128	−0.075	−0.222	0.008	36.41	1.053	0.1083
1,2-Dichloroethane	0.041	0.120	−0.106	−0.131	0.074	29.62	0.331	0.1466
Methyl formate	0.080	0.255	−0.161	−0.333	4.668	25.74	2.554	0.2411
Chloroethane	0.034	0.105	−0.112	−0.227	1.897	24.05	0.961	0.1466
Bromoethane	0.028	0.112	−0.080	−0.217	1.838	26.81	2.514	0.1083
Iodoethane	0.017	0.112	−0.032	−0.304	1.659	29.77	0.103	0.0485
Nitroethane	0.038	0.503	−0.067	−0.377	4.867	33.43	3.169	0.1056
Ethane	0.012	0.071	−0.035	−0.212	0.002	19.09	2.335	0.0462
Ethanol	0.082	0.198	−0.181	−0.339	1.674	22.09	4.625	0.2630
Ethanethiol	0.034	0.103	−0.064	−0.239	1.962	30.98	2.501	0.0977
Ethylamine	0.042	0.143	−0.125	−0.347	1.611	24.92	2.768	0.1669
Dimethylamine	0.041	0.151	−0.126	−0.306	1.366	26.01	1.604	0.1676
Propyne	0.020	0.219	−0.029	−0.229	0.484	26.09	3.868	0.0492
Methoxyflurane	0.159	0.372	−0.132	−0.307	1.654	45.29	5.034	0.2908
1-Chloro-2-propene	0.036	0.131	−0.111	−0.194	1.712	33.58	4.064	0.1466
Propanonitrile	0.024	0.119	−0.064	−0.213	3.366	30.08	5.241	0.0875
Ethyl formate	0.080	0.251	−0.161	−0.333	4.871	34.76	1.689	0.2410
Methyl acetate	0.076	0.305	−0.163	−0.364	1.854	33.96	3.401	0.2382
1,3-Dichloropropane	0.037	0.117	−0.110	−0.179	3.082	37.06	4.822	0.1466
Cyclopropane	0.014	0.108	−0.029	−0.215	0.002	26.03	3.755	0.0433
Propene	0.016	0.117	−0.032	−0.229	0.271	29.74	1.788	0.0482
Propanone	0.032	0.230	−0.128	−0.319	3.269	29.36	2.090	0.1596
Propanoic acid	0.096	0.308	−0.152	−0.370	2.080	32.53	3.703	0.2484
2-Propenol	0.082	0.200	−0.180	−0.330	1.911	31.82	2.071	0.2624
1,2-Dichloropropane	0.041	0.124	−0.106	−0.225	0.522	38.57	7.430	0.1467
Propionaldehyde	0.046	0.187	−0.127	−0.307	2.819	29.28	3.393	0.1724

TABLE 3.7 (*Continued*)

Compound	$q_{max}(Z)$	$q_{max}(A)$	$q_{min}(Z)$	$q_{min}(A)$	$\mu(A)$	α	$\gamma \times 10^{-3}$	T_f
1-Bromopropane	0.029	0.112	−0.080	−0.217	1.824	35.37	3.828	0.1083
2-Bromopropane	0.029	0.116	−0.079	−0.216	2.008	36.43	2.292	0.1083
1-Nitropropane	0.039	0.505	−0.067	−0.381	4.975	42.19	3.200	0.1058
1-Chloropropane	0.035	0.105	−0.112	−0.215	1.838	32.52	4.429	0.1466
2-Chloropropane	0.035	0.109	−0.112	−0.223	2.037	33.17	2.218	0.1466
1-Iodopropane	0.017	0.112	−0.033	−0.304	1.669	38.31	2.438	0.0502
N,*N*-Dimethylformamide	0.056	0.260	−0.117	−0.397	4.158	39.36	3.459	0.1724
2-Nitropropane	0.038	0.495	−0.067	−0.380	5.093	42.31	1.965	0.1056
Propane	0.013	0.076	−0.034	−0.210	0.000	27.87	1.686	0.0469
Methyl ethyl ether	0.031	0.093	−0.189	−0.284	1.451	32.38	1.686	0.2198
1-Propanol	0.082	0.198	−0.181	−0.334	1.595	30.87	6.517	0.2629
2-Propanol	0.082	0.198	−0.181	−0.335	1.732	30.46	4.113	0.2629
2-Methoxyethanol	0.084	0.205	−0.185	−0.337	1.804	35.12	3.669	0.2686
n-Propanethiol	0.034	0.102	−0.063	−0.244	1.874	39.62	5.842	0.0976
Trimethylamine	0.020	0.082	−0.127	−0.263	1.148	35.78	5.999	0.1466
Propylamine	0.042	0.139	−0.125	−0.335	1.721	33.55	2.076	0.1668
Thiophene	0.026	0.552	−0.054	−0.439	0.391	50.83	1.098	0.0793
1-Butyne	0.020	0.219	−0.032	−0.226	0.471	35.02	4.930	0.0524
(*E*)-2-Butenal	0.048	0.217	−0.125	−0.319	3.246	43.67	10.104	0.1724
1-Cyanopropane	0.024	0.117	−0.063	−0.218	3.439	39.23	3.286	0.0877
Isobutyraldehyde	0.046	0.190	−0.126	−0.299	2.712	37.68	2.418	0.1724
Tetrahydrofuran	0.032	0.095	−0.187	−0.290	2.063	37.36	6.888	0.2192
Ethyl acetate	0.076	0.305	−0.162	−0.365	1.836	42.83	2.480	0.2381
Methyl propanoate	0.076	0.307	−0.162	−0.363	1.794	43.23	6.757	0.2385
1,4-Dichlorobutane	0.035	0.115	−0.111	−0.170	2.448	46.15	7.658	0.1466
1-Butene	0.017	0.118	−0.033	−0.226	0.236	38.50	7.887	0.0497
Butanone	0.033	0.228	−0.127	−0.310	3.137	38.29	5.371	0.1599
Isopropyl formate	0.080	0.268	−0.161	−0.363	1.783	42.33	7.898	0.2409

Butanoic acid	0.096	0.311	−0.152	−0.373	2.069	40.97	1.857	0.2483
Propyl formate	0.080	0.254	−0.160	−0.331	4.832	43.41	8.582	0.2408
Butyraldehyde	0.046	0.189	−0.127	−0.305	2.942	38.06	5.404	0.1724
1,4-Dioxane	0.037	0.112	−0.181	−0.270	0.032	42.92	2.849	0.2175
1-Bromobutane	0.029	0.114	−0.079	−0.215	1.787	43.66	8.159	0.1083
1-Chlorobutane	0.035	0.107	−0.112	−0.213	1.816	40.84	7.215	0.1466
1-Nitrobutane	0.039	0.506	−0.067	−0.380	5.065	51.20	6.767	0.1059
2-Bromo-2-methylpropane	0.017	0.090	−0.079	−0.213	2.079	45.25	7.222	0.0954
1-Bromo-2-methylpropane	0.029	0.111	−0.079	−0.215	1.788	43.54	6.234	0.1083
2-Chloro-2-methylpropane	0.018	0.090	−0.112	−0.218	2.092	41.69	6.710	0.1299
2-Chlorobutane	0.035	0.112	−0.112	−0.222	2.084	41.98	6.381	0.1466
1-Iodobutane	0.018	0.114	−0.033	−0.302	1.640	46.49	1.207	0.0507
Morpholine	0.045	0.162	−0.185	−0.293	1.364	46.15	7.220	0.2298
n-Butane	0.014	0.078	−0.033	−0.210	0.011	36.59	4.329	0.0471
Diethyl ether	0.031	0.100	−0.189	−0.283	1.508	41.18	11.771	0.2195
2-Methylpropane	0.015	0.083	−0.033	−0.208	0.008	36.19	6.466	0.0475
2-Methyl-1-propanol	0.082	0.199	−0.181	−0.333	1.612	39.31	5.004	0.2627
2-Butanol	0.082	0.194	−0.181	−0.329	1.773	39.02	3.171	0.2627
2-Methyl-2-propanol	0.082	0.196	−0.181	−0.330	1.771	38.54	5.386	0.2627
Diethyl sulfide	0.019	0.099	−0.062	−0.266	1.689	51.51	3.658	0.0814
Diethyl disulfide	0.022	0.112	−0.044	−0.266	2.570	74.45	11.375	0.0659
2-Ethoxyethanol	0.084	0.202	−0.184	−0.336	1.194	44.24	6.556	0.2683
1-Butanol	0.082	0.198	−0.181	−0.333	1.688	39.21	6.332	0.2628
n-Butanethiol	0.034	0.105	−0.063	−0.240	1.874	48.41	6.643	0.0975
Diethylamine	0.042	0.155	−0.125	−0.312	1.205	44.14	9.235	0.1668
Butylamine	0.042	0.140	−0.125	−0.339	1.765	42.20	11.968	0.1668
2-Chloropyridine	0.056	0.165	−0.090	−0.176	3.200	59.90	4.417	0.1456
3-Chloropyridine	0.040	0.169	−0.098	−0.168	1.941	59.16	5.625	0.1376
Pyridine	0.031	0.154	−0.098	−0.178	2.250	52.06	1.080	0.1290
2-Methylthiophene	0.026	0.550	−0.053	−0.443	0.524	61.95	7.317	0.0787
2-Methylpyrazine	0.035	0.170	−0.093	−0.173	0.540	60.24	4.319	0.1277

TABLE 3.7 (*Continued*)

Compound	$q_{max}(Z)$	$q_{max}(A)$	$q_{min}(Z)$	$q_{min}(A)$	$\mu(A)$	α	$\gamma \times 10^{-3}$	T_f
Cyclopentene	0.018	0.135	−0.028	−0.175	0.185	44.70	2.641	0.0456
2-Methyl-1,3-butadiene	0.018	0.120	−0.030	−0.215	0.205	55.10	8.262	0.0485
Cyclopentanone	0.034	0.237	−0.126	−0.307	3.121	44.65	7.800	0.1607
1,4-pentadiene	0.017	0.119	−0.027	−0.224	0.144	46.16	6.792	0.0442
1-Pentyne	0.020	0.219	−0.033	−0.227	0.434	43.51	6.145	0.0533
1-Cyanobutane	0.024	0.117	−0.063	−0.212	3.418	48.04	10.215	0.0877
Cyclopentane	0.014	0.079	−0.029	−0.157	0.001	42.18	7.638	0.0433
1-Pentene	0.017	0.114	−0.033	−0.228	0.251	46.29	7.064	0.0501
Ethyl propanoate	0.076	0.306	−0.162	−0.364	2.178	52.73	3.471	0.2384
Isobutyl formate	0.081	0.254	−0.160	−0.327	4.869	52.00	9.109	0.2407
2-Methyl-2-butene	0.016	0.111	−0.033	−0.234	0.421	47.80	6.159	0.0484
3-Methyl-1-butene	0.017	0.116	−0.032	−0.223	0.181	45.73	3.660	0.0493
Isopropyl acetate	0.076	0.311	−0.162	−0.367	1.736	51.31	6.361	0.2380
Pentanal	0.046	0.187	−0.126	−0.303	3.066	47.45	2.721	0.1724
Methyl butanoate	0.076	0.302	−0.162	−0.362	1.972	51.57	6.871	0.2386
Pentanoic acid	0.096	0.308	−0.152	−0.368	2.045	50.25	8.195	0.2483
3-Methylbutanoic acid	0.097	0.310	−0.152	−0.370	1.994	49.47	5.847	0.2483
Cyclopentanol	0.082	0.197	−0.180	−0.330	1.636	45.27	3.260	0.2624
3-Methyl-2-butanone	0.033	0.232	−0.127	−0.311	3.225	46.72	3.998	0.1602
3-Pentanone	0.034	0.230	−0.127	−0.305	3.068	47.05	4.360	0.1603
Propyl acetate	0.076	0.303	−0.162	−0.361	1.734	51.26	7.083	0.2379
2-Pentanone	0.033	0.232	−0.127	−0.311	3.089	46.62	7.499	0.1600
2-Methyltetrahydrofuran	0.032	0.099	−0.187	−0.287	1.958	46.15	2.186	0.2193
Tetrahydropyran	0.032	0.093	−0.188	−0.287	1.868	47.24	9.503	0.2193
1-Chloropentane	0.035	0.108	−0.112	−0.214	2.002	50.63	2.420	0.1466
1-Nitropentane	0.039	0.504	−0.067	−0.372	5.125	60.09	6.694	0.1059
1-Bromopentane	0.029	0.116	−0.079	−0.213	1.974	53.46	7.532	0.1083
1-Iodopentane	0.018	0.114	−0.033	−0.306	1.780	56.65	3.208	0.0508

N-Methylmorpholine	0.034	0.104	−0.185	−0.276	1.394	56.08	6.098	0.2185
n-Pentane	0.014	0.079	−0.033	−0.211	0.000	45.48	5.472	0.0475
2-Methyl-1-butanol	0.082	0.200	−0.181	−0.333	1.885	47.54	6.732	0.2627
2-Methyl-2-butanol	0.082	0.196	−0.180	−0.333	1.761	47.35	6.629	0.2626
3-Methyl-1-butanol	0.082	0.197	−0.181	−0.333	1.592	47.98	7.965	0.2628
1-Pentanol	0.082	0.198	−0.181	−0.335	1.680	48.24	4.501	0.2628
2-Pentanol	0.082	0.197	−0.181	−0.336	1.751	47.93	10.873	0.2627
3-Pentanol	0.082	0.197	−0.180	−0.336	1.738	47.94	7.820	0.2626
Methyl *tert*-butyl ether	0.031	0.091	−0.188	−0.281	1.488	48.83	9.901	0.2188
2,2-Dimethylpropane	0.013	0.073	−0.032	−0.204	0.008	44.16	9.093	0.0449
2-Propoxyethanol	0.084	0.198	−0.184	−0.326	1.627	54.01	5.087	0.2680
2-Methylbutane	0.015	0.081	−0.033	−0.209	0.025	44.85	5.038	0.0483
n-Pentylamine	0.042	0.144	−0.125	−0.350	1.638	51.01	6.188	0.1668
1,2,3,4-Tetrachlorobenzene	0.070	0.162	−0.089	−0.109	1.756	85.75	13.511	0.1591
1,2,3,5-Tetrachlorobenzene	0.067	0.170	−0.095	−0.116	0.934	86.22	16.287	0.1621
1,2,4,5-Tetrachlorobenzene	0.057	0.170	−0.089	−0.109	0.027	89.98	13.952	0.1461
1,2,3-Trichlorobenzene	0.064	0.157	−0.091	−0.117	2.290	77.23	10.911	0.1552
1,3,5-Trichlorobenzene	0.043	0.166	−0.096	−0.123	0.038	78.29	15.896	0.1392
1,2,4-Trichlorobenzene	0.054	0.165	−0.097	−0.116	1.129	78.48	9.770	0.1510
1,2-Dichlorobenzene	0.048	0.154	−0.094	−0.120	2.252	69.76	9.929	0.1418
1,3-Dichlorobenzene	0.037	0.158	−0.099	−0.124	1.349	70.40	10.425	0.1364
1,4-Dichlorobenzene	0.035	0.155	−0.100	−0.119	0.011	71.29	11.581	0.1353
3-Cyanopyridine	0.034	0.176	−0.096	−0.182	3.418	69.99	9.551	0.1296
4-Cyanopyridine	0.033	0.169	−0.097	−0.147	1.263	68.50	8.180	0.1293
Bromobenzene	0.025	0.144	−0.070	−0.167	1.662	65.75	3.417	0.0954
Chlorobenzene	0.032	0.145	−0.102	−0.127	1.503	62.94	4.995	0.1337
4-Fluorophenol	0.087	0.223	−0.172	−0.248	1.905	63.28	5.145	0.2582
2-Chlorophenol	0.089	0.223	−0.167	−0.237	2.752	68.37	6.772	0.2559
3-Chlorophenol	0.087	0.224	−0.171	−0.251	2.531	68.66	7.168	0.2580
4-Chlorophenol	0.086	0.223	−0.172	−0.248	1.669	69.11	6.773	0.2585
4-Bromophenol	0.086	0.224	−0.173	−0.247	1.806	72.08	6.486	0.2586

TABLE 3.7 (*Continued*)

Compound	$q_{max}(Z)$	$q_{max}(A)$	$q_{min}(Z)$	$q_{min}(A)$	$\mu(A)$	α	$\gamma \times 10^{-3}$	T_f
2-Iodophenol	0.086	0.225	−0.172	−0.294	2.880	74.42	3.090	0.2583
2-Nitrophenol	0.089	0.582	−0.116	−0.396	7.408	79.66	13.897	0.2555
3-Nitrophenol	0.087	0.564	−0.171	−0.370	7.017	79.67	12.273	0.2579
4-Nitrophenol	0.086	0.574	−0.172	−0.382	6.358	81.95	24.978	0.2584
3-Formalpyridine	0.050	0.230	−0.123	−0.291	1.576	66.76	8.669	0.1724
4-Formalpyridine	0.049	0.221	−0.123	−0.281	2.040	65.59	6.182	0.1724
2-Fluorophenol	0.090	0.222	−0.164	−0.242	3.104	62.84	5.184	0.2544
Fluorobenzene	0.056	0.145	−0.160	−0.167	1.794	57.21	1.172	0.2163
Iodobenzene	0.021	0.143	−0.019	−0.263	1.645	69.24	4.613	0.0396
Nitrobenzene	0.040	0.566	−0.064	−0.374	6.161	73.76	10.772	0.1034
Benzene	0.019	0.130	−0.019	−0.131	0.000	56.20	4.834	0.0385
Phenol	0.086	0.218	−0.173	−0.256	1.404	61.39	4.772	0.2591
4-Chloroaniline	0.045	0.208	−0.119	−0.357	3.303	76.83	12.215	0.1637
2-Nitroaniline	0.048	0.583	−0.114	−0.409	6.240	87.08	17.381	0.1612
3-Nitroaniline	0.046	0.563	−0.118	−0.377	7.211	87.00	21.380	0.1632
4-Nitroaniline	0.045	0.580	−0.118	−0.401	9.807	95.05	48.555	0.1636
3-Chloroaniline	0.046	0.198	−0.118	−0.338	2.732	74.99	11.359	0.1633
2-Chloroaniline	0.047	0.200	−0.114	−0.327	1.950	74.05	11.395	0.1616
Thiophenol	0.037	0.141	−0.057	−0.210	1.392	74.09	25.693	0.0944
2-Methylpyridine	0.031	0.154	−0.098	−0.186	1.959	63.33	6.052	0.1287
3-Methylpyridine	0.031	0.155	−0.098	−0.178	2.388	63.11	7.568	0.1291
4-Methylpyridine	0.031	0.153	−0.098	−0.190	2.666	62.38	4.017	0.1290
N-Acetylpyrrolidine	0.054	0.326	−0.117	−0.323	2.970	68.37	11.277	0.1703
2-Ethylpyrazine	0.035	0.172	−0.093	−0.216	0.679	70.02	6.962	0.1277
Cyclohexene	0.018	0.120	−0.028	−0.167	0.201	55.15	6.373	0.0459
Cyclohexanone	0.034	0.236	−0.126	−0.312	3.248	54.28	6.249	0.1606
1,5-Hexadiene	0.017	0.118	−0.027	−0.228	0.291	55.78	15.058	0.0441
2,3-Dimethyl-1,3-butadiene	0.016	0.114	−0.030	−0.216	0.001	62.89	10.999	0.0462

trans-2-Hexenal	0.048	0.215	−0.125	−0.309	3.897	61.73	10.014	0.1724
1-Hexyne	0.020	0.220	−0.033	−0.229	0.406	51.82	7.837	0.0535
Cyclohexane	0.014	0.082	−0.029	−0.154	0.034	52.21	4.116	0.0433
Methylcyclopentane	0.016	0.084	−0.032	−0.205	0.041	50.78	6.358	0.0480
2-Methyl-1-pentene	0.016	0.110	−0.033	−0.235	0.422	56.80	4.130	0.0489
2-Hexanone	0.033	0.229	−0.127	−0.313	3.208	55.39	10.237	0.1601
Hexanoic acid	0.096	0.308	−0.152	−0.370	2.113	58.46	7.659	0.2483
Isobutyl acetate	0.076	0.303	−0.162	−0.363	2.114	60.97	10.835	0.2377
n-Propyl propanoate	0.076	0.305	−0.162	−0.360	2.282	61.35	2.806	0.2382
1-Hexene	0.017	0.120	−0.033	−0226	0.238	56.13	8.368	0.0502
Isoamyl formate	0.081	0.259	−0.160	−0.359	1.662	59.03	10.380	0.2407
Ethyl butanoate	0.077	0.305	−0.162	−0.365	2.030	61.30	11.850	0.2385
Methyl pentanoate	0.077	0.302	−0.162	−0.361	1.880	60.23	10.643	0.2387
Hexanal	0.046	0.188	−0.126	−0.308	3.029	55.62	8.320	0.1724
4-Methyl-2-pentanone	0.033	0.234	−0.127	−0.304	2.907	54.98	9.865	0.1602
Butyl acetate	0.076	0.304	−0.162	−0.362	1.793	60.18	7.725	0.2378
Cyclohexylamine	0.043	0.141	−0.124	−0.335	1.632	58.35	6.799	0.1663
n-Butylacetamide	0.053	0.299	−0.118	−0.398	4.423	64.73	4.807	0.1702
1-Bromohexane	0.029	0.116	−0.079	−0.211	1.912	61.92	5.668	0.1083
1-Chlorohexane	0.035	0.108	−0.112	−0.212	1.983	59.26	7.322	0.1466
1-Iodohexane	0.018	0.115	−0.033	−0.308	1.777	65.14	3.437	0.0509
N-Methylpiperidine	0.022	0.091	−0.125	−0.271	0.941	60.79	11.564	0.1465
2,2-Dimethylbutane	0.014	0.077	−0.033	−0.210	0.007	52.93	5.173	0.0474
2,3-Dimethylbutane	0.016	0.082	−0.032	−0.208	0.016	53.32	7.243	0.0479
n-Hexane	0.014	0.080	−0.033	−0.213	0.023	54.21	6.905	0.0476
2-Methylpentane	0.015	0.082	−0.033	−0.212	0.032	53.66	6.364	0.0485
3-Methylpentane	0.016	0.079	−0.033	−0.209	0.025	53.66	8.845	0.0487
Diisopropyl ether	0.032	0.086	−0.188	−0.286	1.449	58.59	8.498	0.2193
Di-*n*-propyl ether	0.031	0.091	−0.188	−0.285	1.333	59.88	5.285	0.2194
1-Hexanol	0.082	0.198	−0.181	−0.333	1.650	57.51	10.698	0.2628
Di-*n*-propyl sulfide	0.020	0.099	−0.061	−0.268	1.682	69.41	11.232	0.0812

TABLE 3.7 (*Continued*)

Compound	$q_{max}(Z)$	$q_{max}(A)$	$q_{min}(Z)$	$q_{min}(A)$	$\mu(A)$	α	$\gamma \times 10^{-3}$	T_f
Diisopropyl sulfide	0.021	0.107	−0.060	−0.210	1.727	69.41	6.546	0.0810
4-Methyl-2-pentanol	0.082	0.195	−0.181	−0.333	1.627	56.26	6.011	0.2626
2-Methyl-3-pentanol	0.082	0.197	−0.180	−0.333	1.673	56.35	4.606	0.2625
3-Hexanol	0.082	0.196	−0.180	−0.336	1.679	56.72	10.106	0.2625
2-Butoxyethanol	0.084	0.202	−0.184	−0.332	0.961	61.76	12.449	0.2679
2-Methyl-2-pentanol	0.082	0.196	−0.180	−0.330	1.683	55.98	7.055	0.2625
Di-*n*-Propylamine	0.042	0.157	−0.124	−0.325	1.119	62.24	6.158	0.1666
Hexylamine	0.042	0.145	−0.125	−0.351	1.599	59.87	10.579	0.1668
Diisopropylamine	0.043	0.152	−0.124	−0.301	1.312	61.06	7.151	0.1662
Triethylamine	0.021	0.087	−0.125	−0.270	1.096	62.55	11.990	0.1460
Benzotrifluoride	0.174	0.465	−0.101	−0.171	3.704	65.58	1.847	0.2749
3-Cyanophenol	0.086	0.230	−0.173	−0.249	4.886	79.63	12.410	0.2586
4-Cyanophenol	0.086	0.232	−0.173	−0.237	4.062	81.82	21.662	0.2588
Benzonitrile	0.031	0.148	−0.060	−0.132	4.018	73.92	10.784	0.0909
Benzaldehyde	0.049	0.228	−0.124	−0.306	3.491	71.24	14.447	0.1724
3-Hydroxybenzaldehyde	0.086	0.227	−0.172	−0.300	3.778	76.85	10.995	0.2586
4-Hydroxybenzaldehyde	0.086	0.231	−0.173	−0.307	2.686	78.58	22.377	0.2588
2-Chlorotoluene	0.032	0.145	−0.102	−0.182	1.311	73.22	6.761	0.1339
4-Bromotoluene	0.025	0.144	−0.070	−0.186	2.076	77.60	12.668	0.0955
2-Nitrotoluene	0.040	0.568	−0.064	−0.378	5.865	84.37	15.509	0.1037
3-Nitrotoluene	0.040	0.566	−0.064	−0.374	6.393	84.89	15.586	0.1036
Benzamide	0.058	0.346	−0.115	−0.431	4.204	75.65	9.203	0.1726
4-Chloro-3-methylphenol	0.086	0.222	−0.172	−0.249	1.222	79.31	10.872	0.2584
3-Acetylpyridine	0.040	0.274	−0.123	−0.292	1.842	74.61	11.647	0.1637
4-Acetylpyridine	0.039	0.266	−0.124	−0.287	2.599	73.90	9.753	0.1632
Toluene	0.020	0.132	−0.030	−0.180	0.304	66.95	7.098	0.0491
Benzyl alcohol	0.083	0.203	−0.179	−0.329	1.660	70.28	8.480	0.2620
o-Cresol	0.086	0.217	−0.173	−0.254	1.097	71.97	7.281	0.2590

p-Cresol	0.086	0.217	−0.173	−0.256	1.550	72.75	12.170	0.2590
1,3,5-Cycloheptatriene	0.019	0.126	−0.025	−0.143	0.213	67.12	10.335	0.0441
Methyl phenyl ether	0.032	0.145	−0.182	−0.215	1.447	73.37	6.560	0.2133
2-Methoxyphenol	0.088	0.211	−0.177	−0.236	1.745	79.12	10.881	0.2650
Phenyl methyl sulfide	0.022	0.158	−0.057	−0.345	1.996	81.05	8.733	0.0787
3-Methoxyphenol	0.087	0.219	−0.180	−0.254	1.967	78.41	10.113	0.2666
o-Toluidine	0.045	0.184	−0.119	−0.324	1.670	77.71	14.193	0.1640
p-Toluidine	0.045	0.183	−0.119	−0.327	1.523	79.61	18.539	0.1641
2-Methoxyaniline	0.047	0.189	−0.180	−0.305	1.850	84.34	17.004	0.2263
3-Methoxyaniline	0.045	0.184	−0.181	−0.327	1.635	84.58	14.887	0.2264
4-Methoxyaniline	0.045	0.178	−0.181	−0.320	2.240	86.36	17.904	0.2264
2,3-Dimethylpyridine	0.031	0.153	−0.098	−0.184	2.144	73.99	10.891	0.1285
2,4-Dimethylpyridine	0.031	0.153	−0.098	−0.189	2.363	73.67	9.560	0.1286
2,5-Dimethylpyridine	0.031	0.154	−0.098	−0.178	2.060	75.10	13.426	0.1287
2,6-Dimethylpyridine	0.023	0.143	−0.098	−0.189	1.638	74.54	9.965	0.1209
3,4-Dimethylpyridine	0.031	0.153	−0.098	−0.188	2.839	72.86	10.684	0.1290
3,5-Dimethylpyridine	0.031	0.155	−0.098	−0.178	2.533	74.21	9.996	0.1289
N-Methylaniline	0.044	0.178	−0.120	−0.269	1.495	78.59	19.346	0.1646
2-Ethylpyridine	0.031	0.154	−0.098	−0.200	1.896	72.63	6.491	0.1284
3-Ethylpyridine	0.031	0.157	−0.098	−0.211	2.426	72.20	9.073	0.1291
4-Ethylpyridine	0.031	0.153	−0.098	−0.213	2.751	71.43	9.393	0.1290
1-Methylcyclohexene	0.018	0.119	−0.030	−0.180	0.200	65.19	6.145	0.0484
1-Heptyne	0.020	0.220	−0.033	−0.226	0.414	60.74	6.367	0.0536
1-Heptene	0.017	0.125	−0.033	−0.231	0.274	65.41	3.844	0.0502
Heptanal	0.046	0.184	−0.126	−0.296	2.842	64.80	7.419	0.1724
n-Propyl butanoate	0.077	0.306	−0.162	−0.362	2.136	70.51	14.015	0.2383
Ethyl pentanoate	0.077	0.304	−0.162	−0.364	1.946	70.13	14.571	0.2386
Methyl hexanoate	0.077	0.302	−0.162	−0.360	1.870	69.15	9.940	0.2387
Cycloheptanol	0.082	0.196	−0.180	−0.336	1.727	63.89	8.711	0.2625
Methylcyclohexane	0.016	0.079	−0.032	−0.206	0.019	61.25	9.268	0.0480
2,4-Dimethyl-3-pentanone	0.035	0.237	−0.126	−0.302	2.897	63.65	8.246	0.1608

TABLE 3.7 ***(Continued)***

Compound	$q_{max}(Z)$	$q_{max}(A)$	$q_{min}(Z)$	$q_{min}(A)$	$\mu(A)$	α	$\gamma \times 10^{-3}$	T_f
Isoamyl acetate	0.076	0.305	−0.162	−0.363	1.889	69.92	20.349	0.2378
4-Heptanone	0.034	0.231	−0.127	−0.304	2.932	64.52	5.946	0.1605
trans-2-Heptene	0.017	0.120	−0.033	−0.211	0.146	65.68	2.048	0.0507
2-Heptanone	0.033	0.232	−0.127	−0.305	2.999	63.91	5.343	0.1601
1-Bromoheptane	0.029	0.112	−0.079	−0.212	1.925	70.84	4.348	0.1083
1-Chloroheptane	0.035	0.104	−0.112	−0.210	1.914	67.85	13.968	0.1466
1-Iodoheptane	0.018	0.111	−0.033	−0.308	1.725	73.78	3.795	0.0509
2,3-Dimethylpentane	0.016	0.082	−0.033	−0.211	0.036	62.03	8.533	0.0490
n-Heptane	0.014	0.081	−0.033	−0.211	0.042	62.61	7.383	0.0477
2-Methylhexane	0.015	0.086	−0.033	−0.211	0.010	62.50	7.582	0.0486
3-Methylhexane	0.016	0.085	−0.033	−0.211	0.043	62.27	11.404	0.0489
1-Heptanol	0.082	0.195	−0.181	−0.330	1.816	65.49	7.580	0.2628
2,4-Dimethylpentane	0.015	0.087	−0.033	−0.209	0.017	61.87	6.799	0.0478
3,3-Dimethylpentane	0.015	0.079	−0.033	−0.211	0.013	61.74	12.143	0.0474
2,2-Dimethylpentane	0.015	0.079	−0.033	−0.209	0.022	61.76	8.488	0.0481
n-Heptylamine	0.042	0.143	−0.125	−0.344	1.588	69.03	4.517	0.1668
Styrene	0.020	0.133	−0.026	−0.209	0.067	82.88	23.203	0.0459
Acetophenone	0.038	0.272	−0.125	−0.305	3.372	78.92	12.426	0.1625
4-Methylbenzaldehyde	0.049	0.229	−0.124	−0.309	3.871	83.95	24.660	0.1724
Methyl benzoate	0.082	0.346	−0.160	−0.366	2.393	85.09	17.553	0.2422
Phenyl acetate	0.078	0.310	−0.155	−0.332	1.818	84.71	15.628	0.2329
Ethylbenzene	0.020	0.133	−0.032	−0.211	0.374	76.22	10.764	0.0521
m-Xylene	0.020	0.132	−0.029	−0.180	0.276	77.74	9.245	0.0495
p-Xylene	0.020	0.130	−0.029	−0.178	0.044	78.40	14.709	0.0492
o-Xylene	0.020	0.129	−0.029	−0.181	0.508	77.31	14.297	0.0490
Ethyl phenyl ether	0.032	0.144	−0.181	−0.248	1.715	80.98	8.646	0.2134
2,3-Dimethylphenol	0.086	0.217	−0.173	−0.255	0.885	82.39	10.060	0.2589
2,5-Dimethylphenol	0.086	0.217	−0.173	−0.255	1.443	83.44	15.632	0.2589

2,6-Dimethylphenol	0.086	0.219	−0.173	−0.257	1.372	82.55	10.738	0.2589
3,4-Dimethylphenol	0.086	0.216	−0.173	−0.257	1.368	83.11	12.636	0.2590
3,5-Dimethylphenol	0.086	0.217	−0.173	−0.258	1.579	82.71	11.664	0.2590
4-Ethylphenol	0.086	0.217	−0.173	−0.257	1.583	82.18	12.471	0.2590
2-Phenylethanol	0.082	0.200	−0.180	−0.334	1.631	78.80	8.866	0.2626
2,4-Dimethylphenol	0.086	0.217	−0.173	−0.254	1.169	83.24	14.813	0.2589
3-Ethylphenol	0.086	0.218	−0.173	−0.256	1.794	81.15	14.986	0.2590
2,6-Dimethylaniline	0.045	0.186	−0.119	−0.323	1.720	87.96	14.460	0.1639
2-Isobutylpyrazine	0.035	0.172	−0.093	−0.211	0.659	86.92	14.006	0.1277
trans-2-Octenal	0.048	0.215	−0.125	−0.302	3.838	78.97	16.824	0.1724
1-Octyne	0.020	0.219	−0.033	−0.225	0.554	71.00	16.506	0.0536
n-Propylcyclopentane	0.016	0.086	−0.033	−0.210	0.033	68.78	14.012	0.0496
trans-1,4-Dimethylcyclohexane	0.016	0.084	−0.032	−0.209	0.045	68.95	10.477	0.0480
Octanal	0.046	0.184	−0.126	−0.295	2.839	73.79	8.836	0.1724
n-Pentyl propanoate	0.076	0.305	−0.162	−0.358	1.893	79.27	14.567	0.2381
Ethyl hexanoate	0.077	0.304	−0.162	−0.365	2.025	78.47	9.993	0.2386
Isobutyl isotuanoate	0.077	0.309	−0.162	−0.355	2.111	79.31	8.620	0.2383
cis-1,2-Dimethylcyclohexane	0.016	0.083	−0.032	−0.208	0.008	69.43	9.550	0.0482
Hexyl acetate	0.076	0.309	−0.162	−0.363	1.838	77.67	11.986	0.2378
2-Octanone	0.033	0.235	−0.127	−0.306	2.925	72.53	7.477	0.1601
1-Octene	0.017	0.120	−0.033	−0.231	0.270	74.03	8.874	0.0502
1-Bromooctane	0.029	0.112	−0.079	−0.210	1.887	79.36	23.896	0.1083
n-Octane	0.014	0.079	−0.033	−0.211	0.003	72.07	6.891	0.0477
2,2,4-Trimethylpentane	0.015	0.077	−0.032	−0.207	0.039	69.99	6.764	0.0478
2,3,4-Trimethylpentane	0.016	0.086	−0.032	−0.210	0.010	70.69	7.671	0.0485
Di-*n*-butyl ether	0.032	0.093	−0.188	−0.283	1.312	77.95	8.491	0.2193
3-Methylheptane	0.016	0.086	−0.033	−0.211	0.013	71.39	6.424	0.0490
1-Octanol	0.082	0.197	−0.181	−0.333	1.592	74.68	12.290	0.2628
Di-*n*-butylamine	0.042	0.152	−0.124	−0.310	1.143	80.27	8.606	0.1665
Octylamine	0.042	0.141	−0.125	−0.341	1.700	77.93	13.038	0.1668
Quinoline	0.031	0.156	−0.096	−0.185	2.132	101.67	21.632	0.1269

TABLE 3.7 ***(Continued)***

Compound	$q_{max}(Z)$	$q_{max}(A)$	$q_{min}(Z)$	$q_{min}(A)$	$\mu(A)$	α	$\gamma \times 10^{-3}$	T_f
α-Methylstyrene	0.020	0.132	−0.030	−0.215	0.224	91.29	21.134	0.0498
Ethyl benzoate	0.082	0.347	−0.160	−0.366	2.233	94.47	23.082	0.2420
4-Methylacetophenone	0.038	0.273	−0.125	−0.302	3.562	90.51	21.323	0.1626
Indane	0.020	0.132	−0.027	−0.158	0.518	83.99	11.542	0.0468
1,2,3-Trimethylbenzene	0.020	0.129	−0.029	−0.180	0.556	87.82	16.082	0.0490
1,2,4-Trimethylbenzene	0.020	0.129	−0.029	−0.180	0.277	88.73	15.932	0.0496
1,3,5-Trimethylbenzene	0.020	0.130	−0.029	−0.179	0.058	88.51	1.754	0.0495
Isopropylbenzene	0.020	0.132	−0.032	−0.206	0.292	84.22	7.443	0.0514
2-Ethyltoluene	0.020	0.133	−0.032	−0.210	0.590	86.40	10.717	0.0522
4-Ethyltoluene	0.020	0.136	−0.032	−0.210	0.095	87.72	17.814	0.0522
4-*n*-Propylphenol	0.086	0.219	−0.173	−0.260	1.434	90.42	10.628	0.2590
Propylbenzene	0.020	0.132	−0.033	−0.211	0.381	85.25	14.454	0.0529
3-Phenyl-1-propanol	0.082	0.199	−0.181	−0.331	1.790	88.56	10.524	0.2627
Nonanal	0.046	0.184	−0.126	−0.294	2.839	82.75	8.891	0.1724
5-Nonanone	0.034	0.228	−0.127	−0.304	2.879	82.15	11.605	0.1606
2-Nonanone	0.033	0.232	−0.127	−0.304	2.890	81.08	13.620	0.1601
1-Nonene	0.017	0.117	−0.033	−0.228	0.242	82.24	12.758	0.0502
n-Nonane	0.014	0.085	−0.033	−0.211	0.049	80.23	10.250	0.0477
2,2,5-Trimethylhexane	0.015	0.082	−0.033	−0.213	0.076	79.13	11.122	0.0478
1-Nonanol	0.082	0.197	−0.181	−0.332	1.649	83.62	4.923	0.2628
Naphthalene	0.021	0.135	−0.018	−0.129	0.041	105.58	23.582	0.0387
1-Naphthol	0.086	0.221	−0.173	−0.248	1.100	110.98	28.075	0.2586

2-Naphthol	0.086	0.219	−0.173	−0.254	1.134	111.85	26.540	0.2589
1-Naphthylamine	0.045	0.188	−0.119	−0.327	1.691	117.54	32.456	0.1638
2-Naphthylamine	0.045	0.187	−0.119	−0.329	1.827	121.27	47.521	0.1640
n-Butylbenzene	0.020	0.133	−0.033	−0.211	0.394	94.27	16.118	0.0531
tert-Butylbenzene	0.020	0.133	−0.031	−0.205	0.362	92.58	5.138	0.0509
4-Isopropyltoluene	0.020	0.132	−0.032	−0.205	0.038	95.99	17.185	0.0516
4-*tert*-Butylphenol	0.086	0.217	−0.173	−0.257	1.582	98.82	15.739	0.2590
Isobutylbenzene	0.020	0.133	−0.032	−0.211	0.351	93.70	12.373	0.0521
sec-Butylbenzene	0.020	0.133	−0.033	−0.211	0.345	93.53	14.689	0.0528
n-Pentylcyclopentane	0.016	0.084	−0.033	−0.210	0.026	87.06	10.044	0.0497
2-Decanone	0.033	0.226	−0.127	−0.300	3.046	90.75	11.351	0.1601
n-Decane	0.014	0.084	−0.033	−0.210	0.046	89.47	12.452	0.0477
Decyl alcohol	0.082	0.198	−0.181	−0.331	1.620	92.36	19.566	0.2628
1-Methylnaphthalene	0.021	0.133	−0.029	−0.181	0.321	116.10	24.710	0.0494
n-Pentylbenzene	0.020	0.132	−0.033	−0.211	0.324	102.55	18.948	0.0532
2-Undecanone	0.033	0.227	−0.127	−0.300	3.009	99.16	17.660	0.1601
Biphenyl	0.021	0.136	−0.019	−0.131	0.018	124.31	46.289	0.0393
Acenaphthene	0.021	0.134	−0.023	−0.123	0.649	122.98	25.380	0.0440
2,3-Dimethylnaphthalene	0.021	0.133	−0.029	−0.181	0.695	129.85	37.447	0.0501
2,6-Dimethylnaphthalene	0.021	0.136	−0.029	−0.182	0.435	128.64	40.698	0.0502
1,3-Dimethylnaphthalene	0.021	0.132	−0.029	−0.181	0.382	128.29	31.160	0.0502
1,4-Dimethylnaphthalene	0.021	0.132	−0.029	−0.180	0.034	126.96	29.441	0.0494
1-Ethylnaphthalene	0.021	0.135	−0.032	−0.210	0.294	124.67	24.111	0.0528
n-Hexylbenzene	0.020	0.132	−0.033	−0.210	0.324	111.59	17.689	0.0532
Fluorene	0.021	0.135	−0.020	−0.133	0.449	137.11	62.332	0.0410

TABLE 3.8 Charged Partial Surface Area (CSPA) Descriptors Obtained from AM1 SCRF Quantum Chemical Partial Charges for Structurally Variable Compounds

Compound	PPSA1	PNSA1	FPSA1	FNSA1	WPSA1	WNSA1
Water	72.52	57.31	0.559	0.441	9.42	7.44
Ammonia	106.40	33.49	0.761	0.239	14.88	4.68
Tetrafluoromethane	0.00	196.74	0.000	1.000	0.00	38.71
Tetrachloromethane	0.00	258.55	0.000	1.000	0.00	66.85
Trichloromethane	25.29	210.85	0.107	0.893	5.97	49.79
Tribromomethane	250.60	0.00	1.000	0.000	62.80	0.00
Formaldehyde	100.88	58.90	0.631	0.369	16.12	9.41
Dibromomethane	216.06	2.68	0.988	0.012	47.26	0.59
Dichloromethane	55.35	154.77	0.263	0.737	11.63	32.52
Chloromethane	90.18	95.71	0.485	0.515	16.76	17.79
Bromomethane	89.70	104.68	0.462	0.539	17.44	20.35
Iodomethane	193.56	12.06	0.941	0.059	39.80	2.48
Nitromethane	114.14	91.10	0.556	0.444	23.42	18.70
Methane	130.73	22.12	0.855	0.145	19.98	3.38
Methanol	123.10	52.51	0.701	0.299	21.62	9.22
Methylamine	154.59	30.75	0.834	0.166	28.65	5.70
Tetrachloroethene	259.93	16.80	0.939	0.061	71.93	4.65
Pentachloroethane	19.09	284.82	0.063	0.937	5.80	86.56
Trichloroethene	232.73	16.09	0.935	0.065	57.91	4.00
1,1,1,2-Tetrachloroethane	41.03	237.08	0.148	0.853	11.41	65.93
1,1-Dichloroethene	54.39	181.75	0.230	0.770	12.84	42.92
trans-1,2-Dichloroethene	54.39	180.36	0.232	0.768	12.77	42.34
cis-1,2-Dichloroethene	59.16	173.28	0.255	0.746	13.75	40.28
1,1,2,2-Tetrachloroethane	38.17	246.11	0.134	0.866	10.85	69.96
2,2,2-Trifluoroethanol	78.73	153.72	0.339	0.661	18.30	35.73
Acetonitrile	93.04	110.77	0.457	0.544	18.96	22.58
1,1,1-Trichloroethane	71.57	194.76	0.269	0.731	19.06	51.87
1,1,2-Trichloroethane	64.89	199.08	0.246	0.754	17.13	52.55
Ethene	127.87	53.62	0.705	0.295	23.21	9.73
1,1-Dichloroethane	104.97	137.38	0.433	0.567	25.44	33.29
Acetaldehyde	133.55	62.82	0.680	0.320	26.23	12.34
Acetic acid	119.60	92.29	0.564	0.436	25.34	19.56
1,2-Dibromoethane	89.70	166.61	0.350	0.650	22.99	42.70
1,2-Dichloroethane	93.04	144.46	0.392	0.608	22.10	34.31
Methyl formate	127.64	89.52	0.588	0.412	27.72	19.44
Chloroethane	133.60	79.90	0.626	0.374	28.52	17.06
Bromoethane	131.69	86.96	0.602	0.398	28.79	19.01
Iodoethane	224.18	4.69	0.980	0.021	51.31	1.07
Nitroethane	159.77	85.45	0.652	0.349	39.18	20.95
Ethane	178.92	14.74	0.924	0.076	34.65	2.86
Ethanol	167.47	41.69	0.801	0.199	35.03	8.72
Ethanethiol	159.84	60.83	0.724	0.276	35.27	13.42
Ethylamine	189.90	26.25	0.879	0.121	41.05	5.67
Dimethylamine	187.03	28.20	0.869	0.131	40.26	6.07
Isoflurane	87.74	220.96	0.284	0.716	27.08	68.21
Propyne	115.47	94.15	0.551	0.449	24.20	19.74

TABLE 3.8 (*Continued*)

Compound	PPSA1	PNSA1	FPSA1	FNSA1	WPSA1	WNSA1
Methoxyflurane	96.38	216.01	0.309	0.692	30.11	67.48
1-Chloro-2-propene	131.69	107.96	0.550	0.451	31.56	25.87
Propanonitrile	133.12	92.17	0.591	0.409	29.99	20.77
Ethyl formate	164.38	82.60	0.666	0.334	40.60	20.40
Methyl acetate	174.17	75.99	0.696	0.304	43.57	19.01
1,3-Dichloropropane	133.12	133.93	0.499	0.502	35.55	35.76
Cyclopropane	168.43	28.15	0.857	0.143	33.11	5.53
Propene	170.81	44.23	0.794	0.206	36.73	9.51
Propanone	158.34	62.54	0.717	0.283	34.97	13.81
Propanoic acid	163.30	80.86	0.669	0.331	39.87	19.74
2-Propenol	163.95	70.74	0.699	0.301	38.48	16.60
1,2-Dichloropropane	131.21	139.39	0.485	0.515	35.51	37.72
Propionaldehyde	173.72	52.57	0.768	0.232	39.31	11.90
1-Bromopropane	172.24	86.51	0.666	0.334	44.57	22.38
2-Bromopropane	167.95	87.11	0.659	0.342	42.84	22.22
1-Nitropropane	185.97	80.84	0.697	0.303	49.62	21.57
1-Chloropropane	173.20	78.45	0.688	0.312	43.59	19.74
2-Chloropropane	170.34	74.30	0.696	0.304	41.67	18.18
1-Iodopropane	254.16	10.05	0.962	0.038	67.15	2.66
N,*N*-Dimethylformamide	193.91	58.62	0.768	0.232	48.97	14.80
2-Nitropropane	174.29	85.11	0.672	0.328	45.21	22.08
Propane	204.21	18.10	0.919	0.081	45.40	4.02
Methyl ethyl ether	211.37	37.05	0.851	0.149	52.51	9.20
1-Propanol	204.69	39.97	0.837	0.163	50.08	9.78
2-Propanol	197.05	43.92	0.818	0.182	47.49	10.58
2-Methoxyethanol	198.01	62.41	0.760	0.240	51.57	16.25
n-Propanethiol	200.87	61.33	0.766	0.234	52.67	16.08
Trimethylamine	209.94	30.09	0.875	0.125	50.39	7.22
Propylamine	214.23	30.21	0.876	0.124	52.37	7.38
Thiophene	174.83	69.98	0.714	0.286	42.80	17.13
1-Butyne	157.45	77.36	0.671	0.329	36.97	18.16
(*E*)-2-Butenal	174.79	81.73	0.681	0.319	44.84	20.97
1-Cyanopropane	168.90	95.51	0.639	0.361	44.66	25.25
Isobutyraldehyde	199.56	57.07	0.778	0.222	51.22	14.65
Tetrahydrofuran	216.14	34.08	0.864	0.136	54.08	8.53
Ethyl acetate	214.44	68.58	0.758	0.242	60.69	19.41
Methyl propanoate	215.21	68.51	0.759	0.242	61.06	19.44
1,4-Dichlorobutane	164.61	138.09	0.544	0.456	49.83	41.80
1-Butene	207.07	43.56	0.826	0.174	51.90	10.92
Butanone	208.15	51.02	0.803	0.197	53.95	13.22
Isopropyl formate	205.61	66.13	0.757	0.243	55.87	17.97
Butanoic acid	195.35	79.91	0.710	0.290	53.77	22.00
Propyl formate	187.74	84.39	0.690	0.310	51.09	22.97
Butyraldehyde	200.04	57.10	0.778	0.222	51.44	14.68
1,4-Dioxane	201.83	54.00	0.789	0.211	51.63	13.81
1-Bromobutane	198.49	85.58	0.699	0.301	56.38	24.31
1-Chlorobutane	200.39	74.24	0.730	0.270	55.04	20.39

TABLE 3.8 (*Continued*)

Compound	PPSA1	PNSA1	FPSA1	FNSA1	WPSA1	WNSA1
1-Nitrobutane	208.70	84.62	0.712	0.289	61.22	24.82
2-Bromo-2-methylpropane	195.15	86.21	0.694	0.306	54.91	24.26
1-Bromo-2-methylpropane	192.28	89.28	0.683	0.317	54.14	25.14
2-Chloro-2-methylpropane	198.49	72.06	0.734	0.266	53.70	19.50
2-Chlorobutane	194.19	77.03	0.716	0.284	52.67	20.89
1-Iodobutane	275.51	10.72	0.963	0.038	78.86	3.07
Morpholine	224.73	39.54	0.850	0.150	59.39	10.45
n-Butane	243.34	14.07	0.945	0.055	62.64	3.62
Diethyl ether	248.11	34.36	0.878	0.122	70.08	9.71
2-Methylpropane	230.45	15.41	0.937	0.063	56.66	3.79
2-Methyl-1-propanol	226.16	41.08	0.846	0.154	60.44	10.98
2-Butanol	231.88	40.72	0.851	0.149	63.21	11.10
2-Methyl-2-propanol	220.43	46.89	0.825	0.175	58.93	12.54
Diethyl sulfide	277.56	12.73	0.956	0.044	80.57	3.70
Diethyl disulfide	251.45	91.05	0.734	0.266	86.12	31.18
2-Ethoxyethanol	248.58	45.49	0.845	0.155	73.10	13.38
1-Butanol	234.27	37.95	0.861	0.139	63.77	10.33
n-Butanethiol	223.77	55.54	0.801	0.199	62.50	15.51
Diethylamine	264.33	24.18	0.916	0.084	76.26	6.98
Butylamine	237.13	29.66	0.889	0.111	63.27	7.91
2-Chloropyridine	181.54	106.20	0.631	0.369	52.24	30.56
3-Chloropyridine	172.67	112.77	0.605	0.395	49.29	32.19
Pyridine	138.37	121.74	0.532	0.468	35.99	31.66
2-Methylthiophene	216.63	66.89	0.764	0.236	61.42	18.96
2-Methylpyrazine	157.93	130.48	0.548	0.452	45.55	37.63
Cyclopentene	215.66	37.53	0.852	0.148	54.60	9.50
2-Methyl-1,3-butadiene	188.47	74.39	0.717	0.283	49.54	19.55
Cyclopentanone	207.38	59.19	0.778	0.222	55.28	15.78
1,4-Pentadiene	200.39	63.00	0.761	0.239	52.78	16.59
1-Pentyne	198.01	68.93	0.742	0.258	52.86	18.40
1-Cyanobutane	194.67	95.99	0.670	0.330	56.58	27.90
Cyclopentane	243.34	17.43	0.933	0.067	63.45	4.54
1-Pentene	228.55	42.22	0.844	0.156	61.88	11.43
Ethyl propanoate	255.75	62.42	0.804	0.196	81.37	19.86
Isobutyl formate	213.03	83.21	0.719	0.281	63.11	24.65
2-Methyl-2-butene	229.50	43.56	0.841	0.160	62.67	11.90
3-Methyl-1-butene	229.02	39.54	0.853	0.147	61.51	10.62
Isopropyl acetate	243.63	58.55	0.806	0.194	73.62	17.69
Pentanal	232.21	62.15	0.789	0.211	68.36	18.30
Methyl butanoate	243.82	71.06	0.774	0.226	76.78	22.37
Pentanoic acid	227.24	80.93	0.737	0.263	70.03	24.94
3-Methylbutanoic acid	219.97	82.84	0.726	0.274	66.61	25.08
Cyclopentanol	237.13	40.79	0.853	0.147	65.90	11.34
3-Methyl-2-butanone	216.91	67.39	0.763	0.237	61.67	19.16
3-Pentanone	238.29	53.78	0.816	0.184	69.60	15.71
Propyl acetate	247.73	59.74	0.806	0.194	76.17	18.37
2-Pentanone	229.23	60.67	0.791	0.209	66.45	17.59

TABLE 3.8 (*Continued*)

Compound	PPSA1	PNSA1	FPSA1	FNSA1	WPSA1	WNSA1
2-Methyltetrahydrofuran	236.18	37.65	0.863	0.138	64.67	10.31
Tetrahydropyran	237.13	32.29	0.880	0.120	63.89	8.70
1-Chloropentane	220.91	81.93	0.730	0.271	66.90	24.81
1-Nitropentane	252.73	87.65	0.743	0.258	86.02	29.83
1-Bromopentane	213.28	92.25	0.698	0.302	65.16	28.19
1-Iodopentane	320.73	14.07	0.958	0.042	107.38	4.71
N-Methylmorpholine	257.65	43.93	0.854	0.146	77.70	13.25
n-Pentane	267.19	20.11	0.930	0.070	76.76	5.78
2-Methyl-1-butanol	236.66	47.94	0.832	0.169	67.35	13.64
2-Methyl-2-butanol	242.38	44.58	0.845	0.155	69.55	12.79
3-Methyl-1-butanol	260.99	35.71	0.880	0.120	77.44	10.60
1-Pentanol	258.13	43.92	0.855	0.145	77.97	13.27
2-Pentanol	248.11	48.09	0.838	0.162	73.49	14.24
3-Pentanol	248.11	44.43	0.848	0.152	72.58	13.00
Methyl *tert*-butyl ether	251.45	38.45	0.867	0.133	72.89	11.15
2,2-Dimethylpropane	252.88	26.14	0.906	0.094	70.56	7.29
2-Propoxyethanol	292.48	50.64	0.852	0.148	100.36	17.38
2-Methylbutane	253.36	23.46	0.915	0.085	70.13	6.49
n-Pentylamine	271.96	30.57	0.899	0.101	82.28	9.25
1,2,3,4-Tetrachlorobenzene	282.76	67.19	0.808	0.192	98.95	23.51
1,2,3,5-Tetrachlorobenzene	293.46	67.19	0.814	0.186	105.84	24.23
1,2,4,5-Tetrachlorobenzene	290.67	55.99	0.839	0.162	100.76	19.41
1,2,3-Trichlorobenzene	256.68	65.79	0.796	0.204	82.77	21.21
1,3,5-Trichlorobenzene	272.99	68.59	0.799	0.201	93.25	23.43
1,2,4-Trichlorobenzene	263.66	69.98	0.790	0.210	87.97	23.35
1,2-Dichlorobenzene	161.89	143.38	0.530	0.470	49.42	43.77
1,3-Dichlorobenzene	94.47	211.06	0.309	0.691	28.86	64.48
1,4-Dichlorobenzene	99.24	222.24	0.309	0.691	31.91	71.45
3-Cyanopyridine	105.92	203.52	0.342	0.658	32.78	62.98
4-Cyanopyridine	113.84	191.56	0.373	0.627	34.77	58.50
Bromobenzene	205.29	95.18	0.683	0.317	61.68	28.60
Chlorobenzene	129.78	164.31	0.441	0.559	38.17	48.32
4-Fluorophenol	144.54	150.58	0.490	0.510	42.66	44.44
2-Chlorophenol	132.64	168.91	0.440	0.560	40.00	50.94
3-Chlorophenol	124.79	178.53	0.411	0.589	37.85	54.15
4-Chlorophenol	132.39	179.75	0.424	0.576	41.32	56.11
4-Bromophenol	205.48	110.12	0.651	0.349	64.85	34.75
2-Iodophenol	215.29	100.45	0.682	0.318	67.98	31.72
2-Nitrophenol	135.17	188.22	0.418	0.582	43.71	60.87
3-Nitrophenol	143.00	172.94	0.453	0.547	45.18	54.64
4-Nitrophenol	125.22	216.04	0.367	0.633	42.73	73.72
3-Formalpyridine	140.05	154.06	0.476	0.524	41.19	45.31
4-Formalpyridine	142.15	154.13	0.480	0.520	42.12	45.67
2-Fluorophenol	146.51	147.96	0.498	0.503	43.14	43.57
Fluorobenzene	142.89	136.70	0.511	0.489	39.95	38.22
Iodobenzene	221.78	95.18	0.700	0.300	70.29	30.17
Nitrobenzene	150.40	160.27	0.484	0.516	46.72	49.79

TABLE 3.8 (*Continued*)

Compound	PPSA1	PNSA1	FPSA1	FNSA1	WPSA1	WNSA1
Benzene	160.32	103.58	0.608	0.393	42.31	27.33
Phenol	162.45	121.11	0.573	0.427	46.06	34.34
4-Chloroaniline	153.41	162.96	0.485	0.515	48.54	51.55
2-Nitroaniline	172.38	150.77	0.533	0.467	55.70	48.72
3-Nitroaniline	169.90	155.85	0.522	0.478	55.35	50.77
4-Nitroaniline	174.06	157.72	0.525	0.475	57.75	52.33
3-Chloroaniline	149.63	164.70	0.476	0.524	47.03	51.77
2-Chloroaniline	149.63	157.24	0.488	0.512	45.92	48.25
Thiophenol	205.88	96.58	0.681	0.319	62.27	29.21
2-Methylpyridine	184.17	113.20	0.619	0.381	54.77	33.66
3-Methylpyridine	173.20	115.28	0.600	0.400	49.96	33.25
4-Methylpyridine	178.45	117.24	0.604	0.397	52.76	34.67
N-Acetylpyrrolidine	178.09	129.32	0.579	0.421	54.75	39.75
2-Ethylpyrazine	194.67	113.19	0.632	0.368	59.93	34.85
Cyclohexene	250.02	28.15	0.899	0.101	69.55	7.83
Cyclohexanone	234.95	56.00	0.808	0.193	68.36	16.29
1,5-Hexadiene	242.86	62.33	0.796	0.204	74.12	19.02
2,3-Dimethyl-1,3-butadiene	213.75	70.37	0.752	0.248	60.73	19.99
trans-2-Hexenal	236.34	78.99	0.750	0.251	74.52	24.91
1-Hexyne	225.68	67.38	0.770	0.230	66.14	19.75
Cyclohexane	271.01	7.37	0.974	0.027	75.44	2.05
Methylcyclopentane	278.17	12.73	0.956	0.044	80.92	3.70
2-Methyl-1-pentene	252.88	42.89	0.855	0.145	74.79	12.69
2-Hexanone	256.62	60.43	0.809	0.191	81.36	19.16
Hexanoic acid	268.45	76.32	0.779	0.221	92.55	26.31
Isobutyl acetate	276.95	63.54	0.813	0.187	94.30	21.63
n-Propyl propanoate	285.53	66.73	0.811	0.189	100.58	23.51
1-Hexene	268.15	37.53	0.877	0.123	81.97	11.47
Isoamyl formate	259.90	69.07	0.790	0.210	85.50	22.72
Ethyl butanoate	293.16	53.34	0.846	0.154	101.58	18.48
Methyl pentanoate	269.59	61.43	0.814	0.186	89.24	20.34
Hexanal	256.72	65.00	0.798	0.202	82.59	20.91
4-Methyl-2-pentanone	264.73	48.81	0.844	0.156	83.00	15.30
Butyl acetate	285.15	60.56	0.825	0.175	98.58	20.94
Cyclohexylamine	274.35	18.63	0.936	0.064	80.38	5.46
n-Butylacetamide	267.73	65.05	0.805	0.196	89.10	21.65
1-Bromohexane	247.15	88.52	0.736	0.264	82.96	29.71
1-Chlorohexane	254.79	81.24	0.758	0.242	85.61	27.30
1-Iodohexane	343.94	10.72	0.970	0.030	121.98	3.80
N-Methylpiperidine	289.14	20.53	0.934	0.066	89.54	6.36
2,2-Dimethylbutane	271.49	23.46	0.921	0.080	80.07	6.92
2,3-Dimethylbutane	273.87	26.81	0.911	0.089	82.35	8.06
n-Hexane	296.30	14.07	0.955	0.045	91.96	4.37
2-Methylpentane	283.41	24.13	0.922	0.079	87.16	7.42
3-Methylpentane	276.74	20.11	0.932	0.068	82.15	5.97
Diisopropyl ether	302.02	34.12	0.899	0.102	101.52	11.47
Di-*n*-propyl ether	323.02	23.33	0.933	0.067	111.88	8.08

TABLE 3.8 ***(Continued)***

Compound	PPSA1	PNSA1	FPSA1	FNSA1	WPSA1	WNSA1
1-Hexanol	290.57	42.29	0.873	0.127	96.72	14.08
Di-*n*-propyl sulfide	333.86	15.41	0.956	0.044	116.61	5.38
Diisopropyl sulfide	316.97	23.46	0.931	0.069	107.90	7.99
4-Methyl-2-pentanol	279.12	36.15	0.885	0.115	88.00	11.40
2-Methyl-3-pentanol	274.35	39.13	0.875	0.125	86.00	12.27
3-Hexanol	283.89	41.53	0.872	0.128	92.39	13.52
2-Butoxyethanol	298.21	56.09	0.842	0.158	105.65	19.87
2-Methyl-2-pentanol	276.26	41.00	0.871	0.129	87.64	13.01
Di-*n*-Propylamine	341.62	22.91	0.937	0.063	124.53	8.35
Hexylamine	304.89	26.49	0.920	0.080	101.03	8.78
Diisopropylamine	294.39	26.93	0.916	0.084	94.59	8.65
Triethylamine	303.93	24.49	0.925	0.075	99.82	8.04
Benzotrifluoride	119.28	206.10	0.367	0.633	38.81	67.06
3-Cyanophenol	133.69	185.11	0.419	0.581	42.62	59.01
4-Cyanophenol	132.39	200.87	0.397	0.603	44.12	66.94
Benzonitrile	129.78	185.30	0.412	0.588	40.89	58.38
Benzaldehyde	165.90	143.81	0.536	0.464	51.38	44.54
3-Hydroxybenzaldehyde	163.71	147.86	0.525	0.475	51.01	46.07
4-Hydroxybenzaldehyde	163.01	148.13	0.524	0.476	50.72	46.09
2-Chlorotoluene	168.90	147.34	0.534	0.466	53.41	46.59
4-Bromotoluene	244.89	89.28	0.733	0.267	81.84	29.84
2-Nitrotoluene	193.30	139.87	0.580	0.420	64.40	46.60
3-Nitrotoluene	185.21	148.49	0.555	0.445	61.80	49.55
Benzamide	175.11	141.39	0.553	0.447	55.42	44.75
4-Chloro-3-methylphenol	168.90	157.37	0.518	0.482	55.11	51.34
3-Acetylpyridine	167.68	144.46	0.537	0.463	52.34	45.09
4-Acetylpyridine	168.06	143.59	0.539	0.461	52.38	44.75
Toluene	203.73	100.45	0.670	0.330	61.97	30.56
Benzyl alcohol	198.49	115.66	0.632	0.368	62.35	36.33
o-Cresol	199.73	106.22	0.653	0.347	61.11	32.50
p-Cresol	202.05	115.92	0.635	0.365	64.25	36.86
1,3,5-Cycloheptatriene	207.55	73.72	0.738	0.262	58.38	20.74
Methyl phenyl ether	205.71	111.58	0.648	0.352	65.27	35.40
2-Methoxyphenol	208.09	117.68	0.639	0.361	67.79	38.34
Phenyl methyl sulfide	242.45	91.29	0.727	0.274	80.92	30.47
3-Methoxyphenol	202.40	116.81	0.634	0.366	64.61	37.29
o-Toluidine	214.07	93.92	0.695	0.305	65.93	28.93
p-Toluidine	225.94	96.63	0.700	0.300	72.88	31.17
2-Methoxyaniline	228.64	99.11	0.698	0.302	74.94	32.48
3-Methoxyaniline	226.03	101.70	0.690	0.310	74.08	33.33
4-Methoxyaniline	230.55	105.35	0.686	0.314	77.44	35.39
2,3-Dimethylpyridine	212.80	97.18	0.687	0.314	65.96	30.12
2,4-Dimethylpyridine	220.91	89.03	0.713	0.287	68.47	27.59
2,5-Dimethylpyridine	219.48	102.44	0.682	0.318	70.66	32.98
2,6-Dimethylpyridine	224.25	91.49	0.710	0.290	70.80	28.89
3,4-Dimethylpyridine	199.92	100.36	0.666	0.334	60.03	30.14
3,5-Dimethylpyridine	217.09	95.00	0.696	0.304	67.75	29.65

TABLE 3.8 ***(Continued)***

Compound	PPSA1	PNSA1	FPSA1	FNSA1	WPSA1	WNSA1
N-Methylaniline	217.25	91.30	0.704	0.296	67.03	28.17
2-Ethylpyridine	230.45	93.69	0.711	0.289	74.70	30.37
3-Ethylpyridine	213.75	107.14	0.666	0.334	68.59	34.38
4-Ethylpyridine	215.19	107.53	0.667	0.333	69.44	34.70
1-Methylcyclohexene	282.46	29.49	0.906	0.095	88.11	9.20
1-Heptyne	256.70	63.88	0.801	0.199	82.29	20.48
1-Heptene	292.00	38.20	0.884	0.116	96.42	12.61
Heptanal	307.03	55.33	0.847	0.153	111.26	20.05
n-Propyl butanoate	326.75	66.26	0.831	0.169	128.42	26.04
Ethyl pentanoate	320.64	56.51	0.850	0.150	120.93	21.31
Methyl hexanoate	297.93	68.13	0.814	0.186	109.06	24.94
Cycloheptanol	280.08	34.23	0.891	0.109	88.03	10.76
Methylcyclohexane	288.66	12.06	0.960	0.040	86.81	3.63
2,4-Dimethyl-3-pentanone	266.24	58.05	0.821	0.179	86.34	18.82
Isoamyl acetate	298.61	69.65	0.811	0.189	109.97	25.65
4-Heptanone	293.92	54.03	0.845	0.155	102.27	18.80
trans-2-Heptene	311.57	31.50	0.908	0.092	106.89	10.81
2-Heptanone	289.44	59.92	0.829	0.172	101.12	20.93
1-Bromoheptane	273.40	92.25	0.748	0.252	99.97	33.73
1-Chloroheptane	285.80	86.01	0.769	0.231	106.26	31.98
1-Iodoheptane	378.63	13.40	0.966	0.034	148.43	5.25
2,3-Dimethylpentane	293.43	27.48	0.914	0.086	94.17	8.82
n-Heptane	323.02	16.09	0.953	0.047	109.54	5.45
2-Methylhexane	323.97	22.79	0.934	0.066	112.34	7.90
3-Methylhexane	307.27	25.47	0.924	0.077	102.24	8.47
1-Heptanol	304.89	50.71	0.857	0.143	108.42	18.03
2,4-Dimethylpentane	303.93	24.13	0.927	0.074	99.71	7.92
3,3-Dimethylpentane	297.73	22.12	0.931	0.069	95.23	7.07
2,2-Dimethylpentane	294.39	22.79	0.928	0.072	93.37	7.23
n-Heptylamine	344.01	26.37	0.929	0.071	127.41	9.77
Styrene	195.62	116.21	0.627	0.373	61.00	36.24
Acetophenone	191.44	128.09	0.599	0.401	61.17	40.93
4-Methylbenzaldehyde	202.84	127.71	0.614	0.386	67.05	42.21
Methyl benzoate	206.14	145.96	0.586	0.415	72.58	51.39
Phenyl acetate	218.54	124.12	0.638	0.362	74.89	42.53
Ethylbenzene	234.27	95.61	0.710	0.290	77.28	31.54
m-Xylene	238.56	82.02	0.744	0.256	76.48	26.29
p-Xylene	244.29	95.23	0.720	0.281	82.94	32.33
o-Xylene	229.98	87.62	0.724	0.276	73.04	27.83
Ethyl phenyl ether	237.04	97.32	0.709	0.291	79.26	32.54
2,3-Dimethylphenol	230.71	96.78	0.705	0.296	75.56	31.70
2,5-Dimethylphenol	239.33	103.01	0.699	0.301	81.93	35.26
2,6-Dimethylphenol	236.47	90.79	0.723	0.277	77.39	29.71
3,4-Dimethylphenol	232.11	104.41	0.690	0.310	78.11	35.14
3,5-Dimethylphenol	231.69	101.26	0.696	0.304	77.14	33.71
4-Ethylphenol	234.53	106.06	0.689	0.311	79.88	36.12

TABLE 3.8 ***(Continued)***

Compound	PPSA1	PNSA1	FPSA1	FNSA1	WPSA1	WNSA1
2-Phenylethanol	235.70	112.22	0.678	0.323	82.01	39.04
2,4-Dimethylphenol	249.54	94.25	0.726	0.274	85.79	32.40
3-Ethylphenol	234.30	106.14	0.688	0.312	79.77	36.13
2,6-Dimethylaniline	247.95	85.37	0.744	0.256	82.65	28.46
2-Isobutylpyrazine	254.79	107.53	0.703	0.297	92.31	38.96
Trans-2-octenal	300.45	74.89	0.801	0.200	112.77	28.11
1-Octyne	276.26	87.77	0.759	0.241	100.56	31.95
n-Propylcyclopentane	336.38	20.11	0.944	0.056	119.91	7.17
trans-1,4-Dimethylcyclohexane	313.00	20.11	0.940	0.060	104.26	6.70
Octanal	334.61	53.16	0.863	0.137	129.75	20.61
n-Pentyl propanoate	353.09	56.67	0.862	0.138	144.68	23.22
Ethyl hexanoate	334.00	56.17	0.856	0.144	130.32	21.91
Isobutyl isotuanoate	334.19	63.53	0.840	0.160	132.92	25.27
cis-1,2-Dimethylcyclohexane	303.93	18.10	0.944	0.056	97.87	5.83
Hexyl acetate	347.08	57.53	0.858	0.142	140.43	23.28
2-Octanone	311.67	51.85	0.857	0.143	113.30	18.85
1-Octene	318.25	45.57	0.875	0.125	115.78	16.58
1-Bromooctane	311.09	94.26	0.768	0.233	126.10	38.21
n-Octane	362.14	14.74	0.961	0.039	136.49	5.56
2,3,4-Trimethylpentane	312.04	29.49	0.914	0.086	106.57	10.07
2,2,4-Trimethylpentane	311.57	29.49	0.914	0.087	106.26	10.06
Di-*n*-butyl ether	379.79	28.69	0.930	0.070	155.14	11.72
3-Methylheptane	355.46	21.45	0.943	0.057	133.98	8.08
1-Octanol	348.78	45.64	0.884	0.116	137.57	18.00
Di-*n*-butylamine	383.61	24.98	0.939	0.061	156.74	10.21
Octylamine	358.80	32.89	0.916	0.084	140.54	12.88
Quinoline	173.20	152.81	0.531	0.469	56.46	49.82
α-Methylstyrene	223.30	114.69	0.661	0.339	75.47	38.76
Ethyl benzoate	254.81	124.93	0.671	0.329	96.76	47.44
4-Methylacetophenone	231.70	115.78	0.667	0.333	80.51	40.23
Indane	238.09	85.09	0.737	0.263	76.94	27.50
1,2,3-Trimethylbenzene	256.70	77.70	0.768	0.232	85.84	25.98
1,2,4-Trimethylbenzene	269.10	81.14	0.768	0.232	94.25	28.42
1,3,5-Trimethylbenzene	270.53	79.59	0.773	0.227	94.72	27.87
Isopropylbenzene	259.56	94.68	0.733	0.267	91.94	33.54
2-Ethyltoluene	254.79	88.23	0.743	0.257	87.40	30.26
4-Ethyltoluene	270.53	84.76	0.761	0.239	96.12	30.11
4-*n*-Propylphenol	267.48	96.47	0.735	0.265	97.35	35.11
Propylbenzene	272.44	97.59	0.736	0.264	100.81	36.11
3-Phenyl-1-propanol	255.26	115.46	0.689	0.311	94.63	42.80
Nonanal	361.72	54.66	0.869	0.131	150.61	22.76
5-Nonanone	369.12	50.57	0.880	0.121	154.91	21.23
2-Nonanone	351.27	49.66	0.876	0.124	140.83	19.91
1-Nonene	355.46	48.25	0.881	0.120	143.51	19.48
n-Nonane	367.39	16.09	0.958	0.042	140.88	6.17
2,2,5-Trimethylhexane	357.85	27.48	0.929	0.071	137.89	10.59

TABLE 3.8 (*Continued*)

Compound	PPSA1	PNSA1	FPSA1	FNSA1	WPSA1	WNSA1
1-Nonanol	378.36	42.58	0.899	0.101	159.27	17.92
Naphthalene	186.08	144.87	0.562	0.438	61.58	47.94
1-Naphthol	200.65	159.13	0.558	0.442	72.19	57.25
2-Naphthol	201.67	149.91	0.574	0.426	70.90	52.71
1-Naphthylamine	218.08	147.31	0.597	0.403	79.68	53.83
2-Naphthylamine	228.39	135.41	0.628	0.372	83.09	49.26
n-Butylbenzene	295.82	88.58	0.770	0.230	113.71	34.05
tert-Butylbenzene	277.21	91.61	0.752	0.248	102.24	33.79
4-Isopropyltoluene	301.55	76.88	0.797	0.203	114.11	29.09
4-*tert*-Butylphenol	279.34	108.56	0.720	0.280	108.36	42.11
Isobutylbenzene	294.39	87.65	0.771	0.229	112.47	33.48
sec-Butylbenzene	286.28	84.21	0.773	0.227	106.06	31.20
n-Pentylcyclopentane	407.47	16.76	0.961	0.040	172.86	7.11
2-Decanone	386.30	59.06	0.867	0.133	172.05	26.31
n-Decane	412.72	16.76	0.961	0.039	177.25	7.20
Decyl alcohol	394.11	45.86	0.896	0.104	173.39	20.18
1-Methylnaphthalene	225.68	140.37	0.617	0.384	82.61	51.38
n-Pentylbenzene	332.08	91.09	0.785	0.215	140.53	38.55
2-Undecanone	403.57	52.52	0.885	0.115	184.07	23.96
Biphenyl	231.88	142.77	0.619	0.381	86.88	53.49
Acenaphthene	238.56	127.95	0.651	0.349	87.44	46.90
2,3-Dimethylnaphthalene	258.60	127.60	0.670	0.330	99.87	49.28
2,6-Dimethylnaphthalene	263.85	137.25	0.658	0.342	105.83	55.05
1,3-Dimethylnaphthalene	258.60	125.41	0.673	0.327	99.31	48.16
1,4-Dimethylnaphthalene	248.58	127.45	0.661	0.339	93.48	47.93
1-Ethylnaphthalene	260.04	129.15	0.668	0.332	101.20	50.26
n-Hexylbenzene	358.80	90.39	0.799	0.201	161.17	40.60

atomic charge. The descriptor A_+ is the sum of the surface areas of positively charged atoms multiplied by their corresponding scaled net atomic charge. The descriptor A_{HB} is the sum of the surface areas of hydrogen-bonding hydrogen atoms multiplied by their corresponding scaled net atomic charge. This equation was derived from the theoretical principle that many bulk properties of liquids should be driven by the intermolecular interactions that can be expressed through the molecular surface energy. The latter was presented as a sum of dispersion interactions (proportional to the molecular surface area A), polar interactions (related to the first-order term of the electrostatic interactions A_- and A_+) and hydrogen-bonding interactions (A_{HB}).

In many cases, the electrostatic descriptors are combined with other types of descriptors. For instance, the gas chromatographic retention indexes of 43 mono- and bifunctional molecules on four stationary phases of different polarity were found to be described by a combination of topological indices and quantum chemical CNDO/2 calculated atomic charges [441]. In particular, the ADAPT approach by Jurs and co-workers [474,475] has been successful for the description of a large variety of physical, analytical, and biological properties of compounds. In this approach, the CPSA descriptors discussed above play a central role [444]. Thus, the surface tension for a set of alcohols has been described by the following equation employing topological and CPSA descriptors [476]:

$$\sigma = (27.33 \pm 1.196) + (5.388 \pm 0.491)\log_{10}{}^{2}\kappa + (2.585 \pm 0.392)^{4}\chi^{\nu}_{\text{pc}}$$
$$- (1.592 \pm 0.179)\Delta_c - (0.1581 \pm 0.0216)\text{PNSA1}$$
$$+ (87.52 \pm 15.13)\text{FNSA3} + (0.2939 \pm 0.0420)\text{RSAA}$$
$$R^2 = 0.975 \qquad s = 0.30 \qquad n = 35$$

where ${}^{2}\kappa$ is the Kier–Hall molecular shape index, ${}^{4}\chi^{\nu}_{\text{pc}}$ is the valence-corrected fourth-order path cluster molecular connectivity, Δ_c is the charge separation distance, PNSA1 denotes the partial negative surface area, FNSA3 is the fractional negative surface area, and RSAA is the average surface area of hydrogen bond acceptors. In addition, a 10-parameter equation involving six topological and four CPSA descriptors was developed for the surface tension of a combined data set of alkanes, alcohols, and esters ($R^2 = 0.978$; $s = 0.4$ dyn/cm; $N = 146$).

By applying of CPSA descriptors in combination with a selection of constitutional, topological, and other descriptors, QSPR equations have been developed for the normal boiling points of individual classes of heterocycles. For each single class of heterocycle, satisfactory correlations were obtained. Thus, the best regression for furans and tetrahydrofurans had the squared correlation coefficient $R^2 = 0.969$ [475], for thiophenes $R^2 = 0.974$ [477], for pyrans $R^2 = 0.978$ [477], for pyrroles $R^2 = 0.962$ [477], and for pyridines $R^2 = 0.933$ [478]. Each correlation had standard errors ranging from 8 to 15 K and involved a large number of descriptors. Besides, the correlation models obtained lacked uniformity and involved different, sometimes even specific descriptors for different classes of compounds. Nevertheless, the CPSA descriptors were demonstrated to be useful to predict the boiling point for 268 diverse organic compounds. The respective QSPR equation included 8 parameters four of which are CPSA descriptors [479]:

$$\text{bp} = 153.8 - (0.1102 \pm 0.0064)W + (15.90 \pm 0.26)\text{ID} - (10.04 \pm 1.44)N_{\text{O}}$$
$$- (70.51 \pm 10.72)\bar{\delta}_C + (545.7 \pm 26.4)\text{SCSA} + (1.212 \pm 0.112)\text{PNSA2}$$
$$- (9.353 \pm 0.871)\text{PNSA3} - (514.5 \pm 62.9)\text{FNSA3}$$
$$R^2 = 0.976 \qquad s = 11.85K \qquad n = 268$$

where W is the Wiener index, ID is the molecular identification number, N_{O} is the number of oxygen atoms in the molecule, $\bar{\delta}_C$ is the average charge on carbonyl or cyano carbon atoms, and SCDA is defined as:

$$\mathrm{SCDA} = \frac{q_{\mathrm{acc}} S_{\mathrm{acc}} + q_{\mathrm{don}} S_{\mathrm{don}}}{S_{\mathrm{tot}}} \quad (3.173)$$

and featuring the dipole–dipole intermolecular interactions. Again, a large number of descriptors was required to obtain a satisfactory correlation. However, the combination of the HDSA2 descriptor alone with the cube root of gravitation index has lead to a good two-parameter correlation with the normal boiling points of a set of 298 compounds [87]:

$$\mathrm{bp} = -(170.7 \pm 7.46) + (65.88 \pm 0.86)\sqrt[3]{G_b} + (18470 \pm 540)\mathrm{HDSA2}$$
$$R^2 = 0.9544 \qquad s = 16.15K \qquad n = 298 \qquad F = 3126.5$$

Notably, the 268 compounds used in the previous correlation were a subset of compounds employed in the last correlation. The addition of two further parameters resulted in a correlation with practically the same R^2 as the above eight-parameter correlation.

$$\mathrm{bp} = -(151.3 \pm 6.30) + (67.39 \pm 0.67)\sqrt[3]{G_b} + (21540 \pm 4800)\mathrm{HDSA2}$$
$$+ (140.4 \pm 13.1)\delta^-_{\max} + (17.51 \pm 2.31)N_{\mathrm{Cl}}$$
$$R^2 = 0.9732 \qquad s = 12.41K \qquad n = 298 \quad F = 2700$$

However, because of less parameters, the last correlation is statistically more significant. In this equation, $\delta^-_{\max}$ denotes the largest negative charge in the molecule and N_{Cl} is the number of chlorine atoms.

The CPSA descriptors have been introduced to quantify electrostatic intermolecular interactions in condensed media. Therefore, one of their natural QSPR applications has been the prediction of aqueous solubility of compounds [480–482]. A satisfactory description of the gas solubilities of a set consisting of 406 organic compounds with a large structural variability was obtained using a five-parameter QSPR equation [90]:

$$\log L^w = (2.82 \pm 0.22) + (41.61 \pm 1.11)\mathrm{HDCA2}$$
$$+ (0.13 \pm 0.01)\mathrm{PCWT^E} - (0.17 \pm 0.02)\Delta\varepsilon_{\mathrm{FMO}}$$
$$+ (0.79 \pm 0.06)N_{\mathrm{rings}} + (0.71 \pm 0.02)(2N_{\mathrm{N}} + N_{\mathrm{O}})$$
$$R^2 = 0.9407 \qquad s = 0.53 \qquad n = 406 \qquad F = 1269$$

where the Ostwald solubility coefficient (L^w) is defined as the ratio of the equilibrium concentrations of a gaseous compound in the liquid and in the gas phase, respectively. In the last equation, the most important term involves the charged partial surface area HDCA2 descriptor. In addition, $\Delta\varepsilon_{FMO}$ is the difference in energies of frontier (the highest occupied and the lowest unoccupied) orbitals, N_{rings}, N_N, and N_O are the numbers of rings, nitrogen, and oxygen atoms in the molecule, respectively, and $PCWT^E$ denotes a modified electronic topographical index defined as:

$$PCWT^E = \frac{1}{Q_{min}} \sum_{i<j} \frac{|q_i - q_j|}{r_{ij}^2} \quad (3.174)$$

where q_i and q_j are the Zefirov partial charges on the bonded atoms, i and j, r_{ij} are the respective bond lengths, and Q_{min} is the most negative partial charge in the molecule. The ADAPT approach and CPSA descriptors have been applied also in the development of QSPR equations for the critical temperatures [89,483], for melting points of compounds [484], for refractive indices [485], for gas chromatographic retention times [486–490], for supercritical carbon dioxide solubility of organic compounds [491], for autoignition temperatures of compounds [492,493], and for other properties [494–499].

The Politzer GIPF approach has also been applied in a number of areas. The critical constants of a variety of organic constants were expressed by the following QSPR regression equations [447]:

$$T_c = -530.1 + 74.68\sqrt{S_M} + 152.5\sqrt{(\nu\sigma_{tot}^2)^{0.5}}$$

$$R^2 = 0.826 \qquad s = 60.7 \qquad n = 66$$

$$\bar{V}_c = 43.4 + 74.68 S_M^{1.5}$$

$$R^2 = 0.972 \qquad s = 15.2 \qquad n = 58$$

$$P_c = 61.73 - 0.1764 S_M + 152.5 \frac{\nu\sigma_{tot}^2}{S_M}$$

$$R^2 = 0.828 \qquad s = 4.8 \qquad n = 57$$

where T_c, $\bar{V}_c$, and P_c are the critical temperature, critical volume, and critical pressure of compounds, respectively, and S_M denotes the surface area of a molecule. The octanol/water partition coefficients for a set of 70 compounds has been described using the following equation [447]:

$$\log P = 0.529 + 0.0298 S_M - 0.00912\sigma_-^2 - 0.000849 S_M \Pi$$

$$R^2 = 0.924 \qquad s = 0.437 \qquad n = 70$$

The GIPF descriptors have been useful for the examination of chemical reactivity of compounds [500,501]. Thus, the gas-phase protonation enthalpies of anions were presented as [449]:

$$\Delta H^0_{\text{an}} = -238.00 + 1.09V_{S,\text{min}} + 11.48\bar{I}_{S,\text{min}}$$
$$R^2 = 0.968 \qquad s = 4.6 \qquad n = 9$$

whereas the pK_a of anions in aqueous solutions had the following QSPR description:

$$\text{p}K_a = -9.57 - 0.35V_{S,\text{min}} - 0.27\bar{I}_{S,\text{min}}$$
$$R^2 = 0.925 \qquad s = 4.2 \qquad n = 9$$

The difference in signs of the descriptor terms in the gas phase and in the solution, respectively, is due to the difference in the signs of ΔH^0_{an} and pK_a. However, the results indicate that, as expected, the electrostatic interactions are more efficient in the gas phase. The GIPF electrostatic descriptors have been also successfully applied for the description of the free energies of solvation ΔG_{sol} of compounds in aqueous solution. For a set of 50 compounds, the following QSPR equation was developed [502]:

$$\Delta G_{\text{sol}}(\text{kJ/mol}) = 46.827 + 0.17201V_{\text{min}} - 2.6412 \times 10^{-5}(V_{S,\text{max}} - V_{S,\text{min}})^3$$
$$+ 5.1892 \times 10^{-2} A_S^- \bar{V}_S^- + 9.7042 \times 10^3 (A_S^- \bar{V}_S^-)^{-1}$$
$$R^2 = 0.976 \qquad s = 1.57 \qquad n = 50$$

In this equation, A_S^- denotes the surface area over which the surface potential $V_s(\mathbf{r})$ is negative and $\bar{V}_S^-$ is the average negative potential on the molecular surface.

The Kamlet–Taft empirical acidity and basicity parameters, α_2^H and β_2^H, respectively, have been also successfully correlated with the GIPF descriptors [503]. For the acidity, the following equation was developed:

$$\alpha_2^H = -0.150 + 1.34 \times 10^{-4}\sigma_+^2 V_{S,\text{max}} + 6.40 \times 10^{-3}\nu\sigma_{\text{tot}}^2 + 8.58 \times 10^{-6}S_M^2$$
$$R^2 = 0.966 \qquad n = 20$$

and for the basicity, the QSPR equation was obtained as:

$$\beta_2^H = 0.112 - 0.0275V_{S,\text{min}} - 0.0365\bar{I}_{S,\text{min}} + 0.0164\Pi$$
$$R^2 = 0.970 \qquad n = 18$$

The charge distribution and electrostatic potential in the molecules was calculated using ab initio HF/6-31G* wave functions [504]. As expected, the acidity is determined by the electrostatic potential in the positively charged area and the basicity by the characteristics of minima of potential. The other numerous applications of the GIPF approach include the QSPR treatment of heats of vaporization [503] and sublimation [505], solubilities in supercritical solvents [464,465], diffusion constants [506], buckminsterfullerene solubilities [507], surface tensions [508], liquid and solid densities [508], heats of fusion [508], and lattice energies [509].

The reactivity of compounds in enzymatic reactions has been related to quantum chemically calculated electrostatic descriptors. For example, good correlations were obtained between the CNDO/2 calculated total net ($\sigma + \pi$) atomic and group charges and data on anhydrase inhibition by heterocyclic sulfonamides [510]:

$$\log \Pi_{50} = 8.78 + 37.84 q_{SO_2NH_2}$$

$$R^2 = 0.827 \qquad s = 0.336 \qquad n = 28 \qquad F = 123.2$$

$$\log \Pi\%_{50} = 64.74 + 119.74 q_O$$

$$R^2 = 0.808 \qquad s = 0.364 \qquad n = 28 \qquad F = 109.3$$

where $q_{SO_2NH_2}$ is the charge of the $—SO_2NH_2$ group and q_O is the charge on the oxygen atom of the $—SO_2$ group. The CNDO/2 calculated charge at the most highly charged atom in the molecule, and the shape of the molecule described by the smallest dimension of the molecule, have both been shown to be important descriptors in the quantitative structure–activity relationships of some arylalkylamine and arylalkylamino acid activators of carbonic anhydrase [401]:

$$\log A = 2.208 + 0.480 Q_{max} - 0.0418 L_y$$

$$R^2 = 0.714 \qquad s = 0.043 \qquad n = 19 \qquad F = 20.01$$

where L_y is the length of an ellipsoid of uniform density, with the same principal moments of inertia as the molecule.

The advantage of the use of electrostatic descriptors in QSAR/QSPR correlations arises from their straight relation to physically determined inter- and intramolecular electrostatic interactions. However, because of the operational definition of atomic charges and their conventional nature, the charges for a given set of molecular structures may not be as reliable as necessary for the proper development of structure–property relationships.

3.5 MOLECULAR FIELD APPROACH

In most cases, the adequate description of a chemical or physical property of any compound requires the use of several molecular descriptors, each of which is responsible for individual intra- or intermolecular interactions. The molecular field approach represents a limiting case of a multiparameter QSPR/QSAR, where a spatially continuous molecular property (field created by a molecule) is presented by a large array of numbers. This field is characterized by the respective property values in the predetermined grid points of the three-dimensional space. The actual physical foundation of the field created may be different. In most cases, the electrostatic field created by the charge distribution of the molecule is used for the molecular description. Alternatively, the possible intermolecular van der Waals interactions are frequently accounted for by the introduction of the respective steric fields. In general, all approaches employing molecular fields in three-dimensional space have been referred to as the three-dimensional quantitative structure–activity relationship (3D-QSAR) methods.

Since its introduction in 1988 [511], the comparative molecular field analysis (CoMFA) had become one of the most popular 3D-QSAR methods. A review published 10 years later [512] has given a compilation of 364 CoMFA models developed between 1993 and 1997. An even larger number of publications have been devoted to the development and applications of the CoMFA approach [513]. The molecular field approach has been based on the assumption that in most cases, the biochemical drug–receptor interactions are noncovalent. Consequently, the changes in the biological activity of compounds should be essentially determined by the steric and electrostatic fields of these molecules. To develop the numerical representation of the fields, all molecules under investigation have to be first structurally aligned and the steric or electrostatic fields around them sampled using some probe atoms. For instance, the electrostatic field may be calculated by moving a probe atom with a positive unit charge (+1) or some other charged species on a rectangular grid that encompasses the aligned molecules [514,515]. A CoMFA QSAR table is formed thereafter from the numerical values of the field at each grid point. Because of a large number of columns in this table, it has to be analyzed using special multivariate statistical analysis procedures, such as partial least-squares (PLS) analysis [516,517] together with the cross validation [518].

In principle, the molecular fields applied in CoMFA can be of very different origin. In the introductory work by Cramer et al. [511], the electrostatic field of molecules was calculated from empirical atomic partial charges obtained by the method of Gasteiger and Marsili [367]. The probe atom had the charge +1 and the van der Waals properties of an sp^3 carbon atom. The steric field was represented by the Lennard-Jones potential

$$V_{\mathrm{LJ}}(r) = 4\varepsilon \left[\left(\frac{\sigma}{r} \right)^{12} - \left(\frac{\sigma}{r} \right)^{6} \right] \tag{3.175}$$

where ε is the depth of the minimum in the potential, r is the distance from the atomic center, and σ is the distance at which the potential is equal to zero [519]. In addition to standard electrostatic and steric fields, a number of other fields have been applied either alone or in combination with the standard fields.

One of the earliest attempts to account for the molecular hydrophobicity fields was made by Kellogg and Abraham [520–522]. Their method, called

HINT (Hydropathic INTeraction), proceeded from the CLOGP hydrophobicity fragment constants [523,524] that were further divided into atomic contributions. From these atomic values, the hydrophobic field is calculated at each grid intersection point t as:

$$A_t = \sum_{i=1}^{n} a_i S_i \exp(-r_{it}) \tag{3.176}$$

where a_i is the hydrophobic atom constant for the ith atom in the molecule, S_i is the solvent-accessible surface area for this atom, and r_{it} is the distance between the atom and the grid point. The summation was carried out over all atoms in the molecule. HINT has been shown to give better predictions of experimental log P values than the CLOGP method. An analogous lipophilicity potential has been developed as a combination of fragmental values of lipophilicity and a distance function [525–527]. The molecular lipophilicity potential (MLP) at a given grid point k was expressed as [528]:

$$\mathrm{MLP}_k = \sum_{i=1}^{N} f_i F(r_{ik}) \tag{3.177}$$

In this formula, f_i denotes the lipophilicity increment for fragment i in the molecule, and $F(r_{ik})$ is some function of the distance r_{ik} between the grid point and a given fragment. A number of functions $F(r_{ik})$ have been tested for the distance dependence of the potential. One of the possible forms is a Fermi–Dirac distribution [529] type function:

$$F(r_{ik}) = \frac{1 + \exp(-ab)}{1 + \exp[a(r_{ik} - b)]} \tag{3.178}$$

where a and b are the adjustable parameters. It has been claimed that the MLP encodes the hydrogen-bonding and hydrophobic interactions not adequately described by the steric and electrostatic fields and that it includes an entropy part of interactions [530]. An alternative molecular field approach to describe the hydrophobicity of compounds uses the desolvation free energy fields [531]. In this approach, the linearized Poisson–Boltzmann equation is solved numerically to compute the electrostatic contribution to solvation around a molecule. This field has been suggested for systems where significant resolvation or desolvation takes place during the drug–receptor interactions.

A directional and distance-dependent function has been developed to describe the hydrogen bonding fields [532–534]. Within this approach, a probe that has two hydrogen-bonding donor and two hydrogen-bonding acceptor sites is rotated about the grid point to the energy minimum. The hydrogen-bonding potential energy is calculated at each grid point according to the following function:

$$E_{\mathrm{HB}} = \left[\frac{C}{d^6} - \frac{D}{d^4} \right] \cos m\Theta \tag{3.179}$$

where C and D are the respective interaction parameters, m is usually equal to 1, d is the distance between the atom and the probe, and Θ is the angle defined by the donor, hydrogen, and acceptor atom. It has been suggested, however, that the combination of the electrostatic potential and the electrostatic field strength may give also a satisfactory description of hydrogen bonding [535].

In the case of predominant donor–acceptor interactions between the molecules, the fields derived from the frontier molecular orbitals may be beneficial [536–538]. In this case, the highest occupied (HOMO) or the lowest unoccupied molecular orbital (LUMO) for a given molecule is projected into the CoMFA region, and the electron density on this orbital is recorded as the field characteristics at each grid lattice intersection point. The HOMO field should reflect the Lewis base properties or nucleophilic activity whereas the LUMO field represents the Lewis acid properties or electrophilic activity of a compound.

An alternative approach to molecular fields is based on a comparative analysis of molecular similarity indices (CoMSIA). Three different indices, encoding the steric, electrostatic, and hydrophobic similarity between a probe atom and the molecule at each lattice site were introduced in this method [539]. All similarity indices were calculated at each grid point q according to the following equation:

$$A_k^q(j) = -\sum_{i=1}^{n} w_{\mathrm{probe},k} w_{ik} \exp(-\alpha r_{iq}^2) \tag{3.180}$$

where k is the given property and j denotes a molecule with n atoms. The two coefficients, $w_{\mathrm{probe},k}$ and w_{ik} represent the actual value of the physico-chemical property k of atom i and the value of this property for a probe atom with a charge of $+1$, a radius of 1 Å, and the hydrophobicity of $+1$. The distance between the probe atom at grid point q and atom i in the molecule is denoted by r_{iq}, and α is defined as an attenuation factor. The large α values correspond to steeper distance dependence of the field. On the contrary, with small α values also the remote parts of molecule will be experienced by the probe. It has been thus pointed out that the advantage of CoMSIA descriptors is the generation of more contiguous fields that are easier to interpret.

The molecular fields can also be based on conventional molecular descriptors of different origin. For example, an interesting molecular field approach has been introduced using electrotopological state indices [540]. Two types of field were developed, one of which is based on the electrotopological state indices of nonhydrogen atoms and another using indices for hydrogen atoms (*E*-state and *HE*-state fields, respectively). Kroemer and Hecht [541] have suggested a replacement of CoMFA steric interaction energies with the atom-based indicator vectors. In a way, this approach can be qualified as a three-dimensional analog of the classical Free–Wilson method. A substantial improvement of the results was reported by using the respective field, both alone or in combination with electrostatic fields [541].

The electrostatic molecular field can be obtained from atomic partial charges derived either from some empirical scheme or from the quantum chemical molecular wave function. The semiempirical MNDO-, AM1-, and PM3-calculated Mulliken charges have been widely used for this purpose. Alternatively, the atomic charges may be obtained from fitting the total molecular electrostatic potential. Unfortunately, because of the conceptual arbitrariness of the definition of atomic charges, different methods may result in largely variable values of charges (cf. Table 3.8). The question about the most suitable charges for molecular field analysis is still opened. Kroemer et al. [542] have critically compared the CoMFA charges calculated using 17 different methods. Those included the Gasteiger–Marsili empirical charges, semiempirical MNDO, AM1, and PM3 charges, and ab initio Hartree–Fock charges obtained using STO-3G, 3-21G*, and 6-31G* basis sets. Semiempirical and ab initio partial charges were calculated both by using the Mulliken population analysis and by using the molecular electrostatic potential fitting (ESPFIT charges). The target data set consisted of 37 ligands of the benzodiazepine receptor inverse agonist–antagonist active site, corresponding to different generic structures. It was found that the semiempirical ESPFIT charges had a similar quality compared with the ab initio ESPFIT electron populations. Furthermore, the molecular electrostatic potentials mapped directly onto the CoMFA grid were not superior to the corresponding ESPFIT fields. However, the charges obtained from the Mulliken analysis performed worse than the ESPFIT charges. This result somewhat contradicted an earlier conclusion about the superiority of semiempirical Mulliken charges over the ESPFIT-derived fields [543,544]. The empirical Gasteiger–Marsili charges had the same quality as the semiempirical Mulliken charges. Thus, in overall, the semiempirical quantum chemical charges or even the empirical charges can be considered as sufficient for most CoMFA treatments [512,545].

The most crucial factor in CoMFA analysis is finding the appropriate alignment of a series of molecules to each other and to the field grid lattice. By superimposing the biologically active molecules, the orientation corresponding to the active interaction with the receptor is usually targeted. In the case of availability of the respective crystallographic data, the alignment has often been based on this information [546,547]. It has been pointed out, however, that the active conformation of the substrate molecule in the biological interaction may be substantially different from the solid-state minimum. Therefore, it was not completely unexpected that the alignment rules obtained from a systematic conformational search of thermolysin inhibitors performed significantly better than the experimentally determined active site alignments [548]. The investigated compounds have been also fitted to structurally highly similar reference structures in crystalline complexes [549,550].

A different approach to obtain the appropriate alignment of molecules is based on fitting the experimental activities for some training sets of compounds using different alignments in CoMFA models. The "correct" alignment is thus related to the model with the best performance in the description of the experimental property (activity) studied [551–555].

The fitting of a pharmacophore has also been frequently used to align the molecules with similar structure [556–559]. A pharmacophore is defined as the biologically active fragment or the spatially fixed group of active fragments in a series of molecules [560]. The alignment has also been based on the electrostatic and steric complementarity [561–563]. This type of alignment is usually adequate for a data set that involves structurally closely related molecules. It is, however, much more ambiguous in the case of molecules with diverse structure or if the alignment is based on some structural property other than shape, even for sterically similar molecules. An algorithm has been developed for molecular alignment based on optimization of molecular similarity indices [564]. This Monte-Carlo-based algorithm has been shown more effective and robust than other optimizers applied to the similarity-based alignment problem. The CoMFA quantitative structure–activity relationships derived using these alignments were superior to relationships derived using the more common alignments [564].

The molecular field values in CoMFA grid lattice points create a long vector of numbers for each compound. It is, of course, impossible to employ this whole vector in the development of conventional QSAR using multiple linear regression because the number of independent variables would substantially exceed the number of compounds. A procedure called partial least squares (PLS), which is related to the principal component analysis (PCA), has been, however, applicable to reduce the number of descriptors used in the three-dimensional QSAR [516,517,565–567]. In PLS, the matrix of descriptors, $\boldsymbol{X}$, is decomposed into a product of two matrices $\boldsymbol{T}$ (scores) and $\boldsymbol{B}$ (loadings) and to an additional error term $\boldsymbol{E}$ as:

$$\boldsymbol{X} = \boldsymbol{TB} + \boldsymbol{E} \tag{3.181}$$

The score matrix, $\boldsymbol{T}$, consists of the values of so-called latent variables that are correlated with the biological activity. The elements of the loadings matrix, $\boldsymbol{B}$, describe the contributions of original descriptors into these latent variables. The PLS procedure is carried out subject to several constraints. In addition to the requirement that the latent variables be orthogonal and the elements of $\boldsymbol{E}$ matrix minimal, the covariance between the property studied (biological activity) and the latent variables has to be maximized. The first latent variable accounts most for the variation of the property studied, the second latent variable accounts most for the rest of variation, and so on. The optimum number of latent variables for a given CoMFA relationship is usually determined by using the cross-validation correlation coefficient, or PRESS (*Pre*diction *S*um of *S*quares [568]), statistics.

$$\mathrm{PRESS} = \sum_{i=1}^{N} (Y_{\mathrm{obs},i} - Y_{\mathrm{pred},i})^2 \tag{3.182}$$

$$q^2 = \frac{\mathrm{PRESS}(n = 0) - \mathrm{PRESS}(N)}{\mathrm{PRESS}(n = 0)} \tag{3.183}$$

where $Y_{\mathrm{obs},j}$ is the observed property value for a given compound, i, and N is the total number of compounds in the set. The squared cross-validated correlation coefficient, often denoted as q^2, serves as a quantitative measure of the predictive power of the final CoMFA model. Typically, its value increases with the addition of latent variables, until reaching the maximum value. The set of latent variables corresponding to this maximum determines the "best" description of data. It should be noted that the definition of q^2 is different from that of the cross-validated correlation coefficient in multiple linear regression (R^2_{cv}), and values $q^2 > 0.3$ are considered already statistically significant [569]. A GOLPE (generating optimal linear PLS estimations) algorithm has been developed to improve the consistency and predictive power of 3D-QSAR models by appropriate variable selection [570].

A substantial improvement of the molecular field approach has been made by Richards and co-workers [571]. A tool called SOMFA (self-organizing molecular field analysis) was proposed to reduce the influence of insignificant data and noise in the fields applied. Within the SOMFA methodology, the property (field) value at each grid point is multiplied by the mean centered value of the biological activity correlated. In that case, the compounds having the activity close to the mean value have less weight in the CoMFA PLS correlations than the compounds with the large deviation from this mean value. The respective SOMFA models were shown substantially easier to interpret for the physical insight into the relevant drug–receptor interaction.

As mentioned above, the comparative molecular field approach has been applied extensively in biochemical and biomedical studies [512]. The CoMFA method has been used for the quantitative description of the enzyme and metabolic inhibition activities of compounds [572–583], inhibition of biological transport processes [584–586], neurochemical drug–receptor antagonist and agonist interactions [587–600], antiviral and bacteriostatic activities [601–611], and carcinogenic, mutagenic, and other toxic properties of compounds [612–620].

Typically, the CoMFA treatments do not report equations similar to classical QSPR/QSAR equations. The quality of the CoMFA model is estimated using the cross-validated correlation coefficient or the accuracy of predictions for a test set of compounds. Nevertheless, in some cases such equations are available and can be compared with traditional QSAR results. For example, the measured highest tolerated concentration of benzyldimethylalkylammonium derivatives by the male mice, expressed by LD_{100}, has been correlated using the following equation [621]:

$$\log\left(\frac{1}{\mathrm{LD}_{100}}\right) = 1.940 + 0.038Z^{(1)}_{\mathrm{H_2O}} + 0.038Z^{(2)}_{\mathrm{H_2O}}$$

$$R^2 = 0.944 \qquad R^2_{cv} = 0.892 \qquad s = 0.095 \qquad n = 11$$

where $Z^{(1)}_{\mathrm{H_2O}}$ and $Z^{(2)}_{\mathrm{H_2O}}$ are two CoMFA first latent variables. The quadratic equation developed for the same experimental data using log P as the molecular descriptor

$$\log\left(\frac{1}{\mathrm{LD}_{100}}\right) = 0.200 + 0.939 \log P - 0.107(\log P)^2$$

$$R^2 = 0.864 \qquad s = 0.148 \qquad n = 11$$

has, with the same number of statistical degrees of freedom, substantially worse statistical quality.

Another example concerns the comparative classical and CoMFA QSAR analysis of $\mathrm{pEC_{50}}$, the logarithm of the reciprocal value of the concentration required to stimulate the *N*-acetylglycosamine incorporation into the rice stem borer integument system by 50% of the maximum [622]. For a series of substituted dibenzoylhydrazines,

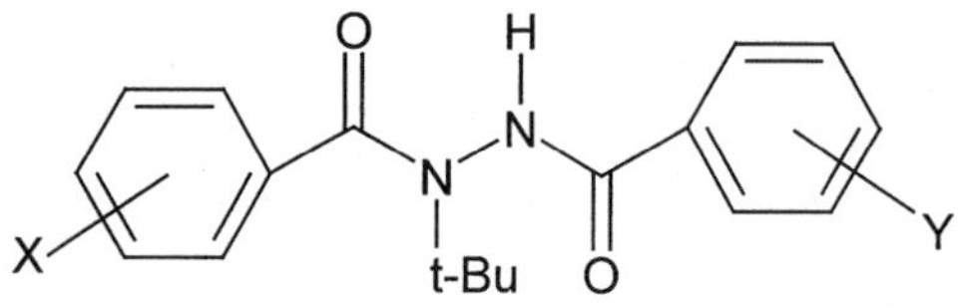

O H N N t-Bu O X Y

the following classical QSAR equation has been reported [623]:

$$\begin{aligned}\mathrm{pEC_{50}} = {} & (4.38 \pm 0.77) + (0.98 \pm 0.26)\log P + (2.89 \pm 1.28)\sigma_I^o(X) \\ & - (0.78 \pm 0.49)\sigma^0(X) - (0.59 \pm 0.54)\Delta V_w^o(X) \\ & - (0.46 \pm 0.36)\Delta V_w^m(X) - (1.06 \pm 0.36)\Delta V_w^p(X) \\ & - (1.00 \pm 0.29)\Delta V_w^o(Y) - (1.25 \pm 0.31)\Delta V_w^m(Y) \\ & - (1.00 \pm 0.29)\Delta V_w^p(Y)\end{aligned}$$

$$R^2 = 0.872 \qquad s = 0.288 \quad n = 37$$

The predictive power of this equation is doubtful because of large errors in multiple regressor coefficients. The same data were correlated, however, with only three CoMFA main latent variables derived from electrostatic and steric fields and the log P descriptor, resulting in the conventional $R^2 = 0.845$ and $s = 0.292$ [622]. Therefore, the CoMFA model would be much more significant.

Furthermore, by appropriate alignment of molecules using the superimposition of four atoms, six structurally rather different ecdysone analogs were included in the CoMFA treatment together with the original set of dibenzoylhydrazines. The resulting QSAR equation involved four CoMFA latent variables and log P and had the best conventional correlation characteristics, $R^2 = 0.900$ and $s = 0.240$. In addition, the comparison of aligned structures allowed the discussion of the possible active regions in the molecules. However, it was pointed out that significant structural variability is necessary in such regions to make justified conclusions about their importance in determining the activity of molecules. In summary, the use of proper alignment of structurally diverse molecules in the CoMFA treatment should give, in principle, additional information about the possible active sites.

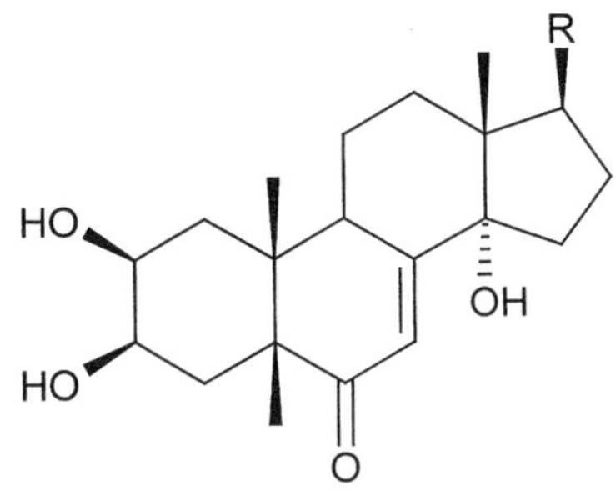

Apart from the biochemical studies, the CoMFA methodology has been also used for the examination of chemical reactivity and physical properties of compounds. The Hammett σ constants that are equivalent to pK_a of meta- and para-substituted benzoic acids, have been described using QSAR regression with seven CoMFA latent variables [543]. The variables were determined from the electrostatic field derived from AM1-calculated partial charges, and the correlation was characterized by $R^2 = 0.953$ and $s = 0.082$. Notably, the data for the same set of 49 compounds was described by the one-parameter equation using the AM1-calculated summary partial charge on the oxygen atoms and hydrogen of the carboxyl group, $\Sigma\, q$:

$$\sigma_{m,p} = (10.62 \pm 0.46) + (24.38 \pm 1.08) \sum q$$
$$R^2 = 0.917 \qquad s = 0.102 \qquad n = 49$$

This result has been interpreted as confirming the validity of CoMFA electrostatic fields. However, it is surprising that so many (seven) latent variables were needed for the CoMFA description of a relatively simple system of substituted benzoic acids. For a limited set of six phenylacetic acids

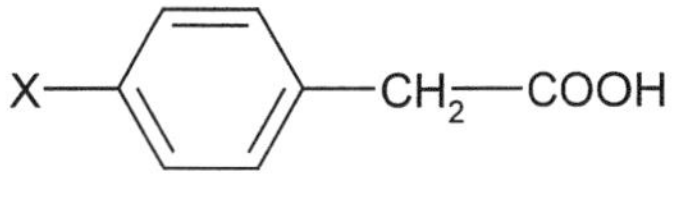

the pK_a were correlated just with a single CoMFA latent variable, $Z1$, as [621]:

$$\mathrm{p}K_a = (4.212 \pm 0.012) + (0.023 \pm 0.002)Z1$$
$$R^2 = 0.982 \qquad R^2_{\mathrm{cv}} = 0.953 \qquad s = 0.029 \qquad n = 6 \qquad F = 221.4$$

Five CoMFA latent variables were involved in the best models developed for the log k of the reaction of substituted benzyl benzenesulfonates

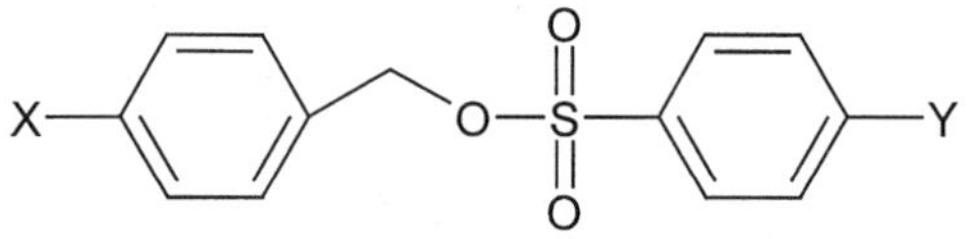

and p-methoxybenzylamine in two solvents (methanol and acetonitrile) [624]. The electrostatic and steric molecular fields were engaged using a C^+ probe atom. The best models had $R^2 = 0.994$ and $s = 0.044$ in the case of the reaction in methanol and $R^2 = 0.998$ and $s = 0.033$ for the reaction in acetonitrile.

An interesting correlation with the molecular electrostatic field parameters has been developed to describe the hydrogen-bonding basicity of heterocyclic compounds [535]. The experimental hydrogen-bonding basicities log K_β measured against p-nitrophenol in 1,1,1,-trichloroethane [625] were correlated using the following equation:

$$pK_\beta = (-6.376 \pm 0.467) - (1.563 \pm 0.368)V_\beta(2.4) + (50.96 \pm 6.23)|\mathbf{F}_\beta(2.4)|$$

$$R^2 = 0.960 \qquad s = 0.171 \qquad n = 23 \qquad F = 242$$

where V_β (2.4) is the electrostatic potential and $|\mathbf{F}_\beta(2.4)|$ is the magnitude of the electric field strength at the distance 2.4 Å from nucleus of the basic nitrogen along the lone-pair axis. The last descriptor quantifies the gradient of the electrostatic potential, and thus it was concluded that this field characteristic might be important in CoMFA analysis.

The importance of the correct orientation of molecules in CoMFA analysis has been illustrated using the data on inhibition of protein synthesis by cephalotaxine esters [626]:

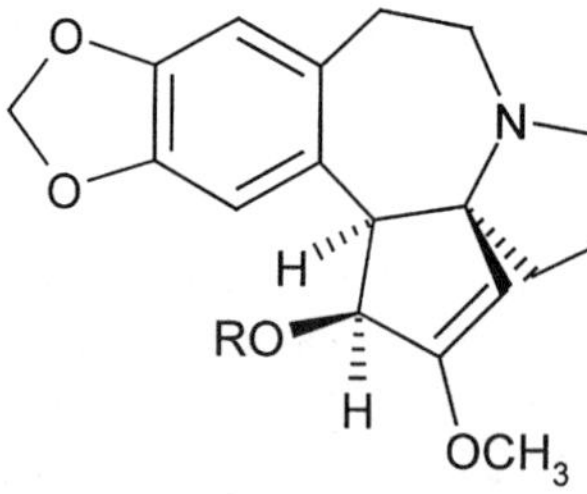

The conventional CoMFA treatment of data resulted in $q^2 = 0.385$. The grid was then divided into 125 small cubic regions of equal size, and the CoMFA treatment was carried out for each of these regions separately. A new CoMFA master region was created by including the small regions that had the q^2 value above some threshold. The final CoMFA model corresponded to the treatment of field in this master region and resulted in significantly better description of data ($q^2 = 0.651$). Similar results had been obtained for other biological activities [626].

3.6 QUANTUM CHEMICAL DESCRIPTORS

The fast progress in computer hardware power and the development of efficient algorithms has assisted in the routine application of molecular quantum mechanical calculations. Depending on size of the molecular system studied, the contemporary semiempirical and ab initio programs can supply realistic quantum chemical molecular characteristics in a relatively short computational time. Alternatively, the quantum chemical calculations, even if not directly applicable for the investigation of complex chemical and biochemical systems, represent an attractive source of new molecular descriptors. These descriptors can express, in principle, almost any electronic and geometric property of molecules and the characteristics of intermolecular interactions. Many recent QSAR/QSPR studies have employed quantum chemical descriptors alone or in combination with conventional descriptors. Three main families of quantum chemically obtained molecular descriptors can be formulated. First, almost any physical property of a molecule can be calculated using the quantum chemical methods and therefore used as a theoretical molecular descriptor. Second, the results of the quantum chemical calculations can be used for the subsequent development of new descriptors based on some physical models or chemical reaction schemes. Finally, the partitioning of quantum chemically calculated energies or other properties can bring up even more molecular descriptors applicable in the development of QSAR/QSPR.

It is generally accepted that quantum chemistry can provide a more accurate and detailed description of electronic distribution than empirical methods. In addition, it has been established that errors due to the approximate nature of quantum chemical methods and the neglect of the solvation may be largely transferable within structurally related series. Thus, the variation or the relative magnitude of calculated descriptors would be meaningful although their absolute values may not be directly applicable. Moreover, the electronic descriptors based on molecular wave function can also be presented for individual atoms or groups, offering the possibility of detailed assessment of active molecular regions or fragments. Much of the work employing quantum chemical descriptors has been done in the field of QSAR rather than QSPR, that is, the descriptors have been correlated with biological activities such as enzyme inhibition activity, carcinogenicity, drug–receptor interactions, and others [627]. However, the quantum chemical descriptors are also reported to correlate the reactivity of organic compounds, octanol/water partition coefficients, gas chromatographic retention indices, and various other physical properties of molecules.

The quantum chemical methodology is based on the solution of the time-independent Schrödinger equation for the stationary states of molecular systems. In QSAR/QSPR descriptor development, the calculations are almost exclusively restricted to the ground states of the molecules. The ab initio model Hamiltonian for a molecular system consisting of n electrons and N nuclei is given as [628]:

$$\hat{H} = -\sum_{i=1}^{m} \frac{\nabla_i^2}{2} - \sum_{a=1}^{N} \frac{\nabla_a^2}{2} - \sum_{i=1}^{m} \sum_{a=1}^{N} \frac{Z_a}{r_{ia}} + \sum_{i}^{m} \sum_{j \neq i}^{m} \frac{1}{r_{ij}} + \sum_{a}^{N} \sum_{b \neq a}^{N} \frac{Z_a Z_b}{R_{ab}} \tag{3.184}$$

where the two first sums represent the kinetic energy operators for electrons and nuclei, respectively, the third term corresponds to the nuclear–electronic attraction energy, and the two last terms describe the electron–electron and nuclear–nuclear electrostatic repulsion energy, respectively. Equation (3.184) provides a complete representation of all nonrelativistic interactions between the nuclei and electrons in a molecule. In most cases, the calculations are carried out in the approximation of fixed nuclei that reduces significantly the computational effort as the Schrödinger equation will be considered only for electrons:

$$\hat{H}_{\text{el}} \Psi_{\text{el}} = E_{\text{el}} \Psi_{\text{el}} \tag{3.185}$$

where E_{el} is the electronic energy of the system that consists the kinetic energy of all electrons, the electron–nuclear electrostatic attraction and electron–electron repulsion energies. In each case, the configuration of nuclei corresponding to the lowest (minimal) value of the total energy of the system

$$E_{\text{tot}} = E_{\text{el}} + \sum_{a<}^{N} \sum_{b}^{N} \frac{Z_a Z_b}{R_{ab}} \tag{3.186}$$

must be established using some geometry optimization routine. This configuration is, in general, different for different methods applied in the solution of the Schrödinger equation. Therefore, it is important to perform the geometry optimization for a series of molecules compared using the identical level of theory [629,630]. Because of the absence of the exact solution of equations of motion for many-particle systems, the solutions of Eq. (3.185) are always approximate and their accuracy depends substantially on the method applied. As a rule, the improvement of the quality of calculations is strongly correlated with the computational effort in terms of time and memory. In fact, practical ab initio calculations are still rather limited by the types of atoms and size of molecules because the computational time is proportional to a high exponential (n^4, n^5, etc.) of the number of electrons in the molecule, n [631]. However, these methods can still be applicable in reasonable time frame after an approximate search on the potential energy surface of the molecule for the global energy minimum has been carried out at lower level of theory. Most ab initio calculations have been based on the orbital approximation (Hartree–Fock method) [632–639]. In general, this method provides better results in terms of closeness to experimental values when larger basis set, that is, the number of atomic orbitals, is employed. According to the variational principle, this rule is strictly valid only for the total electron energy of the molecule. Other electronic properties, particularly those describing electron distribution in the molecule (such as dipole and higher moments of the molecule, partial charges on atoms, or the polarizability of the molecule) are related to the size of the basis set more loosely. For such properties, the use of balanced basis sets is required in the calculations [640–642].

A wide selection of ab initio methods beyond Hartree–Fock, accounting for the electron correlations in the molecule, has been developed and available in various quantum chemical program packages [643–647]. Such methods include the configuration interaction (CI) [648–650], multiconfigurational self-consistent field (MCSCF) [651–653], correlated pair many-electron theory (CPMET) [654] and its various coupled-cluster approximations [655–658], and perturbation theory (e.g., Møller–Plesset theory of various orders) [659–661]. Most of these methods are extremely time consuming, require large CPU memories, and therefore would be impractical for the calculation of extended sets of relatively large molecules (e.g., more than 10 atoms).

As an alternative, the semiempirical quantum chemical methods are relatively inexpensive and can be used for the calculation of molecular descriptors for very large sets of compounds. These methods have been mostly developed within the framework of the molecular orbital theory (SCF MO), using simplifications and approximations in the computational procedure that dramatically reduce the computational time [662–664]. In addition, the molecular integrals necessary for the solution of the molecular Schrödinger equation are approximated from some experimental data on atoms and prototype molecular systems. These estimates can also be considered as the parameters of calculations.

A variety of methods is available for the development of semiempirical quantum chemical descriptors. The earlier methods include the Hückel MO [665] and extended Hückel theory (EHT) [666], complete neglect of differential overlap (CNDO/2) [662], intermediate neglect of differential overlap (INDO) [667,668], modified INDO (MINDO/3) [669], modified neglect of diatomic overlap (MNDO) [670,671], Austin model 1 (AM1) [434], parametric model 3 (PM3) [436,437], and others [672–674]. Because of differences in approximations and parameterizations, the molecular descriptors obtained using different semiempirical methods are not directly comparable. Therefore, only the descriptors obtained by the same method should be applied in the development of each single QSAR/QSPR correlation.

The EHT method is an extension of the Hückel π-electron approximation and treats all valence electrons in a molecule. The overlap integrals are accounted for in the framework of EHT as calculated over the Slater-type atomic functions, but the electronic and electron–nuclear interactions are accounted for deficiently. This method has been mostly applied for the qualitative description of the electronic structure and properties of molecular systems. Because of the neglect of important electronic interactions, the charge distributions deduced are often unrealistic, and thus the EHT has only some historical significance even if available in modern programs.

The CNDO is the simplest and the most approximate method that involves the neglect of both diatomic and single-atom atomic orbital overlap integrals (differential overlap). The improved and much more widely used version is CNDO/2. The CNDO approximation has rather large error due to the neglect of substantial single-atom differential overlap. As a compromise, the INDO and MINDO procedures have been developed in which the one-center overlaps are retained in one-center integrals. The methods based on the correct inclusion of one-center overlap (i.e., neglecting diatomic differential overlap only) are MNDO, AM1, and PM3.

The MNDO and particularly AM1 methods have gained a large popularity in the quantum chemical modeling of larger molecular systems. The advantage of the AM1 method is that in much shorter computational times, the results are obtained that are in reasonable agreement with ab initio calculations or the respective experimental data. The parameterization is available for a variety of atoms including all elements of the second period, halogens, Al, Si, P, S, Sn, Hg, and Pb. In contrast to earlier MNDO, the AM1 method provides a good description also for anions and hydrogen-bonded systems. The MNDO method tends to overestimate electronic repulsion; electronic properties calculated by the MNDO method may well be less reliable than those calculated by the AM1 method. However, the further parameterization of MNDO using substantially larger and variable data has lead to a method of similar quality (MNDO-PM3). The comparison of ab initio and AM1-derived descriptors in the QSAR study has lead to the conclusion that the small basis set ab initio results are frequently inferior to semiempirical calculations [675]. In effect, the semiempirical molecular geometry, charges, and dipole moments are often more reliable than those obtained from low-quality ab initio methods. Moreover, it has been shown that the AM1 method can be used to calculate electronic effects that may be difficult to deal with by ab initio methods [676]. The results produced by different semiempirical methods are generally not comparable, but they often do reproduce similar trends. For example, the electronic net charges calculated by the AM1, MNDO, and INDO methods were found to be quite different in their absolute values but consistent in their trends [677]. The variation in the molecular orbital indices is generally largest for the AM1 and diminishes in the order AM1 > MNDO > INDO. Hence, the AM1 method has been suggested as preferable for QSAR purposes.

Methods based on classical molecular force fields are also capable of minimizing the potential energy of a molecular structure [678–682]. Notably, those methods can be applied in the calculations of thermodynamic functions and dipole moment when combined with some empirical schemes for the calculation of the charge distribution (partial charges) in the molecule. However, only the quantum chemical methods can differentiate between atomic σ and π

charges and calculate HOMO and LUMO energies as well as many other electronic descriptors of potential value to QSAR studies.

The well-defined physical content of the quantum chemically calculated molecular wave function enables the definition of a large number of molecular and local quantities characterizing reactivity, shape, and binding properties of the whole molecule or its fragments. It has been emphasized that their use in the development of a QSAR has two main advantages [683]. First, different compounds and their various fragments and substituents can be directly characterized only using the molecular structure. Second, the proposed mechanism of action can be directly accounted for in terms of chemical reactivity of the compounds under study. As a consequence, the derived QSAR models will include information regarding the nature of the intermolecular forces involved in determining the biological activity of the compounds in question. Unlike the empirical molecular descriptors, there is no experimental error in quantum chemically derived descriptors. There is an inherent error, however, associated with the approximations and simplifications of the theoretical method applied. In most cases the direction but not the magnitude of this error is known. Usually, the computational error can be considered approximately constant for a series of related compounds.

We have seen that many electrostatic descriptors described in the last two paragraphs have been developed from the quantum chemically calculated atomic partial charges in the molecule. The information about the charge distribution in the molecule can be presented also in the form of the electrical moments. Such charge-distribution-related molecular descriptors include the total dipole moment of the molecule (μ),

$$\mu = -\sum_{i=1}^{\text{occ}} \int_{(V)} \phi_i \hat{r} \phi_i \, dv + \sum_{a=1}^{M} Z_a \mathbf{R}_a \tag{3.187}$$

its point-charge (μ_c) and hybridization (μ_h) components, and the absolute value. In Eq. (3.187), the first summation is carried out over all occupied molecular orbitals in the molecule and the second sum involves all nuclei. The quadrupole, octupole, and higher electrical moments of a molecule can also be calculated from the quantum chemical molecular wave function. The dipole moment is often considered as the direct characteristic of the polarity of a molecule.

The polarization of a molecule as a whole by an external electric field can be described in terms of the nth-order susceptibility tensors of the molecule [684]. The first-order term, which is referred to as the polarizability of the molecule α, represents the constant of proportionality between the induced dipole moment μ' and the strength of the external field $\boldsymbol{E}$ [685]:

$$\mu' = \mu + \alpha \boldsymbol{E} + \tfrac{1}{2}\beta \boldsymbol{E}^2 + \cdots \tag{3.188}$$

At higher field strengths, the higher order polarizabilities called superpolarizabilities have to be accounted for (β, γ, etc.). One of the most significant properties of the molecular polarizability is the close relation to the molecular bulk or molar volume. The polarizability values have been also shown to be related to hydrophobicity and thus to the biological activity of compounds [686–688]. Furthermore, the electronic polarizability of molecules shares common features with the electrophilic superdelocalizability [689]. The first-order polarizability tensor contains also information about possible inductive interactions in the molecule [690,691]. The total anisotropy of the polarizability (the second-order term) characterizes the properties of a molecule as an electron acceptor [692].

The polarity and the polarizability of a molecule have been well known to be important for the description of various physico-chemical properties, chemical reactivity and biological activity of molecules. For example, molecular polarity accounts for chromatographic retention on a polar stationary phase [441]. The most obvious and most often used quantity to describe the polarity, the dipole moment of the molecule, reflects only the global polarity of a molecule. Local polarities can be represented by the local dipole moments, calculated for a fragment of a molecule, but these are conceptually difficult to define. First approximations of these quantities can be obtained by considering the atomic charges in the localized regions of the molecule. The local dipole index [439], the differences between net charges on atoms [693], and the topological electronic index [258] have been applied as the charge-based polarity indices. The quadrupole moment tensor can also be used as an index to describe possible electrostatic interactions. However, such tensors depend on the choice of the coordinate system, and therefore the orientation of the congeneric molecular fragment must be the same for all molecules in the series.

The energies of the highest occupied molecular orbital and the lowest unoccupied molecular orbital (HOMO/LUMO energies) are frequently used quantum chemical descriptors [694–698]. It has been recognized that these orbitals play a major role in determining the chemical reactivity and chemical reaction mechanisms. In addition, the difference between the HOMO and LUMO energies has been related to the electronic band gaps in solids and the transition frequencies in the electronic spectra of compounds. The energy localization of frontier molecular orbitals is also important for the description of the molecular charge-transfer complexes. The frontier molecular orbital (FMO) theory of the chemical reactivity is based on the assumption that the formation of a transition state is primarily influenced by the interaction between the frontier orbitals (HOMO and LUMO) of reagents [699]. The FMO theory of reactivity proceeds from the concept of the superdelocalizability, which is an index characterizing the affinity of occupied and unoccupied orbitals in chemical reactions. The superdelocalizability (S_r) on an atom r has been related to the contribution made by atom r to the stabilization energy in the formation of a charge-transfer complex with another molecule [699]. This parameter has been used to characterize molecular interactions [700] and for the comparison of the reactivity and properties of similar atoms in different molecules [439].

A distinction has been made between the electrophilic and the nucleophilic superdelocalizability (or acceptor and donor superdelocalizability), respectively. The former describes the interaction of a compound with the electrophilic center at another reagent. The nucleophilic superdelocalizability characterizes the interaction of a compound with the nucleophilic center at the other reactant. The molecular superdelocalizabilities have been categorized as the dynamic reactivity indices [438]. The static reactivity indices, for example, the partial charges on atoms describe isolated molecules in their ground state. The dynamic indices refer to the transition state of the reaction. Typically, the extreme (maximum and minimum) values of the atomic nucleophilic (N_A), electrophilic (E_A), and one-electron (R_A) Fukui reactivity indices [438] for a given atomic species in the molecule have been used in the QSAR/QSPR developments. These descriptors are defined as:

$$N_A = \sum_{i \in A} c_{i\mathrm{HOMO}}^2 / (1 - \varepsilon_{\mathrm{HOMO}}) \tag{3.189}$$

$$E_A = \sum_{j \in A} c_{j\mathrm{LUMO}}^2 / (\varepsilon_{\mathrm{LUMO}} + 10) \tag{3.190}$$

and

$$R_A = \sum_{i \in A} \sum_{j \in A} c_{i\mathrm{HOMO}} c_{j\mathrm{LUMO}} / (\varepsilon_{\mathrm{LUMO}} - \varepsilon_{\mathrm{HOMO}}) \tag{3.191}$$

where the summations are performed over all atomic orbitals (AO) i, j of a given atom A, $c_{i\mathrm{HOMO}}$ and $c_{j\mathrm{LUMO}}$ denote the ith and jth AO coefficients for the HOMO and LUMO, respectively, and $\varepsilon_{\mathrm{HOMO}}$ and $\varepsilon_{\mathrm{LUMO}}$ are the energies of these orbitals, respectively. The reactivity indices give an estimate of the relative reactivity of the atoms in the molecule for a given series of compounds and relate to the activation energy of the corresponding chemical reaction. In many cases, the extreme (maximum and minimum) values of simplified atomic nucleophilic (N'_A), electrophilic (E'_A) and one-electron (R'_A) reactivity indices for a given atomic species in the molecule have been useful. The latter have been defined as:

$$N'_A = \sum_{i \in A} c_{i\mathrm{HOMO}}^2 \tag{3.192}$$

$$E'_A = \sum_{j \in A} c_{j\mathrm{LUMO}}^2 \tag{3.193}$$

and

$$R'_A = \sum_{i \in A} \sum_{j \in A} c_{i\mathrm{HOMO}} c_{j\mathrm{LUMO}} \tag{3.194}$$

with the same notations as above. Notably, the maximum atomic orbital electronic population for a given atomic species in the molecule ($P_{\mu\mu}$) can be used as another simplified index for the description of the nucleophilicity of a molecule. The Fukui frontier orbital reactivity indices have been extended to groups of atoms in the molecule [701].

According to the Koopmans' theorem [450], the energy of the HOMO is directly related to the ionization potential (IP):

$$\mathrm{IP} = -\varepsilon_{\mathrm{HOMO}} \tag{3.195}$$

The energy of the LUMO is directly related to the electron affinity (EA) of a molecule as:

$$\mathrm{EA} = \varepsilon_{\mathrm{LUMO}} \tag{3.196}$$

The HOMO energy characterizes the susceptibility of the molecule toward the attack by electrophiles whereas the LUMO energy features the susceptibility of the molecule toward the attack by nucleophiles. Both the HOMO and the LUMO energies are important in determining the radical reactivity as both these orbitals are simultaneously engaged in such reactions [699]. The concept of hard and soft nucleophiles and electrophiles has been also directly related to the relative energy of the HOMO/LUMO orbitals. Hard nucleophiles have a low-energy HOMO; soft nucleophiles have a high-energy HOMO; hard electrophiles have a high-energy LUMO; and soft electrophiles have a low-energy LUMO [702].

The HOMO–LUMO gap, that is, the difference in energy between the HOMO and LUMO, has been related to the chemical stability of compounds. Thus, the larger HOMO–LUMO gaps were considered as indicators of higher stability of compounds toward chemical reactions [703]. Whereas this is realistic in the case of radical reactions, the nucleophilic and electrophilic reactivity should primarily depend on the respective individual HOMO and LUMO energies. The HOMO–LUMO gap has also been used as an approximation to the lowest excitation energy of the molecule [439]. This simplified approach, however, neglects the electronic reorganization in the excited state and therefore may often lead to quantitatively incorrect results.

The concept of the activation hardness has been also derived from the HOMO–LUMO energy gap [353,704,705]. The absolute hardness is defined as

$$\eta = (\varepsilon_{\mathrm{LUMO}} - \varepsilon_{\mathrm{HOMO}})/2 \tag{3.197}$$

and the activation hardness as

$$\Delta\eta = \eta_R - \eta_T \tag{3.198}$$

where η_R and η_T stand for the absolute hardness of the reactant and transition state, respectively. The activation hardness is expected to distinguish between the reaction rates at different sites in the molecule, and thus it is relevant for predicting orientation effects. The qualitative definition of hardness is closely related to the polarizability, since a decrease of the energy gap usually leads to easier polarization of the molecule.

It has been suggested that the electron densities on frontier orbitals on atoms provide useful means for the characterization of donor–acceptor interactions between molecules [706,707]. According to the frontier electron reactivity theory, the majority of chemical reactions take place at the position and in the direction where the overlapping of the HOMO and LUMO of the respective reactants is at a maximum. The HOMO or nucleophilic electron density, f_r^N, of the donor molecule and the LUMO or electrophilic electron density, f_r^E, of an acceptor molecule are responsible for the charge transfer. In the framework of the LCAO MO theory, the respective densities can be calculated simply from the respective MO coefficients, c, as:

$$f_r^E = \sum (c_{\mathrm{HOMO},n})^2 \tag{3.199}$$

$$f_r^N = \sum (c_{\mathrm{LUMO},n})^2 \tag{3.200}$$

However, the frontier electron densities can strictly be used only to describe the reactivity of different atoms in the same molecule. To compare the reactivity of different molecules, frontier electron densities have been normalized by the energy of the corresponding orbitals:

$$F_r^E = \frac{f_r^E}{\varepsilon_{\mathrm{HOMO}}} \tag{3.201a}$$

$$F_r^N = \frac{f_r^N}{\varepsilon_{\mathrm{LUMO}}} \tag{3.201b}$$

This logic proceeds from the assumption that the electron density of the frontier MO at an atom characterizes the relative reactivity of that atom within a single molecule. The frontier MO energy, however, reflects the relative reactivity of different molecules. Consequently, the molecules with smaller ionization potentials ($-\varepsilon_{\mathrm{HOMO}}$) are expected to be more reactive as nucleophiles and the molecules with larger electron affinities ($\varepsilon_{\mathrm{LUMO}}$) to be stronger electrophiles.

The self-atom and atom–atom polarizabilities (π_{AA}, π_{AB}) have been also employed to describe chemical reactivity [700]. These quantities are defined on the basis of perturbation theory and represent the effect of a electric field perturbation at one atom on the electronic charge at the same (π_{AA}) or another atom (π_{AB}), respectively. The descriptor π_{AB} is defined as:

$$\pi_{AB} = 4 \sum_i \sum_a \sum_p \sum_r \frac{C_{pi}^A C_{pj}^A C_{ri}^B C_{rj}^B}{\varepsilon_i - \varepsilon_j} \tag{3.202}$$

where the summation is carried out over the MO coefficients C for the AOs p

and r, localized on atoms A and B, respectively, and contributing to MOs i and j, respectively. In the trivial case $A = B$, formula (3.202) results in the self-atom polarizability π_{AA}.

The quantum chemically calculated bond orders contain useful information about the molecular properties and stability [708]. The maximum bond order for a given pair of atomic species A and $B(P_{AB})$ refers, as a rule, to the most stable bond in the molecule. On the contrary, the smallest bond order corresponds to the weakest bond in the molecule. The Mulliken bond order is defined as:

$$P_{AB} = \sum_{i=1}^{\text{occ}} \sum_{\mu \in A} \sum_{\nu \in B} n_i c_{i\mu} c_{j\nu} \tag{3.203}$$

where the first summation is performed over all occupied molecular orbitals (n_i denotes the occupation number of the ith MO), and the two other summations over μ and ν, the atomic orbitals belonging to the different atoms A and B in the molecule. The MO coefficients are denoted as $c_{i\mu}$ and $c_{j\nu}$. The σ-σ bond order ($P_{\sigma\text{-}\sigma}$) for a given pair of atomic species in the molecule can be defined using only the MO coefficients related to the σ orbitals in the molecule. Alternatively, the σ-π and π-π bond orders can be obtained by using the respective MO coefficients in Eq. (3.203). The extreme (maximum and minimum) values of these specific bond orders characterize the molecule as a whole. In parallel to the bond orders, the free valence of a given atomic species or its extreme value in the molecule can be used as molecular stability descriptors. The free valence is defined as [708]:

$$V_{fA} = V_{\max} - P_A \tag{3.204}$$

where $V_{\max}$ is the maximum valency of the given atomic species and

$$P_A = \sum_{B \neq A} P_{AB} \tag{3.205}$$

is the total bond order of a given atom, A.

Many quantum chemical descriptors can be defined proceeding from the different partitionings of the total calculated total energy of the molecule. The total energy of a compound itself calculated by semiempirical quantum chemical methods has been successfully used as a molecular descriptor in a number of cases [709–711]. For example, the total energy has been used as a measure of nonspecific interactions between a solute and stationary phase in the gas chromatography [258]. The difference in heats of formation of conjugated species has been related to at least part of the reaction enthalpy [712–714].

The energy of protonation, calculated from the difference between the total energies of the protonated and neutral forms of the molecule as:

$$\Delta H_{prot} = E_{MH^+} - E_{MH} - E_{H^+} \tag{3.206}$$

has to be considered as a measure of the strength of hydrogen bonds. It has been observed that, in general, the higher protonation energy corresponds to a stronger hydrogen bond. This descriptor has also been used to determine the correct localization of the most favorable hydrogen bond acceptor site in the molecule [715].

The quantum chemically calculated total energy of the molecule (E_{tot}) refers usually to the quantum mechanical standard state for the energy (isolated electrons and nuclei at 0 K).

$$E_{tot} = E_{el} + \sum_{A \neq B} Z_A Z_B / R_{AB} \tag{3.207}$$

where E_{el} denotes the quantum chemically calculated total electronic energy and the second term in the last formula represents the nuclear repulsion in the molecule. In the framework of the self-consistent field (SCF) theory, the total electronic energy of the molecule is calculated as:

$$E_{el} = 2\ \mathrm{Tr}(\mathbf{RF}) - \mathrm{Tr}(\mathbf{RG}) \tag{3.208}$$

where **R** is the first-order density matrix, **F** is the matrix representation of the Hartree–Fock operator and **G** is the matrix representation of the electron repulsion energy. Proceeding from the mathematical structure of the molecular Hamiltonian, various partitionings of the total energy and the total electronic energy of a molecule are possible. Most quantum chemical programs give also the heat of formation of the compound as an output. The quantum chemically calculated heat of formation of the molecule (ΔH_f) gives the energy of the molecule in the thermodynamic standard scale (chemical elements in ideal gas state at 298.15 K and 101,325 Pa).

The electron–electron repulsion energy for a given atomic species (atom A) in the molecule can be calculated as:

$$E_{ee}(A) = \sum_{B \neq A} \sum_{\mu,\nu \in A} \sum_{\lambda,\sigma \in B} P_{\mu\nu} P_{\lambda\sigma} \langle \mu\nu | \lambda\sigma \rangle \tag{3.209}$$

where $P_{\mu\nu}$ and $P_{\lambda\sigma}P_{\mu\nu}$ are the elements of the density matrix and $\langle \mu\nu | \lambda\sigma \rangle$ are the electron repulsion integrals on atomic basis $\{\mu\nu\lambda\sigma\}$. The electron–electron repulsion energy describes the electron repulsion-driven processes in the molecule and may be related to the conformational changes (rotation, inversion) or atomic reactivity in the molecule. As related to a given atom, it may specify the site of a particular chemical activity or conformational change in the molecule.

The nuclear–electron attraction energy for a given atomic species (atom A) in the molecule is defined as:

$$E_{\mathrm{ne}}(A) = \sum_{B} \sum_{\mu,\nu \in A} P_{\mu\nu} \left\langle \mu \left| \frac{Z_B}{R_{iB}} \right| \nu \right\rangle \tag{3.210}$$

The first summation is performed over all atomic nuclei in the molecule (B) whereas the second summation is carried out over all atomic orbitals at a given atom (A). The terms $\langle \mu | Z_B/R_{iB} | \nu \rangle$ denote the nuclear–electron attraction integrals on the given atomic basis. This energy describes the nuclear–electron attraction-driven processes in the molecule and may again be related to the conformational (rotational, inversional) changes or atomic reactivity in the molecule.

Alternatively, the energy of intramolecular electron–electron repulsion between two given atoms (atoms A and B) in the molecule can be calculated as:

$$E_{\mathrm{ee}}(AB) = \sum_{\mu,\nu \in A} \sum_{\lambda,\sigma \in B} P_{\mu\nu} P_{\lambda\sigma} \langle \mu\nu | \lambda\sigma \rangle \tag{3.211}$$

and nuclear–electron attraction energy between given two atoms as:

$$E_{\mathrm{ne}}(AB) = \sum_{\mu,\nu \in A} P_{\mu\nu} \left\langle \mu \left| \frac{Z_B}{R_{iB}} \right| \nu \right\rangle \tag{3.212}$$

with the notations as above. These energies characterize the electrostatic interactions between either the chemically bonded or nonbonded atoms in a molecule. Also, the nuclear repulsion energy between two given atoms (atoms A and B) in the molecule can be calculated as:

$$E_{\mathrm{nn}}(AB) = \frac{Z_A Z_B}{R_{AB}} \tag{3.213}$$

where Z_A and Z_B are the nuclear (core) charges of atoms A and B, respectively, and R_{AB} is the distance between them. This energy describes the nuclear repulsion-driven processes in the molecule and may be again related to the conformational changes or atomic reactivity in the molecule. The extreme values (maximum or minimum values) of the above-discussed energies for a given atomic species or for a pair of given atomic species can be considered as the global characteristics of a compound.

In the framework of LCAO MO theory, the electronic exchange energy between given two atoms (A and B) in the molecule can be found as follows:

$$E_{\mathrm{exc}}(AB) = \sum_{\mu,\nu \in A} \sum_{\lambda,\sigma \in B} P_{\mu\lambda} P_{\nu\sigma} \langle \mu\lambda | \nu\sigma \rangle \tag{3.214}$$

with the notations as above. This energy reflects the change in the Fermi correlation energy between the two electrons localized on atoms A and B, respectively. It can be of importance in determining the conformational changes of the molecule and its spin properties. In the framework of the semiempirical quantum chemical theory, the resonance energy between given two atomic species in the molecule can be defined as:

$$E_R(AB) = \sum_{\mu \in A} \sum_{\nu \in B} P_{\mu\nu}\beta_{\mu\nu} \tag{3.215}$$

where $\beta_{\mu\nu}$ are the resonance integrals on the atomic basis.

The total electrostatic interaction energy between any given two atomic species in the molecule is defined as:

$$E_C(AB) = E_{ee}(AB) + E_{ne}(AB) + E_{nn}(AB) \tag{3.216}$$

whereas the following definition applies for the total interaction energy between two given atomic species in the molecule:

$$E_{tot}(AB) = E_C(AB) + E_{exc}(AB) \tag{3.217}$$

The total energy partitionings corresponding to the molecule as a whole can be also used as the global molecular descriptors. For instance, the total molecular one-center electron–nuclear attraction energy, $E_{ne}(\text{tot})$, and the total molecular one-center electron–electron repulsion energy, $E_{ee}(\text{tot})$, can be defined as:

$$E_{ne}(\text{tot}) = \sum_{A} E_{ne}(A) \tag{3.218}$$

$$E_{ee}(\text{tot}) = \sum_{A} E_{ee}(A) \tag{3.219}$$

respectively. The total intramolecular electrostatic interaction energy would be calculated as:

$$E_C(\text{tot}) = \frac{1}{2} \sum_{A} E_C(A) \tag{3.220}$$

The atomic valence state energies for the given atomic species in the molecule and its various partitionings could also be useful molecular descriptors. The valence state energy characterizes the magnitude of the perturbation experienced by an atom in the molecular environment as compared to the isolated atom.

All of these partitionings consider the total energy of the molecule or its various potential energy components. However, in the framework of the transferable atom equivalent (TAE) method, the descriptors based on the kinetic energy of electrons in the molecule have been developed [716]. Two formulations have been proposed for the electron kinetic energy density, K and G, respectively:

$$K = -\frac{N}{4} \int (\Psi^* \nabla^2 \Psi + \Psi \nabla^2 \Psi^*) \, d\mathbf{r}' \tag{3.221}$$

$$G = \frac{N}{2} \int \nabla \Psi^* \nabla \Psi \, d\mathbf{r}' \tag{3.222}$$

where the integration is carried over all electrons but one. The scalar minimum and maximum for both properties can be derived as the descriptor for the whole molecule, for an atom of a specified kind, or for a molecular fragment. In addition, the surface and three-dimensional spatial representations of these properties can be applied in QSAR/QSPR analysis. Those include the surface integrals of both kinetic energy densities and the rate of the change of K and G normal to the surface, S.

The output of various quantum chemical programs also gives information about the vibrational and rotational energies of the molecule. The quantum mechanically calculated vibrational energies and/or frequencies describe the individual bonds or fragments in the molecule and can be, in some cases, useful as molecular descriptors. These descriptors are calculated on the basis of the total partition function of the molecule Q and its electronic, translational, rotational, and vibrational components [717,718]:

$$Q = Q_{el} Q_{tr} Q_{vib} Q_{rot} \tag{3.223}$$

where

$$Q_{el} = \exp\left(-\frac{E_{el}}{kT}\right) \tag{3.224}$$

$$Q_{tr} = \int_{-\infty}^{\infty} \exp\left(-\frac{p^2}{2mkT}\right) dp \tag{3.225}$$

$$Q_{vib} = \prod_{j=1}^{\alpha} \frac{\exp(-hv_j/2kT)}{1 - \exp(-hv_j/2kT)} \tag{3.226}$$

$$Q_{rot} = \frac{\pi^{1/2}}{\sigma} \prod_{j=1}^{3} \left(\frac{8\pi^2 I_j kT}{h^2}\right)^2 \tag{3.227}$$

In these formulas, E_{el} denotes the electronic energy of the molecule; m and p are the mass and momentum of the molecule, respectively; v_j are the frequencies of normal vibrations of the molecule; α is the number of vibrational degrees of freedom in the molecule; I_j are the principal moments of inertia of the molecule; σ is the symmetry number of the molecule; h is the Planck constant; and kT is the Boltzmann temperature. The following thermodynamic functions that are applicable to a single molecule can be calculated proceeding from the partition function and the quantum chemically calculated total energy of the molecule:

Thermodynamic heat of formation of the molecule at 300 K, ΔH_f^0

Vibrational enthalpy of the molecule at a given temperature T, calculated according to the following formula:

$$H_{\text{vib}} = \frac{1}{2} \sum_{j=1}^{a} h\upsilon_j + \frac{h\upsilon_j \exp(-h\upsilon_j/2kT)}{1 - \exp(-h\upsilon_j/2kT)} \tag{3.228}$$

Translational enthalpy of the molecule at a given temperature T, calculated as:

$$H_{\text{tr}} = \int_{-\infty}^{\infty} \frac{p^2}{2m} e^{-p^2/2mkT} \, dp \tag{3.229}$$

Vibrational entropy of the molecule at a given temperature T, calculated as follows:

$$S_{\text{vib}} = \sum_{j=1}^{\alpha} \left\{ \frac{h\upsilon_j \exp(-h\upsilon_j/2kT)}{kT[1 - \exp(-h\upsilon_j/2kT)]} - \ln[1 - \exp(-h\upsilon_j/2kT)] \right\} \tag{3.230}$$

Rotational entropy of the molecule at a given temperature T, calculated as:

$$S_{\text{rot}} = Nk \ln \left[\frac{\pi^{1/2}}{\sigma} \prod_{j=1}^{3} \left(\frac{8\pi^2 I_j kT}{h^2} \right)^{1/2} \right] \tag{3.231}$$

Internal entropy of the molecule, given by the following sum:

$$S_{\text{int}} = S_{\text{vib}} + S_{\text{rot}} \tag{3.232}$$

Translational entropy of the molecule at temperature T

$$S_{\text{tr}} = \ln \left(\frac{2\pi mkT}{h^2} \right)^{1/2} \frac{Ve^{5/2}}{N_A} \tag{3.233}$$

where V is the volume of the system and N_A is the Avogadro number.

Vibrational heat capacity of the molecule at temperature T

$$c_{\upsilon,\text{vib}} = k \sum_{j=1}^{\alpha} \left(\frac{h\upsilon_j}{kT} \right)^2 \frac{\exp(-h\upsilon_j/2kT)}{1 - \exp(-h\upsilon_j/2kT)} \tag{3.234}$$

An interesting molecular descriptor reflecting the three-dimensional structure of the molecule has been developed at Shell Research Ltd. using the quantum chemically calculated vibrational modes of molecules. This descriptor, called EVA (normal coordinate EigenVAlues) was calculated from the standardized vibrational frequencies allowing the comparison of molecules with different number of atoms [719]. The value of the EVA descriptor at any chosen sampling point, x, along the bounded frequency scale (from 0 to 4000 cm^{-1}) can be determined according to the following equation:

$$EVA_x = \sum_{i=1}^{3N-6} \frac{1}{\sigma\sqrt{2\pi}} \exp\left[-\frac{(x - f_i)^2}{2\sigma^2}\right] \tag{3.235}$$

where the summation is carried out over all vibrational modes in the molecule ($3N - 5$ in the case of linear molecules). The f_i are the individual vibrational frequencies and σ is the fixed standard deviation for all Gaussian functions characterizing the shape of the vibrational peak. Typically, the EVA descriptor has been sampled along the bounded frequency scale using fixed increments, and the resulting data matrix treated using the PLS procedure similarly to the CoMFA approach.

The SPARC (SPARC performs automated reasoning in chemistry) approach has been based on the molecular descriptors derived for specific intramolecular interactions [720–723]. Proceeding from the electronegativities of atoms and the charge distribution in the molecule, the intramolecular interaction is expressed in a SPARC model as a sum of physically distinct terms. Those include the electrostatic and mesomeric field effects, σ induction and π resonance effects, and the intramolecular hydrogen bonding. The overall root-mean-square deviation of SPARC calculated pK_a from the experimental values for more than 3500 organic compounds was only 0.35 logarithmic units [723]. The SPARC approach has been extended to account also for the intermolecular interactions in condensed media. Successful models were developed to describe the refractive index, boiling point, density, and chromatographic retention indices of organic compounds. The advantage of SPARC is that both the intra- and intermolecular interactions are treated using the same physical models that have clear definition. Therefore, an analysis of the respective equations gives a better insight into the nature of physical effects determining the experimentally observable characteristics of chemical compounds in different environments.

A systematic comparison of quantum chemical (AM1), empirical and experimental molecular descriptors has been performed by De Benedetti and coworkers [683]. Fifty theoretical (quantum chemical and others) descriptors, calculated for 50 monosubstituted benzenes, were analyzed using the principal component analysis (PCA) and the partial least squares (PLS) methods. The PCA results revealed that the substituent descriptors cluster into three main groups. The first group involved descriptors reflecting the substituent effects on the aromatic ring such as the net atomic charges, the resonance and field substituent constants, and the substituent-induced chemical shifts. The second group included descriptors describing the bulk properties of the substituents such as Verloop's steric parameters and the molecular refractivity [724,725]. These descriptors clustered together with the polarizability properties of the substituents, the anisotropy of the molecular polarizability, the dispersion terms, and the electrophilic superdelocalizability of substituents. The hydrophobicity parameter was also found close to this cluster. The third group of descriptors included the theoretical and experimental molecular and substituent dipole moments and their squares. The authors also noted that the descriptors are well spread in the principal component space [683].

The PLS analysis was carried out to establish whether the theoretical molecular descriptors bear the same information content as the so-called physicochemical parameters. The conclusions of this analysis can be summarized as follows. First, the substituent-induced chemical shifts, the Hammett substituent constants, and the Taft and Swain–Lupton resonance constants were modeled by the first component. The main contributions to this PLS component originated from the net atomic charges on the carbon atoms, the electrophilic superdelocalizability of the benzene ring, and the energy of the frontier molecular orbitals. The molecular refractivity and the Verloop steric parameters together with the molecular weight and the substituent van der Waals volumes could be modeled by the second PLS component. The main contributions to this component originated mainly from the polarizabilities, dispersion forces, and substituent reactivity indices (superdelocalizabilities). The third PLS component modeled the liphophilic and the lipophobic parameters [726]. The theoretical descriptors contributing to this component were the solvent-accessible surfaces of the substituent, the dipole moments and their square terms, and the HOMO/LUMO energy difference.

Several reviews have been published on the application of quantum chemical descriptors in QSAR [627,698,723]. A summary of the most frequently used quantum chemical descriptors is given in Table 3.9. The numerical values of different quantum chemically calculated molecular descriptors are given in Tables 3.10 and 3.11.

The results of quantum chemical calculations have been used directly by transformation to appropriate molecular descriptors for the correlation of a vast amount of data on chemical, physical, and biological properties of compounds [627].

For instance, the quantum chemical description of the chemical reactivity can be carried out at three levels of theory. First, it is possible, at least for the reactions in the gas phase at moderate temperatures and pressures, to calculate reaction potential surfaces and to localize the global and local minima and saddle points (corresponding to the transition states) on these surfaces. The kinetic or equilibrium parameters of a reaction can be thus calculated directly from the respective partition functions for the reagents, the transition state, and/or the products of the reaction. The theoretical formulation of the free energy of a chemical reaction

$$\mathrm{A} + \mathrm{B} + \cdots \rightleftharpoons \mathrm{P} + \mathrm{R} + \cdots$$

is given as follows

$$\Delta G_r \equiv -RT \ln K_r = \Delta E_r^0 - RT \ln \frac{Q_P Q_R \cdots}{Q_A Q_B \cdots} \tag{3.236}$$

TABLE 3.9 Quantum Chemically Calculated Molecular Descriptors

Notation	Description	Reference
ε_{HOMO}, ε_{LUMO}	Energies of the highest occupied (HOMO) and lowest unoccupied (LUMO) molecular orbitals	*a*
$\varepsilon_{HOMO,A}$, $\varepsilon_{LUMO,A}$	Fraction of HOMO/LUMO energies arising from the atomic orbitals of the atom *A*	*a*
$\varepsilon_{LUMO} - \varepsilon_{HOMO}$	Difference of HOMO and LUMO orbital energies	*a*
η	Absolute hardness	*b*
$\Delta\eta = \eta_R - \eta_T$	Activation hardness, *R* and *T* stand for reactant and transition state	*b*
$q_{A,\sigma}$, $q_{A,\pi}$	π- and σ-electron densities on atom *A*	*c*
$Q_{A,H}$, $Q_{A,L}$	HOMO/LUMO electron densities on atom *A*	*d*
f_r^E	Electrophilic atomic frontier electron densities	*e*
f_r^N	Nucleophilic atomic frontier electron densities	*f*
F_r^E, F_r^N	Indices of frontier electron density	*g*
$S_{E,A}$, $S_{N,A}$	Electrophilic and nucleophilic superdelocalizabilities	*h*
$\Sigma\ S_{E,A}$, $\Sigma\ S_{N,A}$	Sums of electrophilic and nucleophilic superdelocalizabilities	*i*
N_A, E_A, R_A	Nucleophilic, electrophilic, and one-electron Fukui reactivity indices	*h*
ω_{gs}, ω_{vs}	Parr's electrophilicity index	*j*
π_{AA}, π_{AB}	Self-atom polarizabilities and atom–atom polarizabilities	*k*
$\Sigma\ \pi_{AA}$	Sum of self-atom polarizabilities	*i*
P_{AB}	Bond order between atoms *A* and *B*	*l*
$P_{\sigma\sigma}$, $P_{\pi\pi}$, $P_{\sigma\pi}$	σ-σ, π-π, and σ-π bond orders	*m*
B_{AB}	Bond index between atoms *A* and *B*	*n*
V_A	Valency of atom *A*	*o*
E_T	Total energy of the molecule	*p*
E_b	Binding energy of the molecule	*q*
ΔH_f^0	Heat of formation of a compound	*r*
$\Delta(\Delta H_f^0)$	Difference in the heats of formation of an acid and the corresponding anion	*s*
ΔH_{prot}	Energy of protonation, the difference between the total energy of the protonated and neutral forms	*t*
$E_{ee}(A)$	Electron–electron repulsion energy for a given atomic species, *A*	*m*
$E_{ne}(A)$	Electron–nuclear attraction energy for a given atomic species, *A*	*m*
$E_{ee}(AB)$	Electron–electron repulsion energy between two atoms, *A* and *B*	*m*
$E_{nn}(AB)$	Nuclear–nuclear repulsion energy between two atoms, *A* and *B*	*m*
$E_{ne}(AB)$	Electron–nuclear attraction energy between two atoms, *A* and *B*	*m*
$E_{exc}(AB)$	Electronic exchange energy between two atoms, *A* and *B*	*m*
$E_R(AB)$	Resonance energy between two atoms, *A* and *B*	*m*

TABLE 3.9 ***(Continued)***

Notation	Description	Reference
$E_C(AB)$	Total electrostatic interaction energy between two atoms, A and B	*m*
K, G	Electron kinetic energy density, K and G	*u*
EVA	Normal coordinate eigenvalues	*v*

[a]A. Cartier and J.-L. Rivail, *Chemometrics Intell. Lab. Syst.* **1,** 335 (1987).
[b]Z. Zhou and R.G. Parr, *J. Am. Chem. Soc.* **112,** 5720 (1990).
[c]O. Kikuchi, *Quant. Struct.-Act. Relat.* **6,** 179 (1987).
[d]K. Tuppurainen, S. Lötjönen, R. Laatikainen, T. Vartiainen, U. Maran, M. Strandberg, and T. Tamm, *Mutat. Res.* **247,** 97 (1991).
[e]W. Langenaeker, K. Demel, and P. Geerlings, *J. Mol. Struct. (THEOCHEM)* **234,** 329 (1991).
[f]G. Rastelli, F. Fanelli, M.C. Menziani, M. Cocchi, and P.G. DeBenedetti, *J. Mol. Struct. (THEOCHEM)* **251,** 307 (1991).
[g]A. Nakayama, K. Hagiwara, S. Hashimoto, and S. Shimoda, *Quant. Struct.-Act. Relat.* **12,** 251 (1993).
[h]K. Fukui, *Theory of Orientation and Stereoselection*, pp. 34–39. Springer-Verlag, New York, 1975.
[i]M. Cocchi, M.C. Menziani, P.G. De Benedetti, and G. Cruciani, *Chemometrics Intell. Lab. Syst.* **14,** 209 (1992).
[j]R.G. Parr, L. von Szentpály, and S. Liu, *J. Am. Chem. Soc.* **121,** 1922 (1999).
[k]R.E. Brown and A.M. Simas, *Theor. Chim. Acta* **62,** 1 (1982).
[l]C.A. Coulson, *Proc. R. Soc. London, Ser. A* **169,** 413 (1939); R.S. Mulliken, *J. Chem. Phys.* **23,** 1833, 1841 (1955).
[m]J.J.P. Stewart, *MOPAC Program Package 6.0 Manual*, QCPE, Indiana University, Bloomington, IN, No 455, 1990.
[n]K. Wiberg, *Tetrahedron* **24,** 1083 (1968); M. Giambiagi, M.S. de Giambiagi, D.R. Grempel, and C.D. Heyman, *J. Chim. Phys.* **72,** 15 (1975); A.B. Sannigrahi, *Adv. Quantum Chem.* **23,** 301 (1992).
[o]D.A. Armstrong, P.G. Perkins, and J.J.P. Stewart, *J. Chem. Soc., Dalton Trans.*, pp. 838, 2273 (1973); K. Jug, *Theor. Chim. Acta* **51,** 331 (1979); A.B. Sannigrahi, *Adv. Quantum Chem.* **23,** 301 (1992).
[p]K. Osmialowski, J. Halkiewicz, and R. Kaliszan, *J. Chromatogr.* **361,** 63 (1986).
[q]F. Saura-Calixto, A. García-Raso, and M.A. Raso, *J. Chromatogr. Sci.* **22,** 22 (1984).
[r]N. Bodor, Z. Gabanyi, and C.-K. Wong, *J. Am. Chem. Soc.* **111,** 3783 (1989).
[s]C. Gruber and V. Buss, *Chemosphere* **19,** 1595 (1989).
[t]G. Trapani, A. Carotti, M. Franco, A. Latrofa, G. Genchi, and G. Liso, *Eur. J. Med. Chem.* **28,** 13 (1993).
[u]C.M. Breneman and M. Martinov, in *Molecular Electrostatic Potentials: Concepts and Applications* (J.S. Murray and K. Sen, eds.), Vol. 3. Elsevier, Amsterdam, 1996, pp. 143–179.
[v]T.W. Heritage, A.M. Ferguson, D.B. Turner, and P. Willett, in *3D QSAR in Drug Design* (H. Kubinyi, G. Folkers, and Y.C. Martin, eds.), Kluwer/ESCOM, Dordrecht, The Netherlands, 1998, Vol. 2. pp. 381–398.

where

$$\Delta E_r^0 = E_P^0 + E_R^0 + \cdots - E_A^0 - E_B^0 - \cdots \tag{3.237}$$

is the internal energy of reaction evaluated at 0 K as the difference between the quantum mechanically calculated molar zero-point energy levels for reac-

TABLE 3.10 Semiempirical AM1 SCRF Calculated Orbital-Related Molecular Descriptors for Selection of Structurally Variable Compounds

Compound	ε_{HOMO}	ε_{LUMO}	N_C^{max}	E_C^{max}	R_C^{max}	$P_{\sigma\sigma}^{max}$	$P_{\pi\pi}^{max}$
Water	−12.536	4.452	—	—	—	0.9558	0.0000
Ammonia	−10.562	4.398	—	—	—	0.9728	0.0000
Tetrafluoromethane	−15.315	1.162	0.0012	0.0384	0.0000	0.8466	0.1054
Tetrachloromethane	−12.376	−1.119	0.0000	0.0400	0.0000	0.9245	0.0464
Trichloromethane	−11.789	−0.319	0.0082	0.0438	−0.0080	0.9394	0.0391
Tribromomethane	−11.081	−0.757	0.0000	0.0429	0.0000	0.9643	0.0168
Formaldehyde	−10.873	0.684	0.0081	0.0618	0.0000	0.9790	0.9790
Dibromomethane	−10.989	−0.078	0.0017	0.0458	0.0000	0.9733	0.0152
Dichloromethane	−11.417	0.561	0.0094	0.0467	0.0000	0.9552	0.0296
Chloromethane	−11.452	1.485	0.0082	0.0492	0.0000	0.9711	0.0143
Bromomethane	−10.899	0.826	0.0049	0.0487	0.0001	0.9837	0.0093
Iodomethane	−10.596	0.454	0.0031	0.0458	0.0000	0.9862	0.0061
Nitromethane	−12.186	−0.328	0.0000	0.0002	0.0000	0.9584	0.5421
Methane	−13.304	4.657	0.0354	0.0337	−0.0277	0.9867	0.0000
Methanol	−11.174	3.746	0.0115	0.0232	0.0001	0.9627	0.0440
Methylamine	−9.870	3.806	0.0071	0.0172	0.0077	0.9669	0.0439
Pentachloroethane	−11.874	−0.687	0.0079	0.0382	−0.0013	0.9461	0.0428
Trichloroethene	−9.961	−0.063	0.0273	0.0482	0.0347	0.9564	0.8958
1,1,1,2-Tetrachloroethane	−11.794	−0.508	0.0078	0.0419	0.0001	0.9633	0.0395
1,1-Dichloroethene	−10.010	0.337	0.0267	0.0462	0.0362	0.9525	0.9399
trans-1,2-Dichloroethene	−10.023	0.343	0.0266	0.0462	0.0361	0.9524	0.9405
cis-1,2-Dichloroethene	−10.002	0.367	0.0267	0.0460	0.0361	0.9544	0.9384
1,1,2,2-Tetrachloroethane	−11.647	−0.230	0.0075	0.0228	0.0040	0.9431	0.0312
2,2,2-Trifluoroethanol	−11.985	1.359	0.0101	0.0372	0.0001	0.9578	0.0912
Acetonitrile	−12.615	1.391	0.0150	0.0418	0.0033	0.9710	1.9211
1,1,1-Trichloroethane	−12.042	−0.315	0.0007	0.0427	−0.0011	0.9654	0.0370
1,1,2-Trichloroethane	−11.582	0.158	0.0083	0.0382	0.0028	0.9587	0.0309

Ethene	−10.554	1.439	0.0433	0.0437	0.0417	0.9762	1.0257
1,1-Dichloroethane	−11.484	0.517	0.0090	0.0457	0.0043	0.9707	0.0314
Acetaldehyde	−10.852	0.661	0.0138	0.0581	0.0000	0.9726	0.9428
Acetic acid	−11.707	0.890	0.0066	0.0553	0.0001	0.9670	0.8234
1,2-Dibromoethane	−11.014	1.660	0.0038	0.0229	0.0000	0.9666	0.0298
1,2-Dichloroethane	−11.416	0.688	0.0067	0.0237	0.0000	0.9601	0.0268
Methyl formate	−11.484	0.853	0.0024	0.0584	0.0000	0.9632	0.8735
Chloroethane	−11.267	1.367	0.0102	0.0479	0.0002	0.9734	0.0307
Bromoethane	−10.805	0.718	0.0058	0.0480	0.0000	0.9734	0.0300
Iodoethane	−10.524	0.374	0.0036	0.0452	0.0001	0.9770	0.0295
Nitroethane	−11.951	−0.229	0.0099	0.0012	−0.0005	0.9719	0.5455
Ethane	−11.758	4.109	0.0177	0.0155	0.0118	0.9837	0.0265
Ethanol	−10.885	3.570	0.0129	0.0204	−0.0001	0.9747	0.0400
Ethanethiol	−9.004	0.785	0.0028	0.0324	0.0001	0.9836	0.0280
Ethylamine	−9.741	3.648	0.0078	0.0169	0.0082	0.9752	0.0414
Dimethylamine	−9.447	3.478	0.0047	0.0166	−0.0064	0.9667	0.0442
Tetrachloroethene	−9.900	−0.438	0.0245	0.0475	0.0368	0.9438	0.8474
Propyne	−10.756	1.921	0.0377	0.0393	0.0062	0.9628	1.9307
Methoxyflurane	−11.797	−0.103	0.0078	0.0426	0.0043	0.9617	0.0930
1-Chloro-2-propene	−10.292	1.042	0.0368	0.0453	0.0402	0.9682	0.9935
Propanonitrile	−12.076	1.439	0.0160	0.0414	0.0130	0.9745	1.9301
Ethyl formate	−11.373	0.883	0.0036	0.0582	−0.0001	0.9737	0.8729
Methyl acetate	−11.397	1.071	0.0057	0.0538	0.0007	0.9677	0.8329
1,3-Dichloropropane	−11.137	1.284	0.0065	0.0298	0.0010	0.9624	0.0281
Cyclopropane	−11.471	3.139	0.0367	0.0255	0.0169	0.9607	0.0416
Propene	−9.996	1.351	0.0428	0.0425	0.0373	0.9747	0.9903
Propanone	−10.791	0.716	0.0129	0.0546	0.0000	0.9728	0.9169
Propanoic acid	−11.563	0.943	0.0077	0.0550	0.0009	0.9740	0.8307
2-Propenol	−10.444	1.054	0.0380	0.0458	0.0371	0.9691	0.9967
1,2-Dichloropropane	−11.386	0.672	0.0077	0.0263	0.0013	0.9729	0.0284
Propionaldehyde	−10.654	0.757	0.0146	0.0576	0.0009	0.9754	0.9535

TABLE 3.10 (*Continued*)

Compound	ε_{HOMO}	ε_{LUMO}	N_C^{max}	E_C^{max}	R_C^{max}	$P_{\sigma\sigma}^{max}$	$P_{\pi\pi}^{max}$
1-Bromopropane	−10.744	0.756	0.0054	0.0477	0.0005	0.9752	0.0287
2-Bromopropane	−10.744	0.642	0.0049	0.0473	0.0022	0.9737	0.0277
1-Nitropropane	−12.041	−0.314	0.0101	0.0002	0.0004	0.9738	0.5493
1-Chloropropane	−11.225	1.425	0.0095	0.0475	0.0003	0.9751	0.0284
2-Chloropropane	−11.249	1.305	0.0096	0.0466	0.0026	0.9740	0.0278
1-Iodopropane	−10.473	0.410	0.0034	0.0450	0.0002	0.9751	0.0277
N,N-dimethylformamide	−9.732	1.331	0.0031	0.0543	0.0060	0.9671	0.7706
2-Nitropropane	−11.727	−0.095	0.0100	0.0022	0.0011	0.9726	0.5523
Propane	−11.315	3.920	0.0192	0.0161	0.0063	0.9760	0.0244
Methyl ethyl ether	−10.487	3.105	0.0067	0.0205	0.0000	0.9747	0.0487
1-Propanol	−10.862	3.561	0.0126	0.0192	0.0011	0.9757	0.0416
2-Propanol	−10.921	3.476	0.0124	0.0194	0.0039	0.9752	0.0430
2-Methoxyethanol	−10.525	3.140	0.0080	0.0206	0.0010	0.9609	0.0518
n-Propanethiol	−8.924	0.836	0.0028	0.0314	0.0001	0.9837	0.0259
Trimethylamine	−9.172	3.193	0.0041	0.0158	−0.0062	0.9667	0.0428
Propylamine	−9.958	3.690	0.0118	0.0127	0.0036	0.9753	0.0497
Thiophene	−9.210	0.246	0.0230	0.0295	0.0004	0.9580	0.6487
1-Butyne	−10.659	1.942	0.0339	0.0394	0.0208	0.9754	1.9374
(*E*)-2-Butenal	−10.418	−0.171	0.0394	0.0389	0.0334	0.9726	0.9123
1-Cyanopropane	−11.808	1.358	0.0161	0.0409	0.0002	0.9744	1.9294
Isobutyraldehyde	−10.521	0.793	0.0152	0.0577	0.0016	0.9755	0.9647
Tetrahydrofuran	−10.250	3.082	0.0064	0.0210	0.0000	0.9660	0.0441
Ethyl acetate	−11.250	1.129	0.0061	0.0534	0.0012	0.9741	0.8309
Methyl propanoate	−11.341	1.111	0.0082	0.0540	0.0006	0.9743	0.8366
1,4-Dichlorobutane	−11.092	1.308	0.0077	0.0456	0.0005	0.9636	0.0280
1-Butene	−10.055	1.357	0.0423	0.0426	0.0375	0.9756	0.9943
Butanone	−10.605	0.804	0.0138	0.0539	0.0001	0.9752	0.9246
Isopropyl formate	−11.015	1.187	0.0073	0.0570	0.0007	0.9742	0.8512

Butanoic-acid	−11.454	0.966	0.0073	0.0550	0.0002	0.9748	0.8289
Propyl formate	−11.350	0.934	0.0037	0.0577	0.0008	0.9745	0.8731
Butyraldehyde	−10.642	0.761	0.0144	0.0576	0.0011	0.9751	0.9565
1,4-Dioxane	−10.208	2.836	0.0064	0.0123	0.0013	0.9518	0.0467
1-Bromobutane	−10.617	0.773	0.0052	0.0476	0.0000	0.9753	0.0280
1-Chlorobutane	−11.081	1.436	0.0086	0.0473	0.0009	0.9752	0.0270
1-Nitrobutane	−11.846	−0.346	0.0167	0.0003	0.0002	0.9748	0.5505
2-Bromo-2-methylpropane	−10.773	0.595	0.0033	0.0465	0.0005	0.9736	0.0252
1-Bromo-2-methylpropane	−10.698	0.798	0.0054	0.0474	0.0004	0.9755	0.0274
2-Chloro-2-methylpropane	−11.333	1.260	0.0069	0.0453	0.0009	0.9739	0.0254
2-Chlorobutane	−11.223	1.296	0.0098	0.0467	0.0013	0.9751	0.0282
1-Iodobutane	−10.421	0.415	0.0035	0.0449	0.0003	0.9752	0.0258
Morpholine	−9.503	2.990	0.0052	0.0155	0.0030	0.9551	0.0480
n-Butane	−11.339	3.860	0.0136	0.0126	0.0100	0.9758	0.0246
Diethyl ether	−10.447	3.042	0.0073	0.0190	0.0002	0.9748	0.0501
2-Methylpropane	−11.273	3.829	0.0195	0.0159	0.0152	0.9760	0.0228
2-Methyl-1-propanol	−10.841	3.440	0.0125	0.0201	0.0017	0.9757	0.0411
2-Butanol	−10.978	3.540	0.0138	0.0184	0.0037	0.9755	0.0504
2-Methyl-2-propanol	−11.023	3.414	0.0128	0.0188	0.0017	0.9753	0.0441
Diethyl sulfide	−8.466	0.878	0.0019	0.0221	0.0000	0.9737	0.0279
Diethyl disulfide	−9.148	−1.554	0.0038	0.0016	0.0025	0.9728	0.0552
2-Ethoxyethanol	−10.336	3.009	0.0077	0.0180	0.0007	0.9750	0.0532
1-Butanol	−10.650	3.474	0.0106	0.0189	0.0006	0.9760	0.0414
n-Butanethiol	−8.890	0.847	0.0027	0.0320	0.0002	0.9817	0.0253
Diethylamine	−9.303	3.258	0.0050	0.0162	0.0016	0.9755	0.0429
Butylamine	−9.845	3.627	0.0114	0.0121	0.0004	0.9751	0.0491
2-Chloropyridine	−9.933	−0.253	0.0228	0.0324	0.0271	0.9590	0.4911
3-Chloropyridine	−9.858	−0.241	0.0238	0.0326	0.0298	0.9563	0.4913
Pyridine	−9.944	0.123	0.0233	0.0255	0.0250	0.9587	0.4791
2-Methylthiophene	−9.162	0.202	0.0323	0.0295	0.0166	0.9704	0.6420
2-Methylpyrazine	−9.928	−0.305	0.0229	0.0296	0.0272	0.9690	0.4965

TABLE 3.10 (*Continued*)

Compound	$\varepsilon_{\text{HOMO}}$	$\varepsilon_{\text{LUMO}}$	$N_C^{\max}$	$E_C^{\max}$	$R_C^{\max}$	$P_{\sigma\sigma}^{\max}$	$P_{\pi\pi}^{\max}$
Cyclopentene	−9.448	1.284	0.0338	0.0408	0.0375	0.9683	0.9622
2-Methyl-1,3-butadiene	−9.203	0.459	0.0347	0.0314	0.0353	0.9745	0.9553
Cyclopentanone	−10.600	0.754	0.0144	0.0546	0.0001	0.9689	0.9285
1,4-Pentadiene	−9.968	1.215	0.0308	0.0229	0.0208	0.9689	0.9961
1-Pentyne	−10.612	1.947	0.0326	0.0393	0.0104	0.9757	1.9355
1-Cyanobutane	−11.547	1.384	0.0163	0.0407	0.0002	0.9753	1.9306
Cyclopentane	−10.955	3.611	0.0134	0.0058	0.0074	0.9683	0.0210
1-Pentene	−9.938	1.387	0.0413	0.0423	0.0363	0.9757	0.9914
Ethyl propanoate	−11.196	1.137	0.0065	0.0534	0.0009	0.9748	0.8347
Isobutyl formate	−11.343	0.907	0.0060	0.0578	0.0010	0.9747	0.8761
2-Methyl-2-butene	−9.662	1.293	0.0434	0.0417	0.0352	0.9753	0.9636
3-Methyl-1-butene	−10.073	1.384	0.0421	0.0427	0.0415	0.9759	0.9982
Isopropyl acetate	−10.862	1.228	0.0063	0.0527	0.0004	0.9743	0.8304
Pentanal	−10.803	0.609	0.0142	0.0582	0.0000	0.9755	0.9564
Methyl butanoate	−11.222	1.114	0.0076	0.0536	0.0005	0.9751	0.8373
Pentanoic acid	−11.484	0.945	0.0153	0.0551	0.0000	0.9754	0.8317
3-Methylbutanoic acid	−11.506	0.979	0.0077	0.0549	0.0005	0.9752	0.8299
Cyclopentanol	−10.823	3.381	0.0121	0.0149	0.0032	0.9677	0.0459
3-Methyl-2-butanone	−10.519	0.778	0.0141	0.0546	0.0010	0.9751	0.9295
3-Pentanone	−10.464	0.825	0.0136	0.0540	0.0002	0.9753	0.9326
Propyl acetate	−11.080	1.157	0.0079	0.0532	0.0019	0.9750	0.8345
2-Pentanone	−10.561	0.789	0.0137	0.0543	0.0012	0.9750	0.9263
2-Methyltetrahydrofuran	−10.297	3.023	0.0072	0.0210	0.0010	0.9752	0.0449
Tetrahydropyran	−10.164	3.098	0.0063	0.0190	0.0004	0.9668	0.0471
1-Chloropentane	−11.220	1.370	0.0101	0.0474	0.0004	0.9752	0.0264
1-Nitropentane	−11.617	−0.433	0.0169	0.0031	0.0000	0.9746	0.5430
1-Bromopentane	−10.791	0.713	0.0063	0.0478	0.0006	0.9751	0.0262
1-Iodopentane	−10.530	0.370	0.0038	0.0451	0.0001	0.9753	0.0261

N-Methylmorpholine	−9.251	2.780	0.0044	0.0136	0.0008	0.9655	0.0478
n-Pentane	−11.111	3.775	0.0131	0.0094	0.0096	0.9759	0.0245
2-Methyl-1-butanol	−10.859	3.564	0.0121	0.0150	0.0032	0.9752	0.0509
2-Methyl-2-butanol	−10.838	3.428	0.0144	0.0188	0.0015	0.9752	0.0449
3-Methyl-1-butanol	−10.724	3.538	0.0114	0.0170	0.0011	0.9759	0.0418
1-Pentanol	−10.856	3.409	0.0123	0.0176	0.0003	0.9753	0.0412
2-Pentanol	−10.855	3.362	0.0123	0.0178	0.0032	0.9756	0.0422
3-Pentanol	−10.805	3.396	0.0131	0.0192	0.0052	0.9750	0.0438
Methyl *tert*-butyl ether	−10.441	2.976	0.0079	0.0206	0.0005	0.9749	0.0507
2,2-Dimethylpropane	−11.490	3.893	0.0183	0.0165	0.0146	0.9759	0.0215
2-Propoxyethanol	−10.437	2.861	0.0063	0.0180	0.0001	0.9755	0.0431
2-Methylbutane	−11.099	3.837	0.0156	0.0149	0.0075	0.9762	0.0250
n-Pentylamine	−9.730	3.501	0.0078	0.0150	0.0055	0.9752	0.0423
1,2,3,4-Tetrachlorobenzene	−9.741	−0.660	0.0211	0.0343	0.0297	0.9568	0.4734
1,2,3,5-Tetrachlorobenzene	−9.762	−0.685	0.0238	0.0358	0.0322	0.9572	0.4500
1,2,4,5-Tetrachlorobenzene	−9.647	−0.734	0.0160	0.0262	0.0227	0.9560	0.4503
1,2,3-Trichlorobenzene	−9.810	−0.376	0.0264	0.0332	0.0244	0.9563	0.4631
1,3,5-Trichlorobenzene	−9.911	−0.406	0.0277	0.0342	0.0171	0.9531	0.4444
1,2,4-Trichlorobenzene	−9.627	−0.471	0.0222	0.0327	0.0296	0.9547	0.4776
1,2-Dichlorobenzene	−9.643	−0.157	0.0206	0.0277	0.0255	0.9581	0.4684
1,3-Dichlorobenzene	−9.685	−0.160	0.0209	0.0268	0.0221	0.9558	0.4623
1,4-Dichlorobenzene	−9.521	−0.216	0.0225	0.0330	0.0297	0.9538	0.4735
3-Cyanopyridine	−10.332	−0.815	0.0239	0.0307	0.0278	0.9714	1.9117
4-Cyanopyridine	−10.410	−0.734	0.0229	0.0326	0.0191	0.9711	1.9197
Bromobenzene	−9.633	0.048	0.0245	0.0331	0.0285	0.9583	0.4650
Chlorobenzene	−9.581	0.146	0.0257	0.0336	0.0299	0.9582	0.4647
4-Fluorophenol	−9.086	0.051	0.0252	0.0326	0.0143	0.9605	0.4873
2-Chlorophenol	−9.221	0.056	0.0209	0.0284	0.0190	0.9591	0.4689
3-Chlorophenol	−9.331	7.410	0.0286	0.0341	0.0229	0.9592	0.4640
4-Chlorophenol	−9.124	0.085	0.0254	0.0311	0.0229	0.9572	0.4871
4-Bromophenol	−9.192	0.010	0.0252	0.0324	0.0292	0.9568	0.4910

TABLE 3.10 (*Continued*)

Compound	$\varepsilon_{\text{HOMO}}$	$\varepsilon_{\text{LUMO}}$	$N_C^{\max}$	$E_C^{\max}$	$R_C^{\max}$	$P_{\sigma\sigma}^{\max}$	$P_{\pi\pi}^{\max}$
2-Iodophenol	−9.315	−0.025	0.0229	0.0316	0.0165	0.9598	0.4776
2-Nitrophenol	−9.965	−1.136	0.0263	0.0259	0.0150	0.9599	0.5499
3-Nitrophenol	−9.910	−1.271	0.0233	0.0220	0.0176	0.9593	0.5317
4-Nitrophenol	−9.997	−1.174	0.0272	0.0256	0.0235	0.9572	0.5297
3-Formalpyridine	−10.300	−0.802	0.0228	0.0273	0.0237	0.9746	0.9391
4-Formalpyridine	−10.239	−0.751	0.0236	0.0245	0.0182	0.9747	0.9525
2-Fluorophenol	−9.216	0.043	0.0215	0.0271	0.0249	0.9590	0.4572
Fluorobenzene	−9.555	0.156	0.0281	0.0350	0.0308	0.9588	0.4671
Iodobenzene	−9.682	0.049	0.0234	0.0327	0.0294	0.9584	0.4641
Nitrobenzene	−10.544	−1.185	0.0224	0.0228	0.0193	0.9581	0.5288
Benzene	−9.652	0.553	0.0285	0.0303	0.0305	0.9584	0.4602
Phenol	−9.114	0.396	0.0281	0.0304	0.0160	0.9591	0.4740
4-Chloroaniline	−8.442	0.321	0.0246	0.0320	0.0290	0.9565	0.5003
2-Nitroaniline	−9.077	−0.902	0.0239	0.0225	0.0150	0.9595	0.5789
3-Nitroaniline	−9.061	−1.046	0.0205	0.0191	0.0190	0.9588	0.5322
4-Nitroaniline	−9.050	−0.922	0.0267	0.0235	0.0155	0.9573	0.5554
3-Chloroaniline	−8.679	0.258	0.0249	0.0335	0.0144	0.9590	0.4817
2-Chloroaniline	−8.634	0.276	0.0218	0.0309	0.0124	0.9586	0.4849
Thiophenol	−8.468	0.207	0.0143	0.0337	0.0215	0.9828	0.4776
2-Methylpyridine	−9.630	0.162	0.0251	0.0311	0.0283	0.9727	0.4947
3-Methylpyridine	−9.658	0.092	0.0262	0.0286	0.0273	0.9732	0.4915
4-Methylpyridine	−9.955	0.140	0.0241	0.0271	0.0258	0.9735	0.4793
N-Acetylpyrrolidine	−9.096	0.017	0.0359	0.0340	0.0255	0.9657	0.8640
2-Ethylpyrazine	−9.891	−0.276	0.0227	0.0311	0.0276	0.9745	0.5117
Cyclohexene	−9.496	1.322	0.0358	0.0401	0.0381	0.9672	0.9584
Cyclohexanone	−10.405	0.792	0.0129	0.0544	0.0007	0.9685	0.9253
1,5-Hexadiene	−9.880	1.291	0.0295	0.0239	0.0263	0.9690	0.9961
2,3-Dimethyl-1,3-butadiene	−9.129	0.468	0.0304	0.0295	0.0321	0.9744	0.9299

trans-2-Hexenal	−10.509	−0.256	0.0369	0.0390	0.0263	0.9752	0.9248
1-Hexyne	−10.534	1.940	0.0308	0.0392	0.0252	0.9757	1.9323
Cyclohexane	−10.571	3.621	0.0098	0.0066	0.0069	0.9699	0.0230
Methylcyclopentane	−10.896	3.608	0.0122	0.0068	0.0069	0.9754	0.0236
2-Methyl-1-pentene	−9.647	1.286	0.0435	0.0417	0.0353	0.9755	0.9632
2-Hexanone	−10.622	0.723	0.0134	0.0544	0.0002	0.9754	0.9263
Hexanoic acid	−11.292	0.927	0.0120	0.0552	0.0002	0.9756	0.8328
Isobutyl acetate	−11.271	1.098	0.0100	0.0534	0.0001	0.9759	0.8316
n-Propyl propanoate	−11.169	1.084	0.0066	0.0525	0.0017	0.9756	0.8392
1-Hexene	−10.023	1.346	0.0413	0.0426	0.0413	0.9756	0.9940
Isoamyl formate	−11.169	1.112	0.0106	0.0573	0.0020	0.9759	0.8556
Ethyl butanoate	−11.153	1.169	0.0071	0.0532	0.0009	0.9752	0.8344
Methyl pentanoate	−11.171	1.149	0.0074	0.0534	0.0007	0.9756	0.8387
Hexanal	−10.685	0.718	0.0143	0.0574	0.0002	0.9756	0.9503
4-Methyl-2-pentanone	−10.306	0.889	0.0119	0.0535	0.0006	0.9761	0.9285
Butyl acetate	−11.073	1.171	0.0083	0.0529	0.0009	0.9758	0.8339
Cyclohexylamine	−9.757	3.541	0.0123	0.0091	0.0042	0.9664	0.0463
n-Butylacetamide	−10.067	1.245	0.0064	0.0503	0.0024	0.9756	0.7606
1-Bromohexane	−10.782	0.726	0.0062	0.0477	0.0001	0.9758	0.0260
1-Chlorohexane	−11.161	1.390	0.0095	0.0474	0.0010	0.9755	0.0260
1-Iodohexane	−10.529	0.376	0.0038	0.0450	0.0003	0.9757	0.0259
N-Methylpiperidine	−8.929	3.033	0.0043	0.0154	0.0061	0.9672	0.0465
2,2-Dimethylbutane	−11.188	3.754	0.0182	0.0170	0.0147	0.9759	0.0254
2,3-Dimethylbutane	−11.063	3.728	0.0178	0.0168	0.0141	0.9756	0.0232
n-Hexane	−11.034	3.755	0.0126	0.0099	0.0075	0.9759	0.0248
2-Methylpentane	−10.995	3.674	0.0120	0.0137	0.0065	0.9759	0.0249
3-Methylpentane	−10.889	3.802	0.0129	0.0142	0.0100	0.9754	0.0255
Diisopropyl ether	−10.407	2.818	0.0092	0.0179	0.0017	0.9752	0.0460
Di-*n*-propyl ether	−10.381	2.934	0.0054	0.0179	0.0000	0.9754	0.0431
1-Hexanol	−10.864	3.354	0.0122	0.0157	0.0000	0.9758	0.0418
Di-*n*-propyl sulfide	−8.469	0.881	0.0019	0.0217	0.0000	0.9755	0.0247

TABLE 3.10 (*Continued*)

Compound	$\varepsilon_{\text{HOMO}}$	$\varepsilon_{\text{LUMO}}$	$N_C^{\max}$	$E_C^{\max}$	$R_C^{\max}$	$P_{\sigma\sigma}^{\max}$	$P_{\pi\pi}^{\max}$
Diisopropyl sulfide	−8.489	0.812	0.0028	0.0220	0.0005	0.9742	0.0261
4-Methyl-2-pentanol	−10.764	3.492	0.0121	0.0179	0.0065	0.9758	0.0428
2-Methyl-3-pentanol	−10.808	3.393	0.0118	0.0190	0.0048	0.9758	0.0438
3-Hexanol	−10.778	3.407	0.0125	0.0172	0.0056	0.9757	0.0435
2-Butoxyethanol	−10.378	2.715	0.0077	0.0175	0.0016	0.9758	0.0497
2-Methyl-2-pentanol	−10.898	3.438	0.0120	0.0176	0.0021	0.9759	0.0451
Di-*n*-Propylamine	−9.199	3.201	0.0045	0.0165	0.0016	0.9757	0.0439
Hexylamine	−9.723	3.498	0.0077	0.0143	0.0017	0.9756	0.0421
Diisopropylamine	−9.308	3.116	0.0078	0.0153	0.0021	0.9756	0.0467
Triethylamine	−9.008	2.946	0.0066	0.0150	0.0062	0.9754	0.0477
Benzotrifluoride	−10.211	−0.299	0.0234	0.0319	0.0175	0.9581	0.4684
3-Cyanophenol	−9.544	−0.574	0.0261	0.0279	0.0192	0.9715	1.9119
4-Cyanophenol	−9.485	−0.471	0.0270	0.0304	0.0262	0.9714	1.9069
Benzonitrile	−10.042	−0.441	0.0288	0.0276	0.0296	0.9714	1.9104
Benzaldehyde	−9.994	−0.495	0.0303	0.0239	0.0124	0.9739	0.9284
3-Hydroxybenzaldehyde	−9.405	−0.605	0.0254	0.0228	0.0167	0.9740	0.9352
4-Hydroxybenzaldehyde	−9.450	−0.492	0.0277	0.0267	0.0249	0.9742	0.9247
2-Chlorotoluene	−9.423	0.176	0.0235	0.0292	0.0281	0.9733	0.4626
4-Bromotoluene	−9.399	0.701	0.0244	0.0318	0.0301	0.9739	0.4761
2-Nitrotoluene	−10.156	−1.103	0.0253	0.0227	0.0139	0.9708	0.5277
3-Nitrotoluene	−10.211	−1.174	0.0265	0.0227	0.0147	0.9736	0.5319
Benzamide	−9.925	−0.149	0.0295	0.0276	0.0189	0.9637	0.7564
4-Chloro-3-methylphenol	−9.042	0.092	0.0269	0.0342	0.0190	0.9701	0.4743
3-Acetylpyridine	−10.207	−0.578	0.0236	0.0274	0.0253	0.9710	0.9244
4-Acetylpyridine	−10.221	−0.618	0.0233	0.0260	0.0167	0.9698	0.9276
Toluene	−9.338	0.520	0.0292	0.0313	0.0320	0.9740	0.4660
Benzyl alcohol	−9.367	0.459	0.0288	0.0292	0.0307	0.9585	0.4689
o-Cresol	−8.949	0.425	0.0239	0.0254	0.0237	0.9739	0.4604

p-Cresol	−8.892	0.427	0.0276	0.0322	0.0172	0.9748	0.4745
1,3,5-Cycloheptatriene	−8.777	0.172	0.0197	0.0222	0.0233	0.9669	0.8999
Methyl phenyl ether	−9.035	0.455	0.0269	0.0299	0.0162	0.9598	0.4665
2-Methoxyphenol	−8.700	0.360	0.0212	0.0302	0.0126	0.9600	0.4758
Phenyl methyl sulfide	−8.592	−0.046	0.0023	0.0274	0.0004	0.9734	0.4701
3-Methoxyphenol	−8.990	0.365	0.0288	0.0323	0.0149	0.9606	0.4856
o-Toluidine	−8.443	0.583	0.0231	0.0303	0.0131	0.9742	0.4749
p-Toluidine	−8.435	0.621	0.0252	0.0318	0.0290	0.9751	0.4821
2-Methoxyaniline	−8.335	0.522	0.0189	0.0245	0.0212	0.9604	0.4708
3-Methoxyaniline	−8.536	0.602	0.0287	0.0308	0.0229	0.9606	0.4914
4-Methoxyaniline	−8.194	0.507	0.0210	0.0248	0.0167	0.9604	0.4915
2,3-Dimethylpyridine	−9.410	0.142	0.0251	0.0275	0.0266	0.9728	0.4952
2,4-Dimethylpyridine	−9.624	0.185	0.0264	0.0324	0.0290	0.9737	0.4945
2,5-Dimethylpyridine	−9.354	0.135	0.0264	0.0311	0.0305	0.9737	0.5087
2,6-Dimethylpyridine	−9.387	0.217	0.0245	0.0251	0.0235	0.9729	0.4694
3,4-Dimethylpyridine	−9.650	0.120	0.0291	0.0316	0.0280	0.9737	0.4856
3,5-Dimethylpyridine	−9.469	0.095	0.0232	0.0249	0.0235	0.9743	0.4891
N-Methylaniline	−8.569	0.557	0.0214	0.0287	0.0248	0.9649	0.4739
2-Ethylpyridine	−9.566	0.193	0.0245	0.0298	0.0276	0.9748	0.4833
3-Ethylpyridine	−9.648	0.120	0.0264	0.0297	0.0297	0.9749	0.4802
4-Ethylpyridine	−9.966	0.128	0.0254	0.0288	0.0282	0.9746	0.4833
1-Methylcyclohexene	−9.231	1.253	0.0378	0.0394	0.0367	0.9745	0.9303
1-Heptyne	−10.564	1.934	0.0328	0.0393	0.0087	0.9760	1.9329
1-Heptene	−9.940	1.376	0.0412	0.0422	0.0411	0.9758	0.9924
Heptanal	−10.638	0.760	0.0141	0.0570	0.0000	0.9755	0.9583
n-Propyl butanoate	−11.143	1.167	0.0065	0.0532	0.0008	0.9754	0.8364
Ethyl pentanoate	−11.113	1.185	0.0077	0.0531	0.0009	0.9757	0.8333
Methyl hexanoate	−11.132	1.139	0.0066	0.0535	0.0007	0.9756	0.8396
Cycloheptanol	−10.678	3.374	0.0125	0.0142	0.0072	0.9670	0.0424
Methylcyclohexane	−10.739	3.674	0.0105	0.0061	0.0051	0.9756	0.0236
2,4-Dimethyl-3-pentanone	−10.243	0.902	0.0136	0.0544	0.0012	0.9757	0.9437

TABLE 3.10 (*Continued*)

Compound	ε_{HOMO}	ε_{LUMO}	N_C^{max}	E_C^{max}	R_C^{max}	$P_{\sigma\sigma}^{max}$	$P_{\pi\pi}^{max}$
Isoamyl acetate	−11.031	1.148	0.0104	0.0532	0.0009	0.9758	0.8309
4-Heptanone	−10.395	0.856	0.0137	0.0539	0.0012	0.9754	0.9331
trans-2-Heptene	−9.572	1.300	0.0372	0.0401	0.0388	0.9758	0.9608
2-Heptanone	−10.534	0.831	0.0135	0.0538	0.0003	0.9757	0.9308
1-Bromoheptane	−10.812	0.708	0.0058	0.0476	0.0001	0.9758	0.0259
1-Chloroheptane	−10.971	1.382	0.0115	0.0471	0.0001	0.9759	0.0259
1-Iodoheptane	−10.536	0.371	0.0036	0.0449	0.0000	0.9759	0.0259
2,3-Dimethylpentane	−10.814	3.674	0.0141	0.0163	0.0117	0.9757	0.0251
n-Heptane	−10.973	3.722	0.0115	0.0097	0.0007	0.9759	0.0244
2-Methylhexane	−10.971	3.642	0.0117	0.0111	0.0041	0.9758	0.0244
3-Methylhexane	−10.797	3.756	0.0100	0.0120	0.0068	0.9759	0.0256
1-Heptanol	−10.876	3.486	0.0089	0.0118	0.0021	0.9758	0.0489
2,4-Dimethylpentane	−10.827	3.694	0.0126	0.0122	0.0084	0.9761	0.0229
3,3-Dimethylpentane	−10.983	3.720	0.0165	0.0166	0.0102	0.9756	0.0259
2,2-Dimethylpentane	−10.928	3.716	0.0130	0.0148	0.0084	0.9760	0.0247
n-Heptylamine	−9.759	3.463	0.0075	0.0125	0.0023	0.9758	0.0422
Styrene	−9.077	0.107	0.0218	0.0230	0.0244	0.9677	0.9620
Acetophenone	−9.909	−0.290	0.0297	0.0236	0.0177	0.9716	0.9149
4-Methylbenzaldehyde	−9.710	−0.539	0.0276	0.0249	0.0249	0.9739	0.9253
Methyl benzoate	−9.977	−0.365	0.0263	0.0266	0.0189	0.9620	0.8155
Phenyl acetate	−9.435	0.228	0.0276	0.0292	0.0299	0.9666	0.8675
Ethylbenzene	−9.308	0.527	0.0286	0.0311	0.0306	0.9750	0.4749
m-Xylene	−9.188	0.525	0.0246	0.0244	0.0239	0.9744	0.4691
p-Xylene	−9.063	0.486	0.0278	0.0302	0.0311	0.9742	0.4725
o-Xylene	−9.180	0.516	0.0252	0.0257	0.0272	0.9740	0.4702
Ethyl phenyl ether	−9.453	0.338	0.0287	0.0316	0.0319	0.9742	0.4651
2,3-Dimethylphenol	−8.902	0.401	0.0257	0.0293	0.0193	0.9736	0.4681
2,5-Dimethylphenol	−8.826	0.388	0.0244	0.0299	0.0278	0.9743	0.4647

2,6-Dimethylphenol	−8.880	0.393	0.0249	0.0269	0.0185	0.9739	0.4810
3,4-Dimethylphenol	−8.823	0.383	0.0293	0.0328	0.0199	0.9738	0.4665
3,5-Dimethylphenol	−8.988	0.370	0.0309	0.0273	0.0146	0.9744	0.4481
4-Ethylphenol	−8.867	0.433	0.0273	0.0314	0.0136	0.9750	0.4774
2-Phenylethanol	−9.482	0.423	0.0297	0.0316	0.0323	0.9621	0.4651
2,4-Dimethylphenol	−8.766	0.456	0.0252	0.0257	0.0174	0.9744	0.4663
3-Ethylphenol	−9.062	0.365	0.0289	0.0262	0.0177	0.9748	0.4788
2,6-Dimethylaniline	−8.346	0.588	0.0226	0.0238	0.0196	0.9742	0.4569
2-Isobutylpyrazine	−9.922	−0.298	0.0225	0.0291	0.0267	0.9758	0.4969
Trans-2-octenal	−10.534	−0.208	0.0365	0.0386	0.0332	0.9758	0.9325
1-Octyne	−10.612	1.914	0.0317	0.0393	0.0060	0.9758	1.9355
n-Propylcyclopentane	−10.992	3.563	0.0130	0.0096	0.0056	0.9759	0.0248
trans-1,4-Dimethylcyclohexane	−10.672	3.634	0.0095	0.0086	0.0068	0.9759	0.0228
Octanal	−10.636	0.767	0.0141	0.0569	0.0000	0.9758	0.9591
n-Pentyl propanoate	−10.959	1.184	0.0083	0.0535	0.0011	0.9757	0.8414
Ethyl hexanoate	−11.104	1.179	0.0071	0.0532	0.0007	0.9756	0.8328
Isobutyl isotuanoate	−11.074	1.214	0.0091	0.0532	0.0008	0.9755	0.8436
cis-1,2-Dimethylcyclohexane	−10.668	3.649	0.0108	0.0111	0.0095	0.9755	0.0232
Hexyl acetate	−10.988	1.209	0.0103	0.0529	0.0007	0.9759	0.8321
2-Octanone	−10.489	0.869	0.0136	0.0534	0.0011	0.9759	0.9283
1-Octene	−9.929	1.358	0.0406	0.0424	0.0410	0.9758	0.9930
1-Bromooctane	−10.765	0.737	0.0059	0.0475	0.0002	0.9759	0.0257
n-Octane	−11.067	3.640	0.0084	0.0105	0.0001	0.9758	0.0246
2,2,4-Trimethylpentane	−10.701	3.655	0.0128	0.0140	0.0069	0.9759	0.0237
2,3,4-Trimethylpentane	−10.738	3.669	0.0127	0.0149	0.0074	0.9757	0.0237
Di-*n*-butyl ether	−10.381	2.890	0.0054	0.0170	0.0000	0.9756	0.0430
3-Methylheptane	−11.036	3.638	0.0123	0.0126	0.0041	0.9757	0.0251
1-Octanol	−10.756	3.485	0.0116	0.0164	0.0012	0.9761	0.0417
Di-*n*-butylamine	−9.286	3.148	0.0047	0.0147	0.0054	0.9759	0.0431
Octylamine	−9.917	3.489	0.0116	0.0121	0.0011	0.9759	0.0509
Quinoline	−9.177	−0.476	0.0235	0.0222	0.0229	0.9618	0.6663

TABLE 3.10 (*Continued*)

Compound	ε_{HOMO}	ε_{LUMO}	N_C^{max}	E_C^{max}	R_C^{max}	$P_{\sigma\sigma}^{max}$	$P_{\pi\pi}^{max}$
α-Methylstyrene	−9.001	0.120	0.0241	0.0225	0.0228	0.9733	0.9342
Ethyl benzoate	−9.963	−0.273	0.0292	0.0271	0.0193	0.9749	0.8161
4-Methylacetophenone	−9.589	−0.208	0.0282	0.0248	0.0268	0.9720	0.9227
Indane	−9.121	0.529	0.0226	0.0231	0.0242	0.9680	0.4905
1,2,3-Trimethylbenzene	−9.131	0.552	0.0263	0.0267	0.0252	0.9741	0.4616
1,2,4-Trimethylbenzene	−8.956	0.497	0.0280	0.0275	0.0300	0.9748	0.4784
1,3,5-Trimethylbenzene	−9.157	0.564	0.0296	0.0300	0.0225	0.9751	0.4604
Isopropylbenzene	−9.384	0.526	0.0299	0.0314	0.0313	0.9755	0.4656
2-Ethyltoluene	−9.155	0.517	0.0245	0.0241	0.0259	0.9752	0.4575
4-Ethyltoluene	−9.032	0.505	0.0275	0.0305	0.0312	0.9754	0.4863
4-*n*-Propylphenol	−8.906	0.416	0.0278	0.0323	0.0178	0.9758	0.4959
Propylbenzene	−9.312	0.516	0.0285	0.0310	0.0304	0.9759	0.4756
3-Phenyl-1-propanol	−9.367	0.471	0.0287	0.0314	0.0318	0.9633	0.4732
Nonanal	−10.628	0.766	0.0140	0.0569	0.0000	0.9759	0.9597
5-Nonanone	−10.396	0.869	0.0135	0.0537	0.0012	0.9758	0.9320
2-Nonanone	−10.476	0.893	0.0137	0.0534	0.0001	0.9761	0.9284
1-Nonene	−9.981	1.367	0.0405	0.0424	0.0362	0.9758	0.9939
n-Nonane	−10.888	3.655	0.0086	0.0085	0.0011	0.9758	0.0247
2,2,5-Trimethylhexane	−10.978	3.540	0.0138	0.0124	0.0089	0.9763	0.0230
1-Nonanol	−10.878	3.341	0.0113	0.0152	0.0005	0.9757	0.0418
Naphthalene	−8.716	−0.257	0.0181	0.0180	0.0206	0.9612	0.6412
1-Naphthol	−8.444	−0.240	0.0217	0.0201	0.0213	0.9611	0.6355
2-Naphthol	−8.652	−0.341	0.0225	0.0206	0.0227	0.9611	0.6583

1-Naphthylamine	−8.085	−0.170	0.0233	0.0190	0.0201	0.9611	0.6316
2-Naphthylamine	−8.199	−0.177	0.0254	0.0191	0.0230	0.9609	0.6557
n-Butylbenzene	−9.317	0.516	0.0286	0.0313	0.0317	0.9756	0.4742
tert-Butylbenzene	−9.367	0.548	0.0298	0.0313	0.0322	0.9754	0.4723
4-Isopropyltoluene	−9.096	0.498	0.0284	0.0303	0.0315	0.9755	0.4716
4-*tert*-Butylphenol	−8.921	0.432	0.0280	0.0319	0.0171	0.9755	0.4736
Isobutylbenzene	−9.306	0.524	0.0289	0.0315	0.0307	0.9757	0.4696
sec-Butylbenzene	−9.369	0.533	0.0291	0.0305	0.0314	0.9755	0.4826
n-Pentylcyclopentane	−10.972	3.517	0.0122	0.0103	0.0046	0.9758	0.0245
2-Decanone	−10.567	0.806	0.0137	0.0535	0.0002	0.9758	0.9323
n-Decane	−10.930	3.632	0.0110	0.0083	0.0013	0.9760	0.0247
Decyl alcohol	−10.854	3.332	0.0102	0.0149	0.0004	0.9758	0.0415
1-Methylnaphthalene	−8.583	−0.273	0.0195	0.0186	0.0221	0.9738	0.6362
n-Pentylbenzene	−9.368	0.512	0.0292	0.0315	0.0311	0.9758	0.4652
2-Undecanone	−10.575	0.802	0.0139	0.0538	0.0008	0.9758	0.9314
Biphenyl	−8.999	−0.020	0.0154	0.0158	0.0170	0.9585	0.4687
Acenaphthene	−8.491	−0.226	0.0174	0.0176	0.0202	0.9633	0.6459
2,3-Dimethylnaphthalene	−8.578	−0.223	0.0194	0.0179	0.0213	0.9739	0.6349
2,6-Dimethylnaphthalene	−8.518	−0.242	0.0188	0.0182	0.0211	0.9746	0.6392
1,3-Dimethylnaphthalene	−8.510	−0.248	0.0213	0.0181	0.0212	0.9740	0.6345
1,4-Dimethylnaphthalene	−8.546	−0.270	0.0205	0.0186	0.0229	0.9740	0.6305
1-Ethylnaphthalene	−8.595	−0.267	0.0193	0.0186	0.0220	0.9754	0.6343
n-Hexylbenzene	−9.371	0.510	0.0292	0.0315	0.0320	0.9756	0.4649

TABLE 3.11 Semiempirical AM1 SCRE Calculated Molecular Energy Descriptors for Selection of Structurally Variable Compounds

Compound	$\Sigma E_{ee}(A)$	$\Sigma E_{ne}(A)$	$\Sigma E_{el}(AB)$	$\Sigma E_{exc}(AB)$	$\Sigma E_{res}(AB)$
Water	233.60	−555.69	13.49	−10.99	−28.97
Ammonia	152.16	−361.62	17.37	−16.80	−39.70
Tetrafluoromethane	1384.10	−3387.30	35.24	−23.75	−78.68
Tetrachloromethane	1020.40	−2599.50	18.25	−18.60	−43.82
Trichloromethane	794.30	−2012.70	17.83	−19.29	−43.62
Tribromomethane	766.71	−1933.30	13.22	−18.83	−39.78
Formaldehyde	294.37	−720.95	20.23	−21.25	−47.98
Dibromomethane	553.93	−1370.20	14.30	−19.80	−40.72
Dichloromethane	564.55	−1422.10	17.19	−20.02	−43.14
Chloromethane	331.10	−827.67	16.24	−20.79	−42.36
Bromomethane	326.66	−802.53	14.92	−20.73	−41.26
Iodomethane	324.55	−794.50	13.67	−20.62	−40.02
Nitromethane	639.81	−1562.80	45.09	−39.02	−97.41
Methan	91.91	−226.82	15.43	−21.80	−41.95
Methanol	308.88	−750.29	25.87	−27.18	−61.28
Methylamine	231.35	−559.73	29.78	−32.97	−72.55
Tetrachloroethene	1083.70	−2762.90	28.46	−29.88	−70.23
Pentachloroethane	1338.90	−3398.30	31.63	−33.84	−77.45
Trichloroethene	859.21	−2178.40	27.30	−30.30	−68.62
1,1,1,2-Tetrachloroethane	1110.10	−2808.70	30.82	−34.55	−76.91
1,1-Dichloroethene	633.19	−1592.30	26.05	−30.77	−66.99
trans-1,2-Dichloroethene	633.15	−1592.20	26.05	−30.78	−67.00
cis-1,2-Dichloroethene	632.98	−1592.00	26.16	−30.81	−67.16
1,1,2,2-Tetrachloroethane	1112.40	−2811.00	31.04	−34.53	−77.15
2,2,2-Trifluoroethanol	1364.70	−3327.40	52.43	−44.39	−120.09
Acetonitrile	288.05	−706.68	31.40	−37.28	−79.37
1,1,1-Trichloroethane	878.24	−2215.70	29.88	−35.34	−76.29
1,1,2-Trichloroethane	882.19	−2219.90	30.28	−35.29	−76.67
Ethene	168.05	−405.70	23.60	−32.17	−64.14
1,1-Dichloroethane	648.21	−1624.70	29.35	−36.14	−76.08
Acetaldehyde	379.90	−924.67	32.53	−37.50	−81.79
Acetic acid	600.12	−1449.90	43.79	−43.32	−103.57
1,2-Dibromoethane	640.86	−1576.50	26.84	−35.88	−73.69
1,2-Dichloroethane	650.00	−1626.70	29.51	−36.10	−76.18
Methyl formate	591.36	−1442.30	43.61	−43.18	−101.56
Chloroethane	414.16	−1029.40	28.65	−37.01	−75.68
Bromoethane	409.27	−1004.00	27.44	−36.94	−74.54
Iodoethane	406.74	−995.57	26.36	−36.87	−73.41
Nitroethane	722.51	−1764.40	57.39	−55.17	−130.50
Ethane	173.99	−427.57	28.02	−38.09	−75.50
Ethanol	391.23	−951.29	38.31	−43.45	−94.58
Ethanethiol	298.66	−737.54	30.53	−42.13	−82.98
Ethylamine	313.50	−760.57	42.12	−49.22	−105.76
Dimethylamine	310.33	−757.72	42.14	−49.12	−105.22
Propyne	241.06	−580.75	32.77	−42.85	−87.65
Methoxyflurane	1585.30	−3921.60	61.46	−58.52	−145.16

TABLE 3.11 (*Continued*)

Compound	$\Sigma E_{ee}(A)$	$\Sigma E_{ne}(A)$	$\Sigma E_{el}(AB)$	$\Sigma E_{exc}(AB)$	$\Sigma E_{res}(AB)$
1-Chloro-2-propene	488.28	−1206.10	37.04	−47.44	−98.25
Propanonitrile	369.55	−906.93	43.88	−53.56	−112.64
Ethyl formate	674.63	−1644.20	55.96	−59.43	−134.88
Methyl acetate	676.10	−1645.30	56.19	−59.44	−135.69
1,3-Dichloropropane	733.80	−1829.30	41.95	−52.31	−109.36
Cyclopropane	251.64	−607.43	38.03	−48.44	−99.65
Propene	249.67	−605.84	36.33	−48.49	−98.00
Propanone	464.02	−1127.10	45.06	−53.81	−115.62
Propanoic acid	682.03	−1650.60	56.36	−59.58	−136.91
2-Propenol	466.98	−1129.60	46.59	−53.86	−117.12
1,2-Dichloropropane	733.15	−1828.60	41.89	−52.29	−109.34
Propionaldehyde	460.83	−1124.30	45.36	−53.84	−115.37
1-Bromopropane	491.34	−1204.80	40.02	−53.22	−107.95
2-Bromopropane	491.76	−1205.30	39.90	−53.15	−107.68
1-Nitropropane	804.64	−1965.10	70.09	−71.48	−164.08
1-Chloropropane	496.03	−1230.00	41.26	−53.30	−109.12
2-Chloropropane	496.80	−1230.80	41.06	−53.24	−108.89
1-Iodopropane	488.94	−1196.50	38.91	−53.15	−106.78
N,N-Dimethylformamide	599.64	−1454.90	59.62	−65.18	−146.97
2-Nitropropane	805.27	−1966.00	69.66	−71.33	−163.44
Propane	256.10	−628.38	40.60	−54.38	−108.91
Methyl ethyl ether	467.23	−1146.60	50.62	−59.59	−126.83
1-Propanol	472.93	−1151.60	51.05	−59.77	−128.17
2-Propanol	474.41	−1153.00	50.86	−59.73	−128.05
2-Methoxyethanol	685.25	−1671.10	60.96	−64.94	−146.03
n-Propanethiol	380.63	−938.16	43.08	−58.42	−116.38
Trimethylamine	388.95	−955.51	54.38	−65.22	−137.56
Propylamine	396.68	−962.51	54.70	−65.49	−139.21
Thiophene	437.93	−1060.80	41.51	−58.38	−121.20
1-Butyne	322.90	−781.30	45.24	−59.13	−120.92
(*E*)-2-Butenal	535.32	−1301.20	53.51	−64.22	−138.13
1-Cyanopropane	451.51	−1107.50	56.55	−69.87	−146.21
Isobutyraldehyde	542.03	−1324.30	58.02	−70.14	−148.67
Tetrahydrofuran	541.75	−1323.80	59.85	−70.29	−150.86
Ethyl acetate	759.46	−1847.30	68.64	−75.69	−169.10
Methyl propanoate	758.03	−1846.00	68.72	−75.69	−169.01
1,4-Dichlorobutane	817.14	−2031.40	54.56	−68.55	−142.84
1-Butene	331.53	−806.40	48.94	−64.79	−131.43
Butanone	545.10	−1326.90	57.74	−70.13	−149.06
Isopropyl formate	759.92	−1847.90	68.61	−75.62	−168.84
Butanoic-acid	763.81	−1851.00	69.04	−75.90	−170.42
Propyl formate	756.68	−1845.00	68.52	−75.69	−168.28
Butyraldehyde	543.05	−1325.30	57.85	−70.10	−148.70
1,4-Dioxane	753.15	−1842.00	70.12	−75.55	−169.30
1-Bromobutane	573.40	−1405.60	52.51	−69.48	−141.19
1-Chlorobutane	578.13	−1430.80	53.78	−69.57	−142.41
1-Nitrobutane	886.52	−2165.70	82.67	−87.77	−197.50

TABLE 3.11 (*Continued*)

Compound	$\Sigma E_{ee}(A)$	$\Sigma E_{ne}(A)$	$\Sigma E_{el}(AB)$	$\Sigma E_{exc}(AB)$	$\Sigma E_{res}(AB)$
2-Bromo-2-methylpropane	573.58	−1406.00	52.32	−69.36	−140.67
1-Bromo-2-methylpropane	573.29	−1405.40	52.53	−69.48	−141.19
2-Chloro-2-methylpropane	578.66	−1431.40	53.53	−69.48	−142.05
2-Chlorobutane	578.95	−1431.70	53.64	−69.50	−142.27
1-Iodobutane	570.86	−1397.10	51.52	−69.43	−140.14
Morpholine	677.30	−1653.00	73.79	−81.30	−180.63
n-Butane	338.15	−829.13	53.22	−70.67	−142.35
Diethyl ether	550.65	−1348.60	63.05	−75.85	−160.22
2-Methylpropane	338.05	−829.07	53.19	−70.67	−142.23
2-Methyl-1-propanol	554.85	−1352.30	63.65	−76.06	−161.53
2-Butanol	557.12	−1354.40	63.44	−76.02	−161.58
2-Methyl-2-propanol	556.89	−1354.20	63.29	−76.00	−161.23
Diethyl sulfide	465.16	−1140.80	55.12	−74.68	−149.79
Diethyl disulfide	589.64	−1451.10	57.22	−78.48	−156.70
2-Ethoxyethanol	767.55	−1872.00	73.54	−81.23	−179.46
1-Butanol	555.22	−1352.60	63.54	−76.03	−161.45
n-Butanethiol	462.78	−1139.10	55.61	−74.66	−149.70
Diethylamine	574.63	−1159.30	66.92	−81.63	−171.74
Butylamine	478.78	−1163.30	67.24	−81.74	−172.55
2-Chloropyridine	750.01	−1851.30	63.74	−73.82	−164.02
3-Chloropyridine	750.22	−1851.50	64.14	−73.84	−164.42
Pyridine	519.86	−1260.50	62.82	−74.51	−163.00
2-Methylthiophene	517.59	−1259.20	54.66	−74.76	−155.15
2-Methylpyrazine	649.73	−1586.20	74.91	−85.16	−189.37
Cyclopentene	404.28	−981.87	58.41	−75.46	−155.59
2-Methyl-1,3-butadiene	405.24	−982.59	57.41	−75.23	−154.20
Cyclopentanone	617.52	−1501.90	67.41	−80.92	−173.55
1,4-Pentadiene	406.70	−984.18	57.21	−75.19	−153.76
1-Pentyne	404.80	−981.86	57.92	−75.43	−154.41
1-Cyanobutane	533.37	−1308.00	69.27	−86.19	−179.78
Cyclopentane	410.49	−1004.20	62.72	−81.46	−166.50
1-Pentene	413.75	−1007.30	61.54	−81.06	−164.84
Ethyl propanoate	840.50	−2047.10	81.27	−91.98	−202.44
Isobutyl formate	838.33	−2045.30	81.39	−92.02	−201.93
2-Methyl-2-butene	412.73	−1006.10	61.74	−81.12	−165.20
3-Methyl-1-butene	413.21	−1006.80	61.55	−81.08	−164.68
Isopropyl acetate	842.97	−2049.30	80.99	−91.89	−202.36
Pentanal	625.03	−1525.90	70.47	−86.41	−182.22
Methyl butanoate	839.80	−2046.50	81.27	−91.98	−202.31
Pentanoic acid	845.88	−2051.80	81.63	−92.19	−203.90
3-Methylbutanoic acid	846.06	−2052.00	81.61	−92.17	−203.78
Cyclopentanol	628.28	−1528.30	73.01	−86.81	−185.70
3-Methyl-3-butanone	626.98	−1527.60	70.34	−86.38	−182.32
3-Pentanone	626.48	−1527.00	70.46	−86.43	−182.52
Propyl acetate	841.07	−2047.70	81.04	−91.93	−202.21
2-pentanone	627.32	−1527.80	70.33	−86.40	−182.45

TABLE 3.11 (*Continued*)

Compound	$\Sigma E_{ee}(A)$	$\Sigma E_{ne}(A)$	$\Sigma E_{el}(AB)$	$\Sigma E_{exc}(AB)$	$\Sigma E_{res}(AB)$
2-Methyltetrahydrofuran	624.42	−1525.10	72.31	−86.56	−184.14
Tetrahydropyran	623.70	−1524.10	72.73	−86.64	−184.87
1-Chloropentane	660.43	−1631.90	66.28	−85.82	−175.76
1-Nitropentane	968.34	−2366.30	95.11	−104.04	−230.72
1-Bromopentane	655.59	−1606.50	65.13	−85.76	−174.68
1-Iodopentane	653.00	−1598.00	64.06	−85.69	−173.53
N-Methylmorpholine	755.88	−1850.70	86.13	−97.42	−213.15
n-Pentane	420.37	−1030.00	65.75	−86.96	−175.77
2-Methyl-1-butanol	638.27	−1554.30	76.10	−92.26	−194.92
2-Methyl-2-butanol	638.87	−1554.90	75.99	−92.27	−194.74
3-Methyl-1-butanol	636.91	−1553.10	76.16	−92.30	−194.80
1-Pentanol	637.37	−1553.50	76.15	−92.31	−194.92
2-Pentanol	638.64	−1554.70	75.90	−92.26	−194.71
3-Pentanol	638.48	−1554.50	76.00	−92.27	−194.82
Methyl tert-butyl ether	632.92	−1549.50	75.66	−92.13	−193.47
2,2-Dimethylpropane	419.69	−1029.50	65.79	−86.94	−175.44
2-Propoxyethanol	847.54	−2070.70	86.16	−97.57	−212.84
2-Methylbutane	420.01	−1029.70	65.78	−86.93	−175.60
n-Pentylamine	559.88	−1363.00	79.82	−98.05	−205.95
1,2,3,4-Tetrachlorobenzene	1386.40	−3491.60	68.41	−77.81	−175.92
1,2,3,5-Tetrachlorobenzene	1387.50	−3492.80	68.38	−77.77	−175.87
1,2,4,5-Tetrachlorobenzene	1387.50	−3492.80	68.40	−77.77	−175.90
1,2,3-Trichlorobenzene	1158.80	−2903.80	67.05	−78.31	−174.26
1,3,5-Trichlorobenzene	1161.10	−2906.30	67.00	−78.24	−174.18
1,2,4-Trichlorobenzene	1160.00	−2905.10	67.13	−78.29	−174.34
1,2-Dichlorobenzene	930.72	−2315.40	65.76	−78.86	−172.71
1,3-Dichlorobenzene	931.53	−2316.20	65.75	−78.84	−172.72
1,4-Dichlorobenzene	931.76	−2316.50	65.73	−78.82	−172.67
3-Cyanopyridine	709.41	−1733.30	79.10	−90.24	−200.80
4-Cyanopyridine	708.52	−1732.30	78.99	−90.29	−200.66
Bromobenzene	696.09	−1700.00	63.07	−79.31	−169.81
Chlorobenzene	700.96	−1725.10	64.55	−79.49	−171.36
4-Fluorophenol	1002.20	−2437.90	79.70	−86.41	−199.95
2-Chlorophenol	912.49	−2241.70	75.83	−85.31	−192.21
3-Chlorophenol	914.09	−2243.30	75.75	−85.29	−192.28
4-Chlorophenol	914.05	−2243.40	75.79	−85.27	−192.24
4-Bromophenol	909.22	−2218.20	74.26	−85.08	−190.74
2-Iodophenol	905.91	−2209.20	72.76	−84.86	−189.04
2-Nitrophenol	1222.40	−2977.60	103.91	−103.41	−246.92
3-Nitrophenol	1224.00	−2979.60	104.31	−103.44	−246.97
4-Nitrophenol	1224.40	−2979.70	103.84	−103.34	−247.03
3-Formalpyridine	802.04	−1952.30	80.29	−90.38	−203.25
4-Formalpyridine	801.23	−1951.60	80.36	−90.43	−203.05
2-Fluorophenol	1000.50	−2436.10	80.04	−86.50	−200.23
Fluorobenzene	789.17	−1919.50	68.44	−80.64	−179.24
Iodobenzene	693.97	−1692.30	61.69	−79.10	−168.22

TABLE 3.11 (*Continued*)

Compound	$\Sigma E_{ee}(A)$	$\Sigma E_{ne}(A)$	$\Sigma E_{el}(AB)$	$\Sigma E_{exc}(AB)$	$\Sigma E_{res}(AB)$
Nitrobenzene	1011.70	−2462.30	92.59	−97.48	−225.68
Benzene	468.55	−1131.90	63.51	−80.26	−170.26
Phenol	682.33	−1650.90	74.63	−85.98	−191.00
4-Chloroaniline	842.94	−2058.10	78.66	−90.97	−203.99
2-Nitroaniline	1156.30	−2796.30	106.27	−109.07	−259.55
3-Nitroaniline	1153.20	−2794.50	107.36	−109.23	−259.09
4-Nitroaniline	1156.60	−2796.50	106.50	−109.12	−259.73
3-Chloroaniline	841.92	−2057.30	79.10	−91.06	−204.13
2-Chloroaniline	841.19	−2056.80	79.19	−91.05	−204.05
Thiophenol	587.65	−1435.20	66.15	−84.69	−178.65
2-Methylpyridine	600.60	−1459.70	75.43	−90.84	−196.61
3-Methylpyridine	600.63	−1459.70	75.71	−90.88	−196.94
4-Methylpyridine	601.31	−1460.40	75.62	−90.86	−196.90
N-Acetylpyrrolidine	819.31	−1981.30	85.92	−96.96	−217.53
2-Ethylpyrazine	732.27	−1787.40	87.35	−101.39	−222.68
Cyclohexene	484.96	−1180.80	71.41	−91.90	−189.99
Cyclohexanone	699.20	−1702.20	80.07	−97.20	−207.13
1,5-Hexadiene	488.90	−1185.10	69.81	−91.48	−187.21
2,3-Dimethyl-1,3-butadiene	486.42	−1182.30	70.16	−91.55	−187.90
trans-2-Hexenal	698.77	−1702.10	78.65	−96.79	−204.77
1-Hexyne	486.73	−1182.50	70.47	−91.69	−187.73
Cyclohexane	491.40	−1203.60	75.57	−97.76	−200.24
Methylcyclopentane	492.30	−1204.70	75.35	−97.75	−199.86
1-Methyl-1-pentene	494.89	−1207.00	74.26	−97.37	−198.50
2-Hexanone	709.45	−1728.70	82.92	−102.67	−215.79
Hexanoic acid	927.92	−2252.60	94.23	−108.48	−237.29
Isobutyl acetate	922.21	−2247.50	93.86	−108.29	−235.76
n-Propyl propanoate	922.63	−2248.00	93.69	−108.23	−235.63
1-Hexene	495.77	−1208.00	74.10	−97.35	−198.26
Isoamyl formate	922.73	−2248.40	93.86	−108.21	−235.53
Ethyl butanoate	922.63	−2247.90	93.72	−108.25	−235.70
Methyl pentanoate	921.87	−2247.30	93.84	−108.26	−235.71
Hexanal	707.67	−1727.30	83.00	−102.65	−215.56
4-Methyl-2-pentanone	708.89	−1728.00	82.99	−102.70	−215.89
Butyl acetate	923.37	−2248.60	93.67	−108.21	−235.72
Cyclohexylamine	632.69	−1538.40	89.47	−108.82	−230.57
n-Butylacetamide	852.55	−2063.10	96.97	−114.10	−248.08
1-Bromohexane	737.61	−1807.20	77.72	−102.05	−208.10
1-Chlorohexane	742.39	−1832.50	78.83	−102.10	−209.16
1-Iodohexane	735.12	−1798.80	76.64	−101.98	−206.98
N-Methylpiperidine	624.40	−1530.70	89.03	−108.58	−228.88
2,2-Dimethylbutane	501.69	−1230.20	78.44	−103.21	−208.90
2,3-Dimethylbutane	501.94	−1230.40	78.38	−103.20	−208.95
n-Hexane	502.28	−1230.70	78.31	−103.20	−209.05
2-Methylpentane	502.17	−1230.60	78.32	−103.20	−208.95
3-Methylpentane	501.79	−1230.20	78.38	−103.19	−208.97
Diisopropyl ether	715.72	−1751.00	87.94	−108.37	−226.64

TABLE 3.11 (*Continued*)

Compound	$\Sigma E_{ee}(A)$	$\Sigma E_{ne}(A)$	$\Sigma E_{el}(AB)$	$\Sigma E_{exc}(AB)$	$\Sigma E_{res}(AB)$
Di-*n*-propyl ether	713.83	−1749.20	88.24	−108.44	−227.05
1-Hexanol	719.30	−1754.10	88.80	−108.64	−228.52
Di-*n*-propyl sulfide	629.35	−1542.40	80.32	−107.28	−216.68
Diisopropyl sulfide	629.15	−1542.60	80.48	−107.16	−216.39
4-Methyl-2-pentanol	720.11	−1754.90	88.59	−108.55	−228.08
2-Methyl-3-pentanol	720.29	−1755.10	88.57	−108.54	−228.05
3-Hexanol	720.63	−1755.40	88.58	−108.55	−228.23
2-Butoxyethanol	931.07	−2272.90	98.57	−113.79	−246.08
2-Methyl-2-pentanol	720.99	−1755.70	88.53	−108.55	−228.08
Di-*n*-Propylamine	639.27	−1561.20	92.11	−114.21	−238.80
Hexylamine	642.15	−1564.00	92.48	−114.35	−239.51
Diisopropylamine	639.81	−1562.20	91.34	−114.01	−237.47
Triethylamine	636.37	−1558.90	91.45	−113.94	−237.25
Benzotrifluoride	1524.20	−3708.60	89.38	−97.37	−228.59
3-Cyanophenol	872.66	−2124.40	90.95	−101.78	−228.87
4-Cyanophenol	872.24	−2123.80	91.13	−101.81	−229.25
Benzonitrile	659.55	−1606.20	79.73	−95.96	−208.00
Benzaldehyde	752.03	−1825.10	80.79	−96.07	−210.26
3-Hydroxybenzaldehyde	964.94	−2343.10	92.34	−101.94	−231.36
4-Hydroxybenzaldehyde	965.05	−2343.00	91.98	−101.86	−231.35
2-Chlorotoluene	781.98	−1924.60	77.39	−95.84	−205.21
4-Bromotoluene	777.46	−1899.90	75.75	−95.63	−203.55
2-Nitrotoluene	1093.20	−2662.30	105.30	−113.77	−259.48
3-Nitrotoluene	1092.80	−2662.00	105.28	−113.81	−259.38
Benzamide	901.10	−2164.20	95.00	−107.75	−244.03
4-Chloro-3-methylphenol	995.27	−2443.00	88.54	−101.60	−226.03
3-Acetylpyridine	886.14	−2154.60	92.76	−106.68	−236.96
4-Acetylpyridine	885.85	−2154.40	92.90	−106.74	−236.95
Toluene	549.51	−1331.30	76.27	−96.61	−204.05
Benzyl alcohol	765.48	−1853.60	86.54	−101.99	−223.13
o-Cresol	762.63	−1849.70	87.57	−102.36	−224.89
p-Cresol	762.98	−1850.10	87.55	−102.36	−224.91
1,3,5-Cycloheptatriene	550.18	−1332.20	76.08	−96.53	−202.69
Methyl phenyl ether	759.09	−1847.10	86.59	−102.01	−222.78
2-Methoxyphenol	969.59	−2362.90	98.30	−107.92	−243.76
Phenyl methyl sulfide	674.31	−1640.50	77.56	−100.62	−211.16
3-Methoxyphenol	972.78	−2365.70	97.91	−107.81	−243.99
o-Toluidine	688.88	−1662.40	90.88	−108.13	−236.42
p-Toluidine	688.87	−1662.40	90.95	−108.16	−236.51
2-Methoxyaniline	898.00	−2178.20	101.24	−113.50	−254.73
3-Methoxyaniline	899.03	−2178.50	101.08	−113.53	−255.30
4-Methoxyaniline	898.12	−2178.20	101.41	−113.55	−254.99
2,3-Dimethylpyridine	681.42	−1659.10	88.17	−107.15	−230.31
2,4-Dimethylpyridine	681.97	−1659.50	88.19	−107.18	−230.46
2,5-Dimethylpyridine	681.19	−1658.80	88.22	−107.19	−230.40
2,6-Dimethylpyridine	681.40	−1659.00	87.86	−107.14	−230.04
3,4-Dimethylpyridine	681.87	−1659.40	88.51	−107.21	−230.79

TABLE 3.11 ***(Continued)***

Compound	$\Sigma E_{ee}(A)$	$\Sigma E_{ne}(A)$	$\Sigma E_{el}(AB)$	$\Sigma E_{exc}(AB)$	$\Sigma E_{res}(AB)$
3,5-Dimethylpyridine	681.38	−1659.00	88.38	−107.21	−230.62
N-Methylaniline	686.00	−1660.60	90.31	−107.81	−234.61
2-Ethylpyridine	682.16	−1659.90	88.00	−107.13	−230.01
3-Ethylpyridine	682.63	−1660.40	88.35	−107.17	−230.39
4-Ethylpyridine	683.33	−1661.10	88.25	−107.14	−230.34
1-Methylcyclohexene	565.85	−1380.20	84.38	−108.28	−223.96
1-Heptyne	569.06	−1383.50	83.20	−108.02	−221.37
1-Heptene	578.02	−1409.00	86.56	−113.60	−231.51
Heptanal	788.70	−1926.90	95.75	−119.01	−249.17
n-Propyl butanoate	1004.90	−2448.90	106.26	−124.52	−269.13
Ethyl pentanoate	1004.70	−2448.70	106.19	−124.50	−268.98
Methyl hexanoate	1003.90	−2448.10	106.38	−124.52	−269.08
Cycloheptanol	792.20	−1929.50	98.19	−119.30	−252.69
Methylcyclohexane	573.90	−1404.80	88.20	−114.03	−233.88
2,4-Dimethyl-3-pentanone	789.71	−1927.80	95.74	−118.97	−249.12
Isoamyl acetate	1005.00	−2449.00	106.40	−124.54	−269.20
4-Heptanone	790.81	−1928.70	95.65	−119.00	−249.41
Trans-2-heptene	576.98	−1407.80	86.83	−113.67	−232.02
2-Heptanone	791.05	−1928.90	95.65	−119.02	−249.46
1-Bromoheptane	819.55	−2007.80	90.24	−118.33	−241.46
1-Chloroheptane	824.17	−2033.00	91.40	−118.38	−242.50
1-Iodoheptane	817.01	−1999.40	89.23	−118.27	−240.42
2,3-Dimethylpentane	583.76	−1430.90	91.02	−119.46	−242.30
n-Heptane	584.57	−1431.70	90.89	−119.50	−242.47
2-Methylhexane	584.36	−1431.50	90.90	−119.48	−242.37
3-Methylhexane	584.06	−1431.20	91.05	−119.49	−242.46
1-Heptanol	802.40	−1955.90	101.31	−124.88	−261.89
2,4-Dimethylpentane	583.86	−1431.00	90.98	−119.47	−242.28
3,3-Dimethylpentane	583.71	−1430.90	91.02	−119.47	−242.27
2,2-Dimethylpentane	583.63	−1430.90	90.94	−119.44	−242.12
n-Heptylamine	724.07	−1764.60	105.27	−130.70	−273.18
Styrene	623.87	−1508.30	84.51	−107.02	−226.46
Acetophenone	836.15	−2027.40	93.38	−112.40	−244.11
4-Methylbenzaldehyde	833.45	−2024.90	93.51	−112.40	−244.11
Methyl benzoate	1048.70	−2546.00	104.04	−117.96	−263.81
Phenyl acetate	1047.70	−2544.40	105.22	−118.19	−265.13
Ethylbenzene	631.56	−1532.00	88.82	−112.87	−237.43
m-Xylene	630.41	−1530.70	89.02	−112.95	−237.82
p-Xylene	630.28	−1530.60	89.12	−112.98	−237.92
o-Xylene	630.37	−1530.70	89.10	−112.95	−237.87
Ethyl phenyl ether	842.81	−2049.90	98.96	−118.13	−255.66
2,3-Dimethylphenol	843.72	−2049.30	100.17	−118.65	−258.49
2,5-Dimethylphenol	843.67	−2049.20	100.22	−118.68	−258.59
2,6-Dimethylphenol	843.49	−2049.10	100.28	−118.66	−258.55
3,4-Dimethylphenol	844.03	−2049.60	100.12	−118.65	−258.49
3,5-Dimethylphenol	844.47	−2050.00	100.06	−118.66	−258.55

TABLE 3.11 ***(Continued)***

Compound	$\Sigma E_{ee}(A)$	$\Sigma E_{ne}(A)$	$\Sigma E_{el}(AB)$	$\Sigma E_{exc}(AB)$	$\Sigma E_{res}(AB)$
4-Ethylphenol	845.13	−2050.90	99.92	−118.59	−258.08
2-Phenylethanol	848.21	−2055.20	99.17	−118.26	−256.57
2,4-Dimethylphenol	843.38	−2049.00	100.44	−118.74	−258.76
3-Ethylphenol	845.43	−2051.10	100.02	−118.61	−258.28
2,6-Dimethylaniline	769.52	−1861.50	103.68	−124.48	−270.24
2-Isobutylpyrazine	895.85	−2188.50	112.41	−133.94	−289.24
trans-2-Octenal	862.53	−2103.20	104.05	−129.43	−271.89
1-Octyne	651.32	−1584.50	95.51	−124.24	−254.46
n-Propylcyclopentane	656.87	−1606.70	100.60	−130.30	−266.84
trans-1,4-Dimethylcyclohexane	655.78	−1605.50	100.67	−130.28	−266.89
Octanal	870.78	−2127.70	108.41	−135.33	−282.73
n-Pentyl propanoate	1087.10	−2649.80	118.90	−140.80	−302.64
Ethyl hexanoate	1086.80	−2649.50	118.86	−140.79	−302.47
Isobutyl isotuanoate	1085.90	−2648.60	119.05	−140.83	−302.50
cis-1,2-Dimethylcyclohexane	655.73	−1605.30	100.79	−130.29	−267.17
Hexyl acetate	1087.90	−2650.40	118.93	−140.81	−302.74
2-Octanone	873.14	−2129.70	108.15	−135.27	−282.80
1-Octene	660.28	−1610.00	98.73	−129.78	−264.52
1-Bromooctane	901.62	−2208.60	102.68	−134.57	−274.67
n-Octane	666.74	−1632.50	103.58	−135.83	−276.16
2,2,4-Trimethylpentane	665.13	−1631.10	103.61	−135.70	−275.38
2,3,4-Trimethylpentane	665.66	−1631.60	103.59	−135.69	−275.56
Di-*n*-butyl ether	877.81	−2150.60	113.41	−141.02	−293.92
3-Methylheptane	666.57	−1632.30	103.69	−135.81	−276.08
1-Octanol	883.52	−2155.70	113.90	−141.17	−295.18
Di-*n*-butylamine	803.27	−1962.70	117.31	−146.78	−305.58
Octylamine	806.87	−1966.10	117.65	−146.90	−306.32
Quinoline	817.45	−1983.50	102.62	−122.68	−268.50
α-Methylstyrene	705.05	−1708.00	97.30	−123.35	−260.22
Ethyl benzoate	1131.40	−2747.30	116.67	−134.27	−297.33
4-Methylacetophenone	917.08	−2226.70	106.15	−128.77	−277.93
Indane	702.64	−1705.60	98.63	−123.70	−262.03
1,2,3-Trimethylbenzene	711.25	−1730.10	101.77	−129.26	−271.51
1,2,4-Trimethylbenzene	711.15	−1729.90	101.92	−129.31	−271.71
1,3,5-Trimethylbenzene	711.15	−1729.90	101.82	−129.31	−271.64
Isopropylbenzene	713.13	−1732.40	101.42	−129.16	−270.69
2-Ethyltoluene	712.41	−1731.40	101.56	−129.19	−271.18
4-Ethyltoluene	712.30	−1731.30	101.69	−129.25	−271.30
4-*n*-Propylphenol	927.18	−2251.70	112.45	−134.88	−291.43
Propylbenzene	713.69	−1732.90	101.44	−129.17	−270.89
3-Phenyl-1-propanol	930.33	−2255.90	111.81	−134.55	−290.10
Nonanal	952.84	−2328.40	120.87	−151.59	−316.04
5-Nonanone	954.89	−2330.20	120.75	−151.56	−316.17
2-Nonanone	955.25	−2330.50	120.69	−151.56	−316.18
1-Nonene	742.18	−1810.50	111.91	−146.21	−298.54
n-Nonane	748.75	−1833.30	116.03	−152.06	−309.27

TABLE 3.11 ***(Continued)***

Compound	$\Sigma E_{ee}(A)$	$\Sigma E_{ne}(A)$	$\Sigma E_{el}(AB)$	$\Sigma E_{exc}(AB)$	$\Sigma E_{res}(AB)$
2,2,5-Trimethylhexane	747.69	−1832.40	115.96	−151.97	−308.61
1-Nonanol	965.56	−2356.50	126.52	−157.46	−328.69
Naphthalene	767.10	−1856.00	103.24	−128.38	−275.72
1-Naphthol	980.30	−2374.20	114.53	−134.16	−296.73
2-Naphthol	980.67	−2374.70	114.50	−134.15	−296.65
1-Naphthylamine	906.88	−2187.20	118.05	−139.96	−308.53
2-Naphthylamine	906.93	−2187.30	117.94	−139.96	−308.40
n-Butylbenzene	795.83	−1933.70	114.03	−145.46	−304.35
tert-Butylbenzene	794.89	−1932.90	113.98	−145.40	−303.82
4-Isopropyltoluene	793.91	−1931.70	114.19	−145.51	−304.48
4-*tert*-Butylphenol	1008.50	−2451.90	124.94	−151.08	−324.35
Isobutylbenzene	795.65	−1933.60	114.00	−145.42	−304.16
sec-Butylbenzene	795.29	−1933.30	113.87	−145.40	−303.97
n-Pentylcyclopentane	821.15	−2008.40	125.54	−162.82	−333.43
2-Decanone	1037.30	−2531.20	133.40	−167.88	−349.74
n-Decane	830.85	−2034.10	128.36	−168.29	−342.45
Decyl alcohol	1047.70	−2557.40	138.83	−173.67	−361.72
1-Methylnaphthalene	848.05	−2055.40	116.11	−144.73	−309.63
n-Pentylbenzene	877.88	−2134.50	126.82	−161.78	−337.91
2-Undecanone	1119.50	−2732.10	145.96	−184.15	−383.10
Biphenyl	923.73	−2233.70	124.46	−155.14	−332.54
Acenaphthene	919.03	−2228.80	125.98	−155.51	−334.10
2,3-Dimethylnaphthalene	928.93	−2254.70	129.13	−161.14	−343.67
2,6-Dimethylnaphthalene	929.03	−2254.80	128.91	−161.09	−343.47
1,3-Dimethylnaphthalene	928.95	−2254.70	129.04	−161.12	−343.61
1,4-Dimethylnaphthalene	928.94	−2254.70	129.02	−161.10	−343.56
1-Ethylnaphthalene	930.01	−2256.00	128.72	−161.02	−343.02
n-Hexylbenzene	959.86	−2335.10	139.34	−178.05	−371.26
Fluorene	993.35	−2405.50	134.54	−165.95	−357.51

tion components, and Q_A, Q_B, . . . , Q_P, Q_R, . . . are the molar partition functions of reaction components. The individual electronic, rotational, vibrational, and translational contributions to the partition function can be theoretically calculated using the respective statistical mechanical definitions and the quantum mechanically computed rotational and vibrational energetic spectra of molecules. Analogously, the molar Gibbs energy of activation of a bimolecular chemical reaction can be expressed as:

$$\Delta G^{\neq} \equiv -RT \ln k_r = \Delta E_r^{\neq} - RT \ln \frac{Q^{\neq}}{Q_A Q_B} \tag{3.238}$$

where $\Delta E^{\neq} = E^{\neq} - E_A^0 - E_B^0$ is the quantum mechanically calculated activation energy at 0 K, and $Q^{\neq}$ and Q_A, Q_B are the molar partition functions for the transition state and for the reagents A and B, respectively. Eqs. (3.236) to (3.238) have been used for the prediction of the rate of equilibrium constants of reactions, but with variable success. The complications with the theoretical approach described above arise from the limitations in the accuracy of the quantum chemically calculated energies and vibrational spectra and from the ambiguities in the prediction of the energy distribution in the activation complex. In the case of reactions in solutions, additional difficulties come up from the methodological problems related to the proper account for the intermolecular interactions in condensed media. Nevertheless, it has been demonstrated that the calculated free energies of reaction or free energies of activation can be semiquantitatively correlated with the respective experimental data. In this respect, the calculated free energies can be considered as the molecular descriptors used in the framework of the linear free energy relationship or quantitative structure–activity/property relationship approach.

For instance, the Gibbs free energies of activation have been calculated quantum mechanically for the intramolecular S_N2 nucleophilic substitution in the *N*-(ω-aminopropyl)pyridinium cation (reaction I) and for the analogous intermolecular S_N2 reaction between the *N*-propylpyridinium cation and free ammonia (reaction II) [727].

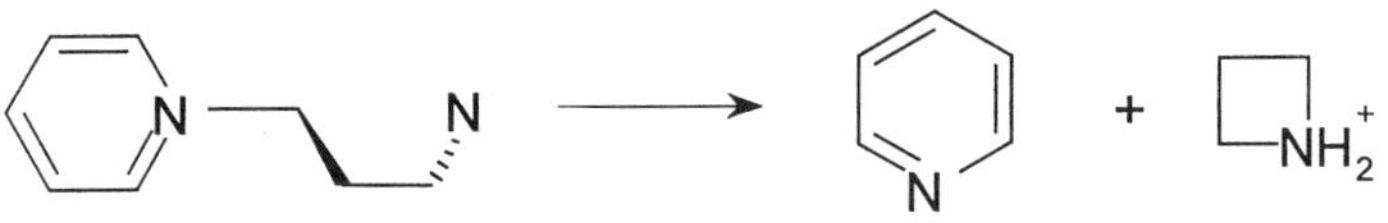

(I)

(II)

The results of theoretical modeling using the semiempirical PM3 and ab initio SCF methods suggested that bimolecular S_N2 reactions involving positively charged substrates with neutral leaving groups should be difficult to induce in the gas phase. Notably, attempts to observe such reactions in the gas phase have so far failed. However, the intramolecular nucleophilic displacement with the cyclic transition state had a much lower predicted activation free energy in the gas phase. In accordance with this prediction, it was confirmed experimentally that the intramolecular S_N2 displacement with cyclic transition states may lead to the formation of charged heterocycles in the gas phase.

The second approach to the use of quantum chemically calculated molecular descriptors in the development of the quantitative relationships to predict the chemical reactivity of compounds is based on the dynamic reactivity indices. The quantum mechanically calculated superdelocalizabilities and simplified Fukui indices have been used to predict the nucleophilic and the electrophilic reactivity of molecules. For instance, the superdelocalizabilities, calculated by the CNDO/2 method, have been directly correlated with the reactivity parameters (Hammett σ constants) of aromatic compounds [700] as follows:

$$\sigma = 8.599 - 1.417S_N^{(C)}$$
$$R^2 = 0.746 \qquad n = 13$$

for simple aromatics

$$\sigma = 3.090 + 1.565S_E^{(C)}$$
$$R^2 = 0.753 \qquad n = 25$$

for substituted benzenes

$$\sigma = 2.817 + 2.500S_E^{(O_1)}$$
$$R^2 = 0.810 \qquad n = 19$$

for substituted benzoic acids, and

$$\sigma = 1.051 + 7.591S_E^{(N)}$$
$$R^2 = 0.748 \qquad n = 19$$

for substituted phenyl amines. It was thus suggested that superdelocalizability apparently includes both the electrostatic and the perturbation effects related to the change of the electron density on the probe atom. The reactivity indices have been applied in correlation of the rate and equilibrium constants of chemical reactions [728–734] and Gibbs free energies of adsorption of ions on electrodes [735,736], and in studies of regioselectivity of reactions [737,738].

The data on chemical reactivity can also be correlated, in principle, with any energy- or charge-distribution-related molecular descriptor obtained from the quantum chemical calculations. The gas-phase acidity of substituted benzoic acids has been shown to be related linearly either to the AM1 calculated net charges on oxygen atoms, q_O, to the energy of the highest occupied molecular orbital (HOMO) of the corresponding benzoate anions, or to the difference in heats of formation between the acids and their anions [676].

$$\Delta(\Delta G^0) = -359.217 - 316.431 \sum q_O$$
$$R^2 = 0.962 \qquad s = 0.853 \qquad n = 14 \qquad F = 302.8$$
$$\Delta(\Delta G^0) = 81.891 + 17.328E_{HOMO}$$
$$R^2 = 0.956 \qquad s = 0.953 \qquad n = 14 \qquad F = 261.8$$
$$\Delta(\Delta G^0) = -0.396 + 0.887\Delta(\Delta H_f^0)$$
$$R^2 = 0.970 \qquad s = 0.758 \qquad n = 14 \qquad F = 387.2$$

A similar statistical quality of these equations indicates the existence of a close relationship between the three descriptors, for this set of compounds. Therefore, one has to be careful in jumping to quick conclusions about the possible interactions that influence the property by using the results from a single correlation. In the above case, the HOMO is obviously located on the oxygen atoms of the carboxyl group, and the change in the enthalpies of formation is due to the redistribution of charge on this group.

The AM1-calculated energy differences between the acids and conjugate bases and the anion HOMO energies have also been correlated satisfactorily with the experimental (condensed phase) acidity of phenols, and aromatic and aliphatic acids and related properties [711,739]. The best correlation employed four descriptors and included calculated atomic charge densities:

$$pK_a = 33.74d_{11} - 13.01d_1 + 0.16\Delta(\Delta H_f^0) + 0.12E_{\mathrm{HOMO}}$$

$$R^2 = 0.880 \qquad s = 1.01 \qquad n = 183$$

where d_1 and d_{11} are the atomic partial charges on acid group oxygen atoms in the neutral acid and in the anion, respectively.

The frontier molecular orbital energies and the related hardness–softness parameters have been commonly used in the study of the chemical reactivity of compounds [740–743]. The activation hardness, calculated from the Hückel MO theory, has been introduced as a descriptor for the prediction orientation effects [705]. The activation hardness, defined as the difference between the absolute hardness of the reactant and the transition state, was successfully used to predict orientations in electrophilic aromatic substitution. It was concluded that the reaction coordinate follows the path that minimizes the change in the HOMO–LUMO gap, or hardness. The relative hardness, η_r, defined relative to some hypothetical acyclic structure, was correlated with different aromaticity scales for a series of annulenes and radialenes [353]. The graphical relationships were presented for the correlations with the different scales based on the resonance energy per π electron [744–746]. It was concluded that the relative hardness, defined from the difference of frontier molecular orbital energies, could be another quantitative scale of aromaticity.

Quantum chemical descriptors have been widely used for the quantitative structure–activity relationship prediction of the biological activity of compounds. The atomic partial charges, HOMO–LUMO energies, frontier orbital electron densities, superdelocalizabilities, and other quantum chemically calculated molecular descriptors have been shown to correlate with various biological activities.

Significant correlations were found between Ames TA100 mutagenicity and electron affinity (EA) or the AM1-calculated LUMO energy (i.e., the stability of the corresponding anion radical) for a series of chlorofuranones [706,747,748]:

$$\ln(\text{TA100}) = 21.40 - 12.53\text{EA}$$

$$R^2 = 0.913 \qquad n = 5$$

$$\ln(\text{TA100}) = -13.39 - 14.23E_{\text{LUMO}}$$

$$R^2 = 0.918 \qquad s = 1.309 \qquad n = 5$$

The correlations observed suggest a reaction mechanism in which chlorofuranones act as electron acceptors in the interaction with DNA. In general, the participation of frontier orbitals in mutagenic activity seems to be essential, but often it is masked by the hydrophobicity.

In a QSAR study on the mutagenicity of quinolines,

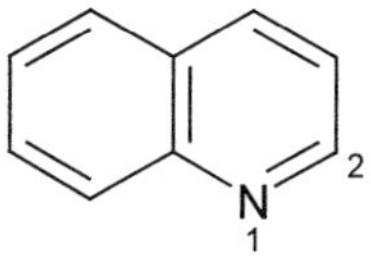

2
N
1

the best correlation was obtained with the AM1-calculated net atomic charges on a carbon atom (q_2) and the hydrophobic parameter (log P) [749]:

$$\ln(\text{TA100}) = -5.39 - 45.76q_2 + 1.14 \log P$$

$$R^2 = 0.726 \qquad s = 0.565 \qquad n = 21 \qquad F = 11.9$$

Notably, the HOMO and LUMO energies and electron densities were also correlating this property. The involvement of the net atomic charge on the carbon atom in the 2-position (q_2) suggests that this might be the site for activity. In addition, linear correlations have been established between the calculated HOMO or LUMO energy and the mutagenicity of aromatic and heteroaromatic nitro-compounds [713,750–753], aromatic and heteroaromatic amines [754], aryltriazenes and heterocyclic triazenes [714], and benzanthracenes [755]. A significant correlation has been obtained between the inhibition potency of indanone-benzylpiperidine inhibitors of acetylcholinesterase and the MNDO-calculated HOMO energy [756]:

$$-\log(\text{IC}_{50}) = -757.52 + 2.21C_4 - 162.9E_{\text{HOMO}} - 8.85E_{\text{HOMO}}^2 - 6.65\mu + 1.18\mu^2$$

$$R^2 = 0.882 \qquad s = 0.25 \qquad n = 16 \qquad F = 14.8$$

where C_4 is the HOMO out-of-plane π-orbital coefficient of the ring carbon atom, and μ is the total dipole moment of the compound.

The treatment of the antirhinoviral activity of 9-benzylpurines has resulted in QSAR correlation equations involving Hückel MO-generated electronic parameters and empirical substituent constants of substituents [707]. The following equation has been obtained using the LUMO energy and the total π-electron energy (E_{π}^{T}) of compounds as the quantum chemical descriptors:

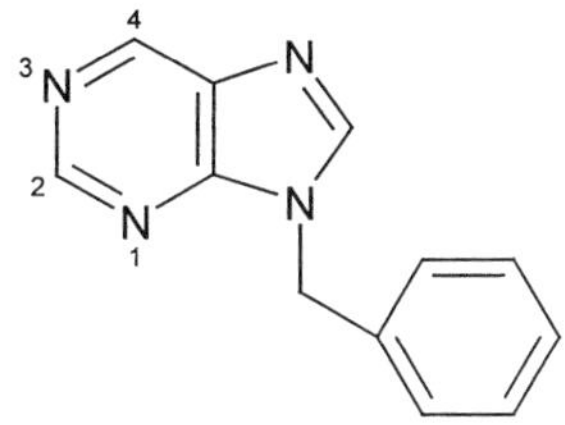

$$-\log(\mathrm{IC}_{50}) = 6.044 + 2.056R_2 + 0.873F_4 - 0.289\pi_4 - 0.094E_{\pi}^{T} - 2.323E_{\mathrm{LUMO}}$$

$$R^2 = 0.684 \quad s = 0.503 \quad n = 50 \quad F = 19.0$$

where R and F are the Swain–Lupton resonance and field parameters, respectively, and π is the hydrophobicity substituent constant at a given position in the purine ring. However, it was established that various serotypes of rhinovirus behave differently in terms of the electronic parameters that inhibit their action.

The pharmacological activity of the para-, ortho-, and meta-monoderivatives of 1,4-dihydropyridine was correlated with quantum chemical AM1 parameters such as net atomic charges, hybridization dipole moment, frontier orbital densities, HOMO energy, and the rotational energy barrier of the phenyl ring [757]. The QSAR treatment of all monosubstituted derivatives generated the following equation:

$$\log\left(\frac{1}{\mathrm{IC}_{50}}\right) = -17.4 + 0.56\pi - 3.4E_{\mathrm{HOMO}} - 0.49L_m - 3.40B_{1,p}$$

$$R^2 = 0.828 \quad s = 0.61 \quad n = 35 \quad F = 35.85$$

where π is the Hansch's hydrophobicity constant, and $B_{1,p}$ and L_m are the Verloop Sterimol parameters for the substituents in the para- and meta-position, respectively [758].

Both the hydrophobicity and the LUMO energy were found to determine the activity of a series of alkyl-substituted phenols

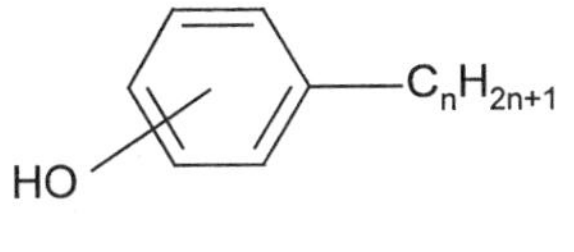

against Chinese hamster V76 tumor cells. The following quadratic equation was obtained [759]:

$$\log\left(\frac{1}{\mathrm{ED}_{50}}\right) = (0.818 \pm 0.062) + (0.278 \pm 0.132)\log P - (0.017 \pm 0.001)(\log P)^2 - (3.485 \pm 0.458)E_{\mathrm{LUMO}}$$

$$R^2 = 0.714 \qquad s = 0.227 \qquad n = 29 \qquad F = 20.74$$

The presence of E_{LUMO} in this equation shows that the activity of these compounds does not depend only on their ability to penetrate the cell membrane but also upon their ability to participate as electron acceptors in the interaction with the receptor. Another antitumor activity, the toxicity of substituted phenols against L1210 leukemia cells, has been related to the HOMO–LUMO energy gap, ΔE_{FMO} [760]:

$$\log\left(\frac{1}{\mathrm{IC}_{50}}\right) = (26.58 \pm 3.30) + (0.25 \pm 0.05)\log P - (2.50 \pm 0.37)\Delta E_{\mathrm{FMO}}$$

$$R^2 = 0.903 \qquad s = 0.176 \qquad n = 26 \qquad R^2_{\mathrm{cv}} = 0.874$$

This equation indicates that the toxicity increases in parallel with the decrease in the HOMO–LUMO energy gap. This result is natural as the smaller gap of frontier orbitals is usually related to higher radical reactivity.

Another QSAR study on the series of 8-substituted xanthines

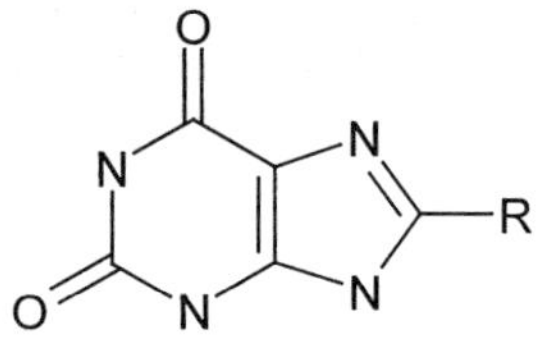

has provided correlation equations between classical and MNDO-calculated parameters and the affinity toward A_1 and A_2 adenosine receptors [761]:

$$\mathrm{p}K_i(A_1) = -17.02 + 0.57\pi_8 - 2.06I_7 + 1.54I_{13} - 59.21q_1$$

$$R^2 = 0.828 \qquad s = 0.57 \qquad n = 37 \qquad F = 37.66$$

$$\mathrm{p}K_i(A_2) = 5.90 + 0.41\pi_1 + 0.66\pi_8 - 0.47I_7 + 45.39q_3 + 61.57S_{N(8)}$$

$$R^2 = 0.689 \qquad s = 0.59 \qquad n = 38 \qquad F = 14.17$$

where π is the substituent hydrophobicity parameter, I is the substituent indicator variable, S_N is the donor superdelocalizability, and q is the net atomic charge at the given position. According to the polyelectronic perturbation theory by Klopman and Hudson [762], the drug–receptor interactions are under either charge or orbital control. Thus, the net atomic charges may be considered as the characteristics of electrostatic interactions, while the donor superdelocalizability characterizes the covalent component of the interaction.

The CNDO/2-calculated molecular polarizability (α) and the HOMO–LUMO energy difference (ΔE_{FMO}) have been successfully correlated with a number of biological activities (ethanol inhibition, rate of oxidative metabolism, and acute toxicity) for a series of 20 nitriles [703]:

$$-\log K_i = 0.95 - 1.25 \frac{\alpha}{\Delta E_{FMO}}$$

$$R^2 = 0.578 \quad s = 0.282 \quad n = 13 \quad F = 15.3$$

$$\log(\text{EtOH/glucose}) = 0.227 - 0.006\alpha$$

$$R^2 = 0.624 \quad s = 0.046 \quad n = 20 \quad F = 30.5$$

$$-\log(\text{LD}_{50}) = 0.43 - 0.03\alpha$$

$$R^2 = 0.757 \quad s = 0.199 \quad n = 13 \quad F = 42.1$$

$$-\log(\text{LD}_{50}) = 0.43 - 1.69 \frac{\alpha}{\Delta E_{FMO}}$$

$$R^2 = 0.757 \quad s = 0.199 \quad n = 13 \quad F = 42.4$$

The $\alpha/\Delta E_{FMO}$ parameter is an orbital energy weighted polarizability term and thus implies that the acute toxicity of nitriles is a function of molecular size/polarity and electronic activation energy. Various other quantum chemical descriptors have been used for the QSPR study of acute toxicity of compounds [763–767]. An interesting application of quantum chemical indices involves the assessment of biodegradability of environmental pollutants [768,769].

An index of frontier orbital electron density has been derived from the results of semiempirical molecular orbital calculations (MNDO-PM3) as:

$$R(1) \equiv 100 \frac{f_r(1)}{-E_{HOMO}} \tag{3.239}$$

where $f_r(1)$ is the HOMO electron density on a given (sulfur) atom. This descriptor correlated the fungicidal activity of Δ^3-1,2,4-thiadiazolines as [770]:

$$\text{pEC}_{50} = 1.91 + 2.14R(1) - 0.18R(1)^2$$

$$R^2 = 0.774 \quad s = 0.21 \quad n = 17 \quad F = 24.76$$

The acute toxicity of soft electrophiles such as substituted benzenes, phenols, and anilines has been correlated with the soft electrophilicity for aromatics calculated as the MNDO average superdelocalizability as follows [771]:

$$\log\left(\frac{1}{\text{LC}_{50}}\right) = -1.49 + 0.56 \log P + 13.7S_{av}^N$$

$$R^2 = 0.81 \quad s = 0.44 \quad n = 114 \quad F = 238.7$$

The average acceptor superdelocalizability S_{av}^{N} was obtained by averaging the sum of individual superdelocalizabilities S_i^N over the atoms involved in the π bonds. The hydrophobicity (log P) and soft electrophilicity descriptors were shown orthogonal for the 114 compounds studied.

The electrophilic superdelocalizability at the 6-position of purine derivatives was found as the main factor determining the activity of these compounds against murine solid adenocarcinoma CA 755. The following QSAR equation was developed [772]:

$$\log\left(\frac{1}{C}\right) = (3.69 \pm 0.14) + (0.51 \pm 0.14)S_6^E + (0.24 \pm 0.14)\pi_6$$

$$R^2 = 0.846 \qquad s = 0.265 \qquad n = 17 \quad F = 39.67$$

where π_6 denotes the hydrophobic constant for the substituent at 6-position of purine, and S_6^E is the electrophilic superdelocalizability at the same position. It was concluded from this result that the charge transfer from this position to the biomacromolecule is an important electronic process related to the activity of compounds. Alternatively, the antitumor activity of some 3,5-disubstituted *N*-formylheteroaromatic thiosemicarbazones

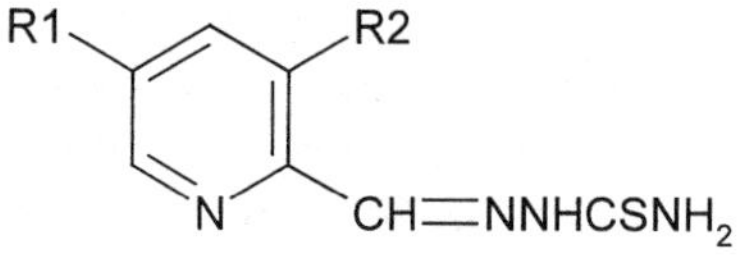

R1 R2 N CH=NNHCSNH$_2$

was correlated with the nucleophilic superdelocalizability on carbon atoms as [773]:

$$\log\left(\frac{1}{\mathrm{IC}_{50}}\right) = -2.5 + 18.6S_C^N$$

$$R^2 = 0.835 \qquad n = 10$$

In this case, the biological target was ribonucleoside diphosphate reductase, and it was speculated using the result that before the compound interacts with enzyme, it should form a complex with the Fe(II) ion.

The nucleophilic superdelocalizability on the terminal piperazine nitrogen atom in a series of compounds

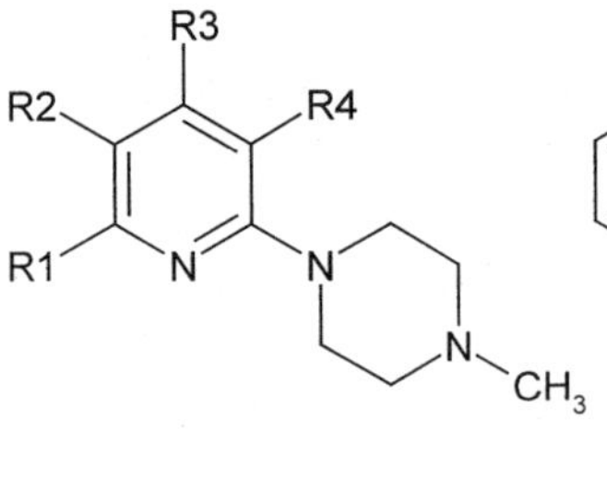

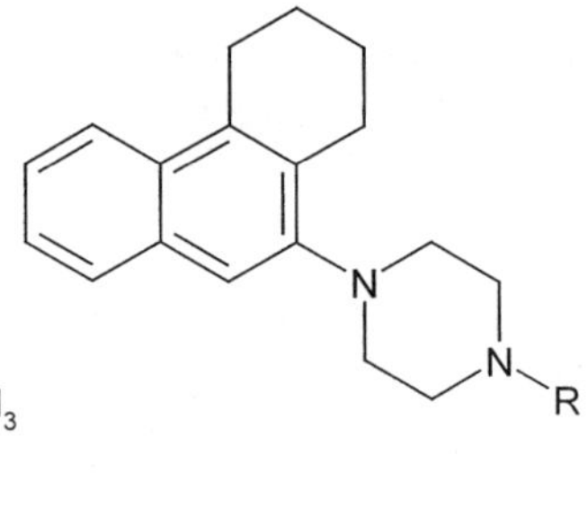

R3 R2 R4 R1 N N N CH$_3$ N N R

was found to correlate the central nervous system 5-HT_3 binding affinities as [774]:

$$pK_i = (6.763 \pm 0.207) + (271.6 \pm 0.14)S_N^N$$
$$R^2 = 0.704 \qquad s = 0.596 \qquad n = 16 \qquad F = 33.2$$

The elimination of outliers resulted in significant improvement of this correlation:

$$pK_i = (6.7612 \pm 0.159) + (334.6 \pm 36.0)S_N^N$$
$$R^2 = 0.887 \qquad s = 0.411 \qquad n = 13 \qquad F = 86.5$$

The 5-$HT1_A$ serotoninergic receptor binding affinities of a series of arylpiperazine derivatives were satisfactorily correlated using two-parameter equations [775]:

$$pK_i = (7.95 \pm 0.28) + (3.18 \pm 0.39)V_{\mathrm{dif}} - (0.569 \pm 0.149)\mu_h$$
$$R^2 = 0.8106 \qquad s = 0.193 \qquad n = 29 \qquad F = 55.64$$
$$pK_i = (10.92 \pm 0.31) + (3.76 \pm 0.39)V_{\mathrm{dif}} - (1.915 \pm 0.574)E_{NN}(\mathrm{N—H})$$
$$R^2 = 0.7934 \qquad s = 0.201 \qquad n = 29 \qquad F = 49.91$$
$$pK_i = (7.02 \pm 0.18) + (3.79 \pm 0.39)V_{\mathrm{dif}} - (1050 \pm 317)P_{\sigma\pi}(\mathrm{N—H})$$
$$R^2 = 0.7925 \qquad s = 0.202 \qquad n = 29 \qquad F = 49.66$$

where V_{dif} is the intersection of the inner and outer volumes of the ligands with respect to the reference volume of the supermolecule, μ_h denotes the AM1-calculated hybridization component of the dipole moment, E_{NN}(N—H) is the nuclear repulsion energy of the N—H bond in the protonated compound and $P_{\sigma\pi}$(N—H) is the σ-π bond order in the same compound. These results indicate that apart from the steric influence, the electronic factor connected with the piperazine nitrogen in protonated form has also important role in ligand–receptor interaction.

The quantum chemical molecular descriptors have been used for the QSPR prediction of chromatographic properties of compounds [776]. For instance, the total energy (E_T) or binding energy (E_b) for a homologous series of esters were shown to give linear correlations with the gas chromatographic retention indexes related both to nonpolar and polar stationary phases [709]. The energies E_T and E_b have been calculated using the CNDO/2 method. In another study, a two-parameter regression equation was derived that satisfactorily describes the retention index $I_{\mathrm{OV\text{-}101}}$ of structurally different polar solutes on a relatively nonpolar stationary phase [693]. The equation involves the CNDO/2-calculated total energy (E_{tot}) and the original submolecular polarity parameter, Δ:

$$I_{\text{OV-101}} = 301.88 - 11.66E_{\text{tot}} - 1016.80\Delta$$
$$R^2 = 0.865 \qquad s = 67.45 \qquad n = 22$$

The last parameter was defined as the largest difference between the electronic charges of any two atoms in the molecule. According to the authors, the total energy represents a bulk measure of the solute's ability to participate in non-specific interactions with the stationary phase, while the submolecular polarity parameter is a measure of the solute's ability to take part in polar solute–stationary phase interactions. The topological electronic index, in combination with the quantum chemically calculated total energy, has provided even a better regression equation for retention data of a diverse set of aliphatic and heterocyclic amines [258]:

$$I_{\text{OV-101}} = 160.6 - 13.39E_{\text{tot}} - 149.6T^{E}$$
$$R^2 = 0.962 \qquad s = 35.29 \qquad n = 22$$

Good correlations have been obtained with nonpolar stationary phase retention times, t_R, and response factors (RF) for a set of 152 diverse organic compounds using a combination of conventional and AM1-calculated quantum chemical descriptors [777]:

$$t_R = 26.5 - 6.9N_{\text{CH}} - 0.87\frac{S_{\text{tot}}}{N} + 0.046\alpha + 0.019\text{MW}$$
$$- 21.55V_{\text{H,min}} + 0.93\text{AOP}_{\text{max}}$$
$$R^2 = 0.959 \qquad s = 0.515 \qquad n = 152$$
$$\text{RF} = -2.33 + 10.96\text{RW}_{\text{Ceff}} - 0.0003E^{T}_{\text{EE}} - 1.16\text{RN}_{\text{Ceff}}$$
$$-0.21b_{\text{C,min}} + 3.22V_{\text{H,min}} - 0.03\mu_h$$
$$R^2 = 0.892 \qquad s = 0.054 \qquad n = 152$$

where N_{CH} is the relative number of C—H bonds, S_{tot} is the total entropy of the molecule at 300 K, N is the number of atoms in the molecule, α is the polarizability, MW is the molecular weight, $V_{\text{H,min}}$ is the minimum valency of a H atom, AOP_{max} is the maximum atomic orbital electronic population, RW_{Ceff} is the relative weight of "effective" C atoms, E^{T}_{EE} is the total molecular one-center electron–electron repulsion energy, RN_{Ceff} is the relative number of "effective" C atoms, μ_h is the total hybridization component of the molecular dipole, and $b_{\text{C,min}}$ is the minimal total bond order of a C atom. The most significant quantum chemical descriptors correlated with the retention times were the molecular polarizability (characterizing molecular bulk and dispersion interaction with the media) and the minimum valency of a hydrogen atom (related to hydrogen-bonding interactions). The gas-chromatographic response factors were correlated with the minimum total bond order of a carbon atom and the total molecular one-center electron–electron repulsion. The last descriptor reflects the inclination of the thermally cracked products to undergo "chemiionization" in the flame ionization detector of the chromatograph.

The involvement of quantum chemical descriptors has enabled obtaining successful QSPR correlations for the gas-chromatographic retention data, boiling points, melting points, and flash points of substituted pyridines [778,779]. For example, the following six-parameter correlation equation was derived for boiling points:

$$\text{bp} = -247.4 + 0.278G_i + 19.25\mu_{\text{char}} + 1713.2S_{\text{FHA}} + 136.1b_{\text{N,min}} - 26.62S_{\text{RN}} + 2503.4\bar{f}^{N}_{r,\text{N}}$$

$$R^2 = 0.943 \qquad s = 14.5 \qquad n = 152 \qquad F = 214.5$$

where G_i is the gravitation index, μ_{char} is the total point-charge component of the molecular dipole, S_{FHA} is the fractional hydrogen acceptors surface area, $b_{\text{N,min}}$ is the minimal total bond order of a nitrogen atom, S_{RN} is the relative negatively charged surface area, and $\bar{f}^{N}_{r,\text{N}}$ is the average nucleophilic reactivity index for a nitrogen atom. All the descriptors involved, except the gravitation index, were calculated by the AM1 method. Charged surface areas were calculated from net atomic charges provided by AM1 method. The flash point (fp) for a limited set of substituted pyridines was described by the following best QSPR equation [779]:

$$\text{fp} = 1968.0 + 1.971\alpha + 344.5\text{FHASA} - 38.75\,{}^{1}\text{CIC}_{\text{av}} - 738.9V_{\text{min}}(N) + 181.5R(YZ) - E^{\text{min}}_{\text{ee}}(\text{C—N})$$

$$R^2 = 0.837 \qquad s = 16.7 \qquad n = 126 \qquad F = 97.6$$

where α is the α polarizability of the molecule, FHASA is the fractional hydrogen-bonding acceptor surface area, ${}^{1}\text{CIC}_{\text{av}}$ is the average complementary information content of first order, $V_{\text{min}}(N)$ denotes minimal valency of a N atom in the molecule, $R(YZ)$ has been calculated as the ratio of the shadow of the molecule to the rectangle defined by the maximum dimensions of the molecule in YZ plane, and $E^{\text{min}}_{\text{ee}}(\text{C—N})$ is the minimal electron–electron repulsion for a C—N bond.

A general QSPR treatment of densities has been carried out for a data set incorporating 303 individual structures from a wide cross section of classes of organic compounds (containing C, H, N, O, S, F, Cl, Br, and I) [780]. For the normal densities of liquids, ρ^{20}, the following two-parameter correlation was obtained:

$$\rho^{20} = -(0.398 \pm 0.022) + (0.942 \pm 0.008)\rho_R + (0.0894 \pm 0.0040)E_{\text{el}}$$

$$R^2 = 0.9749 \qquad R^2_{\text{cv}} = 0.9736 \qquad s = 0.046 \qquad n = 303 \qquad F = 5937.3$$

The intrinsic density ρ_R was calculated as the ratio of the molecular weight M_w and molecular volume M_{vdW} calculated from the overlapped van der Waals spheres of the atoms in the molecule. The total molecular electrostatic interaction E_{el} involves the different interaction energies, and this descriptor is calculated as:

$$E_{\text{el}} = \frac{[E_{\text{ee}}(AB) + E_{\text{ne}}(AB) + E_{\text{nn}}(AB)]}{N} \tag{3.240}$$

where $E_{\text{ee}}(AB)$ denotes electron–electron repulsion between atoms A and B, and $E_{\text{ne}}(AB)$ and $E_{\text{nn}}(AB)$ are the energies of nuclear–electron attraction and nuclear–nuclear repulsion, respectively. The first descriptor accounts successfully for most of the variation of hydrocarbon liquids. The electrostatic energy quantifies the volume change due to the interatomic electrostatic attraction and repulsion.

The quantum chemically calculated descriptors have been useful in the QSPR studies of complex technological processes. For example, successful QSPR equations have been developed for two characteristics of rubber vulcanization, the maximum rate of cure (mxr) and the time to scorch (ts2). For a set of 23 Zn complexes of sulfenamides and sulfenimides, the following regressions were obtained [781]:

$$\begin{aligned} \text{mxr} = {} & -(198.0 \pm 16.2) - (0.0691 \pm 0.0052)E_{\text{ee}}^{\max}(\text{S—Zn}) \\ & + (0.259 \pm 0.023)E_{\text{ee}}^{\max}(\text{C—N}) + 28.7 \pm 1.90)E_{\text{exch}}^{\max}(\text{N—H}) \\ & + (8.02 \pm 2.09)^0\text{CIC}_{\text{av}} \\ & \qquad R^2 = 0.9673 \qquad R_{\text{cv}}^2 = 0.9490 \qquad n = 23 \end{aligned}$$

$$\begin{aligned} \text{ts2} = {} & (88.5 \pm 7.00) - (4.68 \pm 0.47)E_{\text{exch}}^{\max}(\text{C—C}) \\ & - (6.09 \pm 3.51)E_{\text{exch}}^{\max}(\text{N—H}) - (169.0 \pm 19.2)V_{\min}(N) \\ & + (0.0116 \pm 0.0021)\text{PNSA2} \\ & \qquad R^2 = 0.9617 \qquad R_{cv}^2 = 0.9403 \qquad n = 23 \end{aligned}$$

where $E_{\text{ee}}^{\max}(\text{S—Zn})$ and $E_{\text{ee}}^{\max}(\text{C—N})$ are the maximum electron–electron repulsion energies for the S—Zn and C—N bonds, respectively, $E_{\text{exch}}^{\max}(\text{N—H})$ and $E_{\text{exch}}^{\max}(\text{C—C})$ are the maximum exchange energies for the N—H and C—C bonds, respectively, $^0\text{CIC}_{\text{av}}$ is the average complementary information content of the zeroth order, and PNSA2 denotes the total charge-weighted partial negative surface area of the molecule. The implications of each descriptor involved were discussed in view of the possible mechanism of vulcanization [781]. The quantum chemical descriptors have also played an important role in the QSPR presentation of the refractive indices of organic compounds and polymers [782,783], and in correlation of glass transition temperatures of polymers [784,785].

3.7 SOLVATIONAL DESCRIPTORS

The solvent and environmental effects on the chemical structure and properties can be also accounted for using theoretically calculated molecular descriptors within the framework of QSAR/QSPR methodology. This approach proceeds from the construction of certain physical models for the description of different solvation phenomena. The respective mathematical apparatus is then developed and applied for the calculation of a variety of properties of molecules in solution used thereafter as the molecular descriptors in the QSAR/QSPR treatment of experimental data. The solvation process can be considered as consisting of several steps. First, a cavity has to be created in the solvent to accommodate the solute molecule. Second, the solute is immersed into the cavity. This process is accompanied by the electronic and orientational polarization of the solvent and the dispersion interaction between the solute and solvent molecules. Finally, hydrogen bonds or donor–acceptor complexes are formed between the solute and solvent molecules. Theoretically, the cavity formation effects can be treated using the statistical mechanical methods whereas the polarization and dispersion effects have to be described using the combination of statistical and quantum mechanics. A proper description of the hydrogen bonding in condensed media requires also the use of advanced quantum mechanical methods. Hence, we proceed with an overview of physical models of solvation that can be used as the basis for the development of theoretical solvent effect descriptors.

The following definition of the Gibbs free energy of solvation has been given [786]:

$$\Delta G_{\text{sol}} = W(\text{M}|\text{S}) + p\,\Delta V + RT\left(\frac{q_{\text{rot},g}q_{\text{vib},g}}{q_{\text{rot},s}q_{\text{vib},s}}\right) - RT\ln\left(\frac{n_{M,g}\Lambda_{M,g}^{3}}{n_{M,s}\Lambda_{M,s}^{3}}\right) \tag{3.241}$$

where $q_{\text{rot},g}$, $q_{\text{vib},g}$, $q_{\text{rot},s}$, and $q_{\text{vib},s}$ are the microscopic partition functions of rotation and vibration of the solute in the gas phase and in the solution, respectively; $n_{M,g}$ and $n_{M,s}$ are the number densities of the solute in the corresponding phases; and $\Lambda_{M,g}$ and $\Lambda_{M,s}$ are the respective translational partition functions. The last three terms in this equation are usually small and can be neglected if no significant changes are expected in the rotational and vibrational energy levels of molecules when transferred from the gas phase to solution. However, if the rotational, vibrational, or translational motions of the solute molecule are substantially restricted in a condensed medium, these terms may become significant [787]. Such restrictions may arise from the specific bonding with the solvent or because of the formation of solvational cages around the solute molecule. In principle, each of the two last terms in Eq. (3.241) can be then computed and applied as the solvational molecular descriptor.

In most cases, the free energy of solvation has been thus approximated by the coupling work of the solute with solvent, expressed by the following equation [787]:

$$W(\mathrm{M}|\mathrm{S}) = \Delta G_{\mathrm{el}} + \Delta G_{\mathrm{cav}} + \Delta G_{\mathrm{disp}} + \Delta G_{\mathrm{rep}} + \Delta G_{\mathrm{HB}} \tag{3.242}$$

This work is related to the transfer of a solute M from a fixed position in the ideal-gas phase to a fixed position in solvent S, at constant temperature T, pressure p, and the chemical composition. In the last equation, ΔG_{el} denotes the solvent electrostatic polarization term, ΔG_{cav} is the term representing the work needed to create the cavity for the solute in the liquid, ΔG_{disp} and ΔG_{rep} are the solute–solvent dispersion and exchange–repulsion interaction energy, respectively, and ΔG_{HB} is the specific hydrogen-bonding energy between the solute and solvent. A variety of theoretical models has been developed for each individual term in the last equation. The corresponding mathematical expressions can be effectively used to calculate the solvational molecular descriptors for the QSAR/QSPR treatment.

The main contribution to the macroscopic solvent effects on the polar molecules and ions rises from the electrostatic interaction between the charge distribution of the solute and the reaction field created in the solvent by this charge distribution. The respective solvent electrostatic polarization term can be calculated using either classical electrostatics or quantum theory [788]. The classical theory of electrostatic solvation originates from Kirkwood and Onsager [789–791]. By assuming the statistical averaging over all solute–solvent intermolecular distances and configurations, the solvent can be represented as a homogeneous polarizable dielectric continuum. The energy of electrostatic interaction between the charge distribution of a solute molecule and the dielectric medium is given by the following general equation [792]:

$$E_{\mathrm{el}} = -\frac{1}{8\pi}\int_V \mathbf{E}_S \cdot \mathbf{E}_0(\varepsilon - 1)\, dV \tag{3.243}$$

where $\mathbf{E}_0$ is the electrostatic field created by the molecular charge distribution in vacuo and $\mathbf{E}_S$ is the modified field in the presence of dielectric medium. According to Eq. (3.243), the dielectric constant ε of the medium is a continuous function of the space coordinates, and therefore both the interior of the molecule and the surrounding medium are treated by the same equation. Due to mathematical difficulties, there is, however, no analytical solution to the integral in the last equation in the general case. Moreover, even the numerical integration over the full space V presents a difficult mathematical task. Therefore, the electrostatic equation has to be simplified by using some approximations. In most cases, the volume integral in Eq. (3.243) is transformed into a surface integral over some boundary S, using the Gauss divergence theorem.

The respective electrostatic interaction energy between the charge distribution inside the closed surface and the homogeneous medium with dielectric constant ε outside the surface is then given as:

$$E_{\mathrm{el}} = \frac{\varepsilon - 1}{8\pi} \int_S \Phi_S \cdot \mathbf{E}_0 \mathbf{n} \, dS \tag{3.244}$$

where S is the boundary surface, $\mathbf{n}$ the unit outward normal vector on S, and the reaction potential Φ_S corresponds to the field $\mathbf{E}_S = -\mathrm{grad}\Phi_S$. It has been assumed that the dielectric constant inside the cavity is equal to that of vacuum and outside the cavity has a constant value, usually the macroscopic dielectric constant of the medium studied. Equation (3.244) has been solved in the case of different shapes of molecular surfaces. In the framework of the classical Kirkwood and Onsager reaction field theory [788–790], the solute molecule is represented by a set of point charges fixed inside of sphere of a given radius. The electrostatic equation (3.244) can then be solved analytically by applying appropriate boundary conditions inside and outside the sphere. The resulting energy of the electrostatic interaction between the solute charge distribution and the surrounding dielectric medium is given by the following infinite expansion:

$$E_{\mathrm{el}} = \frac{1}{2} \sum_{i,j} e_i e_j \sum_{1=0}^{\infty} \left[\frac{(l+1)(1-\varepsilon)}{\varepsilon(l+1)+l} \right] \frac{\mathbf{r}_i^l \mathbf{r}_j^l}{a_0^{2l+1}} P_l(\cos\theta_{ij}) \tag{3.245}$$

where e_i and e_j are the charges inside the sphere at positions $\mathbf{r}_i$ and $\mathbf{r}_j$, respectively, and Θ_{ij} is the angle at the center of the sphere between vectors $\mathbf{r}_i$ and $\mathbf{r}_j$, and $P_1(\cos\theta_{ij})$ are the Legendre polynomials of lth order. The summation is carried out over all nuclei and electrons of the solute. By expressing the Legendre polynomials as the products of the respective spherical harmonics of order m $(-l \leq m \leq l)$, the last equation can be rewritten as follows [793]:

$$E_{\mathrm{el}} = -\frac{1}{2} \sum_{l=0}^{\infty} \sum_{m=-l}^{l} \mathbf{R}_l^m \cdot \mathbf{M}_l^m \tag{3.246}$$

where

$$\mathbf{R}_l^m = f_l \mathbf{M}_l^m \tag{3.247}$$

and

$$f_l = \frac{(l+1)(\varepsilon-1)}{(l+1)\varepsilon+l} \frac{1}{a_0^{2l+1}} \tag{3.248}$$

In these equations, $\mathbf{M}_l$ and $\mathbf{R}_l$ represent the lth-order electrical momentum of the compound and the corresponding reaction field created in the dielectric medium. The first term (l = 0) in the infinite expansion (3.245) gives the interaction of the ionic charge of the solute with the respective reaction field

$$E_{\text{born}} = \frac{1 - \varepsilon}{2\varepsilon} \frac{Q^2}{a_0} \tag{3.249}$$

where Q is the value of the ionic charge (Born term). The next term (l = 1) corresponds to the interaction of the total dipole moment of the molecule with the corresponding reaction field

$$E_{\text{Onsager}} = \frac{(1 - \varepsilon)}{(2\varepsilon + 1)} \frac{\boldsymbol{\mu}^2}{a_0^3} \tag{3.250}$$

where $\boldsymbol{\mu}$ is the dipole moment of the solute (Onsager dipolar term). Both these terms are applicable as the solvational descriptors. In fact, we have seen above that the dielectric constant dependent factors in these equations have been used as the empirical solvent effect scales. However, depending on the charge distribution in the system studied, the interaction of higher electrical moments with the corresponding reaction field may become also significant. Hence, the terms corresponding to higher electrical moments of order 2^l (quadrupole, octupole, hexadecapole, etc.) can also be calculated and applied as molecular descriptors.

The shape of the solute molecule may deviate considerably from the perfect sphere. The next approximation is the ellipsoidal cavity for which the closed formulas are still available for the calculation of the charge and dipolar terms of the electrostatic interaction with the reaction field. The charge term for an ellipsoidal cavity with main semiaxes $\boldsymbol{a}$, $\boldsymbol{b}$, and $\boldsymbol{c}$ is given as:

$$E_{\text{Born}}^{\text{ell}} = \frac{1 - \varepsilon}{2\varepsilon} \frac{Q^2}{\boldsymbol{abc}} \tag{3.251}$$

whereas in the respective dipolar term [791]

$$E_{\text{Onsager}}^{\text{ell}} = \boldsymbol{R\mu} \tag{3.252}$$

the reaction field $\boldsymbol{R}$ is presented by a tensor

$$\boldsymbol{R} = \begin{pmatrix} R_a & 0 & 0 \\ 0 & R_b & 0 \\ 0 & 0 & R_c \end{pmatrix} \tag{3.253}$$

with the diagonal terms R_i defined as:

$$R_i = \frac{3A_i(1 - A_i)(1 - \varepsilon)\mu_i}{abc[\varepsilon + (1 - \varepsilon)A_i]} \tag{3.254}$$

where A_i denote the standard ellipsoidal shape factor integrals [791], and μ_i are the dipole moment components along the main semiaxes, a, b and c, of the ellipsoid. Different definitions can be used for the orientation of the ellipsoid in the molecular frame. For instance, the ellipsoid semiaxes may be taken collinear with the axes of the solute dipole polarizability tensor and their lengths proportional to the respective eigenvalues [794]. Another definition proceeds from the inertia tensor of the van der Waals solid, that is, a solid or uniform density composed of interlocking van der Waals spheres [795]. The ellipsoidal surface has been defined also in terms of the best fitting of a given molecular electrostatic isopotential surface [796].

For a cavity of arbitrary shape, the electrostatic energy of interaction between the charge distribution inside the cavity and the respective reaction field can be calculated numerically. According to the classical electrostatics, the potential Φ_S can be calculated at any point of the space in terms of the apparent charge distribution σ on the cavity surface. The latter is composed from two terms,

$$\Phi_S = \Phi_M + \Phi_\sigma \tag{3.255}$$

the first of which (Φ_M) corresponds to the electrostatic potential created by the charge distribution of the solute and the second (Φ_σ) represents the reaction potential of the solvent. The reaction potential can be expressed using the apparent charge distribution on the surface of the cavity $\sigma(\mathbf{s})$ as:

$$\Phi_\sigma(\mathbf{r}) = \int_S \frac{\sigma(\mathbf{s})}{|\mathbf{r} - \mathbf{s}|} d^2\mathbf{s} \tag{3.256}$$

where S is the cavity surface and the $\mathbf{s}$ vector defines a point on S. The last integral can be approximated numerically using the boundary element method:

$$\Phi_\sigma(\mathbf{r}) = \sum_k \frac{q_k}{|\mathbf{r} - \mathbf{s}_k|} \tag{3.257}$$

where the charge q_k is localized at some internal point of the surface, $\mathbf{s}_k$. This charge has been usually found as the product of the charge density at a given point and the respective surface element

$$q_k = \Delta S_k \sigma(\mathbf{s}_k) \tag{3.258}$$

The success of the boundary element method depends critically upon the selection of appropriate boundary surface for the solute cavity [797,798].

The energy of electrostatic solvation can also be calculated using the quantum chemical extension of the electrostatic polarization theory of solvation. The respective molecular descriptors should be more adequate, particularly in the systems having substantial charge redistribution due to solvent reaction field. Within this approach, the interaction energy between the quantum mechanically treated solute and the surrounding dielectric continuum is accounted for by additional terms in the solute Hamiltonian. The self-consistent reaction field (SCRF) method has been frequently used to solve the respective Schrödinger equation for a solute molecule in a polarizable dielectric medium [799–803]. According to this method, the electrostatic interaction energy, E_{el}, between the solute and its reaction field is given as the following difference:

$$E_{el} = E_{rf} - E_0 \tag{3.259}$$

where E_0 and E_{rf} are the quantum chemically calculated total energies for the isolated solute molecule and for the solute molecule in the dielectric medium, respectively. Proceeding from the classical expression for the electrostatic solvation energy of a solute molecule in a dielectric medium in the dipole–dipole interaction approximation, the total energy of the solute is given as follows:

$$E_{rf} = E^0 - \tfrac{1}{2}\,\Gamma[\langle\psi|\hat{\boldsymbol{\mu}}|\psi\rangle\langle\psi|\hat{\boldsymbol{\mu}}|\psi\rangle + 2\boldsymbol{\mu}_{nuc}\langle\psi|\hat{\boldsymbol{\mu}}|\psi\rangle + \boldsymbol{\mu}^2_{nuc}] \tag{3.260}$$

where $E^0 = \langle\psi|\hat{\mathbf{H}}^0|\psi\rangle$, $\hat{\mathbf{H}}^0$ is the Hamiltonian for the reaction field unperturbed solute molecule and ψ is the molecular electronic wave function. In the last equation, $\hat{\boldsymbol{\mu}}$ denotes the quantum mechanical operator of the total dipole moment of the molecule

$$\hat{\boldsymbol{\mu}} = \hat{\boldsymbol{\mu}}_{el} + \boldsymbol{\mu}_{nuc} = \sum_{i=1}^{n} e_i\hat{\mathbf{r}}_i + \sum_{a=1}^{N} Z_a\mathbf{R}_a \tag{3.261}$$

The two terms correspond to the electronic dipole moment operator and the nuclear component of the dipole moment (for nuclei with charges Z_a at fixed positions $\mathbf{R}_a$), respectively. In Eq. (3.260), the factor Γ describes the intensity of the reaction field response of the solvent. This factor depends both on the dielectric constant of the solvent and the size and shape of the solute cavity. In the case of the simplest, spherical model for the solute cavity

$$\Gamma = \frac{2(\varepsilon - 1)}{(2\varepsilon + 1)a_0^3} \tag{3.262}$$

where a_0 is the cavity radius and ε is the dielectric constant of medium. In the case of the ellipsoidal cavity with principal axes $\boldsymbol{a}$, $\boldsymbol{b}$, and $\boldsymbol{c}$, the factor Γ becomes a tensor (3.253). However, the general form of Eq. (3.260) remains the same. In the case of ionic solutes, the factor Γ has to be modified to maintain the gauge invariance of Eq. (3.260). For positive ions (cations) with the ionic charge Q

$$\Gamma_{\text{cation}} = \Gamma \frac{n_e}{n_e + Q} \tag{3.263a}$$

and for anions with the negative charge $-Q$

$$\Gamma_{\text{anion}} = \Gamma \frac{n_e - Q}{n_e} \tag{3.263b}$$

where n_e is the number of electrons in the ionic solute. In addition, the ionic solvation term, calculated according to the classical Born formula (3.249), has to be accounted for in the electrostatic solvation energy for ionic solutes.

$$E_{\text{el}} = -\frac{1}{2}\Gamma\langle\hat{\boldsymbol{\mu}}\rangle^2 - \frac{(\varepsilon - 1)Q^2}{2\varepsilon a_0} \tag{3.264}$$

The expectation value of the dipole moment in Eq. (3.260) depends, of course, on the molecular wave function in the given environment. Therefore, due to the interaction of the solute electron distribution with the solvent, it has to be calculated during each self-consistent field cycle of solving the respective Schrödinger equation. The self-consistent reaction field (SCRF) method to find the solution of this equation was first developed by Tapia and Goscinski [799]. By describing the electronic wave function of a solute molecule Ψ as a proper spin-projected antisymmetrized product of molecular (or atomic) orbitals,

$$\Psi = O_s[\phi_1, \ldots, \phi_n] \tag{3.265}$$

and recalling that the dipole moment operator is an one-electron operator, the following orbital equations can be obtained [804]

$$\mathbf{f}(k)\phi_i(k) = \varepsilon_i\phi_i(k) \tag{3.266}$$

with

$$\mathbf{f}(k) = \mathbf{f}_0(k) - \Gamma\langle\Psi|\hat{\boldsymbol{\mu}}|\Psi\rangle\hat{\boldsymbol{\mu}}_{\text{el}}(k) \tag{3.267}$$

where $\mathbf{f}_0(k)$ is the usual Fock operator for the isolated molecule, ε_i is the energy of molecular orbital ϕ_i, and $\hat{\boldsymbol{\mu}}_{\text{el}}(k)$ is the electronic part of the dipole moment operator. The last equation is solved iteratively, using the usual self-consistent field (SCF) procedure and the expectation value of the total dipole moment from the previous SCF cycle. The SCRF approach has been further developed by Tapia and others [805–808] into a form of the generalized self-consistent reaction field theory (GSCRF). According to this theory, the effective Hamiltonian of the solute in the solvent has the following form:

$$\hat{\mathbf{H}}_s = \hat{\mathbf{H}}_s^0 + \int d\mathbf{r}\ \Omega_s(\mathbf{r}) \left[V_m^0(\mathbf{r}) + \int d\mathbf{r}'\ G(\mathbf{r}, \mathbf{r}')\Omega_s(\mathbf{r}') \right] \quad (3.268)$$

where $\Omega_s(\mathbf{r})$ is the solute charge density operator defined as:

$$\Omega_s(\mathbf{r}) = -\sum_i \delta(\mathbf{r} - \mathbf{r}_i) + \sum_a Z_a \delta(\mathbf{r} - \mathbf{R}_a) \quad (3.269)$$

and $\mathbf{r}_i$ stands for the ith electron position vector operator, $\mathbf{R}_a$ is the position vector of the ath nucleus with the charge Z_a in the solute, and $\delta(\mathbf{r})$ is the Dirac delta function. The first term in square brackets in Eq. (3.268), $V_m^0(\mathbf{r})$, represents the electrostatic potential created by the solvent in the absence of the solute, and the second, integral term corresponds to the reaction potential response function of the polarizable solvent. Together these terms produce the reaction field potential applying to the solvent molecule in the polarizable dielectric medium. Notably, an important part of the Hamiltonian (3.268) is the solute charge density that can be represented using different approximations of which the multipolar expansion has been mostly applied. By using the distributed multipole model, it is possible to obtain the GSCRF equations for the molecules of a complex shape [805]. A crucial step in the general self-consistent reaction field procedure is the estimation of the solvent charge density needed to obtain the response function $G(\mathbf{r}, \mathbf{r}')$ and the reaction potential. The use of Monte Carlo or molecular dynamics simulations of the system consisting of the solute and surrounding solvent molecules has been proposed to find the respective solvent static and polarization densities. The GSCRF methodology has been widely used for the modeling of large biomolecules in aqueous solutions [809–812].

In the quantum chemical calculation of the electrostatic solvation energy, the solvent part of the solute–solvent system can also be treated not as a dielectric continuum but as the distribution of the polarizable point dipoles, interacting with each other. The respective direct reaction field (DRF) Hamiltonian of the solute–solvent system is given by the following formula [813–815]:

$$\hat{\mathbf{H}}_{\mathrm{DRF}} = \hat{\mathbf{H}}^0 - \frac{1}{2} \sum_{i,j} \sum_{p,q} \mathbf{F}_{\mathbf{ip}}^{+} \boldsymbol{\alpha}_{\mathbf{pq}} \mathbf{F}_{\mathbf{jq}} \quad (3.270)$$

where indexes i and j correspond to the solute particles (electrons and nuclei) and p and q run over the external polarizable points; $\mathbf{F}_{\mathbf{ip}}$ is the field created by ith particle at the pth point and factor $\boldsymbol{\alpha}_{\mathbf{pq}}$ represents the induced dipole created at qth point by a field applied at pth point. The respective Schrödinger equation can be solved immediately, without the iterative adjustment of the solvent charge distribution and the respective reaction field potential. It has been argued [787] that the use of operator $\hat{\mathbf{H}}_{DRF}$ does not correspond to the correct introduction of electrostatic polarization effects that are properly defined only when the average distributions are used. While the GSCRF approach uses the average reaction field model, the DRF method proceeds from the direct reaction field obtained as the linear solute–solvent interaction operator, proportional to the square of the electric field operator. The additional energy contributions can be interpreted as due to the dispersion interaction between the solute and solvent molecules [816]. The DRF method has been combined with the continuum approach by dividing the space around the solute into two regions. The immediate surrounding of the solute is treated using the direct reaction field whereas the more distant interactions are described by the macroscopic dielectric properties of the solvent [817,818].

The polarizable continuum model (PCM) for the calculation of electrostatic solvation effects is based on the classical electrostatic boundary element method discussed above [Eq. (3.257)] [819–826]. Within the PCM model, the solute energy in a polarizable medium is calculated by solving the respective Schrödinger equation:

$$(\hat{\mathbf{H}}^0 + \hat{V}_{PCM})\Psi = E\Psi \tag{3.271}$$

where the reaction field potential $\hat{V}_{PCM}$ is expressed as the sum over the small surface elements ΔS_k on the boundary between the solute and solvent:

$$\hat{V}_{PCM} = \sum_k \frac{q_k}{|\mathbf{r} - \mathbf{s}_k|} \tag{3.272}$$

The surface charges are usually defined as

$$q_k = \Delta S_k \sigma(\mathbf{s}_k) \tag{3.273}$$

where $\sigma(\mathbf{s}_k)$ is the charge density at a given point. The charges on the boundary surface are created by the electrostatic polarization of the dielectric medium due to the potential derived from the charge distribution of the solute and from other (induced) charges on the surface. The induced surface charge is evaluated iteratively at each step of the SCF procedure to solve the Schrödinger equation (3.271). It has been noticed that a simultaneous iteration of the surface charge with the Fock procedure reduces substantially the computation time without the loss in the precision of calculations [827].

The advantage of the PCM method is that it is applicable to the solute cavity of practically any shape in the solution. However, it is not clear how precisely should the molecular cavity be defined bearing in mind the classical (quasi-macroscopic) representation of the solvent. It is difficult to imagine that the solvent, for example, the water molecules, can produce the electrical polarization corresponding to the statistically average distribution in the macroscopic liquid at infinitely small regions on the cavity surface. However, it is conceivable that larger chemical groups in the molecules may possess their own reaction field created by their charge distribution and the reaction fields of other groups in the solute molecule. A respective multicavity self-consistent reaction field (MCa SCRF) has been proposed [828] for the description of rotationally flexible molecules in condensed dielectric media. The COSMO (conductor-like screening model) approach has also been developed to calculate the electrostatic solvation energy for more realistic cavity shapes of the solutes [829,830]. This approach proceeds from the assumption that the solute has ideal dielectric screening by the surrounding conductor-like medium. This corresponds to the infinite value of the dielectric constant of the medium and should be, therefore, primarily applicable for the high dielectric constant solvents (e.g., water).

Finally, the virtual charge methods can be used for the calculation of the electrostatic polarization energy of the solute–solvent interaction. In the formulation of this model by Klopman [831], an imaginary particle called solvaton was associated to each atom A of the solute. The charge of the solvaton, Q_A^s, was taken equal in magnitude but opposite in sign to the Mulliken partial charge q_A on the atom A. The interaction energy was then calculated between all partial charges and solvatons. However, the interactions between the partial charges themselves and solvatons themselves were neglected. The virtual charge approach has been incorporated into the SCF procedure by modifying the molecular Hamiltonian $\hat{\mathbf{H}}$ as [832,833]

$$\hat{\mathbf{H}} = \hat{\mathbf{H}}_0 - \frac{\varepsilon - 1}{\varepsilon} \sum_{s=1}^{N} \sum_{n=1}^{N} \frac{Q_s Z_n}{r_{sn}} \tag{3.274}$$

where $\hat{\mathbf{H}}_0$ is the Hamiltonian for the isolated molecule, ε is the dielectric permittivity of the solvent, Q_s is the charge of the sth solvaton and Z_n—the charge of the nth atomic nucleus in the molecule. The distance between the solvaton and the atomic nucleus is denoted as r_{sn} and the summation in the last formula is carried out over all pairs of nuclei and solvatons.

A similar generalized Born theory has been used in the AM1-SMx solvational models [834–838]. In the original AM1-SM1 method, the total molecular solvation energy was divided into the solute–solvent electrostatic and inductive polarization terms, the standard-state free energy of cavity creation, the solute–solvent dispersion interaction, and an empirical part of the free energy of nuclear motion. The polarization term was described using the generalized Born formula as:

$$G_p = -\frac{1}{2}\left(1 - \frac{1}{\varepsilon}\right) \sum_{k=1}^{\mathrm{N}} \sum_{k'=1}^{\mathrm{N}} q_k q_{k'} \gamma_{kk'} \tag{3.275}$$

where the double summation is performed over all atomic partial charges q_k in the solute molecule, ε is the relative dielectric permittivity of the solvent, and

$\gamma_{kk'}$ is the Coulomb integral between two centers k and k', parameterized for the interactions with the solvent.

A number of methods have been developed that account simultaneously for the solvent effects and electron correlation effects on the molecular electronic structure. Those include the multiconfigurational self-consistent reaction field (MC SCRF) method [839–842] and various applications of the quantum mechanical perturbation theory [843–847].

One of the most critical aspects of electrostatic polarization theory of solvation is the applicability of the bulk dielectric constant in the respective equations. The validity of this assumption has been examined by the theoretical calculations of the dielectric medium effects on the tautomeric equilibrium between acetylacetone (**1**) and its enol form (**2**)

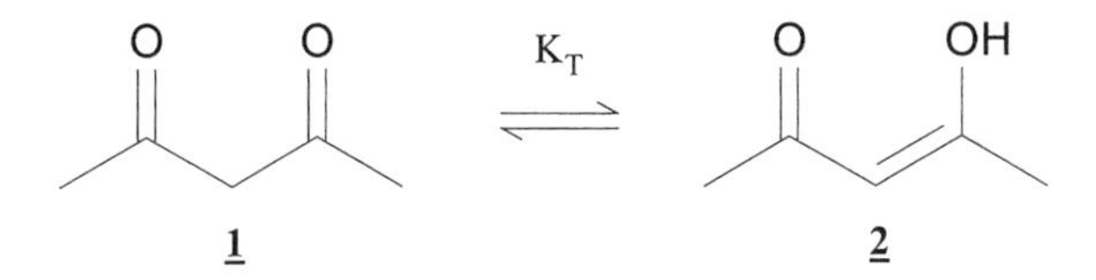

The experimental equilibrium constant of this reaction has been precisely measured using bromatometry and NMR spectroscopy in the dioxane–water mixtures having a large variation of the dielectric constant, from pure dioxane (ε = 2.209) to water (ε = 78.4) [848]. It is obvious that the solute–solvent dispersion energy, the free energy of cavity formation, and the hydrogen-bonding energy should be very similar for the two tautomeric forms. Therefore, the theoretical tautomeric equilibrium constant can be expressed through the following equation:

$$\log K_T = \frac{\Delta G_T}{2.3RT} \approx \frac{E_{\mathrm{el}}}{2.3RT} = \frac{\delta\ \Delta H}{2.3RT}$$

where ΔG_T is the free energy of the tautomeric equilibrium and $\delta\ \Delta H$ is the difference in the calculated heats of formation of the tautomeric species participating in this equilibrium. An excellent agreement between the calculated and experimental log K_T values in different media was obtained using the self-consistent reaction field method [849]. Within the spherical cavity model

$$\log K(\mathrm{calc}) = 0.004 + 0.911 \log K(\mathrm{exp})$$

$$R^2 = 0.979 \qquad s = 0.081 \qquad n = 11$$

The slope of this regression is close to one that confirms the applicability of the macroscopic dielectric constant of the medium.

Another component of the free energy of solvation is the dispersion energy between the solute and solvent. This energy can also effectively be used as the theoretical molecular descriptor that reflects the solvent effects on chemical reactivity and physical properties of compounds in solutions. Several approaches have been developed for the calculation of solute–solvent dispersion energy in condensed media. The pair-potential approach is based on the discrete representation of the pairs of solvent and solute molecules or some fragments of them. The dispersion potential is expressed as a truncated asymptotic expansion in powers of $1/r$, the reciprocal of the distance between the interacting entities [787]:

$$U_{ms}(\text{disp}) = \sum_{k=6,8,10} d_{ms}^{(k)} r_{ms}^{-k} \tag{3.276}$$

where the indexes m and s correspond to the structural entities (atoms, bonds, chemical groups) belonging to the solute and solvent molecules, respectively. The expansion coefficient $d_{ms}^{(6)}$ may be approximated, for instance, by the well-known London formula

$$d_{ms}^{(6)} = -\frac{3}{2}\alpha_m \alpha_s \frac{\bar{I}_m \bar{I}_s}{\bar{I}_m + \bar{I}_s} \tag{3.277}$$

where α_M and α_S are the isotropic polarizabilities for interacting molecules, and $\bar{I}_m$ and $\bar{I}_s$ are the mean excitation energies of these molecules. In the first approximation, the dispersion potential is limited to the dipole–dipole term, and the mean excitation energies are approximated by the respective ionization potentials for the solute and solvent molecules. Thus, when the solute molecule is surrounded by a small cluster of solvent molecules, the dispersion energy is given by the following formula [850]:

$$E_{\text{disp}} = -\frac{x}{4}\frac{\bar{I}_M \bar{I}_S}{\bar{I}_M + \bar{I}_S} \sum_{u=1}^{B_M} \sum_{v=1}^{B_S} \{r_{uv}^{-6}\,\text{Tr}[\mathbf{T}_{uv}\mathbf{A}_u\mathbf{T}_{uv}\mathbf{A}_v]\} \tag{3.278}$$

where B_M and B_S are the numbers of bonds in the solute and the solvent molecule, respectively, and $\mathbf{T}_{uv}$ is the tensor

$$\mathbf{T}_{uv} = 3\frac{\mathbf{r}_{uv}}{r_{uv}} \otimes \frac{\mathbf{r}_{uv}}{r_{uv}} - \mathbf{1} \tag{3.279}$$

where r_{uv} and $\mathbf{r}_{uvv}$ are the distance and the radius-vector between the bonds u and v, respectively, and $\mathbf{A}_u$ is the polarizability tensor for bond u. The factor x in Eq. (3.278) has been introduced to achieve the agreement between the molecule–molecule pair dispersion potential and a simpler expression derived on the basis of the assumption that the dispersion energy between two molecules may be reduced to the sum of independent atom–atom contributions [851].

Tomasi and Persico [787] have developed a scheme that reduces the spatial representation of the dispersion interaction into a surface representation of this interaction. Therefore, the concept of the molecular cavity was introduced into the framework of the pair–pair dispersion potential approach. The corresponding average dispersion–repulsion energy of a dense solute–solvent system was derived as:

$$E_{\text{disp-rep}} = \int \cdots \int U(\Omega) g(\Omega)\, d\Omega \tag{3.280}$$

where Ω stands for the set of all coordinates of the molecules involved, $g(\Omega)$ is the distribution function, and $U(\Omega)$ is expressed as a sum of two-body dispersion–repulsion potentials. In the case of the fixed geometry of the solute molecule, the last integral can be transformed as:

$$E_{\text{disp-rep}} = n_S \sum_{s \in S} N_s \sum_{m \in M} \sum_{k} \int_S \mathbf{A}_{ms}^{(k)} \mathbf{n}_\sigma \, d\sigma \tag{3.281}$$

where $\mathbf{n}_\sigma$ is the outer normal to the surface S at the position σ and $\mathbf{A}_{ms}^{(k)}$ are the auxiliary functions. The integral in the last equation may be calculated numerically using an appropriate partitioning (tessellation) of the surface.

A quantum mechanical method of calculation of the solute–solvent dispersion energy has been developed by Rösch and Zerner [852] using Abe's theory [853,854]. According to this method, the dispersion energy, E_{disp}, for a solute molecule in a spherical cavity is given as:

$$E_{\text{disp}} = -\frac{2}{3} \frac{1}{a_S^3 a_M^3} \sum_{J \neq I} \sum_{K \neq O} \frac{(\mu_{IJ}^M)^2 (\mu_{KO}^S)^2}{E_K^S - E_O^S + E_J^M - E_I^M} \tag{3.282}$$

where the superscript S refers to the solvent molecule and the superscript M to the solute molecule. Thus, μ_{IJ}^M and μ_{KO}^S are the transition dipoles between the respective states of the solute (I and J) and the solvent (K and O) molecules. In Eq. (3.282), E_K^S, E_O^S and E_J^M, E_I^M denote the energies of the Kth and Oth state of the solvent and of the Jth and Ith state of the solute molecule, respectively. The a_S and a_M are the cavity radii for the solvent and solute molecules, respectively.

Another term of the free energy of solvation is the free energy of the cavity formation, which is predominantly of entropic nature. The concept of cavity formation in the process of dissolution of a compound in the pure liquid was introduced by Eley [855], following a suggestion by Susskind and Kasarnowski [856]. Several theoretical equations based on the shape and size of the solute and various parameters of the solvent have been derived for the calculation of ΔG_{cav}. The simplest approach proceeds from the abstraction of the surface tension on the boundary between the solute cavity and the solvent. In this case, the free energy of cavity formation is simply proportional to the surface of the solute cavity, S_M:

$$\Delta G_{\text{cav}} = \sigma S_M \tag{3.283}$$

where σ is the surface tension of the solvent. This formula has also been extended to account for the size of the solvent molecule as:

$$\Delta G_{\text{cav}} = \sigma S_M - RT \ln(1 - V_S n_S) \tag{3.284}$$

where V_S is the intrinsic volume of a solvent molecule and n_S is the number density of the solvent. It has been suggested to use different surface tension values for different atomic types in the solute molecule [857]. Thus,

$$\Delta G_{\text{cav}} = C + \sum_i \sigma_i A_i \tag{3.285}$$

where A_i are the solvent-accessible surface areas of different atoms in the solute molecule and C and σ_i are empirically adjustable parameters. Another theoretical construction for the calculation of the free energy of cavity formation proceeds from the theory of microscopic curved surfaces. According to this theory [858,859],

$$\Delta G_{\text{cav}} = k_s^g(V_S|V_M)\sigma S \tag{3.286}$$

where S is the area of the cavity and $k_s^g(V_S|V_M)$ is a correction factor, characteristic of the solvent and depending on the ratio of molecular volumes of the solvent and solute. This factor has been approximated by the following formula:

$$k_s^g(V_S|V_M) = 1 + \left(\frac{V_S}{V_M}\right)^{2/3} [k_s^g(1) - 1] \tag{3.287}$$

where $k_s^g(1)$ is estimated from the solubility of a given solute in a given solvent. The main deficiency of this approach is connected with the introduction of additional empirical information, often not readily available. It has been also suggested to estimate the free energy of cavity formation from the data on isothermal compressibility, β_T, as follows [860–862]:

$$\Delta G_{\text{cav}} = \frac{V_{\text{cav}}}{\beta_T} + C \tag{3.288}$$

where V_{cav} is the volume of the cavity and C is a constant term.

A different, basically entropic approach to the calculation of the free energy of cavity formation proceeds from the scaled particle theory (SPT) [863–866]. The free energy of the formation of a spherical cavity in a solvent, ΔG_{cav}, can be calculated proceeding from the SPT as:

$$\Delta G_{\text{cav}} = RT\left\{1 - \ln(1-y) + \left(\frac{3y}{1-y}\right)\frac{a_M}{a_S} + \left[\frac{3y}{1-y} + \frac{9}{2}\left(\frac{y}{1-y}\right)^2\right]\left(\frac{a_M}{a_S}\right)^2\right\} \tag{3.289}$$

where

$$y = \frac{4\pi\rho a_S^2}{3} \tag{3.290}$$

is the reduced number density of the solvent. In the last formulas, a_M and a_S denote the intrinsic radii of the solute and solvent molecules, respectively, and ρ is the number density of the solvent. The SPT cavity formation energy can be calculated also for the ellipsoidal cavities [867] and for the diluted solutions of arbitrarily shaped solutes [868].

In many cases, the free energy of solvation has been calculated for aqueous solutions. However, both the ΔG_s and its components can also be found for nonaqueous solutions [21,869–874]. Within the SM5.4 quantum mechanical solvation models [23,838], the free energy of solvation in any solvent is expressed by the following equation:

$$\Delta G_s = \Delta G_{\text{ENP}} + G_{\text{CDS}} \tag{3.291}$$

The first term is computed as the sum of two contributions

$$\Delta G_{\text{ENP}} = \Delta E_{\text{EN}} + G_p \tag{3.292}$$

where ΔE_{EN} is the change in the total energy of solute upon transformation from the optimized gas-phase structure to the solution-phase structure. The second contribution, G_P, arises from the screening of the solvent charges by solvent dielectric field, the solute–solvent electrostatic interactions, and the solvent–solvent dielectric polarization cost. The calculation of ΔE_{EN} can be carried out in the frame of any quantum chemical method as:

$$\Delta G_{\text{EN}} = (E_e^{(s)} - E_e^{(g)}) + \frac{1}{2}\left(\sum_{a\neq b}\frac{Z_a Z_b}{R_{ab}^{(s)}} - \sum_{a\neq b}\frac{Z_a Z_b}{R_{ab}^{(g)}}\right) \tag{3.293}$$

where $E_e^{(g)}$ and $E_e^{(s)}$ are the quantum mechanically calculated total electronic energies for the solute in the gas phase and in solution, respectively, and $R_{ab}^{(g)}$ and $R_{ab}^{(s)}$ correspond to equilibrium internuclear distances in these two media. The electronic energy depends directly upon the electron density, and thus the redistribution of charge due to the solvation of the solute molecule is taken into account by the last equation. In SM5.4 solvation models, the polarization term, G_P, is derived using the following modified generalized Born equation:

$$G_p = -\frac{1}{2}\left(1 - \frac{1}{\varepsilon}\right) \sum_{k,k'} q_k q_{k'} \left[r_{kk'}^2 + \alpha_k \alpha_{k'} \exp\left(-\frac{r_{kk'}^2}{d_{kk'} \alpha_k \alpha_{k'}} \right) \right]^{-1/2} \quad (3.294)$$

In this equation, ε is the bulk dielectric constant of the solvent; q_k is the partial charge on atom k; $r_{kk'}$ is the distance between atoms k and k'; α_k is the fixed Coulomb radius of atom k; and $d_{kk'}$ is an empirical constant optimized for a given pair of atoms. As a whole, the ΔE_{ENP} term is computed using the self-consistent reaction field method [875]. The atomic partial charges were derived using a special class IV mapping of NDDO Mulliken charges [22].

The second contribution to the free energy of solvation, G_{CDS}, represents within the SM5.4 models the short-range solute–solvent and solvent–solvent interactions such as dispersion, cavity formation, and solvent structural reorganization. It has been divided into two components as:

$$G_{\mathrm{CDS}} = G_{\mathrm{CD}} + G_{\mathrm{CS}} \quad (3.295)$$

$$\left[G_{\mathrm{CD}} = \sum_k \sigma_k A_k(\{R_Z\}, R_S^{\mathrm{CD}}) \right] \quad (3.296)$$

and

$$G_{\mathrm{CS}} = \sigma^{\mathrm{CS}} \sum_k A_k(\{R_Z\}, R_S^{\mathrm{CS}}) \quad (3.297)$$

where σ_k is the geometry-dependent cavity-dispersion atomic surface tension of atom k, and σ^{CS} represents the cavity-structural surface tension of the whole molecule. In the last equation, $A_k(\{R_Z\}, R_S^{\mathrm{CD}})$ and $A_k(\{R_Z\}, R_S^{\mathrm{CS}})$ denote the solvent-accessible surface areas of atom k, calculated using the same set of solute atomic radii $\{R_Z\}$ but different solvent effective radii for the cavity-dispersion and cavity-structural effects, R_S^{CD} and R_S^{CS}, respectively. The van der Waals radii values by Bondi [876] were used for the solute radii whereas the solvent effective radii and surface tensions were the empirically adjusted parameters of SM5.4 models. The models have been developed for more than 90 solvents using the experimentally measured free energies of solvation for altogether 206 solutes [23]. The parameters have been derived for both the solutes and solvents containing H, C, N, O, F, S, Cl, Br, and I atoms. Different sets of parameters have been reported [22,23,877] based on the AM1 [434], PM3 [436], and MNDO/d [878–880] semiempirical theories, respectively. The SM5.4 models have been successfully used for the prediction of partition coefficients of compounds and for the study of conformational equilibria in solutions.

In principle, any component of the theoretically calculated free energy of solvation as well as the total free energy, ΔG_s, is applicable as a solvational molecular descriptor. The use of individual components is advantageous for the interpretation of the obtained relationships as each of them corresponds to a clearly defined physical model [881–885]. A list of common theoretical solvational descriptors is given in Table 3.12. In Tables 3.13, 3.14, and 3.15, the numerical values of individual components of ΔG_s are presented for aqueous and nonaqueous solutions.

An original approach called TLSER (theoretical linear solvation energy descriptors) for the QSAR/QSPR account for the solvent effects has been proposed by Famini [886]. The TLSER methodology was developed as a theoretical extension of the empirical Kamlet–Taft–Abraham LSER method [504,887]. The linear solvatochromic relation (LSER) descriptors have been shown to be successful in correlating a wide range of chemical and physical properties involving solute–solvent interactions as well as biological activities of compounds. However, the LSER descriptors have been derived from the Ultraviolet/Visible solvatochromic spectral shifts of indicator dyes or some other experimental data. Thus, their ability to make a priori predictions is limited because of the lack of descriptors for compounds for which the necessary

TABLE 3.12 Theoretical Solvational Descriptors

Notation	Description	Reference
ΔG_{el}	Free energy of solvent electrostatic polarization	*a*
ΔG_{cav}	Free energy of the solute cavity formation	*b*
ΔG_{disp}	Solute–solvent dispersion interaction energy	*c*
ΔG_{rep}	Solute–solvent exchange–repulsion interaction energy	*d*
ΔG_{ENP}	Solute–solvent electrostatic interaction	*e*
G_{CDS}	Solute–solvent short-range interactions	*e*
ε_B	Covalent contribution to Lewis basicity of the solvent	*f*
q_-	Electrostatic contribution to Lewis basicity of the solvent	*f*
ε_A	Covalent contribution to Lewis acidity of the solvent	*f*
q_+	Electrostatic contribution to Lewis acidity of the solvent	*f*
V_{mc}	Molecular van der Waals volume	*g*
π_1	Ratio of the molecular polarizability and volume	*h*
δ_H^2	Hildebrand's solubility parameter	*f*

[a] R.J. Abraham, B.D. Hudson, M.W. Kermode, and J.R. Mines, *J. Chem. Soc., Faraday Trans.* **84,** 1911 (1988); M. Karelson and G.H.F. Diercksen, in *Problem Solving in Computational Molecular Science*, S. Wilson, G.H.F. Diercksen (Eds.) pp. 215–248. Kluwer Acad. Publ., Dordrecht, The Netherlands, 1997.
[b] R.A. Pierotti, *Chem. Rev.* **76,** 717 (1976).
[c] T. Abe, Y. Amako, T. Nishioka, and H. Azumi, *Bull. Chem. Soc. Jpn.* **39,** 845 (1966); M.J. Huron and P. Claverie, *Chem. Phys. Lett.* **4,** 429 (1969); N. Rösch and M.C. Zerner, *J. Phys. Chem.* **98,** 5817 (1994).
[d] J. Tomasi and M. Persico, *Chem. Rev.* **94,** 2027 (1994).
[e] D.J. Giesen, G.D. Hawkins, D.A. Liotard, C.J. Cramer, and D.G. Truhlar, *Theor. Chim. Acta* **98,** 85 (1997).
[f] G.R. Famini, *Using Theoretical Descriptors in Quantitative Structure-Activity Relationships*, V.CRDEC-TR-085. U.S. Army Chemical Research, Development and Engineering Center, Aberdeen Proving Ground, MD, 1989.
[g] A.J. Hopfinger, *J. Am. Chem. Soc.* **102,** 7126 (1980).
[h] H.A. Kurtz, J.J.P. Stewart, and K.M. Dieter, *J. Comput. Chem.* **11,** 82 (1990).

TABLE 3.13 AM1 SCRF Calculated Electrostatic Solvation Energies (E_{el}) INDO/1-Calculated Dispersion Energies (E_{disp}), and SPT Cavity Formation Free Energies (ΔG_{cav}) of Some Organic Compounds in Aqueous Solution (kcal/mol) [849]

Compound	$E_{el}(S^a)$	$E_{el}(E^b)$	$E_{disp}(S)$	$E_{disp}(E)$	$\Delta G_{cav}(S)$	$\Delta G_{cav}(E)$
Methane	0.00	0.00	−2.75	−2.75	4.22	4.22
Ethane	0.00	0.00	−3.55	−3.40	4.95	4.96
n-Hexane	0.00	0.00	−4.99	−4.95	7.34	7.50
Cyclohexane	0.00	0.00	−5.30	−4.79	6.74	6.31
Ethene	0.00	0.00	−3.64	−3.55	4.81	4.84
Ethyne	0.00	0.00	−3.50	−3.95	4.53	4.60
Propyne	−0.05	−0.01	−4.40	−4.99	5.11	5.19
Benzene	0.00	0.00	−6.49	−6.52	6.18	6.35
Methyl fluoride	−1.06	−0.95	−3.76	−3.70	4.28	4.24
Methyl chloride	−0.91	−0.61	−4.29	−3.92	4.60	4.66
Methyl bromide	−0.67	−0.51	−4.06	−3.78	5.10	5.11
Methyl iodide	−0.53	−0.43	−4.13	−3.99	5.31	5.33
Dimethyl ether	−0.78	−o.70	−3.56	−3.20	5.69	5.24
Dimethyl sulphide	−0.61	−0.66	−4.39	−4.32	5.68	5.73
Methanol	−0.92	−1.06	−3.34	−3.30	4.46	4.47
Ethanol	−1.95	−1.98	−4.26	−3.80	5.02	5.03
Glycol	−1.78	−3.06	−4.96	−4.44	5.08	5.11
Ethanal	−2.23	−1.97	−4.34	−3.81	5.09	4.92
Acetone	−2.27	−2.04	−4.86	−4.34	5.69	5.33
Acetic acid	−1.08	−0.82	−4.92	−3.93	5.02	5.33
Methyl formate	−0.66	−0.65	−4.16	−3.88	5.29	5.32
Acetonitrile	−2.81	−1.89	−4.21	−3.71	4.94	4.85
Pyridine	−0.92	−0.73	−6.78	−6.39	5.93	6.10
Dimethyl amine	−0.69	−0.88	−4.48	−3.75	5.16	5.24
Trifluoromethane	−0.97	−0.73	−5.05	−4.88	4.69	4.70
Tetrafluoromethane	0.00	0.00	−6.76	−6.76	4.17	4.17
Ethylene diamine	−1.39	−1.20	−5.13	−4.27	5.47	5.52
Methyl amine	−0.88	−0.99	−3.67	−3.54	4.72	4.73
Acetamide	−0.12	−0.09	−5.05	−4.45	5.20	5.23
Methyl mercaptane	−1.87	−1.78	−3.95	−4.25	5.06	5.08

[a] For the spherical cavity.
[b] For the ellipsoidal cavity.

experimental information is missing. Although the respective tables of parameters and predictive relationships to help in their estimation are available, the LSER descriptors for new compounds are not as easily found. Attempts to correlate computationally derived structural and electronic descriptors with the solvatochromic parameters have met with only moderate success [888].

To circumvent the limitations of LSER parameters, the computationally derived TLSER descriptors attempt to maintain the same QSPR relationships for

TABLE 3.14 Individual Components of Free Energy of Solvation in Aqueous Solutions (ΔG_{ENP} and G^0_{CDS}) Calculated Using the SM5.4/A Model and SM5.4/A-Calculated and Experimental Total Free Energy of Solvation [ΔG^0_S(calc) and ΔG^0_S(exp)], respectively (kcal/mol) [885a]

Compound	ΔG_{ENP}	G^0_{CDS}	ΔG^0_S(calc)	ΔG^0_S(exp)
Methane	−0.1	2.1	2.0	2.0
n-Butane	−0.2	1.7	1.4	2.1
n-Octane	−0.1	2.2	2.1	2.9
2,2-Dimethyl propane	−0.2	2.0	1.8	2.5
2,2,4-Trimethyl pentane	−0.1	2.5	2.4	2.9
Cyclopentane	−0.6	1.4	0.8	1.2
cis-1,2-Dimethylcyclohexane	−0.2	1.9	1.8	1.6
1-Butene	−0.8	2.4	1.6	1.4
Cyclopentene	−1.4	1.6	0.2	0.6
(*E*)-2-pentene	−0.9	2.0	1.2	1.3
Ethyne	−2.3	2.8	0.5	0.0
1-Hexyne	−2.4	2.6	0.2	0.3
Benzene	−3.1	2.1	−1.0	−0.9
m-Xylene	−3.2	2.4	−0.8	−0.8
Naphthalene	−4.9	2.6	−2.3	−2.4
1,2-Ethanediol	−6.4	−2.3	−8.8	−9.3
1-Butanol	−3.7	−0.7	−4.4	−4.7
Cyclopentanol	−3.1	−0.8	−3.9	−5.5
p-Cresol	−6.7	0.2	−6.5	−6.1
1-Octanol	−3.5	−0.2	−3.7	−4.1
Tetrahydrofuran	−2.9	−0.6	−3.6	−3.5
1,4-Dioxane	−3.9	−2.3	−6.2	−5.1
Methyl propyl ether	−2.1	0.7	−1.4	−1.7
1,2-Dimethoxyethane	−3.3	−0.3	−3.6	−4.8
Anisole	−4.4	1.4	−3.0	−1.0
Butanal	−4.3	1.0	−3.3	−3.2
Benzaldehyde	−6.1	1.7	−4.4	−4.0
Octanal	−4.2	1.6	−2.6	−2.3
Butanone	−4.6	1.0	−3.6	−3.6
Cyclopentanone	−4.3	0.4	−3.9	−4.7
3-Pentanone	−4.3	1.3	−3.0	−3.4
Methyl phenyl ketone	−6.3	1.6	−4.7	−4.6
2-Octanone	−4.3	1.6	−2.7	−2.9
Ethanoic acid	−8.0	1.5	−6.5	−6.7
Pentanoic acid	−7.0	1.8	−5.2	−6.2
Methyl methanoate	−6.0	2.1	−3.8	−2.8
Methyl butanoate	−4.8	1.9	−2.9	−2.8
Butyl ethanoate	−5.3	2.2	−3.1	−2.6
Methyl octanoate	−4.7	2.5	−2.2	−2.0
2-Propen-1-ol	−4.3	−0.1	−4.4	−5.1
2-Methoxyethanol	−5.4	−1.8	−7.2	−6.8
Butenyne	−3.2	3.1	−0.1	0.0
m-Hydroxybenzaldehyde	−9.5	−0.2	−9.7	−9.5

TABLE 3.14 (*Continued*)

Compound	ΔG_{ENP}	G^0_{CDS}	ΔG^0_S(calc)	ΔG^0_S(exp)
Methylamine	−2.5	−3.2	−5.8	−4.6
Dimethylamine	−2.0	−2.7	−4.7	−4.3
Propylamine	−1.8	−2.6	−4.4	−4.4
Trimethylamine	−1.8	−1.3	−3.2	−3.2
Pyrrolidine	−1.9	−3.1	−5.1	−5.5
Diethylamine	−1.1	−1.6	−2.8	−4.1
N-Methylpiperazine	−3.0	−4.1	−7.1	−7.8
N,*N*-Dimethylpiperazine	−2.6	−2.5	−5.1	−7.6
Dipropylamine	−1.1	−1.8	−2.9	−3.7
Pyridine	−4.8	−0.1	−5.0	−4.7
2-Ethylpyrazine	−4.7	−1.0	−5.7	−5.5
2,5-Dimethylpyridine	−5.1	0.7	−4.4	−4.7
Ethanonitrile	−6.1	1.8	−4.3	−3.9
Benzonitrile	−5.1	2.2	−3.0	−4.1
Nitroethane	−9.2	5.0	−4.1	−3.7
2-Nitropropane	−8.6	5.2	−3.4	−3.1
2-Methyl-1-nitrobenzene	−8.1	5.2	−2.8	−3.6
(*E*)-*N*-Methylacetamide	−6.9	−0.8	−7.8	−10.0
2-Methoxyethanamine	−3.3	−3.5	−6.9	−6.5
Morpholine	−3.6	−4.2	−7.8	−7.2
N-Methylmorpholine	−3.3	−2.5	−5.7	−6.3
Hydrazine	−3.7	−3.2	−6.9	−9.3
Ethanethiol	−1.3	0.3	−1.0	−1.3
Thiophenol	−3.5	0.4	−3.1	−2.6
Hydrogen sulfide	−1.1	0.3	−0.8	−0.7
Dimethyl sulfide	−1.3	−0.4	−1.7	−1.5
Thioanisole	−3.9	0.7	−3.3	−2.7
Dimethyl disulfide	−2.4	0.4	−2.0	−1.8
Diethyl disulfide	−2.3	0.9	−1.4	−1.6
Fluoromethane	−1.8	2.5	0.7	−0.2
1,1-Difluoroethane	−2.8	2.8	0.0	−0.1
Fluorobenzene	−3.1	2.9	−0.2	−0.8
Dichloromethane	−1.8	0.1	−1.6	−1.4
1,1,2-Trichloroethane	−1.8	−0.1	−1.9	−2.0
2-Chloropropane	−1.5	1.1	−0.4	−0.3
3-Chloropropene	−1.5	1.7	0.2	−0.6
Trichloroethene	−0.7	0.3	−0.4	−0.4
Chlorobenzene	−2.8	1.6	−1.2	−1.1
p-Dichlorobenzene	−2.0	1.1	−1.0	−1.0
Bromomethane	−1.5	0.5	−1.0	−0.8
2-Bromopropane	−1.6	0.7	−0.9	−0.5
Bromobenzene	−2.8	1.1	−1.7	−1.5
Diiodomethane	−0.7	−1.3	−1.9	−2.5
1-Iodobutane	−0.9	0.4	−0.5	−0.3
Iodobenzene	−2.6	0.9	−1.6	−1.7
1-Bromo-1-chloro-2,2,2-trifluoroethane	−1.4	1.8	0.4	−0.1

TABLE 3.14 (*Continued*)

Compound	ΔG_{ENP}	G^0_{CDS}	ΔG^0_S(calc)	ΔG^0_S(exp)
1-Bromo-1,2,2,2-tetrafluoroethane	−1.8	2.8	1.0	0.5
1-Chloro-2,2,2-trifluoroethane	−2.8	2.7	−0.1	0.1
1,1,1-Trifluoropropan-2-ol	−6.0	1.4	−4.6	−4.2
bis-(2-Chloroethyl)sulfide	−2.0	−0.8	−2.9	−3.9
2,2,2-Trifluoroethyl vinyl ether	−3.7	3.8	0.0	−0.1
p-Bromophenol	−6.5	−0.7	−7.2	−7.1

TABLE 3.15 Individual Components of Free Energy of Solvation in Chloroform Solutions (ΔG_{ENP} and G_{CDS}) Calculated Using the SM5.4/A Model and SM5.4/A-Calculated and Experimental Total Free Energy of Solvation [ΔG_S(calc) and ΔG_S(exp)], respectively (kcal/mol) [838]

Compound	ΔG_{ENP}	G_{CDS}	ΔG_S(calc)	ΔG_S(exp)
Toluene	−3.0	−2.4	−5.4	−5.5
m-Xylene	−3.6	−2.4	−6.1	−5.9
Ethanol	−0.1	−3.5	−3.6	−3.9
1-Butanol	−2.1	−2.9	−4.9	−5.3
o-Cresol	−2.3	−5.2	−7.6	−7.6
1-Heptanol	−4.6	−2.8	−7.4	−7.5
Tetrahydropyran	−3.3	−2.2	−5.5	−5.8
Ethyl phenyl ether	−4.0	−3.2	−7.2	−7.2
Ethanal	0.3	−3.6	−3.3	−3.7
Acetophenone	−3.2	−4.6	−7.9	−7.8
Pentanoic acid	−1.0	−5.5	−6.5	−6.6
Propyl ethanoate	−2.5	−4.1	−6.5	−6.4
Methyl pentanoate	−3.0	−3.6	−6.6	−6.7
p-Hydroxybenzaldehyde	−1.8	−7.7	−9.5	−10.3
Dimethylamine	−2.5	−1.6	−4.1	−3.7
Propylamine	−3.4	−1.4	−4.8	−4.7
Trimethylamine	−2.7	−1.6	−4.2	−3.9
Piperidine	−4.7	−1.1	−5.8	−6.4
Pyridine	−2.7	−3.5	−6.3	−6.5
Aniline	−3.5	−3.7	−7.2	−7.3
2-Ethylpyrazine	−4.3	−3.5	−7.8	−7.7
Ethanonitrile	−0.2	−4.5	−4.7	−4.4
Nitrobenzene	−1.3	−6.4	−7.7	−7.8
Ethanamide	−1.3	−6.0	−7.3	−7.1
Diethyl sulfide	−4.0	−1.1	−5.1	−6.4
Thioanisole	−5.5	−0.8	−6.3	−6.0
p-Dichlorobenzene	−4.4	−1.6	−6.0	−6.3
Iodobenzene	−4.7	−2.0	−6.7	−6.6

a given property. However, each of the steric, polarizability, and hydrogen-bonding terms was derived using the semiempirical molecular orbital methods, permitting a much greater degree of a priori prediction oncc a correlation is derived, than does the LSER. Like the LSER, TLSER uses a single set of descriptors and each of them describes a single, independent molecular event or characteristics:

$$P = P_0 + aV_{\mathrm{mc}} + b\pi_1 + c\varepsilon_B + dq_- + e\varepsilon_A + fq_+ \qquad (3.298)$$

where V_{mc} is the molecular van der Waals volume calculated according to the Hopfinger's method [889], and the polarizability term π_1 is derived as the ratio of the polarization and total volume of the solute [890]. The hydrogen-bonding effects were separated into donor and acceptor components. The covalent contribution to Lewis basicity, ε_B, is calculated as the difference in energy between the lowest unoccupied molecular orbital (E_{LUMO}) of standard solvent (water) and the highest occupied molecular orbital (E_{HOMO}) of solute. The electrostatic contribution to basicity, q_-, is simply the most negative atomic partial charge in the solute molecule. Analogously, the hydrogen-bonding donating ability is divided into two components: ε_A is the energy difference between the E_{HOMO} of water and E_{LUMO} of solute, whereas q_+ is the most positive charge on a hydrogen atom in the solute molecule. The TLSER parameters for a representative set of compounds are given in Table 3.16.

The TLSER regressions have been developed for over 80 solute/solvent-based properties. These properties, like the LSER before it, run the gamut of physical, chemical, spectral, and toxicological properties. A number of TLSER correlations have been reported for different chemical and physical properties of substances in solution, including adsorption on charcoal, Hafkensheid retention indices, octanol/water partition coefficients, hydrolysis rates of organophosphorus compounds, and pK_a of organic acids [891]. For instance, the acidity of substituted acetic acids XCH_2COOH in aqueous solutions has been described by the following excellent QSPR equation [892]:

$$\mathrm{p}K_a(\mathrm{H_2O}) = -16.441 - (30.22 \pm 7.05)\varepsilon_B + (68.69 + 0.6)q_-$$

$$R^2 = 0.986 \qquad s = 0.209 \qquad n = 14 \qquad F = 389$$

The correlation with the covalent basicity descriptor, ε_B, was interpreted as arising from the specific solvation of the carboxyl group by water. The higher basicity of this group enhances the solvation and thus assists the release of acidic carboxyl proton. The largest negative charge, q_-, is localized on the carbonyl oxygen of acetic acids. The decrease of this charge by electronegative substituents results in weakening of the adjacent O—H bond and the parallel increase of acidity of compounds. Interestingly, different TLSER terms were found to determine the acidity of substituted acetic acids in the gas phase. The following correlation was obtained [892]:

TABLE 3.16 TLSER Descriptors for Some Solvents [894]

Compound	V_{mc}	π_1	ε_B	q_-	ε_A	q_+	δ_H^2
Water	0.1933	0.0581	0.1237	0.3255	0.1237	0.1628	2.2970
Methanol	0.3647	0.0860	0.1314	0.3291	0.1402	0.1803	0.8586
Formamide	0.4090	0.0965	0.1371	0.3553	0.1667	0.1572	1.5134
Ethanol	0.5423	0.0927	0.1326	0.3235	0.1429	0.1800	0.6782
N-Methylformamide	0.5790	0.1006	0.1381	0.3437	0.1682	0.1557	0.6150
Nitromethane	0.4740	0.1093	0.1302	0.3342	0.1817	0.0500	0.6632
Acetonitrile	0.4529	0.0937	0.1177	0.1145	0.1622	0.0209	0.5766
Dimethylsulfoxide	0.7209	0.1046	0.1475	0.7196	0.1734	0.0525	0.7063
Propanone	0.6441	0.0972	0.1381	0.2867	0.1715	0.0232	0.3791
Dimethyl formamide	0.7693	0.1042	0.1441	0.4698	0.1649	0.0576	0.5812
Tetramethylene sulfoxide	1.0129	0.1192	0.1368	0.6788	0.1942	0.0814	0.7489
Dimethyl acetamide	0.9634	0.1026	0.1452	0.4656	0.1658	0.0285	0.4879
N-Methylpyrrolidine	1.0311	0.1053	0.1516	0.4541	0.1495	0.0169	0.5339
Hexamethylphosphoramide	1.8436	0.1107	0.1456	0.6518	0.1874	0.0096	0.3071
Trichloromethane	0.7540	0.1114	0.1165	0.1130	0.1849	0.0876	0.3711
Tetrachloromethane	0.9058	0.1172	0.1132	0.0704	0.1912	0.0000	0.3088
Benzene	0.8463	0.1204	0.1517	0.0594	0.1744	0.0593	0.3506
Dimethoxymethane	0.8076	0.0978	0.1369	0.3304	0.1474	0.0190	0.5905
Dioxane	0.8598	0.1045	0.1379	0.3275	0.1480	0.0364	0.4184
Dichloromethane	0.6045	0.1036	0.1207	0.1605	0.1773	0.0555	0.4088
Ethoxyethane	0.9035	0.0996	0.1365	0.3423	0.1455	0.0072	0.2351
Benzonitrile	0.9984	0.1274	0.1474	0.0865	0.1833	0.0699	0.5142
Tetrahydrofuran	0.7889	0.1021	0.1378	0.3270	0.1471	0.0217	0.3615
Di(2-methoxyethyl)ether	1.4171	0.1024	0.1157	0.3572	0.1502	0.0128	0.4000

$$pK_a(\text{gas phase}) = 665.33 - (273.54 \pm 79.35)\pi_1 - (1504.1 \pm 161.9)q_+$$

$$R^2 = 0.922 \qquad s = 1.434 \qquad n = 14 \qquad F = 65$$

The negative contribution of the dipolarity/polarizability term, π_1, reflects the stabilization of the conjugated anion of acids by bulky and polarizable groups that leads to the respective enhancement of the acidity of compounds. The most positive atom in a substituted acetic acid is the acidic hydrogen and, as expected, the increase of its charge, q_+, leads to the lowering of the pK_a of acids. The TLSER descriptors have also been successfully employed for the correlation of the gas-phase acidities of hydrocarbons and nitrohydrocarbons [893].

In other solvents, including various alcohols and 1,2-dimethoxyethane, the pK_a of substituted acetic acids correlated with the molecular van der Waals volume V_{mc}, the electrostatic basicity q_-, and covalent acidity ε_A terms of TLSER. The descriptor ε_A is a natural descriptor for the acidity of compounds whereas the volume effects reflect the necessity to create a cavity for the acid in the solvent before the cleavage of the O—H bond. Thus, the TLSER analysis gives a consistent picture of solvent effects determining the acid–base properties of compounds in different media.

The TLSER descriptors have enabled to obtain excellent correlations of the solvent effects on the rate constants of decarboxylation of 3-carboxybenzisoxazoles to form 2-cyanophenolates in the presence of base [894].

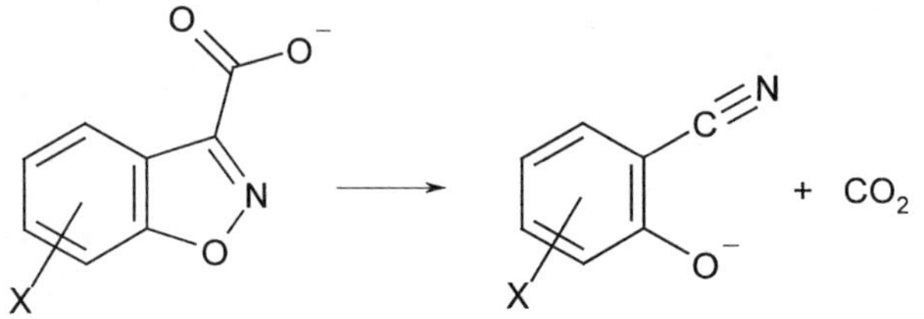

For instance, the correlations for anionic compounds involved only two descriptors, the electrostatic basicity q_- and the electrostatic acidity q_+. For the data corresponding to the aqueous solutions, the following TLSER equation was obtained:

$$\log k = (30.99 \pm 8.64) - (73.33 \pm 14.89)q_- + (27.58 \pm 5.62)q_+$$

$$R^2 = 0.978 \qquad s = 0.195 \qquad n = 7 \qquad F = 89.6$$

Similar correlations were obtained for the rate constants measured in methanol and in nitromethane. Apparently, the reactivity of anionic compounds is primarily determined by the local polarity at the reaction center. The correlations for more extensive sets of electroneutral substrates involved also other TLSER descriptors. The detailed analysis of descriptors involved and the signs and size of the respective regression coefficients enabled to elucidate the possible mechanism of the reaction [894].

The TLSER equations have been also developed for the gas chromatographic retention indices of different classes of compounds [895–897] and solvent effects on the vibrational frequencies in the infrared (IR) spectra of compounds [898]. More extensively, the TLSER descriptors have been used for the correlation and physical interpretation of a variety of biological activities of chemical compounds, including nonspecific toxicities [899], activities of some local anesthetics [900], opiate receptor activity of some fenantyl-like compounds [901], and acetylcholinesterase inhibition [901]. The cytochrome P450-mediated acute toxicity of nitriles has been described by the following TLSER equation [902]:

$$\log\left(\frac{1}{\mathrm{LD}_{50}}\right) = 10.38 - 0.0104V_{\mathrm{mc}} - 0.186\pi_1 - 0.541\varepsilon_A - 2.51q_-$$

$$R^2 = 0.810 \qquad s = 0.288 \qquad n = 23 \qquad F = 18.3$$

The cavity and polarization effects determine the effectiveness of the toxic

TABLE 3.17 Correlation between Molecular Descriptors Calculated for Isolated Molecules and for Molecules in Dielectric Medium With ε = 80, respectively

Descriptors	N	$R^2 < 0.95$	$R^2 < 0.99$
Geometrical	12	1	4
Quantum chemical energy	162	1	4
Quantum chemical charge	176	9	35
Molecular orbital related	64	0	5
All	414	11	48

agent via a mechanism that includes the hydrogen abstraction, hydroxyl addition, and dissociation of hydroxylated nitrile.

The theoretical molecular descriptors, including the quantum chemical descriptors have been obtained, as a rule, for the isolated molecules. However, we have seen that the molecular wave function and the respective electronic and structural properties of molecules may be significantly affected by the solvation. Thus, it is important to estimate the size of such effects on the numerical values of different descriptors. The respective analysis has been performed for a set of 406 compounds incorporating a wide variety of organic molecules [903]. In this analysis, the pair correlation coefficients were calculated between two sets of data for each descriptor. One of them included the values calculated for the isolated molecules and another for the molecules in a condensed dielectric medium. This medium was characterized within the self-consistent reaction field model by the dielectric constant of water ($\varepsilon = 80$). In Table 3.17, the distribution of different types descriptors is given according to this pair correlation coefficient. Notably, the intercorrelation between the two types is very high ($R^2 > 0.99$) for most descriptors. Obviously, the perturbation by dielectric medium is still similar in the case of different compounds, and thus the linear relationships obtained with one set of descriptors will remain valid if another set is applied. This result is rather important as validating the use of quantum chemical descriptors for isolated molecules in different condensed media.

TABLE 3.18 Correlation between Molecular Descriptors Calculated for Molecules in full-trans and in the Folded Optimum Conformation, respectively

Descriptors	N	$R^2 < 0.5$	$R^2 < 0.95$	$R^2 < 0.99$
Geometrical	15	4	7	0
Quantum chemical energy	120	9	27	86
Quantum chemical charge	141	3	19	47
Molecular orbital related	64	11	37	20
All	330	27	81	153

The molecular descriptors can also depend on the conformation of the molecule. Whereas the constitutional and topological descriptors are insensitive to the conformational change, most of geometrical and charge-distribution-related descriptors should be affected by these changes. To examine the conformational effects on the molecular descriptors, an analogous analysis as described above for the solvent-dependent descriptors has been carried out [903]. For the following series of compounds:

$$CH_3(CH_2)_nCH_3$$

$$CH_3(CH_2)_nOH$$

$$CH_3(CH_2)_nNH_2$$

$$CH_3(CH_2)_nCOOH$$

$$CH_3(CH_2)_nCN$$

$$HO(CH_2)_nOH$$

$$H_2N(CH_2)_nNH_2$$

$$HOOC(CH_2)_nCOOH$$

$$NC(CH_2)_nCN$$

where $n = 2, \ldots, 5$, two sets of molecular descriptors were calculated. The first of them corresponded to the full-trans conformation of the hydrocarbon skeleton of compounds, whereas the other was calculated for the molecules in the folded conformational minimum. The data on the distribution of the pair correlation coefficients between these two sets of descriptors is given in Table 3.18. It is interesting to note that the conformational change of molecule has larger influence on the values of molecular descriptors than the change of the medium. Therefore, the choice of the "correct" conformation of the molecule from which the molecular descriptors are calculated is vital for the development of legitimate QSAR equations. Richards and co-workers have also pointed out the importance of statistical weighting over the conformations at given temperature, using the respective Boltzmann factors [121,146].

REFERENCES

1. S.M. Free and J.W. Wilson, *J. Med. Chem.* **7,** 395 (1964).
2. E.L. Plummer, *Rev. Comput. Chem.* **1,** 119 (1990).
3. T. Fujita and T. Ban, *J. Med. Chem.* **14,** 158 (1971).
4. R.D. Cramer, III, G. Redl, and C. E. Berkoff, *J. Med. Chem.* **17,** 533 (1974).
5. K.C. Chu, R.J. Feldmann, M.B. Shapiro, G.F. Hazard, and R.I. Geran, *J. Med. Chem.* **18,** 539 (1975).

6. L. Hodes, G.F. Hazard, R.I. Geran, and S. Richman, *J. Med. Chem.* **20,** 469 (1977).
7. L. Hodes, *J. Chem, Inf. Comput. Sci.* **21,** 128 (1981).
8. K. Enslein and P.N. Craig, *Toxicol. Environ. Health* **10,** 521 (1982).
9. G. Klopman, *J. Am. Chem. Soc.* **106,** 7315 (1984).
10. G. Klopman, M.R. Frierson, and H.S. Rosenkrantz, *Mutat. Res.* **1,** 228 (1990).
11. C. Hansch, A. Vittoria, C. Silipo, and P.Y.C. Jow, *J. Med. Chem.* **18,** 546 (1975).
12. F.R. Quinn and J.A. Beisler, *J. Med. Chem.* **24,** 251 (1981).
13. F.R. Quinn, Z. Neiman, and J.A. Beisler, *J. Med. Chem.* **24,** 636 (1981).
14. S.P. Gupta, *Chem. Rev.* **94,** 1507 (1994).
15. W.A. Denny, B.F. Cain, G.J. Atwell, C. Hansch, A. Pathananickal, and A. Leo, *J. Med. Chem.* **25,** 276 (1982).
16. *Handbook of Chemistry and Physics*, R.C. Weast (ed.) p. F-112. CRC Press, Cleveland, OH, 1974.
17. R.H. Rohrbaugh and P.C. Jurs, *Anal. Chim. Acta* **199,** 99 (1987).
18. C.J. Cramer and D.G. Truhlar, *J. Am. Chem. Soc.* **113,** 8305 (1991).
19. C.J. Cramer and D.G. Truhlar, *Science* **256,** 213 (1992).
20. J. Tomasi and M. Persico, *Chem. Rev.* **94,** 2027 (1994).
21. D.J. Giesen, J.W. Storer, C.J. Cramer, and D.G. Truhlar, *J. Am. Chem. Soc.* **117,** 1057 (1995).
22. G.D. Hawkins, C.J. Cramer, and D.G. Truhlar, *J. Phys. Chem.* **100,** 19824 (1996).
23. D.J. Giesen, G.D. Hawkins, D.A. Liotard, C.J. Cramer, and D.G. Truhlar, *Theor. Chim. Acta* **98,** 85 (1997).
24. M.L. Connolly, *J. Appl. Crystallogr.* **16,** 548 (1983).
25. B. Lee and F.M. Richards, *J. Mol. Biol.* **55,** 379 (1971).
26. R.B. Hermann, *J. Phys. Chem.* **76,** 2754 (1972).
27. C.J. Alden and S.-H. Kim, *J. Mol. Biol.* **132,** 411 (1979).
28. M.L. Connolly, *J. Appl. Crystallogr.* **16,** 548 (1983).
29. P.A. Bash, N. Pattabiraman, C. Huang, T.E. Ferrin, and R. Langridge, *Science* **222,** 1325 (1983).
30. T.J. Richmond, *J. Mol. Biol.* **178,** 63 (1984).
31. M.L. Connolly, *J. Appl. Crystallogr.* **18,** 499 (1985).
32. F.M. Richards, *Methods Enzymol.* **115,** 440 (1985).
33. N.L. Max, *J. Mol. Graphics* **6,** 210 (1988).
34. P. Mezey, *Rev. Comput. Chem.* **1,** 265 (1990).
35. J.L. Pascual-Ahuir and E. Silla, *J. Comput. Chem.* **11,** 1047 (1990).
36. H. Wang and C. Levinthal, *J. Comput. Chem.* **12,** 868 (1991).
37. F. Eisenhaber and P. Argos, *J. Comput. Chem.* **14,** 1272 (1993).
38. B.S. Duncan and A.J. Olson, *Biopolymers* **33,** 219 (1993).
39. W. Heiden, T. Goetze, and J. Brickmann, *J. Comput. Chem.* **14,** 246 (1993).
40. S.M. Le Grand and J.K.M. Merz, *J. Comput. Chem.* **14,** 349 (1993).
41. J.-D. Boissonat, O. Devillers, J. Duquesne, and M. Yvinec, *J. Mol. Graphics* **12,** 61 (1994).

42. V. Gogonea and E. Osawa, *Supramol. Chem.* **3,** 303 (1994).
43. J.L. Pascual-Ahuir, E. Silla, and I. Tuñón, *J. Comput. Chem.* **15,** 1127 (1994).
44. S. Leicester, J. Finney, and R. Bywater, *J. Math. Chem.* **16,** 315 (1994).
45. F. Eisenhaber, P. Lijnzaad, P. Argos, C. Sander, and M. Scharf, *J. Comput. Chem.* **16,** 273 (1995).
46. B.S. Duncan and A.J. Olson, *J. Mol. Graphics* **13,** 250 (1995).
47. B.S. Duncan and A.J. Olson, *J. Mol. Graphics* **13,** 258 (1995).
48. R.J. Zauhar, *J. Comput.-Aided Mol. Des.* **9,** 149 (1995).
49. D.R. Flower, *J. Mol. Graphics* **15,** 238 (1997).
50. C.S. Pomelli, *J. Mol. Graphics* **15,** 393 (1997).
51. R. Fraczkiewicz and W. Braun, *J. Comput. Chem.* **19,** 319 (1998).
52. F.M. Richards, *Annu. Rev. Biophys. Bioeng.* **6,** 151 (1977).
53. J.L. Finney, *J. Mol. Biol.* **119,** 415 (1978).
54. W. Brostow, J.P. Dussault, and B. Fox, *J. Comput. Phys.* **29,** 81 (1978).
55. B.J. Gellatly and J.L. Finney, *J. Mol. Biol.* **161,** 305 (1982).
56. R. Pavani and G. Ranghino, *Comput. Chem.* **6,** 1335 (1982).
57. J.J. Muller, *J. Appl. Crystallogr.* **16,** 74 (1983).
58. A. Gavezzotti, *J. Am. Chem. Soc.* **105,** 5220 (1983).
59. M. Yu. Pavlov and B.A. Fedorov, *Biopolymers* **22,** 1507 (1983).
60. M.L. Connolly, *J. Am. Chem. Soc.* **107,** 1118 (1985).
61. K.D. Gibson and H.A. Scheraga, *Mol. Phys.* **62,** 12475 (1987).
62. J. Higo and N. Go, *J. Comput. Chem.* **10,** 376 (1989).
63. R.S. Bohacek and W.C. Guida, *J. Mol. Graphics* **7,** 113 (1989).
64. E. Silla, F. Villar, O. Nilsson, J.L. Pascual-Ahuir, and O. Tapia, *J. Mol. Graphics* **8,** 168 (1990).
65. F. Colonna-Cesari and C. Sander, *Biophys. J.* **57,** 1103 (1990).
66. C.E. Kundrot, J.W. Ponder, and F.M. Richards, *J. Comput. Chem.* **12,** 402 (1991).
67. M. Petitjean, *J. Comput. Chem.* **15,** 507 (1994).
68. M. Petitjean, *J. Comput. Chem.* **16,** 80 (1995).
69. M.L. Connolly, *J. Math. Chem.* **15,** 339 (1994).
70. V. Gogonea and E. Osawa, *J. Comput. Chem.* **16,** 817 (1995).
71. L.M. Rellick and W.J. Becktel, *Methods Enzymol.* **259,** 377 (1995).
72. M.L. Connolly, *J. Am. Chem. Soc.* **107,** 1118 (1985).
73. C. Hansch and T.E. Klein, *Acc. Chem. Res.* **19,** 392 (1986).
74. I.D. Kuntz, *Science* **257,** 1078 (1992).
75. L.M. Balbes, S.W. Mascarella, and D.B. Boyd, *Rev. Comput. Chem.* **5,** 337 (1994).
76. P.J. Whittle and T.L. Blundell, *Annu. Rev. Biophys. Biomol. Struct.* **23,** 349 (1994).
77. I.D. Kuntz, J.M. Blaney, S.J. Oatley, R. Langridge, and T.E. Ferrin, *J. Mol. Biol.* **161,** 269 (1982).
78. B.K. Shoichet and I.D. Kuntz, *J. Mol. Biol.* **221,** 327 (1991).

79. A.R. Leach and I.D. Kuntz, *J. Comput. Chem.* **13,** 730 (1992).
80. I.D. Kuntz, E.C. Meng, and B.K. Shoichet, *Acc. Chem. Res.* **27,** 117 (1994).
81. T. Ooi, M. Oobatake, G. Nemethy, and H.A. Scheraga, *Proc. Natl. Acad. Sci. U.S.A.* **84,** 3086 (1987).
82. I. Tuñón, E. Silla, and J.L. Pascual-Ahuir, *Protein Eng.* **5,** 715 (1992).
83. W. Heiden, G. Moeckel, and J. Brickmann, *J. Comput.-Aided Mol. Des.* **7,** 503 (1993).
84. V.V. Krishnan and M. Cosman, *J. Biomol. NMR Res.* **12,** 177 (1998).
85. Y.S. Prabhakar, S.P. Gupta, and A. Ray, *J. Pharmacobio-Dyn.* **9,** 61 (1986).
86. A.R. Katritzky, V.S. Lobanov, and M. Karelson, *CODESSA User's Manual*. University of Florida, Gainesville, 1994.
87. A.R. Katritzky, L. Mu, V.S. Lobanov, and M. Karelson, *J. Phys. Chem.* **100,** 10400 (1996).
88. A.R. Katritzky, V.S. Lobanov, and M. Karelson. *J. Chem. Inf. Comput. Sci.* **38,** 28 (1998).
89. A.R. Katritzky, L. Mu, and M. Karelson, *J. Chem. Inf. Comput. Sci.* **38,** 293 (1998).
90. A.R. Katritzky, L. Mu, and M. Karelson, *J. Chem. Inf. Comput. Sci.* **36,** 1162 (1996).
91. A. Verloop, *Drug Des.* **3,** 133 (1976).
92. A. Verloop, W. Hoogenstraaten, and J. Tipker, *Drug Des.* **7,** 165 (1976).
93. J. Tipker and A. Verloop, *ACS Monogr.* **255,** 279 (1984).
94. A. Verloop and J. Tipker, in *QSAR in Drug Design and Toxicology* (D. Hadzi and B. Jorman-Blazic, eds.), p. 97. Elsevier, Amsterdam, 1987.
95. A. Verloop, *The STERIMOL Approach to Drug Design*. Dekker, New York, 1987.
96. C. Hansch and A. Leo, *Exploring QSAR,* ACS Prof. Ref. Book, American Chemical Society, Washington, DC, 1995. p. 80.
97. R.A. Coburn, M. Weirzba, M.J. Suto, A.J. Solo, A.M. Triggle, and D.J. Triggle, *J. Med. Chem.* **31,** 2103 (1988).
98. G.A. Arteca, *Rev. Comput. Chem.* **9,** 191 (1996).
99. Y.C. Martin, M. G. Bures, and P. Willett, *Rev. Comput. Chem.* **1,** 213 (1990).
100. G.M. Downs and P. Willett, *Rev. Comput. Chem.* **7,** 1 (1995).
101. M.V. Volkenstein, *Configurational Statistics of Polymeric Chains*. Wiley-Interscience, New York, 1963.
102. C. Tanford, *Physical Chemistry of Macromolecules*. Wiley, New York, 1961.
103. J. Rudnick and G. Gaspari, *J. Phys. A* **19,** L191 (1986).
104. J. Rudnick and G. Gaspari, *Science* **237,** 384 (1986).
105. R. Todeschini, M. Lasgani, and E. Marengo, *J. Chemometrics* **8,** 263 (1994).
106. R. Todeschini, P. Gramatica, R. Provenzani, and E. Marengo, *Chemometrics Intell. Lab. Syst.* **27,** 221 (1995).
107. R. Todeschini and P. Gramatica, in *3D QSAR in Drug Design* (H. Kubinyi, G. Folkers, and Y.C. Martin, eds.), Vol. 2, pp. 355–380. Kluwer/ESCOM, Dordrecht, The Netherlands, 1998.

108. L.B. Kier and L.H. Hall, *Molecular Connectivity in Structure-Activity Analysis.* Research Studies Press, Letchworth, UK, 1986.
109. R. Todeschini, P. Gramatica, R. Provenzani, and E. Marengo, *Chemometrics Intell. Lab. Syst.* **27,** 221 (1995).
110. R. Todeschini and P. Gramatica, *Quant. Struct.-Act. Relat.* **16,** 120 (1997).
111. R. Todeschini, C. Bettiol, G. Giurin, P. Gramatica, P. Miana, and E. Argese, *Chemosphere* **33,** 71 (1996).
112. S. Shapiro and B. Guggenheim, *Quant. Struct-Act. Relat.* **17,** 327 (1998).
113. S. Shapiro and B. Guggenheim, *Quant. Struct-Act. Relat.* **17,** 338 (1998).
114. G. Crippen and T. Havel, *Distance Geometry and Molecular Conformation.* Research Studies Press, Taunton, Somerset, England, 1988.
115. J.M. Blaney and J.S. Dixon, *Annu Rep. Med. Chem.* **26,** 281 (1991).
116. J.M. Blaney and J.S. Dixon, *Rev. Comput. Chem.* **5,** 299 (1994).
117. A.K. Ghose and G.M. Crippen, *Mol. Pharm.* **37,** 725 (1990).
118. P.I. Nagy, J. Tokarski, and A.J. Hopfinger, *J. Chem. Inf. Comput. Sci.* **34,** 1190 (1994).
119. K.-B. Rhyu, H.C. Patel, and A.J. Hopfinger, *J. Chem. Inf. Comput. Sci.* **35,** 771 (1995).
120. M. Randić, *J. Chem. Inf. Comput. Sci.* **35,** 373 (1995).
121. A.C. Good and W.G. Richards, in *3D QSAR in Drug Design,* (H. Kubinyi, G. Folkers, and Y.C. Martin, eds.), Vol. 2, pp. 321–338. Kluwer/ESCOM, Dordrecht, The Netherlands, 1998.
122. H. Kubinyi, in *3D QSAR in Drug Design,* (H. Kubinyi, G. Folkers, and Y.C. Martin, eds.), Vol. 2, pp. 225–252. Kluwer/ESCOM, Dordrecht, The Netherlands, 1998.
123. P.G. Mezey, *Int. J. Quantum Chem., Quantum Biol. Symp.* **12,** 113 (1986).
124. P.G. Mezey, *J. Comput. Chem.* **8,** 462 (1987).
125. G.A. Arteca, W.B. Jammal, and P.G. Mezey, *J. Comput. Chem.* **9,** 608 (1988).
126. P.D. Walker, G.A. Arteca, and P.G. Mezey, *J. Comput. Chem.* **12,** 220 (1990).
127. J. Cioslowski and E.D. Fleischmann, *J. Am. Chem. Soc.* **113,** 64 (1991).
128. D.H. Rouvray, *J. Chem. Inf. Comput. Sci.* **34,** 446 (1994).
129. S.C. Basak and G.D. Grunwald, *J. Chem. Inf. Comput. Sci.* **35,** 366 (1994).
130. L.H. Hall and L.B. Kier, *J. Chem. Inf. Comput. Sci.* **35,** 1074 (1995).
131. A.C. Good, T.J.A. Ewing, D.A. Gschwend, and I.D. Kuntz, *J. Comput.-Aid. Mol. Des.* **9,** 1 (1995).
132. S.K. Kearsley, S. Sallamack, E.M. Fluder, J.D. Andose, R.T. Mosley, and R.P. Sheridan, *J. Chem. Inf. Comput. Sci.* **36,** 118 (1996).
133. R.P. Sheridan, M.D. Miller, D.J. Underwood, and S.K. Kearsley, *J. Chem. Inf. Comput. Sci.* **36,** 128 (1996).
134. J. Mestres, D.C. Rohrer, and G.M. Maggiora, *J. Mol. Graphics* **15,** 114 (1997).
135. M. Snarey, N.K. Terrett, P. Willett, and D.J. Wilton, *J. Mol. Graphics* **15,** 372 (1997).
136. P. Ertl, *J. Mol. Graphics* **16,** 11 (1998).
137. N.C. Perry and V.J. van Geerestein, *J. Chem. Inf. Comput. Sci.* **32,** 607 (1992).

138. P.M. Dean and P.-L. Chau, *J. Mol. Graphics* **5,** 152 (1987).
139. F.E. Blaney, P. Finn, R.W. Phippen, and M. Wyatt, *J. Mol. Graphics* **5,** 98 (1993).
140. R. Carbó, L. Leyda, and M. Arnau, *Int. J. Quantum Chem.* **17,** 1185 (1980).
141. R. Carbó and L. Domingo, *Int. J. Quantum Chem.* **32,** 517 (1987).
142. E.E. Hodgkin and W.G. Richards, *Int. J. Quantum Chem., Quantum Biol. Symp.* **14,** 105 (1987).
143. P.E. Bowen-Jenkins and W.G. Richards, *Int. J. Quantum Chem.* **30,** 763 (1986).
144. C. Burt, P. Huxley, and W.G. Richards, *J. Comput. Chem.* **11,** 1139 (1990).
145. A.M. Richard, *J. Comput. Chem.* **12,** 959 (1991).
146. C. Burt and W.G. Richards, *J. Comput.-Aided Mol. Des.* **4,** 231 (1990).
147. A.M. Meyer and W.G. Richards, *J. Comput.-Aided Mol. Des.* **5,** 426 (1991).
148. M. Hahn, *J. Chem. Inf. Comput. Sci.* **37,** 80 (1997).
149. C.A. Reynolds, C. Burt, and W.G. Richards, *Quant. Struct.-Act. Relat.* **11,** 34 (1992).
150. G. Downs and P. Willett, *Rev. Comput. Chem.* **7,** 1 (1995).
151. M. Hahn, *J. Chem. Inf. Comput. Sci.* **37,** 80 (1997).
152. S. Namasivayam and P.M. Dean, *J. Mol. Graphics* **4,** 46 (1986).
153. P.-L. Chau and P.M. Dean, *J. Mol. Graphics* **5,** 97 (1987).
154. P.M. Dean, P. Callow, and P.-L. Chau, *J. Mol. Graphics* **6,** 28 (1988).
155. M. Manaut, F. Sanz, J. Jose, and M. Milesi, *J. Comput.-Aided Mol. Des.* **5,** 371 (1991).
156. F. Sanz, M. Manaut, J. Rodriguez, E. Lozoya, and E. Lopez-de-Brinao, *J. Comput.-Aided Mol. Des.* **7,** 337 (1993).
157. A.C. Good, E.E. Hodgkin, and W.G. Richards, *J. Chem. Inf. Comput. Sci.* **32,** 188 (1992).
158. E.E. Hodgkin and W.G. Richards, *J. Chem. Soc., Chem. Commun.*, p. 1342 (1986).
159. A.C. Good and W.G. Richards, *J. Chem. Inf. Comput. Sci.* **33,** 112 (1993).
160. F.A. Cotton, *Chemical Applications of Group Theory.* Wiley-Interscience, New York, 1971.
161. A.T. Balaban, ed., *Chemical Applications of Graph Theory.* Academic Press, London, 1976.
162. E. Ash, S.E. Chubb, S.E. Ward, S.M. Welford, and P. Willett, *Communication, Storage and Retrieval of Chemical Information.* Ellis Horwood, Chichester, 1985.
163. P. Willett, *Similarity and Clustering in Chemical Information Systems.* Research Studies Press, Letchworth, UK, 1987.
164. S. Fujita, *Symmetry and Combinatorial Enumeration in Chemistry.* Springer-Verlag, Berlin, 1991.
165. R. Carlson, *Design and Optimization in Organic Synthesis.* Elsevier, London, 1992.
166. S. El-Basil and M. Randić, *Adv. Quant Chem.* **24,** 239 (1992).
167. R. Carbó, B. Calabuig, L. Vera, and E. Besalú, *Adv. Quant. Chem.* **25,** 253 (1994).

168. L.B. Kier and L.H. Hall, *Molecular Connectivity in Chemistry and Drug Reasearch.* Academic Press, New York, 1976.
169. D. Bonchev, *Information Theoretic Indices for Characterization of Chemical Structure.* Wiley-Interscience, New York, 1983.
170. L.B. Kier and L.H. Hall, *Molecular Connectivity in Structure-Activity Analysis.* Wiley, New York, 1986.
171. R. Kaliszan, *Quantitative Structure-Chromatographic Retention Relationships.* Wiley, New York, 1987.
172. R.E. Merrifield and H.E. Simmons, *Topological Methods in Chemistry.* Wiley, New York, 1989.
173. D.H. Rouvray, ed., *Computational Chemical Graph Theory.* Nova Science Publishers, New York, 1990.
174. M.I. Stankevich, I.V. Stankevich, and N.S. Zefirov, *Russ. Chem. Rev. (Engl. Transl.)* **57,** 191 (1988).
175. S.S. Basak, G.J. Niemi, and G.D. Veith, *J. Math. Chem.* **4,** 185 (1990).
176. H. Wiener, *J. Am. Chem. Soc.* **69,** 17 (1947).
177. A.T. Balaban, *J. Chem. Inf. Comput. Sci.* **32,** 23 (1992).
178. M. Randić, *Chem. Phys. Lett.* **211,** 478 (1993).
179. M. Randić, X. Guo, and H.J. Krishnapriyan, *J. Chem. Inf. Comput. Sci.* **33,** 709 (1993).
180. I. Lukovits and W. Linert, *J. Chem. Inf. Comput. Sci.* **34,** 899 (1994).
181. D.J. Klein, I. Lukovits, and I. Gutman, *J. Chem. Inf. Comput. Sci.* **35,** 50 (1995).
182. I. Lukovits, *Comput. Chem.* **19,** 27 (1995).
183. W. Linert, F. Renz, K. Kleestorfer, and I. Lukovits, *Comput. Chem.* **19,** 395 (1995).
184. I. Lukovits, *Rep. Mol. Theory* **1,** 127 (1990).
184a. M. Diudea and B. Pârv, *J. Chem. Inf. Comput. Sci.* **35,** 1015 (1995).
185. I. Gutman, B. Rusič, N. Trinajstić, and C.F. Wilcox, Jr., *J. Chem. Phys.* **62,** 3399 (1975).
186. M. Gordon and G.R. Scantlebury, *Trans. Faraday Soc.* **60,** 605 (1964).
187. H. Hosoya, *Bull. Chem. Soc. Jpn.* **44,** 2332 (1971).
188. Y. Gao and H. Hosoya, *Bull. Chem. Soc. Jpn.* **61,** 3093 (1988).
189. L. Buydens and D.L. Massart, *Anal. Chem.* **53,** 1990 (1981).
190. L. Buydens, D. Coomans, M. Vanbelle, D.L. Massart, and R. Van der Driesche, *J. Pharm. Sci.* **72,** 1327 (1983).
191. A. Hermann and P. Zinn, *J. Chem. Inf. Comput. Sci.* **35,** 551 (1995).
192. A.T. Balaban, *Theor. Chim. Acta* **53,** 355 (1979).
193. R.E. Merrifield and H.E. Simmons, *Topological Methods in Chemistry.* New York, 1989.
194. D. Bonchev, A.T. Balaban, and M. Randić, *Int. J. Quantum Chem.* **19,** 61 (1981).
195. M. Randić, *J. Am. Chem. Soc.* **97,** 6609 (1975).
196. E. Estrada, *J. Chem. Inf. Comput. Sci.* **35,** 1022 (1995).
197. A.T. Balaban, *Pure Appl. Chem.* **55,** 199 (1983).
198. A.T. Balaban, *Chem. Phys. Lett.* **89,** 399 (1981).

199. A.T. Balahan and P. Filip, *Math. Chem.* **16,** 163 (1984).
200. A.T. Balaban, *Math. Chem.* **21,** 115 (1986).
201. D. Bonchev and N. Trinajstić, *J. Chem. Phys.* **67,** 4517 (1977).
202. H.P. Schultz, *J. Chem. Inf. Comput. Sci.* **29,** 227 (1989).
203. W.R. Müller, K. Szymanski, J.V. Knop, and N. Trinajstić, *J. Chem. Inf. Comput. Sci.* **30,** 160 (1990).
204. H.P. Schultz, E.B. Schultz, and T.P. Schultz, *J. Chem. Inf. Comput. Sci.* **30,** 27 (1990).
205. J.V. Knop, W.R. Müller, K. Szymanski, and N. Trinajstić, *J. Chem. Inf. Comput. Sci.* **31,** 83 (1991).
206. H.P. Schultz, E.B. Schultz, and T.P. Schultz, *J. Chem. Inf. Comput. Sci.* **32,** 69 (1992).
207. H.P. Schultz and T.P. Schultz, *J. Chem. Inf. Comput. Sci.* **32,** 364 (1992).
208. H.P. Schultz and T.P. Schultz, *J. Chem. Inf. Comput. Sci.* **33,** 231 (1993).
209. H.P. Schultz and T.P. Schultz, *J. Chem. Inf. Comput. Sci.* **33,** 863 (1993).
210. H.P. Schultz and T.P. Schultz, *J. Chem. Inf. Comput. Sci.* **34,** 1151 (1994).
211. H.P. Schultz and E.B. Schultz, and T.P. Schultz, *J. Chem. Inf. Comput. Sci.* **35,** 864 (1995).
212. H.P. Schultz and E.B. Schultz, and T.P. Schultz, *J. Chem. Inf. Comput. Sci.* **36** 996 (1996).
213. B. Mohar, in *MATH/CHEM/COMP 1988* (A. Graovac, ed.), pp. 1–8. Elsevier, Amsterdam, 1989.
214. N. Trinajstić, D. Babić, S. Nikolić, D. Plavić, D. Amić, and Z. Mihalić, *J. Chem. Inf. Comput. Sci.* **34,** 277 (1994).
215. M. Barysz, G. Jashari, R.S. Lall, V.K. Srivastava, and N. Trinajstić, *Stud. Phys. Theor. Chem.* **28,** 222 (1983).
216. A.T. Balaban, *J. Chem. Inf. Comput. Sci.* **34,** 398 (1994).
217. A.T. Balaban, D. Bonchev, and W.A. Seitz, *J. Mol. Struct. (THEOCHEM)* **280,** 253 (1993).
218. O. Ivanciuc, T.S. Balaban, and A.T. Balaban, *J. Math. Chem.* **12,** 309 (1993).
219. D. Plavsić, S. Nikolić, N. Trinajstić, and Z. Mihalić, *J. Math. Chem.* **12,** 235 (1993).
220. D.J. Klein and M. Randić, *J. Math. Chem.* **12,** 85 (1993).
221. A.T. Balaban and C. Catana, *J. Comput. Chem.* **14,** 155 (1993).
222. P.A. Filip, S. Balaban, and A.T. Balaban, *J. Math. Chem.* **1,** 61 (1987).
223. A.T. Balaban, *Stud. Phys. Theor. Chem.* **51,** 159 (1987).
224. L.B. Kier and L.H. Hall, *Eur. J. Med. Chem.* **12,** 307 (1977).
225. L.B. Kier and L.H. Hall, *J. Pharm. Sci.* **70,** 583 (1981).
226. L.H. Hall and L.B. Kier, *Bull. Environ. Contam. Toxicol.* **32,** 354 (1984).
227. L.B. Kier and L.H. Hall, *Pharm. Res.* **6,** 497 (1989).
228. L.B. Kier and L.H. Hall, *Quant. Struct.-Act. Relat.* **10,** 134 (1991).
229. L.B. Kier, *Quant. Struct.-Act. Relat.* **4,** 109 (1985).
230. L.B. Kier, *Quant. Struct.-Act. Relat.* **5,** 1 (1986).

231. L.B. Kier, *Quant. Struct.-Act. Relat.* **6,** 8 (1987).
232. L.B. Kier, *Acta Pharm. Jugosl.* **36,** 171 (1986).
233. L.B. Kier, in *Computational Chemical Graph Theory* (D. H. Rouvray, ed.), pp. 152–174, Nova Science Publishers, New York, 1990.
234. L.H. Hall and L.B. Kier, *Rev. Comput. Chem.* **2,** 367 (1991).
235. W. Fisanick, K. Criss, and A. Rusink, III, *Tetrahedron Comput. Methodol.* **3,** 635 (1990).
236. C-W. von der Lieth, K. Stumpf-Nothof, and U. Prior, *J. Chem. Inf. Comput. Sci.* **36,** 711 (1996).
237. H. Bandemer and M. Otto, *Microchim. Acta* **2,** 93 (1987).
238. A.T. Balaban and M.V. Diudea, *J. Chem. Inf. Comput. Sci.* **33,** 421 (1993).
239. M.V. Diudea, *J. Chem. Inf. Comput. Sci.* **34,** 1064 (1994).
240. M.V. Diudea, M. Topan, and A. Graovac, *J. Chem. Inf. Comput. Sci.* **34,** 1072 (1994).
241. M.V. Diudea, O.V. Minailiuc, and A.T. Balaban, *J. Comput. Chem.* **12,** 527 (1991).
242. L. Pogliani, *J. Phys. Chem.* **99,** 925 (1995).
243. L. Pogliani, *J. Chem. Inf. Comput. Sci.* **36,** 1082 (1996).
244. L. Pogliani, *J. Phys. Chem.* **100,** 18065 (1996).
245. L. Pogliani, *Croat. Chim. Acta* **70,** 803 (1997).
246. L. Pogliani, *J. Chem. Inf. Comput. Sci.* **36,** 104 (1999).
247. L. Pogliani, *J. Phys. Chem. A* **103,** 1598 (1999).
248. L.B. Kier, *J. Pharm. Sci.* **69,** 807 (1980).
249. S.C. Basak, D.K. Harriss, and V.R. Magnuson, *J. Pharm. Sci.* **73,** 429 (1984).
250. D. Bonchev and N. Trinajstić, *J. Chem. Phys.* **67,** 4517 (1977).
251. S. Basak, in *Practical Applications of Quantitative Structure-Activity Relationships (QSAR) in Environment Chemistry and Toxicology* (W. Karcher and J. Devillers, eds.), pp. 83–103. ECSC, EEC, EAEC, Brussels and Luxembourg, 1990.
252. L.B. Kier and L.H. Hall, *Pharm. Res.* **7,** 801 (1990).
253. L.H. Hall, B.K. Mohney, and L.B. Kier, *J. Chem. Inf. Comput. Sci.* **31,** 76 (1991).
254. L.H. Hall, B.K. Mohney, and L.B. Kier, *Quant. Struct.-Act. Relat.* **10,** 43 (1991).
255. L.B. Kier and L.H. Hall, *J. Math. Chem.* **7,** 229 (1991).
256. L.H. Hall, B.K. Mohney, and L.B. Kier, *Quant. Struct.-Act. Relat.* **12,** 44 (1993).
257. L.H. Hall and G.E. Kellogg, *Molconn-Z: Software Package for Molecular Topology Analysis, User's Guide, Version, 3.15.* eduSoft, LC, Ashland, VA, 1997.
258. K. Osmialowski, J. Halkiewicz, and R. Kaliszan, *J. Chromatogr.* **361,** 63 (1986).
259. E. Estrada, *J. Chem. Inf. Comput. Sci.* **35,** 708 (1995).
260. J. Gálvez, R. Garcia, M.T. Salabert, and R. Soler, *J. Chem. Inf. Comput. Sci.* **34,** 520 (1994).
261. Z. Mihalić, S. Nikolić, and N. Trinajstić, *J. Chem. Inf. Comput. Sci.* **32,** 28 (1992).
262. M. Randić, *Int. J. Quantum Chem., Quantum Biol. Symp.* **28,** 60 (1988).
263. B. Bogdanov, S. Nikolić, and N. Trinajstić, *J. Math. Chem.* **3,** 299 (1989).

264. E. Estrada, *J. Chem. Inf. Comput. Sci.* **35,** 31 (1995).
265. E. Estrada and A. Ramirez, *J. Chem. Inf. Comput. Sci.* **36,** 837 (1996).
266. D.R. Armstrong, P.G. Perkins, and J.J.P. Stewart, *J. Chem. Soc., Dalton Trans.,* p. 838 (1973).
267. X.J. Yan, Z.H. Xu, and Q. Dong, *Thermochim. Acta* **194,** 289 (1992).
268. D.A. Morales and O. Arujo, *J. Math. Chem.* **13,** 95 (1993).
269. M. Randić, *New J. Chem.* **20,** 1001 (1996).
270. D. Plavsić, N. Trinajstić, D. Amić, and M. Soskić, *New J. Chem.* **20,** 1075 (1998).
271. E.W. Pitzer, *Tribol. Trans.* **36,** 417 (1993).
272. A. Pyka, *JPC, J. Planar Chromatogr.-Mod. TLC* **9,** 181 (1996).
273. D. Bonchev and O. Mekenyan, *Z. Naturforsch., A* **35A,** 739 (1980).
274. D. Bonchev, O. Mekenyan, and O.E. Polansky, *Z. Naturforsch., A* **36A,** 643 (1981).
275. M. Guo, L. Xu, H. Li, and C.Y. Hu, *Anal. Sci.* **12,** 291 (1996).
276. L.F. Li, X.Z. You, and T.M. Yao, *Main Group Met. Chem.* **20,** 97 (1997).
277. C. Mercier, O. Mekenyan, J.E. Dubois, and D. Bonchev, *Eur. J. Med. Chem.* **26,** 575 (1991).
278. J. Szymanowski, A. Borowiakresterna, and A. Voelkel, *J. Chem. Technol. Biotechnol.* **62,** 233 (1995).
279. A. Pyka, *JPC, J. Planar Chromatogr.-Mod. TLC* **9,** 52 (1996).
280. A. Pyka, *JPC, J. Planar Chromatogr.-Mod. TLC* **11,** 61 (1998).
281. E.A. Smolenskii, *Russ. J. Phys. Chem. (Engl. Transl.)* **38,** 1288 (1964).
282. H. Hosoya, *Bull. Chem. Soc. Jpn.* **44,** 2332 (1971).
283. D. Bonchev, D. Kamenski, and V. Kamenska, *Bull. Math. Biol.* **38,** 119 (1976).
284. O. Mekenyan, D. Bonchev, and N. Trinajstić, *Int. J. Quantum Chem.* **18,** 369 (1980).
285. H. Narumi and H. Hosoya, *Bull. Chem. Soc. Jpn.* **58,** 1778 (1985).
286. Y. Gao and H. Hosoya, *Bull. Chem. Soc. Jpn.* **61,** 3093 (1988).
287. M. Randić, *J. Mol. Struct. (THEOCHEM)* **79,** 45 (1991).
288. L.F. Li and X.Z. You, *Thermochim. Acta* **225,** 85 (1993).
289. I. Lukovits, *Int. J. Quantum Chem., Quantum Biol. Symp.* **19,** 217 (1992).
290. A. Voelkel and J. Szymanowski, *J. Chem. Technol. Biotechnol.* **56,** 279 (1993).
291. A. Kozera and J. Sliwiok, *JPC, J. Planar Chromatogr.-Mod. TLC* **8,** 450 (1995).
292. A. Pyka, *JPC, J. Planar Chromatogr.-Mod. TLC* **9,** 215 (1996).
293. D. Bonchev, O. Mekenyan, G. Protić, and N. Trinajstić, *J. Chromatogr.* **176,** 149 (1979).
294. B. Bogdanov, *Croat. Chim. Acta* **63,** 671 (1991).
295. E.J. Kupchik, *J. Chromatogr.* **630,** 223 (1993).
296. A. Pyka, *JPC, J. Planar Chromatogr.-Mod. TLC* **7,** 108 (1994).
297. A. Bemgard, A. Colmsjo, and K. Wranskog, *Anal. Chem.* **104,** 4288 (1994).
298. R. Gautzsch and P. Zinn, *Chromatographia* **43,** 163 (1996).
299. E. Brendle and E. Papirer, *J. Colloid Interface Sci.* **194,** 207 (1997).
300. E. Brendle and E. Papirer, *J. Colloid Interface Sci.* **194,** 217 (1997).

301. A. Jurić, M. Gagro, S. Nikolić, and N. Trinajstić, *J. Math. Chem.* **11,** 179 (1992).
302. O.G. Mekenyan, T.W. Schultz, G.D. Veith, and V. Kamenska, *J. Appl. Toxicol.* **16,** 355 (1996).
303. Q.G. Huang, W.L. Song, and L.S. Wang, *Chemosphere* **35,** 2849 (1997).
304. L.B. Kier, L.H. Hall, W.J. Murray, and M. Randić, J. Pharm. Sci. **64,** 1971 (1975).
305. J.W. Murray, L.B. Kier, and L.H. Hall, *J. Pharm. Sci.* **64,** 1978 (1975).
306. L.B. Kier, W.J. Murray, and L.H. Hall, *J. Med. Chem.* **18,** 1272 (1975).
307. W.J. Murray, L.B. Kier, and L.H. Hall, *J. Med. Chem.* **19,** 573 (1976).
308. R. Sarkar, A.B. Roy, and P.K. Sarkar, *Math. Biosci.* **39,** 299 (1978).
309. S.K. Ray, S.C. Basak, C. Raychaudhury, A.B. Roy, and J.J. Ghosh, *Indian J. Chem., Sect. B* **20B,** 894 (1981).
310. S.K. Ray, S.C. Basak, C. Raychaudhury, A.B. Roy, and J.J. Ghosh, *Arzneim.-Forsch.* **32,** 322 (1982).
311. S.K. Ray, S.C. Basak, C. Raychaudhury, A.B. Roy, and J.J. Ghosh, *Arzneim.-Forsch.* **33,** 352 (1983).
312. S.C. Basak, D.K. Harriss, and V.R. Magnuson, *J. Pharm. Sci.* **73,** 429 (1984).
313. L.B. Kier, *Quant. Struct.-Act. Relat.* **4,** 109 (1985).
314. I. Lukovits, *J. Chem. Soc., Perkin Trans. 2,* p. 1667 (1988).
315. A. Vera, M. Montes, J.L. Usero, and J. Casado, *J. Pharm. Sci.* **81,** 791 (1992).
316. E.P. Jaeger, P.C. Jurs, and T.R. Stouch, *Eur. J. Med. Chem.* **28,** 275 (1993).
317. D. Bonchev, C.F. Mountain, W.A. Seitz, and A.T. Balaban, *J. Med. Chem.* **36,** 1562 (1993).
318. S. Nikolić, M. Medicsarić, and J. Matijevicsosa, *Croat. Chem. Acta* **66,** 151 (1993).
319. S. Mendiretta and A.K. Madan, *J. Chem. Inf. Comput. Sci.* **34,** 867 (1994).
320. B. Llorente, N. Rivero, R. Carrasco, R.S. Martinez, and E. Estrada, *Quant. Struct.-Acta. Relat.* **13,** 419 (1994).
321. A.K. Srivastava, N. Mishra, D.K. Gupta, and A.A. Khan, *J. Indian Chem. Soc.* **72,** 107 (1995).
322. E. Estrada and A. Ramirez, *J. Chem. Inf. Comput. Sci.* **36,** 833 (1996).
323. T. Langer and R.D. Hoffmann, *Quant. Struct.-Act. Relat.* **17,** 211 (1998).
324. L.H. Hall and T.A. Vaughn, *Med. Chem. Res.* **7,** 407 (1998).
325. A.R. Katritzky and E.V. Gordeeva, *J. Chem. Inf. Comput. Sci.* **33,** 835 (1993).
326. Z. Mihalić, S. Nikolić, and N. Trinajstić, *J. Chem. Inf. Comput. Sci.* **32,** 28 (1992).
327. D.E. Needham, I.-C. Wei, and P.G. Seybold, *J. Am. Chem. Soc.* **110,** 4186 (1988).
328. I. Lukovits and W. Linert, *J. Chem. Inf. Comput. Sci.* **34,** 899 (1994).
329. M. Randić, *Chem. Phys. Lett.* **211,** 478 (1993).
330. L.H. Hall and C.T. Story, *J. Chem. Inf. Comput. Sci.* **36,** 1004 (1996).
331. L.H. Hall and L.B. Kier, *J. Chem. Inf. Comput. Sci.* **35,** 1039 (1995).
332. R.D. Cramer, III, *J. Am. Chem. Soc.* **102,** 1837 (1980).
333. I. Motoc and A.T. Balaban, *Rev. Roum. Chim.* **26,** 593 (1981).

334. J.K. Labanowski, I. Motoc, and R.A. Dammkoehler, *Comput. Chem.* **15,** 47 (1991).
335. D.H. Rouvray, *J. Mol. Struct. (THEOCHEM)* **185,** 187 (1989).
336. M. Randić and P.G. Seybold, *SAR QSAR Environ. Res.* **1,** 77 (1993).
337. M. Randić, *J. Chem. Inf. Comput. Sci.* **31,** 311 (1991).
338. Y. Michotte and D.L. Massart, *J. Pharm. Sci.* **66,** 1630 (1977).
339. J.S. Millership and A.D. Woolfson, *J. Pharm. Pharmacol.* **30,** 483 (1978).
340. M. Randić, *J. Chromatogr.* **161,** 1 (1978).
341. L.B. Kier and L.H. Hall, *J. Pharm. Sci.* **68,** 120 (1979).
342. A. Sablijić, *J. Chromatogr.* **319,** 1 (1985).
343. D. Amić, D. Davidović-Amić, and N. Trinajstić, *J. Chem. Inf. Comput. Sci.* **35,** 136 (1995).
344. M. Randić, *New J. Chem.* **15,** 517 (1991).
345. M. Randić, *J. Chem. Inf. Comput. Sci.* **31,** 311 (1991).
346. E. Estrada and A. Ramirez, *J. Chem. Inf. Comput. Sci.* **36,** 833 (1996).
347. G. Rücker and C. Rücker, *J. Chem. Inf. Comput. Sci.* **33,** 683 (1993).
348. K.J. Rossife, *Chem. Rev.* **96,** 3201 (1996).
349. J.E. Amoore, *Nature (London)* **233,** 270 (1971).
350. J.C. Dearden, *Food Qual. Preferemce* **5,** 81 (1994)
351. H. Boelens and P. Punter, *3rd ECRO Congr. Eur. Chemoreception Organ.*, Pavia, Italy, 1978, pp. 1–9.
352. D. Amić, D. Davidović-Amić, A. Jurić, B. Lučić and N. Trinajstić, *J. Chem. Inf. Comput. Sci.* **35,** 1034 (1995).
353. Z. Zhou and G.R. Parr, *J. Am. Chem. Soc.* **111,** 7371 (1989).
354. I. Gutmann, M. Milun, and N. Trinajstić, *J. Am. Chem. Soc.* **99,** 1692 (1977).
355. J. Gálvez, R. Garcia-Domenoch, V. De Julián-Ortiz, and R. Soler, *J. Chem. Inf. Comput. Sci.* **34,** 1198 (1994).
356. K. Osmialowski and R. Kaliszan, *Quant. Struct.-Act. Relat.* **10,** 125 (1991).
357. G.G. Cash and J.J. Breen, *Chemosphere* **24,** 1607 (1992).
358. S.C. Basak, L.J. Monsrud, M.E. Rosen, C.M. Frane, and V.R. Magnuson, *Acta Pharm. Jugosl.* **36,** 81 (1986).
359. G.G. Cash and J.J. Breen, *J. Chem. Inf. Comput. Sci.* **33,** 275 (1993).
360. J. Gálvez, R. Garcia-Domenoch, V, De Julián-Ortiz, and R. Soler, *J. Chem. Inf. Comput. Sci.* **35,** 272 (1995).
361. H. Wiener, *J. Am. Chem. Soc.* **69,** 2636 (1947).
362. H. Wiener, *J. Chem. Phys.* **15,** 766 (1947).
363. A. Goel and A.K. Madan, *J. Chem. Inf. Comput. Sci.* **35,** 504 (1995).
364. A. Goel and A.K. Madan, *J. Chem. Inf. Comput. Sci.* **35,** 510 (1995).
365. S.C. Basak, G.J. Niemi, and G.D. Veith, *J. Math. Chem.* **4,** 185 (1990).
366. J. Gasteiger and M. Marsili, *Tetrahedron Lett.,* p. 3181 (1978).
367. J. Gasteiger and M. Marsili, *Tetrahedron* **36,** 3219 (1980).
368. R.T. Sanderson, *Science* **144,** 670 (1951).
369. R.T. Sanderson, *Chemical Periodicity.* Reinhold, New York, 1960.

370. M. Marsili and J. Gasteiger, *Croat. Chem. Acta* **53,** 601 (1980).
371. M.D. Guillen and J. Gasteiger, *Tetrahedron* **39,** 1331 (1983).
372. J. Gasteiger and M.D. Guillen, *J. Chem. Res., Miniprint,* p. 2611 (1983).
373. N.S. Zefirov, M.A. Kirpichenok, F.F. Izmailov, and M.I. Trofimov, *Dokl. Akad. Nauk SSSR* **296,** 883 (1987).
374. M.A. Kirpichenok and N.S. Zefirov, *Zh. Org. Khim.* **23,** 4 (1987).
375. J. Mullay, *J. Am. Chem. Soc.* **106,** 6842 (1984).
376. J. Mullay, *J. Am. Chem. Soc.* **108,** 1770 (1986).
377. L.H. Hall and L.B. Kier, *Tetrahedron* **33,** 1953 (1977).
378. D. Bonchev and L.B. Kier, *J. Math. Chem.* **9,** 75 (1992).
379. D. Bonchev and E.V. Gordeeva, *J. Chem. Inf. Comput. Sci.* **35,** 383 (1995).
380. I. Gutman, D. Bonchev, E.V. Gordeeva, and W.A. Seitz, *J. Chem. Inf. Comput. Sci.* **35,** 894 (1995).
381. D. Bonchev, L.B. Kier, and O. Mekenyan, *Int. J. Quantum Chem.* **46,** 635 (1993).
382. S.W. Benson, *J. Am. Chem. Soc.* **97,** 704 (1975).
383. V.A. Palm and N.V. Palm, *Org. React.* **10,** 391 (1973).
384. M.M. Karelson and V.A. Palm, *Org. React.* **11,** 663 (1975).
385. M.M. Karelson, *Org. React.* **11,** 679 (1975).
386. J.E. Huheey, *Inorganic Chemistry: Principles of Structure and Reactivity.* Harper & Row, New York, 1978.
387. R.G. Parr and L.J. Bartolotti, *J. Am. Chem. Soc.* **104,** 3801 (1982).
388. R.J. Abraham and B. Hudson, *J. Comput. Chem.* **6,** 173 (1985).
389. R.J. Abraham and P.E. Smith, *J. Comput. Chem.* **9,** 288 (1988).
390. W.J. Mortier, K. Van Genechten, and J. Gasteiger, *J. Am. Chem. Soc.* **107,** 829 (1985).
391. W.J. Mortier, S.K. Ghosh, and S. Shankar, *J. Am. Chem. Soc.* **108,** 4315 (1986).
392. W.J. Mortier, *Struct. Bonding (Berlin)* **66,** 125 (1987).
393. J. Cioslowski and M. Martinov, *J. Chem. Phys.* **102,** 749 (1995).
394. J. Cioslowski and A. Nanayakkara, *J. Chem. Phys.* **99,** 5151 (1993).
395. L. Uytterhoeven and W.J. Mortier, *J. Chem. Soc., Faraday Trans.* **88,** 2747 (1992).
396. K.T. No, J.A. Grant, and H.A. Scheraga, *J. Phys. Chem.* **94,** 4732 (1990).
397. K.T. No, J.A. Grant, M.S. Jhon, and H.A. Scheraga, *J. Phys. Chem.* **94,** 4740 (1990).
398. T.K. Ghanty and S.K. Ghosh, *J. Mol. Struct. (THEOCHEM)* **95,** 83 (1992).
399. J.A. Pople and D.L. Beveridge, *Approximate Molecular Orbital Theory.* McGraw-Hill, New York, 1970.
400. R.S. Mulliken, *J. Chem. Phys.* **23,** 1833 (1955).
401. B.W. Clare and C.T. Supuran, *J. Pharm. Sci.* **83,** 768 (1994).
402. P. Politzer and R.R. Harris, *Tetrahedron* **27,** 1567 (1971).
403. R.W.F. Bader, *Acc. Chem. Res.* **18,** 9 (1985).
404. R.W.F. Bader, *Atoms in Molecules. A Quantum Theory.* Claredon Press, Oxford, 1990.

405. C.E. Shannon, *Bell Sys. Tech. J.* **27,** 379 (1948).
406. C.E. Shannon, *Bell Sys. Tech. J.* **27,** 623 (1948).
407. W. Weaver and C.E. Shannon, *The Mathematical Theory of Communication,* University of Illinois Press, Urbana, 1949.
408. R.F. Bader, *Acc. Chem. Res.* **8,** 34 (1975).
409. H. Lischka, *J. Am. Chem. Soc.* **99,** 353 (1977).
410. K.B. Wiberg, *J. Am. Chem. Soc.* **102,** 1229 (1980).
411. M.M. Franel, R.F. Hout, and W.J. Hehre, *J. Am. Chem. Soc.* **106,** 563 (1984).
412. A. Warshel, *Acc. Chem. Res.* **14,** 284 (1981).
413. S.L. Price and A.J. Stone, *J. Chem. Phys.* **86,** 2859 (1987).
414. P.A. Kollman, *J. Am. Chem. Soc.* **99,** 4875 (1977).
415. P.A. Kollman, *J. Am. Chem. Soc.* **100,** 2974 (1978).
416. F.A. Momany, *J. Phys. Chem.* **82,** 592 (1978).
417. P.H. Smit, J.L. Derissen, and F.B. Van Duijneveldt, *Mol. Phys.* **37,** 521 (1979).
418. S.R. Cox and D.E. Williams, *J. Comput. Chem.* **2,** 304 (1981).
419. U.C. Singh and P.A. Kollman, *J. Comput. Chem.* **5,** 129 (1984).
420. A.E. Reed, R.B. Weinstock, and F. Weinhold, *J. Chem. Phys.* **83,** 735 (1985).
421. L.E. Chirlian and M.M. Francl, *J. Comput. Chem.* **8,** 894 (1987).
422. D.E. Williams, *J. Comput. Chem.* **9,** 745 (1988).
423. A.E. Reed, F. Weinhold, and L.A. Curtiss, *Chem. Rev.* **88,** 899 (1988).
424. C.M. Breneman and K.B. Wiberg, *J. Comput. Chem.* **11,** 361 (1990).
425. D.E. Williams, *Biopolymers* **29,** 1367 (1990).
426. D.E. Williams, *Rev. Comput. Chem.* **2,** 21 (1991).
427. K.M. Merz, *J. Comput. Chem.* **11,** 749 (1992).
428. R. Montagnini and J. Tomasi, *J. Mol. Struct. (THEOCHEM)* **279,** 131 (1993).
429. G. Rauhut and T. Clark, *J. Comput. Chem.* **14,** 503 (1993).
430. C.I. Bayly, P. Cieplak, W.D. Cornell, and P.A. Kollman, *J. Phys. Chem.* **97,** 10269 (1993).
431. B. Beck, G. Rauhut, and T. Clark, *J. Comput. Chem.* **15,** 1064 (1994).
432. M.M. Francl, C. Carey, L.E. Chirlian, and D.M. Gange, *J. Comput. Chem.* **17,** 367 (1996).
433. J.W. Storer, D.J. Giesen, C.J. Cramer, and D.G. Truhlar, *J. Comput.-Aided Mol. Des.* **9,** 87 (1995).
434. M.J.S. Dewar, E.G. Zoebisch, E.F. Healy, and J.J.P. Stewart, *J. Am. Chem. Soc.* **107,** 3902 (1985).
435. M.J. Frisch, G.W. Trucks, M. Head-Gordon, P.M.W. Gill, M.W. Wong, J.B. Foresman, B.G. Johnson, H.B. Schlegel, M.A. Robb, E.S. Repolgle, R. Gomperts, J.L. Andres, K. Raghavachari, J.S. Binkley, J.J.P. Stewart, and J.A. Pople, *GAUSSIAN92,* Gaussian, Inc., Pittsburgh, PA, 1992.
436. J.J.P. Stewart, *J. Comput. Chem.* **10,** 209 (1989).
437. J.J.P. Stewart, *J. Comput. Chem.* **10,** 221 (1989).
438. R. Franke, *Theoretical Drug Design Methods.* Elsevier, Amsterdam, 1984.
439. O. Kikuchi, *Quant. Struct.-Act. Relat.* **6,** 179 (1987).

440. N. Bodor, Z. Gabanyi, and C.-K. Wong, *J. Am. Chem. Soc.* **111,** 3783 (1989).
441. L. Buydens, D. Massart, and P. Geerlings, *Anal. Chem.* **55,** 738 (1983).
442. G. Klopman and L.D. Iroff, *J. Comput. Chem.* **2,** 157 (1981).
443. M.A. Ordorica, M.L. Velázquez, J.G. Ordonica, J.L. Escobar, and P.A. Lehmann, *Quant. Struct.-Act. Relat.* **12,** 246 (1993).
444. D.T. Stanton and P.C. Jurs, *Anal. Chem.* **62,** 2323 (1990).
445. D.T. Stanton, L.M. Egolf, P.C. Jurs, and M.G. Hicks, *J. Chem. Inf. Comput. Sci.* **32,** 306 (1992).
446. J.S. Murray, T. Brinck, P. Lane, K. Paulsen, and P. Politzer, *J. Mol. Struct. (THEOCHEM)* **307,** 55 (1995).
447. J.S. Murray and P. Politzer, in *Quantitative Treatments of Solute/Solvent Interactions* (P. Politzer and J. . Murray, eds.), pp. 243–289. Elsevier, Amsterdam, 1994.
448. P. Sjöberg, J.S. Murray, T. Brinck, and P. Politzer, *Can. J. Chem.* **68,** 1440 (1990).
449. T. Brinck, J.S. Murray, and P. Politzer, *Int. J. Quantum Chem.* **48,** 73 (1993).
450. T.C. Koopmans, *Physica (Amsterdam)* **1,** 104 (1933).
451. T. Brinck, J. S. Murray, P. Politzer, and R. E. Carter, *J. Org. Chem.* **56,** 2934 (1991).
452. J. S. Murray, T. Brinck, and P. Politzer, *J. Mol. Struct. (THEOCHEM)* **255,** 271 (1992).
453. M. Haberlein, J. S. Murray, T. Brinck, and P. Politzer, *Can. J. Chem.* **70,** 2209 (1992).
454. Scrocco and J. Tomasi, *Adv. Quantum Chem.* **11,** 115 (1978).
455. P. Politzer and D.G. Truhlar, eds., *Chemical Applications of Atomic and Molecular Electrostatic Potentials.* Plenum, New York, 1981.
456. P. Politzer and J.S. Murray, *Rev. Comput. Chem.* **2,** 273 (1991).
457. R.K. Rathak and S.R. Gadre, *J. Chem. Phys.* **93,** 1770 (1993).
458. J.S. Murray, P. Lane, T. Brinck, and P. Politzer, *J. Phys. Chem.* **94,** 844 (1990).
459. J.S. Murray, and P. Politzer, *J. Org. Chem.* **56,** 6715 (1991).
460. J.S. Murray and P. Politzer, *J. Chem. Res., Synop.*, p. 110 (1993).
461. P. Politzer and K.C. Daiker, in *The Force Concept in Chemistry* (B. M. Deb, ed.), Chapter 6. Van Nostrand-Reinhold, New York, 1981.
462. T. Brinck, S. Ranganathan, and P. Politzer, *J. Org. Chem.* **56,** 3736 (1991).
463. T. Brinck, J.S. Murray, and P. Politzer, *Mol. Phys.* **76,** 609 (1992).
464. P. Politzer, P. Lane, J.S. Murray, and T. Brinck, *J. Phys. Chem.* **96,** 7938 (1992).
465. P. Politzer, J.S. Murray, P. Lane, and T. Brinck, *J. Phys. Chem.* **97,** 729 (1993).
466. J.S. Murray, P. Lane, T. Brinck, and P. Politzer, *J. Phys. Chem.* **97,** 5144 (1993).
467. J.S. Murray, P. Lane, T. Brinck, K. Paulsen, M.E. Grice, and P. Politzer, *J. Phys. Chem.* **97,** 9369 (1993).
468. R. Todeschini and P. Gramatica, in *3D QSAR in Drug Design* (H. Kubinyi, G. Folkers, and Y.C. Martin, eds.), Vol. 2, p. 355. Kluwer/ESCOM, Dordrecht, The Netherlands, 1998.
469. R. Todeschini, G. Moro, R. Boggia, L. Bonati, U. Cosentino, M. Lasagni, and D. Pitea, *Chemometrics Intell. Lab. Syst.* **36,** 65 (1997).

470. R.J. Abraham, L. Griffiths, and P. Loftus, *J. Comput. Chem.* **3,** 407 (1982).
471. S.L. Dixon and P.C. Jurs, *J. Comput. Chem.* **14,** 1460 (1993).
472. N. Bodor and M.-J. Huang, *J. Comput. Chem.* **12,** 1182 (1991).
473. S. Grigoras, *J. Comput. Chem.* **11,** 493 (1990).
474. D.T. Stanton and P.C. Jurs. *Anal. Chem.* **61,** 1328 (1989).
475. D.T. Stanton and P.C. Jurs, *J. Chem. Inf. Comput. Sci.* **31,** 301 (1991).
476. D.T. Stanton and P.C. Jurs, *J. Chem. Inf. Comput. Sci.* **32,** 109 (1992).
477. D.T. Stanton, L.M. Egolf, P.C. Jurs, and M.G. Hicks, *J. Chem. Inf. Comput. Sci.* **32,** 306 (1992).
478. L.M. Egolf and P.C. Jurs, *J. Chem. Inf. Comput. Sci.* **33,** 616 (1993).
479. L.M. Egolf, M.D. Wessel, and P.C. Jurs, *J. Chem. Inf. Comput. Sci.* **34,** 947 (1994).
480. T.M. Nelson and P.C. Jurs, *J. Chem. Inf. Comput. Sci.* **34,** 601 (1994).
481. J.M. Sutter and P.C. Jurs, *J. Chem. Inf. Comput. Sci.* **36,** 100 (1996).
482. B.E. Mitchell and P.C. Jurs, *J. Chem. Inf. Comput. Sci.* **38,** 489 (1998).
483. L.M. Egolf and P.C. Jurs, *J. Chem. Inf. Comput. Sci.* **34,** 947 (1994).
484. A.R. Katritzky, U. Maran, M. Karelson, and V.S. Lobanov, *J. Chem. Inf. Comput. Sci.* **37,** 913 (1997).
485. A.R. Katritzky, S. Sild, and M. Karelson, *J. Chem. Inf. Comput. Sci.* **38,** 840 (1998).
486. C.G. Georgakopoulos, J.C. Kiburis, and P.C. Jurs, *Anal. Chem.* **63,** 2012 (1991).
487. C.G. Georgakopoulos, O.G. Tsika, J.C. Kiburis, and P.C. Jurs, *Anal. Chem.* **63,** 2025 (1991).
488. M.D. Needham and P.C. Jurs, *Anal. Chim. Acta* **258,** 183 (1992).
489. M.D. Needham, K.C. Adams, and P.C. Jurs, *Anal. Chim. Acta* **258,** 199 (1992).
490. J.M. Sutter, T.A. Peterson, and P.C. Jurs, *Anal. Chim. Acta* **342,** 113 (1997).
491. H.L. Engelhardt and P.C. Jurs, *J. Chem. Inf. Comput. Sci.* **37,** 478 (1997).
492. L.M. Egolf and P.C. Jurs, *Ind. Eng. Chem. Res.* **31,** 1798 (1992).
493. B.E. Mitchell and P.C. Jurs, *J. Chem. Inf. Comput. Sci.* **37,** 538 (1997).
494. P.A. Edwards, L.S. Anker, and P.C. Jurs, *Chem. Senses* **16,** 447 (1991).
495. M.D. Wessel, J.M. Sutter, and P.C. Jurs, *Anal. Chem.* **68,** 4237 (1996).
496. B.E. Mitchell and P.C. Jurs, *J. Chem. Inf. Comput. Sci.* **38,** 200 (1998).
497. E.P. Jaeger, T.R. Stouch, and P.C. Jurs, *Eur. J. Med. Chem.* **28,** 275 (1993).
498. D.T. Stanton, W.J. Murray, and P.C. Jurs, *Quant. Struct.-Act. Relat.* **12,** 239 (1993).
499. Lu Xu, J.W. Ball, S.L. Dixon, and P.C. Jurs, *Quant. Struct.-Act. Relat.* **13,** 841 (1994).
500. J.S. Murray, T. Brinck, and P. Politzer, *Int. J. Quantum Chem., Quantum Biol. Symp.* **18,** 91 (1991).
501. J.S. Murray and P. Politzer, *J. Chem. Res., Synop.* **18,** 110 (1992).
502. J.S. Murray, F. Abu-Awwad, and P. Politzer, *J. Phys. Chem. A* **103,** 1853 (1999).
503. R.W. Taft, J.S. Murray, and P. Politzer, in *Quantitative Treatments of Solute/Solvent Interactions* (P. Politzer and J.S. Murray, eds.), pp. 55–82. Elsevier, Amsterdam, 1994.

504. M.H. Abraham, P.L. Grellier, D.V. Prior, R.W. Taft, J.J. Morris, P.J. Taylor, C. Lauren, M. Berthelot, R.M. Doherty, M.J. Kamlet, J.-L.M. Abboud, K. Sraidi, and J. Guiheneuf, *J. Am. Chem. Soc.* **10,** 8534 (1988).
505. P. Politzer, J.S. Murray, M.E. Grice, M. DeSalvo, and E. Miller, *Mol. Phys.* **91,** 923 (1997).
506. P. Politzer, J.S. Murray, and P. Flodmark, *J. Phys. Chem.* **100,** 5538 (1996).
507. J.S. Murray, S.G. Gagarin, and P. Politzer, *J. Phys. Chem.* **99,** 12081 (1995).
508. J.S. Murray, T. Brinck, and P. Politzer, *Chem. Phys.* **204,** 289 (1996).
509. P. Politzer, J.S. Murray, and P. Flodmark, *J. Phys. Chem. A* **102,** 1018 (1998).
510. P.G. De Benedetti, M.C. Menziani, M. Cocchi, and C. Frassineti, *Quant. Struct.-Act. Relat.* **6,** 51 (1987).
511. R.D. Cramer, III, D.E. Patterson, and J.D. Bunce, *J. Am. Chem. Soc.* **110,** 5959 (1988).
512. K.H. Kim. G. Greco, and E. Novellino, in *3D QSAR in Drug Design* (H. Kubinyi, G. Folkers, and Y.C. Martin eds.), Vol. 3, pp. 257–316. Kluwer/ESCOM, Dordrecht, The Netherlands, 1998.
513. K.H. Kim, in *3D QSAR in Drug Design* (H. Kubinyi, G. Folkers, Y.C. Martin, eds.), Vol. 3, pp. 317–338. Kluwer/ESCOM, Dordrecht, The Netherlands, 1998.
514. K.H. Kim, *J. Comput.-Aided Mol. Des.* **7,** 71 (1993).
515. A.K. Debnath, C. Hansch, K.H. Kim, and Y.C. Martin, *J. Med. Chem.* **36,** 1007 (1993).
516. S. Wold, C. Albano, W.J. Dunn, III, U. Edlund, K. Esbensen, P. Geladi, P.S. Hellberg, E. Johansson, W. Lindberg, and M. Sjöström, in *CHEMOMETRICS: Mathematics and Statistics in Chemistry* (B. Kowalski, ed.), pp. 17 ff. Reidel, Dordrecht, The Netherlands, 1984.
517. S. Wold, A. Ruhe, H. Wold, and W.J. Dunn, III, *SIAM J. Sci. Stat. Comput.* **5,** 735 (1984).
518. R.D. Cramer, J.D. Bunce, and D.E. Patterson, *Quant. Struct.-Act. Relat.* **7,** 18 (1988).
519. I.N. Levine, *Physical Chemistry,* p. 731. McGraw-Hill, New York, 1978.
520. G.E. Kellogg, S.F. Semus, and D.J. Abraham, *J. Comput.-Aided Mol. Des.* **5,** 545 (1991).
521. G.E. Kellogg and D.J. Abraham, in *3D QSAR in Drug Design* (H. Kubinyi, ed.), pp. 506–522. ESCOM, Leiden, The Netherlands, 1993.
522. G.E. Kellogg and D.J. Abraham, *HINT Software,* eduSoft, LC, Ashland, VA, 1993.
523. A. Leo, *Chem. Rev.* **93,** 1281 (1993).
524. C. Hansch and A. Leo, *Exploring QSAR,* Chapter 5. American Chemical Society, Washington, DC, 1995.
525. P. Gaillard, P.-A. Carrupt, and B. Testa, *J. Mol. Graphics* **12,** 73 (1994).
526. P. Gaillard, P.-A. Carrupt, B. Testa and A. Boudon, *J. Comput.-Aided Mol. Des.* **8,** 83 (1994).
527. B. Testa, P.-A. Carrupt, P. Gaillard, F. Billois, and P. Weber, *Pharm. Res.* **13,** 335 (1996).
528. P.-A. Carrupt, B. Testa, and P. Gaillard, *Rev. Comput. Chem.* **11,** 241 (1997).

529. P.W. Atkins, *Quanta: A Handbook of Concepts,* 2nd ed., p. 121. Oxford University Press, Oxford, 1991.
530. P. Gaillard, P.-A. Garrupt, B. Testa, and P. Schambel, *J. Med. Chem.* **39,** 126 (1996).
531. C.L. Waller and G.R. Marshall, *J. Med. Chem.* **36,** 2390 (1993).
532. K.H. Kim, G. Greco, E. Novellino, C. Silipo, and A. Vittoria, *J. Comput.-Aided Mol. Design* **7,** 263 (1993).
533. K.H. Kim, *Med. Chem. Res.* **1,** 259 (1991).
534. K.H. Kim, *Quant. Struct.-Act. Relat.* **12,** 232 (1993).
535. P.W. Kenny, *J. Chem. Soc., Perkin Trans. 2,* p. 199 (1994).
536. C.L. Waller and G.R. Marshall, *J. Med. Chem.* **36,** 2390 (1993).
537. A. Poso, K. Tuppurainen, and J. Gynther, *J. Mol. Struct. (THEOCHEM)* **304,** 255 (1994).
538. C. Navajas, A. Poso, K. Tuppurainen, and J. Gynther, *Quant. Struct.-Act. Relat.* **15,** 189 (1996).
539. G. Klebe, U. Abraham, and T. Mietzner, *J. Med. Chem.* **37,** 4130 (1994).
540. G.E. Kellogg, L.B. Kier, P. Gaillard, and L.H. Hall, *J. Comput.-Aided Mol. Des.* **10,** 51 (1996).
541. R.T. Kroemer and P. Hecht, *J. Comput.-Aided Mol. Des.* **9,** 205 (1995).
542. R.T. Kroemer, P. Hecht, and K.R. Liedl, *J. Comput. Chem.* **17,** 1296 (1996).
543. K.H. Kim and Y.C. Martin, *J. Org. Chem.* **56,** 2723 (1991).
544. G. Folkers, A. Merz, and D. Rognan, in *3D QSAR in Drug Design* (H. Kubinyi, ed.), pp. 603–605. ESCOM, Leiden, The Netherlands, 1993.
545. U. Norinder, in *3D QSAR in Drug Design* (H. Kubinyi, G. Folkers, and Y.C. Martin, eds.), Vol. 3, pp. 25–39. Kluwer/ESCOM, Dordrecht, The Netherlands, 1998.
546. G. Klebe and U. Abraham, *J. Med. Chem.* **36,** 70 (1993).
547. C.L. Waller, T.I. Oprea, A. Giolitti, and G.R. Marshall, *J. Med. Chem.* **36,** 4152 (1993).
548. S.A. DePriest, D. Mayer, C.B. Naylor, and G.R. Marshall, *J. Am. Chem. Soc.* **115,** 5372 (1993).
549. W. Brandt, T. Lehmann, C. Willkomm, S. Fittkau, and A. Barth, *Int. J. Pept. Res.* **46,** 73 (1995).
550. R. T. Kroemer, P. Ettmayer, and P. Hecht, *J. Med. Chem.* **38,** 4917 (1995).
551. A.M. Gamper, R.H. Winger, K.R. Liedl, C.A. Sotriffer, J.M. Varga, R.T. Kroemer, and B.M. Rode, *J. Med. Chem.* **39,** 3882 (1996).
552. S.J. Cho and A. Tropsha, *J. Med. Chem.* **38,** 1060 (1995).
553. R.T. Kroemer and P.A. Hecht, *J. Comput.-Aided Mol. Des.* **9,** 396 (1995).
554. R.X. Wang, Y. Gao, L. Liu, and L.H. Lai, *J. Mol. Model.* **4,** 276 (1998).
555. R.T. Kroemer, P.A. Hecht, S. Guessregen, and K.R. Liedl, in *3D QSAR in Drug Design,* (H. Kubinyi, G. Folkers, and Y.C. Martin, eds.), Vol. 3, kpp. 41–56. Kluwer/ESCOM, Dordrecht, The Netherlands, 1998.
556. Y.C. Martin, M.G. Bures, E.A. Danaher, J. DeLazzer, I. Lico, and P.A. Pavlik, *J. Comput.-Aided Mol. Des.* **7,** 83 (1993).

557. R. Bureau, J.C. Lancelot, J. Prunier, and S. Rault, *Quant. Struct.-Act. Relat.* **15,** 373 (1996).
558. A.E. Medvedev, A.S. Ivanov, A.V. Veselovsky, V.S. Skvortsov, and A.I. Archakov, *J. Chem. Inf. Comput. Sci.* **36,** 664 (1996).
559. F. Corelli, F. Manetti, A. Tafi, G. Campiani, V. Nacci, and M. Botta, *J. Med. Chem.* **40,** 125 (1997).
560. T. Oprea and C.L. Waller, *Rev. Comput. Chem.* **11,** 127 (1997).
561. J.A. Calder, J.A. Wyatt, D.A. Frenkel, and J.E. Casida, *J. Comput.-Aided Mol. Des.* **7,** 45 (1993).
562. C.L. Waller and G.R. Marshall, *J. Med. Chem.* **36,** 2390 (1993).
563. J. P. Horwitz, I. Massova, T.E. Wiese, B.H. Besler, and T.H. Corbett, *J. Med. Chem.* **37,** 781 (1994).
564. M.F. Parretti, R.T. Kroemer, J.H. Rothman, and W.G. Richards, *J. Comput. Chem.* **18,** 1344 (1997).
565. B. Kowalski, R. Gerlach, and H. Wold, in *Systems under Indirect Observation* (K. Jörenskog and H. Wold, eds.), pp. 191–206. North-Holland, Amsterdam, 1982.
566. S. Wold, A. Ruhe, H. Wold, and W. Dunn, *SIAM J. Sci. Stat. Comput.* **5,** 735 (1984).
567. P. Geladi and B.R. Kowalski, *Anal. Chim. Acta* **185,** 1 (1986).
568. R.H. Myers, *Classical and Modern Regression with Applications.* PWS-KENT Publ. Co., Boston, 1990.
569. A. Agarwal, P.P. Pearson, E.W. Taylor, H.B. Li, T. Dahlgren, M. Herslof, Y. Yang, G. Lambert, D.L. Nelson, J.W. Regan, and A.R. Martin, *J. Med. Chem.* **36,** 4006 (1993).
570. M. Baroni, G. Constantino, G. Cruciani, D. Riganelli, R. Valigi, and S. Clementi, *Quant. Struct.-Act. Relat.* **12,** 9 (1993).
571. D.D. Robinson, P.J. Winn, P.D. Lyne, and W.G. Richards, *J. Med. Chem.* **42,** 573 (1999).
572. M.C. Nicklaus, G.W.A. Milne, and T.R. Burke, *J. Comput.-Aided Mol. Des.* **6,** 487 (1992).
573. S.A. DePriest, D. Mayer, C.B. Naylor, and G.R. Marshall, *J. Am. Chem. Soc.* **115,** 5372 (1993).
574. G. Greco, E. Novellino, M. Pellechia, C. Silipo, and A. Vittoria, *J. Comput.-Aided Mol. Des.* **8,** 97 (1994).
575. K. Raghawan, J.K. Buolamwini, M.R. Fesen, Y. Pommier, K.W. Kohn, and J.N. Weinstein, *J. Med. Chem.* **38,** 890 (1995).
576. U. Thull, S. Kneubuhler, P. Gaillard, P.-A. Carrupt, B. Testa, C. Altomare, A. Carotti, P. Jenner, and K.S.P. McNaught, *Biochem. Pharmacol.* **50,** 869 (1995).
577. S.J. Cho, M.L.S. Garsia, J. Bier, and A. Tropsha, *J. Med. Chem.* **39,** 5064 (1996).
578. J.P. Jones, M. He, W.F. Trager, and A.E. Rettie, *Drug. Metab. Dispos.* **24,** 1 (1996).
579. K.S.P. McNaught, U. Thull, P.-A. Carrupt, C. Altomare, S. Cellamare, A. Carotti, B. Testa, P. Jenner, and C.D. Marsden, *Biochem. Pharmacol.* **51,** 1503 (1996).
580. W. Tong, E.R. Collantes, Y. Chen, and W.J. Welsh, *J. Med. Chem.* **39,** 380 (1996).

581. S. Kim, M.W. Chi, C.N. Yoon, and H.C. Sung, *J. Biochem. Mol. Biol.* **31,** 459 (1998).
582. M. Recanatini and A. Cavalli, *Bioorg. Med. Chem.* **6,** 377 (1998).
583. A.K. Debnath, *J. Chem. Inf. Comput. Sci.* **38,** 761 (1998).
584. A.M. Davis, N.P. Gensmantel, E. Juhansson, and D.P. Marriott, *J. Med. Chem.* **37,** 963 (1994).
585. F. Corelli, F. Manetti, A. Tafi, G. Campiani, V. Nacci, and M. Botta, *J. Med. Chem.* **40,** 125 (1997).
586. P.W. Swaan, F.C. Szoka, and S. Oie, *J. Comput.-Aided Mol. Des.* **11,** 581 (1997).
587. S. Rault, R. Bureau, J.C. Pilo, and M. Robba, *J. Comput.-Aided Mol. Des.* **6,** 553 (1992).
588. D.A. Loughney and C.F.A. Schwender, *J. Comput.-Aided Mol. Des.* **6,** 569 (1992).
589. J.A. Calder, J.A. Wyatt, D.A. Frenkel, and J.E. Casida, *J. Comput.-Aided Mol. Des.* **7,** 45 (1993).
590. C.L. Walker and G.R. Marshall, *J. Med. Chem.* **36,** 2390 (1993).
591. J.S. Tokarski and A.J. Hopfinger, *J. Med. Chem.* **37,** 3639 (1994).
592. J.D. Burke, W.J. Dunn, III, and A.J. Hopfinger, *J. Med. Chem.* **37,** 3768 (1994).
593. Y.C. Martin, C.T. Lin, and J. Wu, in *3D QSAR in Drug Design* (H. Kubinyi, ed.), pp. 643–660. ESCOM, Leiden, The Netherlands, 1993.
594. A. Morreale, E. Galvez-Ruano, I. Iriepa-Canalda, and D.B. Boyd, *J. Med. Chem.* **41,** 2029 (1998).
595. S.M. Bromidge, S. Dabbs, D.T. Davies, D.M. Duckworth, I.T. Forbes, P. Ham, G.E. Jones, F.D. King, D.V. Saunders, S. Starr, K.M. Thewlis, P.A. Wyman, F.E. Blaney, C.B. Naylor, F. Bailey, T.P. Blackburn, V. Holland, G.A. Kennett, G.J. Riley, and M.D. Wood, *J. Med. Chem.* **41,** 159 (1998).
596. R.T. Kroemer, E. Koutsilieri, P. Hecht, K.R. Liedl, P. Riederer, and J. Kornhuber, *J. Med. Chem.* **41,** 393 (1998).
597. R.E. Wilcox, T. Tseng, M.Y.K. Brusniak, B. Ginsburg, R.S. Pearlman, M. Teeter, C. DuRand, S. Starr, and K.A. Neve, *J. Med. Chem.* **41,** 4385 (1998).
598. Y. Takeuchi, E.F.B. Shands, D.D. Beusen, and G.R. Marshall, *J. Med. Chem.* **41,** 3609 (1998).
599. D.A. Winkler and F.R. Burden, *Quant. Struct.-Act. Relat.* **17,** 224 (1998).
600. R. Xu, M.K. Sim, and M.L. Go, *Chem. Pharm. Bull.* **46,** 231 (1998).
601. G.D. Diana, P. Kowalczyk, A.M. Treasurywala, R.C. Oglesby, D.C. Peaver, and F.J. Dutko, *J. Med. Chem.* **35,** 1002 (1992).
602. C.L. Waller, T.I. Oprea, A. Giolitti, and G.R. Marshall, *J. Med. Chem.* **36,** 4152 (1993).
603. G.D. Diana, T.J. Nitz, J.P. Mallamo, and A. Treasurywala, *Antiviral Chem. Chemother.* **4,** 1 (1993).
604. T.I. Oprea, C.L. Waller, and G.R. Marshall, *J. Med. Chem.* **37,** 2206 (1994).
605. J.P. Horwitz, I. Massova, T.E. Wiese, B.H. Besler, and T.H. Corbett, *J. Med. Chem.* **37,** 781 (1994).
606. S.R. Krystek, J.T. Hunt, Jr., P.D. Stein, and T.R. Stouch, *J. Med. Chem.* **38,** 659 (1995).

607. S.J. Cho, A. Tropsha, M. Suffness, Y.C. Cheng, and K.H. Lee, *J. Med. Chem.* **39,** 1383 (1996).
608. S.W. Carrigan, P.C. Fox, M.E. Wall, M.C. Wani, and J.P. Bowen, *J. Comput.-Aided Mol. Des.* **11,** 71 (1997).
609. N.S. Zefirov, V.A. Palyulin, and E.V. Radchenko, *Dokl. Akad. Nauk SSSR* **352,** 630 (1997).
610. Y.L. Li, A.D. MacKerell, M.J. Egorin, M.F. Ballesteros, D.M. Rosen, Y.Y. Wu, D.A. Blamble, and P.S. Callery, *Cancer Res.* **57,** 234 (1997).
611. J.R. Woolfrey, M.A. Avery, and A.M. Doweyko, *J. Comput.-Aided Mol. Des.* **12,** 165 (1998).
612. C.L. Waller and J.D. McKinney, *J. Med. Chem.* **35,** 3660 (1992).
613. A.K. Debnath, C. Hansch, K.H. Kim, and Y.C. Martin, *J. Med. Chem.* **36,** 1007 (1993).
614. G. Caliendo, C. Fattorusso, G. Greco, E. Novellino, E. Persisutti, and V. Santagada, SAR QSAR Environ. Res. **4,** 21 (1994).
615. A. Poso, K. Tuppurainen, and J. Gynther, *J. Mol. Struct. (THEOCHEM)* **304,** 255 (1994).
616. A. Poso, K. Tuppurainen, and J. Gynther, *J. Mol. Graphics* **12,** 70 (1994).
617. F. Briens, R. Bureau, S. Rault, and M. Robba, *Ecotoxicol. Environ, Saf.* **31,** 37 (1995).
618. K.H. Kim, *Med. Chem. Res.* **7,** 45 (1997).
619. Y. Kamano, K. Kotake, H. Hashima, M. Inoue, H. Morita, K. Takeya, H. Itokawa, N. Nandachi, T. Segawa, A.Yukita, K. Saitou, M. Katsuyama, and G.R. Pettit, *Bioorg. Med. Chem.* **6,** 1103 (1998).
620. I.K. Pajeva and M. Wiese, *Quant. Struct.-Act. Relat.* **17,** 301 (1988).
621. K.H. Kim, in *Classical and Three-Dimensional QSAR in Agrochemistry* (C. Hansch, and T. Fujita, eds.), Chapter 23. American Chemical Society, Washington, DC, 1995.
622. Y. Nakagawa, B. Shimizu, N. Oikawa, M. Akamatsu, K. Nishimura, N. Kurihara, T. Ueno, and T. Fujita, in *Classical and Three-Dimensional QSAR in Agrochemistry* (C. Hansch and T. Fujita, eds.), Chapter 23. American Chemical Society, Washington, DC, 1995.
623. Y. Nakagawa, Y. Soya, K. Nakai, N. Oilawa, K. Nishimura, N. Kurihara, T. Ueno, and T. Fujita, *Pestic. Sci.* **43,** 339 (1995).
624. S.-E. Yoo and O.J. Cha, *J. Comput. Chem.* **16,** 449 (1995).
625. M.H. Abraham, P.P. Duce, D.V. Prior, D.G. Barrett, J.J. Morris, and P.J. Taylor, *J. Chem. Soc., Perkin Trans. 2*, p. 1355 (1989).
626. A. Tropsha and S.J. Cho, in *3D QSAR in Drug Design* (H. Kubinyi, G. Folkers, and Y.C. Martin, eds.), Vol. 3, pp. 57–69. Kluwer/ESCOM, Dordrecht, The Netherlands, 1998.
627. M. Karelson, V.S. Lobanov, and A.R. Katritzky, *Chem. Rev.* **96,** 1027 (1996).
628. J.N. Murrell, S.F. Kettle, and J.M. Tedder, *The Chemical Bond.* Wiley, Chichester, 1985.
629. W.J. Hehre, L. Radom, P. von R. Schleyer, and J.A. Pople, *Ab Initio Molecular Orbital Theory.* Wiley, New York, 1986.

630. P. Jøgensen and J. Simons, eds., *Geometrical Derivatives of Energy Surfaces and Molecular Properties.* Reidel, Dordrecht, The Netherlands, 1986.
631. E.R. Davidson, *Rev. Comput. Chem.* **1,** 373 (1990).
632. D.R. Hartree, *Proc. Cambridge Philos. Soc.* **24,** 89, 111, 426 (1928).
633. V. Fock, *Z. Phys.* **61,** 126 (1930).
634. G.G. Hall, *Proc. R. Soc. London, Ser. A* **205,** 541 (1951).
635. C.C. Roothaan, *Rev. Mod. Phys.* **23,** 69 (1951).
636. P.-O. Löwdin, *Phys. Rev.* **97,** 1509 (1955).
637. C.C. Roothaan, *Rev. Mod. Phys.* **32,** 179 (1960).
638. R. McWeeny and G.H.F. Diercksen, *J. Chem. Phys.* **49,** 4852 (1968).
639. P.-O. Löwdin and I. Mayer, *Adv. Quant. Chem.* **24,** 79 (1992).
640. S. Wilson, in *Methods in Computational Molecular Physics* (G.H.F. Diercksen and S. Wilson, eds.), pp. 71–93. Reidel, Dordrecht, The Netherlands, 1983.
641. D. Feller and E. Davidson, *Rev. Comput. Chem.* **1,** 1 (1990).
642. S. Wilson, in *Problem Solving in Computational Molecular Science* (S. Wilson and G. H. F. Diercksen, eds.), pp. 109–158. Kluwer Acad. Publ., Dordrecht, The Netherlands, 1997.
643. R. McWeeny and B.T. Sutcliffe, *Methods of Molecular Quantum Mechanics.* Academic Press, London, 1969.
644. P. Čarsky and M. Urban, *Ab Initio Calculations. Methods and Applications in Chemistry.* Springer-Verlag, Berlin, 1980.
645. S. Wilson, *Electron Correlation in Molecules.* Clarendon Press, Oxford, 1984.
646. K. Andersson, M.P. Fülsher, R. Lindh, P.-Å. Malmqvist, J. Olsen, and B. Roos, in *Methods and Techniques in Computational Chemistry: METECC-94* (E. Clementi, ed.), Vol. B, Chapter 7, STEF, Cagliari, 1994.
647. R.J. Bartlett and J.F. Stanton, *Rev. Comput. Chem.* **5,** 65 (1994).
648. I. Shavitt, in *Methods of Electronic Structure Theory* (H. F. Schaefer, ed.), pp. 189–276. Plenum, New York, 1977.
649. G.H.F. Diercksen and S. Wilson, eds., *Methods in Computational Molecular Physics.* Reidel, Dordrecht, The Netherlands, 1983.
650. M. Dupuis, ed., *Supercomputer Simulations in Quantum Chemistry*, Lect. Notes Chem., Vol. 44. Springer-Verlag, Berlin, 1986.
651. J.D. Goddard, N.C. Handy, and H.F. Schaefer, *J. Chem. Phys.* **71,** 1525 (1979).
652. R. Krishnan, H.B. Schlegel, and J.A. Pople, *J. Chem. Phys.* **72,** 4654 (1980).
653. J. Jortner, and B. Pullman, eds., *Perspectives in Quantum Chemistry.* Kluwer Acad. Publ., Dordrecht, The Netherlands, 1989.
654. S.A. Kucharski and R.J. Bartlett, *Adv. Quantum Chem.* **18,** 281 (1986).
655. J. Čižek, *J. Chem. Phys.* **45,** 4256 (1966).
656. J.A. Pople, R. Krishnan, H.B. Schlegel, and J.S. Binkley, *Int. J. Quantum Chem.* **14,** 545 (1978).
657. R.J. Bartlett, *J. Phys. Chem.* **93,** 1697 (1989).
658. R.J. Bartlett, in *Modern Electronic Structure Theory* (D. R. Yarkony, ed.), p. 1047. World Scientific Publishing Co., Singapore, 1995.
659. C. Møller and M.S. Plesset, *Phys. Rev.* **46,** 618 (1934).

660. J.A. Pople, J.S. Binkley, and R. Seeger, *Int. J. Quantum Chem., Symp.* **10,** 1 (1976).
661. J. Gauss and D. Cremer, *Adv. Quantum Chem.* **23,** 205 (1992).
662. J.A. Pople and D.L. Beveridge, *Approximate Molecular Orbital Theory.* Mc-Graw-Hill, New York, 1970.
663. M.J.S. Dewar, *Science* **187,** 1037 (1975).
664. J.J.P. Stewart, *Rev. Comput. Chem.* **1,** 45 (1990).
665. A. Streitwieser, *Molecular Orbital Theory for Organic Chemists.* Wiley, New York, 1959.
666. R. Hoffman, *J. Chem. Phys.* **39,** 1397 (1963).
667. J.A. Pople, D.L. Beveridge, and P.A. Dobosh, *J. Chem. Phys.* **47,** 2026 (1967).
668. W.P. Anderson, T. Cundari, R. Drago, and M.C. Zerner, *Inorg. Chem.* **29,** 1 (1990).
669. R.C. Bingham, M.J.S. Dewar, and D.H. Lo, *J. Am. Chem. Soc.* **97,** 1285 (1975).
670. M.J.S. Dewar and W. Thiel, *J. Am. Chem. Soc.* **99,** 4899 (1977).
671. M.J.S. Dewar and W. Thiel, *J. Am. Chem. Soc.* **99,** 4907 (1977).
672. S. Diner, J.P. Malrieu, F. Jordan, and M. Gilbert, *Theor. Chim. Acta* **15,** 100 (1969).
673. D.N. Nanda and K. Jug, *Theor. Chim. Acta* **57,** 95 (1980).
674. W. Thiel, *Tetrahedron* **44,** 7393 (1988).
675. K.M. Gough, K. Belohorcova, and K.L.E. Kaiser, *Sci. Total Environ.* **142,** 179 (1994).
676. T. Sotomatsu, Y. Murata, and T. Fujita, *J. Comput. Chem.* **10,** 94 (1989).
677. P.G. De Benedetti, *J. Mol. Struct. (THEOCHEM)* **256,** 231 (1992).
678. U. Burkett and N.L. Allinger, *Molecular Mechanics.* American Chemical Society, Washington, DC, 1982.
679. A.R. Leach, *Rev. Comput. Chem.* **2,** 1 (1991).
680. J.P. Bowen and N.L. Allinger, *Rev. Comput. Chem.* **2,** 81 (1991).
681. U. Dinur and A.T. Hagler, *Rev. Comput. Chem.* **2,** 99 (1991).
682. I. Pettersson and T. Liljefors, *Rev. Comput. Chem.* **9,** 1 (1996).
683. M. Cocchi, M.C. Menziani, P.G. De Benedetti, and G. Cruciani, *Chemometrics Intell. Lab. Syst.* **14,** 209 (1992).
684. H.A. Kurtz, J.J.P. Stewart, and K.M. Dieter, *J. Comput. Chem.* **11,** 82 (1990).
685. P.W. Atkins, *Quanta.* Oxford University Press, Oxford, 1991.
686. A. Cammarata, *J. Med. Chem.* **10,** 525 (1967).
687. A. Leo, C. Hansch, and C. Church, *J. Med. Chem.* **12,** 766 (1969).
688. C. Hansch and E. Coats, *J. Pharm. Sci.* **59,** 731 (1970).
689. D.V.F. Lewis, *J. Comput. Chem.* **8,** 1084 (1987).
690. H. Sklenar and J. Jäger, *Int. J. Quantum Chem.* **16,** 467 (1979).
691. A.C. Gaudio, A. Korolkovas, and Y. Takahata, *J. Pharm. Sci.* **83,** 1110 (1994).
692. A. Cartier and J.-L. Rivail, *Chemometrics Intell. Lab. Syst.* **1,** 335 (1987).
693. K. Osmialowski, J. Halkiewicz, A. Radecki, and R. Kaliszan, *J. Chromatogr.* **346,** 53 (1985).

694. S.N. Bushelev and N.F. Stepanov, *Z. Naturforsch.* **44,** 212 (1989).

695. Y. Kawakami and A.J. Hopfinger, *Chem. Res. Toxicol.* **3,** 244 (1990).

696. M.N. Ramos and B. de Barros Neto, *J. Comput. Chem.* **11,** 569 (1990).

697. B.W. Clare, *J. Med. Chem.* **33,** 687 (1990).

698. B.W. Clare, *Theor. Chim. Acta* **87,** 415 (1994).

699. K. Fukui, *Theory of Orientation and Stereoselection.* Springer-Verlag, New York, 1975.

700. R.E. Brown and A.M. Simas, *Theor. Chim. Acta* **62,** 1 (1982).

701. J.M. Pieres, W.B. Floriano, and A.C. Gaudio, *J. Mol. Struct. (THEOCHEM)* **389,** 159 (1997).

702. I. Fleming, *Frontier Orbitals and Organic Chemical Reactions.* Wiley, New York, 1976.

703. D.F.V. Lewis, C. Ioannides, and D.V. Parke, *Xenobiotica* **24,** 401 (1994).

704. Z. Zhou and R.G. Parr, *J. Am. Chem. Soc.* **112,** 5720 (1990).

705. R.G. Pearson, *J. Org. Chem.* **54,** 1423 (1989).

706. K. Tuppurainen, S. Lötjönen, R. Laatikainen, T. Vartiainen, U. Maran, M. Strandberg, and T. Tamm, *Mutat. Res.* **247,** 97 (1991).

707. Y.S. Prabhakar, *Drug Des. Delivery* **7,** 227 (1991).

708. A.B. Sannigrahi, *Adv. Quantum Chem.* **23,** 30 (1992).

709. F. Saura-Calixto, A. García-Raso, and M.A. Raso, *J. Chromatogr. Sci.* **22,** 22 (1984).

710. M. Bodor, Z. Gabanyi, and C.-K. Wong, *J. Am. Chem. Soc.* **111,** 3783 (1989).

711. C. Gruber and V. Buss, *Chemosphere* **19,** 1595 (1989).

712. A.J. Shusterman, *CHEMTECH* **21,** 624 (1991).

713. A.K. Debnath, R.L. Lopez de Compadre, A.J. Shusterman, and C. Hansch, *Environ. Mol. Mutagen.* **19,** 53 (1992).

714. A.K. Debnath, A.J. Shusterman, R.L. Lopex de Compadre, and C. Hansch, *Mutat. Res.* **305,** 63 (1994).

715. G. Trapani, A. Carotti, M. Franco, A. Latrofa, G. Genchi, and G. Liso, *Eur. J. Med. Chem.* **28,** 13 (1993).

716. C.M. Breneman and M. Martinov, in *Molecular Electrostatic Potentials: Concepts and Applications* (J. S. Murray and K. Sen, eds.), Vol. 3, pp. 143–179, Elsevier, Amsterdam, 1996.

717. D.A. McQuarrie, *Statistical Thermodynamics.* Harper & Row, New York, 1973.

718. A.I. Akhiezer and S.V. Peltminskii, *Methods of Statistical Physics.* Pergamon, Oxford, 1981.

719. T.W. Heritage, A.M. Ferguson, D.B. Turner, and P. Willett, in *3D QSAR in Drug Design* (H. Kubinyi, G. Folkers, and Y.C. Martin, eds.), Vol. 2, pp. 381–398. Kluwer/ESCOM, Dordrecht, The Netherlands, 1998.

720. S.W. Karickoff, V.K. McDaniel, C.M. Melton, A.N. Vellino, D.E. Nute, and L.A. Carreira, *Environ. Toxicol. Chem.* **10,** 1405 (1991).

721. S.H. Hilal, L.A. Carreira, C.M. Melton, and S.W. Karickoff, *Quant. Struct.-Act. Relat.* **12,** 389 (1993).

722. S.H. Hilal, L.A. Carreira, C.M. Melton, and S.W. Karickoff, *J. Chromatogr.* **662,** 269 (1994).
723. S.H. Hilal, L.A. Carreira, and S.W. Karickoff, in *Quantitative Treatments of Solute/Solvent Interactions* (P. Politzer and J.S. Murray, eds), pp. 291–353. Elsevier, Amsterdam, 1994.
724. C. Ebert, P. Linda, S. Alunni, S. Clementi, G. Cruciani, and S. Santini, *Gazz. Chim. Ital.* **120,** 29 (1990).
725. B. Skagerberg, D. Bonelli, S. Clementi, G. Cruciani, and C. Ebert, *Quant. Struct.-Act. Relat.* **8,** 32 (1989).
726. H. Van De Waterbeemd and B. Testa, *Adv. Drug Res.* **16,** 85–225 (1987).
727. A.R. Katritzky, R.D. Burton, Ming Qi, P.A. Shipkova, C.H. Watson, Z. Dega-Szafran, J.R. Eyler, M. Karelson, U. Maran, and M.C. Zerner, *J. Chem. Soc., Perkin Trans. 2*, p. 825 (1998).
728. W. Langenaeker, K. Demel, and P. Geerlings, *J. Mol. Struct. (THEOCHEM)* **80,** 329 (1991).
729. M. Mestechkin and L. Sivakova, *J. Mol. Struct. (THEOCHEM)* **74,** 117 (1991).
730. G. Del Re, *Theor. Chim. Acta* **85,** 109 (1993).
731. E.N. Grigorieva, S.S. Panchenko, V.Y. Korobkov, and I.V. Kalechitz, *Fuel Proc. Technol.* **41,** 39 (1994).
732. A. Tuulmets and M. Karelson, *Proc. Estonian Acad. Sci, Chem.* **43,** 51 (1994).
733. T.R. Ghanty and S.K. Ghosh, *J. Phys. Chem. A* **101,** 5022 (1997).
734. F. Gilardoni, J. Weber, H. Chermette, and T.R. Ward, *J. Phys. Chem. A* **102,** 3607 (1998)
735. M.G. Väärtnõu, and M.M. Karelson, *Russ J. Electrochem.* **27,** 1366 (1991).
736. M. Väärtnõu, *J. Electroanal. Chem.* **353,** 247 (1993).
737. G.P. Mishra and A.B. Sanniraghi, *J. Mol. Struct. (THEOCHEM)* **361,** 63 (1996).
738. K. Choco, W. Langenaeker, G. Van de Woude, and P. Geerlings, *J. Mol. Struct. (THEOCHEM)* **362,** 305 (1996).
739. A.K. Debnath and C. Hansch, *Environ. Mol. Mutagen.* **20,** 140 (1992).
740. A. Tachibana, S. Kawauchi, K. Nakamura, and H. Inaba, *Int. J. Quant. Chem.* **57,** 673 (1996).
741. F. Mendez, M.D. Romero, F. DeProft, and P. Geerlings, *J. Org. Chem.* **63,** 5774 (1998).
742. R.K. Roy, S. Krishnamurti, P. Geerlings, and S. Pal, *J Phys. Chem. A* **102,** 3746 (1998).
743. P. Fuentealba, *J. Mol. Struct. (THEOCHEM)* **433,** 113 (1998).
744. B.A. Hess, Jr. and L.J. Schaad, *J. Am. Chem. Soc.* **93,** 305 (1971).
745. J.-I. Aihara, *J. Am. Chem. Soc.* **98,** 2750 (1976).
746. I. Gutman, M. Milun, and N. Trinajstić, *J. Am. Chem. Soc.* **99,** 1692 (1977).
747. K. Tuppurainen, S. Lötjönen, R. Laatikainen, and T. Vartiainen, *Mutat. Res.* **266,** 181 (1992).
748. K. Tuppurainen, *J. Mol. Struct. (THEOCHEM)* **306,** 49 (1994).
749. A.K. Debnath, R.L. Lopez de Compadre, and C. Hansch, *Mutat. Res.* **280,** 55 (1992).

750. G. Klopman, D.A. Tonucci, M. Holloway, and H.S. Rosendranz, *Mutat. Res.* **126,** 139 (1984).

751. A.K. Debnath, R.L. Lopez de Compadre, G. Debnath, A.J. Shusterman, and C. Hansch, *J. Med. Chem.* **34,** 786 (1991).

752. A.K. Debnath and C. Hansch, *Mol. Mutagen.* **19,** 53 (1992).

753. A.K. Debnath and C. Hansch, *Mol. Mutagen.* **20,** 140 (1992).

754. A.K. Debnath, G. Debnath, A.J. Shusterman, and C. Hansch, *Environ. Mol. Mutagen.* **19,** 37 (1992).

755. D.F.V. Lewis and D.V. Parke, *Mutat. Res.* **328,** 207 (1995).

756. M.G. Cardozo, Y. Iimura, H. Sugimoto, Y. Yamanishi, and A.J. Hopfinger, *J. Med. Chem.* **35,** 584 (1992).

757. A.C. Gaudio, A. Korolkovas, and Y. Takahata, *J. Pharm. Sci.* **83,** 1110 (1994).

758. R.A. Coburn, M. Weirzba, M.J. Suto, A.J. Solo, A.M. Triggle, and D.J. Triggle, *J. Med. Chem.* **31,** 2103 (1988).

759. H. Itokawa, N. Totsuka, K. Nakahara, M. Maezuru, K. Takeya, M. Kondo, M. Inamatsu, and H. Morita, *Chem. Pharm. Bull.* **37,** 1619 (1989).

760. L. Zhang, H. Gao, C. Hansch, and C.D. Selassie, *J. Chem. Soc., Perkin Trans. 2*, p. 2553 (1997).

761. I.A. Doichinova, R.N. Natcheva, and D.N. Mihailova, *Eur. J. Med. Chem.* **29,** 133 (1994).

762. G. Klopman and R.F. Hudson, *Theor. Chim. Acta* **8,** 165 (1967).

763. S. Sixt, J. Altschuh, and R. Bruggemann, *Chemoshpere* **30,** 2397 (1995).

764. H.J.M. Vehaar, L. Eriksson, M. Sjöström, G. Schüürmann, W. Seinen, and J.L. M. Hermens, *Quant. Struct.-Act Relat.* **13,** 133 (1994).

765. L. Eriksson, S. Rannar, and M. Sjöström, *Environmetrics* **5,** 197 (1994).

766. S. Karabunarliev, O.G. Mekenyan, W. Karcher, C.L. Russom, and S.P. Bradbury, *Quant. Struct.-Act. Relat.* **15,** 302 (1966).

767. E.U. Ramos, W.H.J. Waes, H.J.M. Vehaar, and J.L.M. Hermens, *Environ. Sci. Pollut. Res.* **4,** 83 (1997).

768. L. Eriksson, J. Jonsson, and M. Tysklind, *Environ. Toxicol. Chem.* **14,** 209 (1995).

769. B.M. Berger and N.L. Wolfe, *Environ. Toxicol. Chem.* **15,** 1500 1996).

770. A. Nakayama, K. Hagiwara, S. Hashimoto, and S. Shimoda, *Quant. Struct.-Act. Relat.* **12,** 251 (1993).

771. G.D. Veith and O.G. Mekenyan, *Quant. Struct.-Act. Relat.* **12,** 349 (1993).

772. O.G. Mekenyan, D. Bonchev, D.H. Rouvray, D. Petichev, and I. Bangov, *Eur. J. Med. Chem.* **26,** 305 (1991).

773. S. Miertuš, J. Miertušova, and P. Filipovic, in *QSAR in Toxicology and Xenobiochemistry* (M. Tichy, ed.), p. 143. p. 143. Elsevier, Amsterdam, 1985.

774. A. Capelli, M. Anzini, S. Vomero, L. Mennuni, F. Makovec, E. Doucet, M. Hamon, G. Bruni, M.R. Romeo, M.C. Menziani, P.G. De Benedetti, and T. Langer, *J. Med. Chem.* **41,** 728 (1998).

775. M.C. Menziani, P.G. De Benedetti, and M. Karelson, *Bioorg. Med. Chem.* **6,** 535 (1998).

776. J.M. Luco, L.J. Yamin, and H.F. Ferretti, *J. Pharm. Sci.* **184,** 903 (1995).

777. A.R. Katritzky, E.S. Ignatchenko, R.A. Barcock, V.S. Lobanov, and M. Karelson, *Anal. Chem.* **11,** 1799 (1994).
778. R. Murugan, M.P. Grendze, J.E. Toomey, Jr., A.R. Katritzky, V.S. Lobanov, P. Rachwal, and M. Karelson, *CHEMTECH* **24,** 17 (1994).
779. A.R. Katritzky, V.S. Lobanov, M. Karelson, R. Murugan, M.P. Grendze, and J.E. Toomey, *Rev. Roum. Chim.* **41,** 851 (1996).
780. M. Karelson and A. Perkson, *Comput. Chem.* **23,** 49 (1999).
781. F. Ignatz-Hoover, A.R. Katritzky, V.S. Lobanov, and M. Karelson, *Rubber Chem. Technol.* (1999) (in press).
782. A.R. Katritzky, S. Sild, and M. Karelson, *J. Chem. Inf. Comput. Sci.* **38,** 840 (1998).
783. A.R. Katritzky, S. Sild, and M. Karelson, *J. Chem. Inf. Comput. Sci.* **38,** 1171 (1998).
784. A.R. Katritzky, P. Rachwal, K.W. Law, M. Karelson, and V.S. Lobanov, *J. Chem. Inf. Comput. Sci.* **36,** 879 (1996).
785. A.R. Katritzky, S. Sild, V. Lobanov, and M. Karelson, *J. Chem. Inf. Comput. Sci.* **38,** 300 (1998).
786. A. Ben-Naim, *Solvation Thermodynamics.* Plenum, New York, 1987.
787. J. Tomasi and M. Persico, *Chem. Rev.* **94,** 2027 (1994).
788. M. Schaefer and M. Karplus, *J. Phys. Chem.* **100,** 1578 (1996).
789. J.G. Kirkwood, *J. Chem. Phys.* **7,** 911 (1939).
790. L. Onsager, *J. Am. Chem. Soc.* **58,** 1486 (1936).
791. C.J.F. Böttcher and P. Bordewijk, *Theory of Electric Polarization*, 2nd ed., Vol. II. Elsevier, Amsterdam, 1978.
792. R.J. Abraham, B.D. Hudson, M.W. Kermode, and J.R. Mines, *J. Chem. Soc., Faraday Trans.* **184,** 1911 (1988).
793. J.-L. Rivail and D. Rinaldi, *Chem. Phys.* **18,** 233 (1976).
794. D. Rinaldi, M.F. Ruiz-Lopez, and J.-L. Rivail, *J. Chem. Phys.* **78,** 834 (1983).
795. D. Rinaldi, J.-L. Rivail, and N. Rguini, *J. Comput. Chem.* **13,** 675 (1992).
796. J.-L. Rivail, B. Terryn, D. Rinaldi, and M.F. Ruiz-Lopez, *J. Mol. Struct. (THEOCHEM)* **166,** 319 (1988).
797. J.L. Pascual-Ahuir and E. Silla, *J. Comput. Chem.* **11,** 1047 (1990).
798. E. Silla, L. Tuñón, and J.L. Pascual-Ahuir, *J. Comput. Chem* **12,** 1077 (1991).
799. O. Tapia and O. Goscinski, *Mol. Phys.* **29,** 1653 (1975).
800. M.M. Karelson, *Org. React.* **17,** 357 (1980).
801. M.M. Karelson, *Org. React.* **17,** 366 (1980).
802. M.W. Wong, M.J. Frisch, and K.B. Wiberg, *J. Am. Chem. Soc.* **113,** 4776 (1991).
803. M.W. Wong, K.B. Wiberg, and M.J. Frisch, *J. Chem. Phys.* **95,** 8991 (1991).
804. M. Karelson and M.C. Zerner, *J. Phys. Chem.* **92,** 6949 (1992).
805. O. Tapia, *J. Mol. Struct. (THEOCHEM)* **226,** 59 (1991).
806. O. Tapia, F. Colonna, and J. Ángyán, *J. Chem. Phys. Phys.-Chim. Biol.* **87,** 875 (1990).
807. J. Ángyán, *J. Math. Chem.* **10,** 93 (1992).
808. G. Jansen, F. Colonna, and J. Ángyán, *Int. J. Quantum Chem.* **58,** 251 (1996).

809. O. Tapia and G. Johannin, *J. Chem. Phys.* **75,** 3624 (1981).
810. J. Ángyán, M. Allavena, M. Picard, A. Potier, and O. Tapia, *J. Chem. Phys.* **77,** 4723 (1982).
811. O. Tapia, in *Theoretical Models of Chemical Bonding* (Z.B. Maksic, ed.), Vol. 4. Springer-Verlag, Berlin, 1990.
812. J. Ángyán and C. Chipot, *Int. J. Quantum Chem.* **52,** 17 (1994).
813. B.T. Thole, P.T. van Duijnen, *Chem. Phys.* **71,** 211 (1982).
814. B.T. Thole, P.T. van Duijnen, *Biophys. Chem.* **18,** 53 (1983).
815. P.T. van Duijnen, A.H. Juffer, and H. Dijkman, *J. Mol. Struct. (THEOCHEM)* **260,** 195 (1992).
816. A.H. deVries, and P.T. van Duijnen, *Int. J. Quantum Chem.* **57,** 1067 (1996).
817. H. Dijkman and P.T. van Duijnen, *Int. J. Quantum Chem., Quantum Biol. Symp.* **18,** 49 (991).
818. P.T. van Duijnen and A.H. deVries, *Int. J. Quantum Chem., Quantum Chem. Symp.* **29,** 531 (1995).
819. S. Miertuš, E. Scrocco, and J. Tomasi, *Chem. Phys.* **55,** 117 (1981).
820. J.L. Pascual-Ahuir, E. Silla, J. Tomasi, and R. Bonaccorsi, *J. Comput. Chem.* **8,** 778 (1987).
821. H. Hoshi, M. Sakurai, Y. Inoue, and R. Chûjô, *J. Chem. Phys.* **87,** 1107 (1987).
822. M.L. Drummond, *J. Chem. Phys.* **88,** 5014 (1988).
823. R.J. Zauhar and R.S. Morgan, *J. Comput. Chem.* **11,** 603 (1990).
824. B. Wang, and G.P. Ford, *J. Chem. Phys.* **97,** 4162 (1992).
825. T. Fox, N. Rösch, and R.J. Zauhar, *J. Comput. Chem.* **14,** 253 (1993).
826. F.J. Luque, M. Orozco, P.K. Bhadane, and S.R. Gadre, *J. Chem. Phys.* **100,** 6718 (1994).
827. M.C. Zerner, private communication (1997).
828. M. Karelson, T. Tamm, and M.C. Zerner, *J. Phys. Chem.* **97,** 11901 (1993).
829. A. Klamt and G. Schüürmann, *J. Chem. Soc., Perkin Trans. 2*, p. 799 (1993).
830. A. Klamt, *J. Phys. Chem.* **99,** 2224 (1995).
831. G. Klopman, *Chem. Phys. Lett.* **1,** 200 (1967).
832. H.A. Germer, *Theor. Chim. Acta* **34,** 145 (1974).
833. R. Constanciel and O. Tapia, *Theor. Chim. Acta* **48,** 75 (1978).
834. C.J. Cramer and D.G. Truhlar, *J. Am. Chem. Soc.* **113,** 8305 (1991).
835. C.J. Cramer and D.G. Truhlar, *J. Am. Chem. Soc.* **113,** 8552 (1991).
836. C.J. Cramer and D.G Truhlar, *J. Comput. Chem.* **13,** 1089 (1992).
837. C.J. Cramer and D.G. Truhlar, *Science* **256,** 213 (1992).
838. D.J. Giesen, C.C. Chambers, C.J. Cramer, and D.G. Truhlar, *J. Phys. Chem. B* **101,** 2061 (1997).
839. K.V. Mikkelsen, H. Ågren, H.J. Aa. Jensen, and T. Helgaker, *J. Chem. Phys.* **89,** 3086 (1988).
840. H. Ågren, C. Medina-Llanos, K.V. Mikkelsen, and H.J. Aa. Jensen, *Chem. Phys. Lett.* **153,** 322 (1988).
841. C. Medina-Llanos, H. Ågren, K.V. Mikkelsen, and H.J. Aa. Jensen, *J. Chem. Phys.* **90,** 6422 (1989).

842. K.V. Mikkelsen and H. Ågren, *J. Phys. Chem.* **94,** 6220 (1990).
843. J.-L. Rivail, *C. R. Sceances Acad. Sci, Ser. 2*, **311,** 307 (1990).
844. J.-L. Rivail, D. Rinaldi, and M.F. Ruiz-Lopez, in *Theoretical and Computational Models for Organic Chemistry* (S.J. Formosinho, I.G. Csizmadia, and L.G. Arnaut, eds), pp. 77–92. Kluwer Acad Publ., Dordrecht, The Netherlands, 1991.
845. F.J. Olivares del Valle and J. Tomasi, *Chem. Phys.* **150,** 139 (1991).
846. F.J. Olivares del Valle and M.A. Aguilar, *J. Comput. Chem.* **13,** 115 (1992).
847. F.J. Olivares del Valle and M.A. Aguilar, *J. Mol. Struct. (THEOCHEM)* **280,** 25 (1993).
848. C.H. Lochmüller, T. Maldacker, and M. Cefola, *Anal. Chim. Acta* **48,** 139 (1969).
849. M. Karelson, *Adv. Quantum Chem.* **28,** 141 (1997).
850. M.J. Huron and P. Claverie, *Chem. Phys. Lett.* **4,** 429 (1969).
851. A.J. Pertsin and A.I. Kitaigorodsky, *The Atom-Atom Pair Potential Method.* Springer, Berlin, 1986.
852. N. Rösch and M.C. Zerner, *J. Phys. Chem.* **98,** 5817 (1994).
853. T. Abe, Y. Amako, T. Nishioka, and H. Azumi, *Bull. Chem. Soc. Jpn.* **39,** 845 (1966).
854. A.T. Amos and B.L. Burrows, *Adv. Quantum Chem.* **7,** 289 (1973).
855. D.D. Eley, *Trans. Faraday Soc.* **35,** 1281 (1939).
856. B. Susskind and I. Kasarnowski, *Z. Anorg. Allg. Chem.* **214,** 385 (1933).
857. T. Simonson and A.T. Brünger, *J. Phys. Chem.* **98,** 4683 (1994).
858. O. Sinanoglu, *Chem. Phys. Lett.* **1,** 283 (1967).
859. O. Sinanoglu, in *Molecular Interactions* (H. Ratajczak and W.J. Orville-Thomas, eds.), p. 283. Wiley, New York, 1982.
860. D.G. Oakenfull and D.E. Fenwick, *J. Phys. Chem.* **78,** 1759 (1974).
861. J.P.M. Postma, J.J.C. Berendsen, and J.R. Haak, *Symp. Faraday Soc.* **17,** 55 (1982).
862. V. Gogonea and E. Osawa, *J. Mol. Struct. (THEOCHEM)* **311,** 305 (1994).
863. H. Reiss, H.L. Frisch, and J.L. Lebowitz, *J. Chem. Phys.* **31,** 369 (1959).
864. H.L. Frisch, *Adv. Chem. Phys.* **6,** 229 (1963).
865. H. Reiss, *Adv. Chem. Phys.* **9,** 1 (1966).
866. R.A. Pierotti, *Chem. Rev.* **76,** 717 (1976).
867. R.M. Gibbons, *Mol. Phys.* **17,** 81 (1969).
868. M. Irisa, K. Nagayama, and F. Hirata, *Chem. Phys. Lett.* **207,** 430 (1993).
869. R. Bonaccorsi, F. Floris, and J. Tomasi, *J. Mol. Liquids* **47,** 25 (1990).
870. R. Bonaccorsi, F. Floris, F. Palla, and J. Tomasi, *Theor. Chim. Acta* **162,** 213 (1990).
871. D.J. Giesen, C.J. Cramer, and D.G. Truhlar, *J. Phys. Chem.* **99,** 7137 (1995).
872. D. Sitkoff, N. Ben-Tal, and B. Honig, *J. Phys. Chem.* **100,** 2744 (1996).
873. F.J. Luque, Y. Zhang, C. Alemán, M. Bachs, J. Gao, and M. Orozco, *J. Phys. Chem.* **100,** 4269 (1996).
874. F. J. Luque, C. Alemán, M. Bachs, and M. Orozco, *J. Comput. Chem.* **17,** 806 (1996).

875. O. Tapia, in *Quantum Theory of Chemical Reactions* (R. Daudel, A. Pullman, L. Salem, and A. Veillard, eds), p. 25. Reidel, Drodrecht, The Netherlands, 1980.
876. A. Bondi, *J. Phys. Chem.* **68,** 441 (1964).
877. G.D. Hawkins, C.J. Cramer, and D.G. Truhlar, *J. Phys. Chem. B* **102,** 3257 (1998).
878. W. Thiel and A.A. Voityuk, *Int. J. Quantum Chem.* **44,** 807 (1992).
879. W. Thiel and A.A. Voityuk, *Theor. Chim. Acta* **93,** 315 (1996).
880. W. Thiel and A.A. Voityuk, *J. Phys. Chem.* **100,** 616 (1996).
881. M.M. Karelson, A.R. Katritzky, and M.C. Zerner, *Int. J. Quantum Chem.* **S20,** 521 (1986).
882. M. Karelson, T. Tamm, A.R. Katritzky, and M.C. Zerner, *Int. J. Quantum Chem.* **37,** 1 (1990).
883. C.J. Cramer, *J. Org. Chem.* **57,** 7034 (1992).
884. C.J. Cramer and D.G. Truhlar, *J. Am. Chem. Soc.* **115,** 8810 (1993).
885. E.V. Patterson and C.J. Cramer, *J. Phys. Org. Chem.* **11,** 232 (1998).
885a. C.C. Chambers, G.D. Hawkins, C.J. Cramer, and D.G. Truhlar, *J. Phys. Chem.* **100,** 16385 (1996).
886. G.R. Famini, *Using Theoretical Descriptors in Quantitative Structure-Activity Relationships*, V.CRDEC-TR-085. U.S. Army Chemical Research, Development and Engineering Center, Aberdeen Proving Ground, MD, 1989.
887. M.J. Kamlet, J.-L.M. Abboud, M.H. Abraham, and R.W. Taft, *J. Org. Chem.* **48,** 2877 (1983).
888. D.F.V. Lewis, *J. Comput. Chem.* **8,** 1084 (1987).
889. A.J. Hopfinger, *J. Am. Chem. Soc.* **102,** 7126 (1980).
890. H.A. Kurtz, J.J.P. Stewart, and K.M. Dieter, *J. Comput. Chem.* **11,** 82 (1990).
891. G.R. Famini, C.A. Penski, and L.Y. Wilson, *J. Phys. Org. Chem.* **5,** 395 (1992).
892. A.D. Headley, S.D. Starnes, L.Y. Wilson, and G.R. Famini, *J. Org. Chem.* **59,** 8040 (1994).
893. A.H. Lowrey, G.R. Famini, and L.Y. Wilson, *J. Chem. Soc., Perkin Trans. 2*, p. 1381 (1997).
894. G.R. Famini and L.Y. Wilson, *J. Chem Soc., Perkin Trans. 2*, p. 1641 (1994).
895. W.H. Donovan and G.R. Famini, *J. Chem. Soc., Perkin Trans. 2*, p. 83 (1996).
896. G.R. Famini and L.Y. Wilson, *Network Sci.* **2**(1) (1996).
897. C.M. Breneman and M. Rhem, *J. Comput. Chem.* **18,** 182 (1997).
898. J.B.F.N. Engberts, G.R. Famini, A. Perjéssy, and L.Y. Wilson, *J. Phys Org. Chem.* **11,** 261 (1998).
899. L.Y. Wilson and G.R. Famini, *J. Med. Chem.* **34,** 1668 (1991).
900. G.R. Famini, R.J. Kassel, J.W. King, and L.Y. Wilson, *Quant. Struct.-Act. Relat.* **10,** 344 (1991).
901. G.R. Famini, W.P. Ashman, A.P. Mickiewicz, and L.Y. Wilson, *Quant. Struct.-Act. Relat.* **11,** 162 (1992).
902. G.R. Famini, L.Y. Wilson, and S.C. DeVito, in *Biomarkers of Human Exposure to Pesticides* (M.A. Saleh, J.N. Blancato, and C.H. Nauman, eds.), p. 22. American Chemical Society, Washington, DC, 1994.
903. M. Karelson, S. Sild, and U. Maran, *Mol. Simul.* (1999) (submitted for publication).

4

METHODS FOR DEVELOPMENT OF QSAR/QSPR

4.1 INTRODUCTION

The practical application of molecular descriptor scales in the development of quantitative structure–activity/property relationship (QSAR/QSPR) models is not a trivial task. First, as demonstrated throughout this book, an extremely large number of molecular descriptors of different complexity and of different conceptuality have been derived and suggested over the years. At the same time, no strict rules have been established or even proposed for the selection of "correct" descriptors from the myriad available. In most cases, the choice has been based on the chemical intuition of researchers or on tradition. For instance, in many QSAR models the biological activity of a series of congeneric compounds has been correlated with the water–octanol partition coefficient, log *P*. This, of course, demonstrates that the solubility in different biological environments is an important factor determining the biological activity of a compound. However, the actual physico-chemical interaction between the drug molecule and a biological receptor, as related to other specific molecular descriptors, may remain hidden. Thus, the models that predict better activity of compounds by improving their solubility may be practical for a series of structurally similar compounds but not so useful if structurally divergent compounds are considered. In fact, in the drug design or in the development of new materials, the target is usually a completely new molecular structure. The series of congeneric compounds are commonly patented for given purposes, and thus it would be of high interest to design a new, structurally divergent compound that exhibits a similar or, preferably, a better performance than the known compounds.

Another difficulty in the selection of QSAR/QSPR descriptors arises from the nonstandardization of scales. The empirically derived induction, resonance, or steric substituent constants or empirical solvent effect scales possess intrinsic inaccuracy originating from the error of the respective experimental measurements. On the other hand, the quantum chemical methods applied for the calculation of molecular-orbital- (MO)-related and charge-distribution-related molecular descriptors are often based on different semiempirical parameterizations or on the use of different basis sets in ab initio calculations. Naturally, a descriptor that is constructed using different experimental or theoretical methods for different compounds cannot be used in the development of a single QSAR/QSPR model. The choice of a particular descriptor scale is usually based on practical considerations, for example, the availability of descriptors for all compounds studied or how similar is the process or phenomenon studied to that used for the definition of the scale. A systematic approach to the selection of descriptor scales in QSAR/QSPR models is based on the statistical discrimination among the very large number of descriptor scales. In the present chapter, we will contemplate various approaches used to develop the "best" QSAR/QSPR equations in large descriptor spaces.

Finally, the quantitative structure property/activity relationships can be developed within different mathematical models generally related to multivariate statistical analysis. The first and most widespread model is a (multiple) linear equation obtained by the least-squares regression of experimental data against a set of preselected descriptors (or a single descriptor). In some cases, the physical or chemical models known for the phenomenon studied predict certain mathematically nonlinear (e.g., logarithmic or exponential) forms of dependence between the experimental data and molecular descriptors. The respective QSAR/QSPR equations can be found using the nonlinear least-squares regression technique. Another group of models has been developed using factor analysis or principal component analysis. The advantage of these methods is that they evade the multicollinearity problem inherent to the linear regression methods. At the same time, the interpretation of the respective QSAR/QSPR equations is impeded by the formal nature of factors or principal components applied. An alternative to more classical multiple linear regression and principal component regression methods is the partial least-squares (PLS) regression technique [1–5]. Also, the modern artificial intelligence methods including the neural networks, genetic algorithms, and other global optimization methods have been applied to develop the QSAR/QSPR models [6–9]. Therefore, we shall give in the following a short overview of different mathematical methods used for the development of QSAR/QSPR equations.

4.2 LINEAR AND MULTIPLE LINEAR REGRESSION METHODS

4.2.1 General Comments

The linear and multiple regression methods have been used throughout the classical QSAR/QSPR work. In each case, the investigator has chosen a molecular descriptor or a set of descriptors that is expected to reflect the physical or chemical interactions determining the molecular property or the characteristics of the phenomenon studied. As mentioned above, the choice of the descriptors has been usually based on chemical intution, tradition, or simply on the availability of descriptor. Nevertheless, we would like to emphasize five principles that should assist the selection of proper molecular descriptors for QSAR/QSPR.

1. The maximum number of experimental data studied (preferably all of them) should have complementary original descriptor values.
2. The descriptor values should be obtained from the same source and, preferably, as measured using the same experimental method or calculated using the same software.
3. The number of descriptors in multiple regression models should be minimized, without the loss of information as reflected by the respective statistical criteria (F- and t-test values, etc.).
4. In multiple linear regression models, the descriptors employed should be statistically orthogonal.
5. Provided that the other criteria are similar, the chemical or physical nature of the descriptor selected should be closest to the property or phenomenon studied.

Indeed, in practical life it may be difficult to follow all the listed principles. However, the neglect of several of them may lead to useless equations of little or missing predictive power.

4.2.2 Preliminary Evaluation of Data

Before proceeding with the actual development of QSAR/QSPR regression equations, it is highly advisable to examine the statistical quality of starting data, both the data to be correlated (dependent variable) and the descriptors used in the correlation (independent variables). We have distinguished the one-dimensional and the two-dimensional analysis in such preliminary treatment of data [10–14].

In the one-dimensional analysis, it is advisable to assess the accordance of the data to the normal distribution. A special caution should be taken in the subsequent regression procedure if the values of the property studied or a descriptor do not follow the normal distribution. This can be evaluated, for instance, using the statistical measures, asymmetry and excess. We shall proceed with the following statistical characteristics adjacent to a set of data on a given property or on a descriptor, O:

$$\text{Mean value } \bar{O} = \frac{1}{n}\sum_{i=1}^{n} O_i \tag{4.1}$$

where n denotes the number of the individual values of the given property (or descriptor) O_i;

$$\text{Dispersion of data } \sigma = \frac{1}{n-1}\sum_{i=1}^{n} (O_i - \bar{O})^2 \tag{4.2}$$

or the standard deviation of data

$$S = \sqrt{\sigma} \tag{4.3}$$

the variation coefficient

$$Q = 100S/\bar{O} \tag{4.4}$$

with the same notations as above.

The one-dimensional data arrays of a property or of a descriptor can be further separated into N_c classes according to the Stuergess method and the asymmetry

$$\rho = \sum_{j=1}^{N_c} n_j(O_k - \bar{O})^3/nS^3 \tag{4.5}$$

and the excess

$$\varepsilon = \sum_{j=1}^{N_c} n_j(O_k - \bar{O})^4/nS^4 - 3 \tag{4.6}$$

of the data distribution calculated. The individual data O_k belong to the jth class and n_j is their total number in the jth class. For a perfectly symmetrical distribution of data, asymmetry $\rho = 0$. Situation $\rho < 0$ corresponds to the shift of data to smaller values and situation $\rho > 0$ to the shift of data to larger values of the data distribution (Fig. 4.1). The excess characterizes the "sharpness" of the distribution curve. Excess value $\varepsilon = 0$ corresponds to a perfect normal distribution of data; values $\varepsilon > 0$ refer to a sharper peak of the distribution whereas excess values $\varepsilon < 0$ correspond to a less sharper peak than the normal distribution (Fig. 4.2). The large deviations of excess and asymmetry values from zero indicate serious distortion of the data distribution that may cause complications in their (multiple) regression treatment. For instance, the data clustered into two groups with significantly different values have large negative excess of the distribution. Effectively, such data represent just two points in

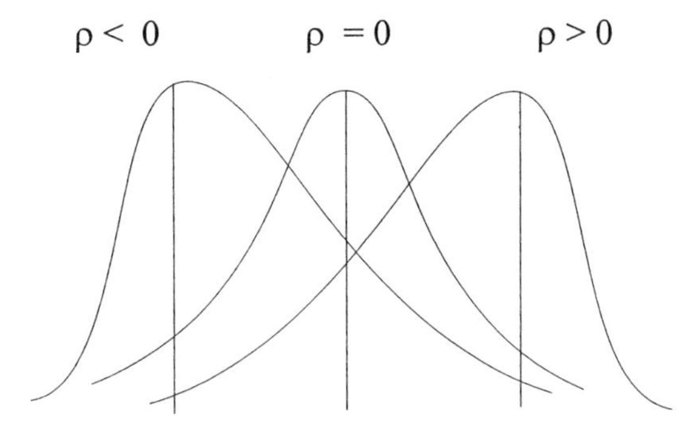

Figure 4.1 Perfectly symmetrically distribution of data ($\rho = 0$), and distributions with negative ($\rho < 0$) and positive ($\rho > 0$) asymmetry.

the regression line, and the effective number of statistical degrees of freedom is much smaller than formally derived from the number of data points.

In the case of serious deviations from the normal distribution, the measurement of more data or inclusion of more compounds in the set studied will be required. If the distribution of a descriptor does not correspond to a normal distribution according to either of these criteria, it is advisable to discard this descriptor in the subsequent development of QSAR/QSPR equations.

For a set of different descriptors, it is necessary to carry out a two-dimensional analysis of data, that is, to calculate the linear pair correlation coefficient between every two descriptor scales, X and Z, defined as:

$$R = \frac{1}{n} \sum_{i=1}^{n} \frac{X_i Z_i}{\sigma_X \sigma_Z} \tag{4.7}$$

where σ_X and σ_Z denote the dispersion of these two descriptor scales, respectively. If the correlation coefficient between two descriptors is statistically sig-

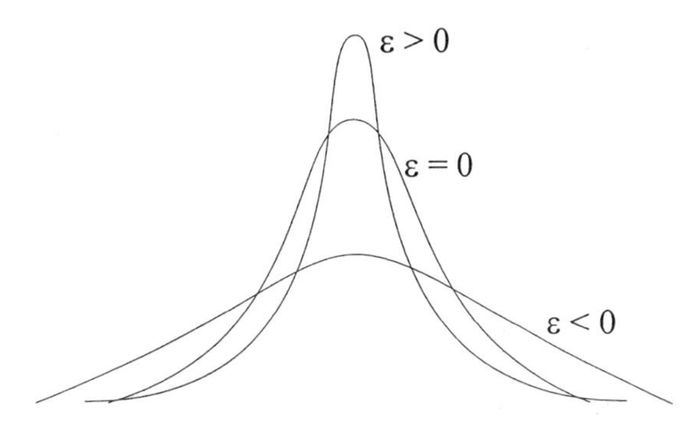

Figure 4.2 Perfectly normal distribution of data ($\varepsilon = 0$), and distributions with negative ($\varepsilon < 0$) and positive ($\varepsilon > 0$) excess values.

nificant (e.g., $R > 0.9$), both of these descriptors should not be used simultaneously in the subsequent multiple linear regression analysis.

4.2.3 Linear Regression

The one-parameter linear regression treatment between the user-defined property (dependent variable) P and a single descriptor x_1 (independent variable) is performed according to the linear equation

$$P = a_0 + a_1 x_1 \tag{4.8}$$

where a_0 and a_1 are the intercept of the regression model and the slope of the regression line, respectively. These coefficients are found subject to the so-called least-squares conditions:

$$\frac{\partial}{\partial a_0}\left[\sum_{i=1}^{n}(P_i - P_i(\text{calc}))^2\right] = \frac{\partial}{\partial a_0}\left[\sum_{i=1}^{n}(P_i - a_0 - a_1 x_{1i})^2\right] = 0 \tag{4.9}$$

$$\frac{\partial}{\partial a_1}\left[\sum_{i=1}^{n}(P_i - P_i(\text{calc}))^2\right] = \frac{\partial}{\partial a_1}\left[\sum_{i=1}^{m}(P_i - a_0 - a_1 x_{1i})^2\right] = 0 \tag{4.10}$$

where P_i(calc) are the values of the property calculated according to (4.8). We would emphasize that the dependent and the independent variable are not equivalent in the last two equations. In fact, the coefficients determined using the least-squares method for the inverse relationship

$$x_1 = a_0' + a_1' P \tag{4.11}$$

have, in general, no relation to the slope and intercept of the original regression. As an example, let us consider a set of four data points with the following values:

	P	x
1	1.0	2.0
2	3.0	2.0
3	3.0	4.0
4	5.0	4.0

The intercept of the respective linear relationship (4.8) obtained using the least-squares method is $a_0 = 0.0$ and the slope of the regression line is $a_1 = 1.0$ (Fig. 4.3.*a*). The inverse relationship (4.11), however, has the least-squares intercept $a_0' = 1.66$ and the least-squares slope $a_1' = 0.5$ (Fig. 4.3.*b*). If one interprets these two formulas just as algebraic equations, the coefficients for the first equation (3.8) result in the coefficient values $a_0' = 0.0$ and $a_1' = 1$ for the second

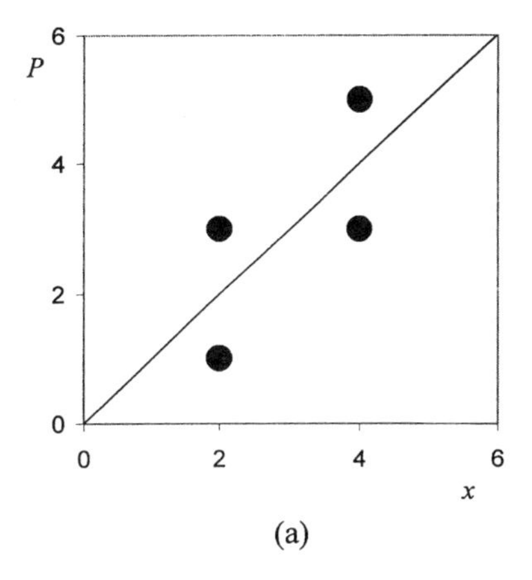

(a)

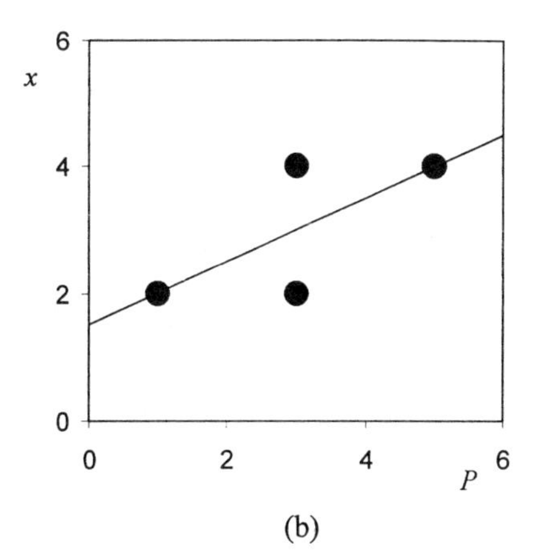

(b)

Figure 4.3 Least-squares estimates of regression line for direct and inverse linear relationships, (3.8) and (3.11), respectively.

equation. Thus, the direct and reverse least-squares linear equations do not follow, in general, the rules of linear algebra. Consequently, for the proper interpretation of regression coefficients, it is important that the causal relationship between the molecular property and the descriptor used is predetermined. The quality of the regression, that is, how good it is for the overall description of the relationship between the molecular property studied and the molecular descriptor, can be evaluated using several statistical characteristics. Since the simple linear regression is a special case of the multiple linear regression, we shall examine it in the general case.

4.2.4 Multiple Linear Regression

A multiparameter linear regression treatment between the user-defined property (dependent variable) P and a set of user-chosen descriptors (independent variables) x_k involves the determination of the regression coefficients a_0, a_1, a_2, ... in the following equation:

$$P = a_0 + \sum_k a_k x_k \tag{4.12}$$

under the condition

$$\frac{\partial}{\partial a_k}\left[\sum_{i=1}^{n} (P_i - P_i(\text{calc}))^2\right] = \frac{\partial}{\partial a_k}\left[\sum_{i=1}^{n}\left(P_i - a_0 - \sum_k a_k x_{ki}\right)^2\right] = 0 \tag{4.13}$$

In Eq. (4.13), i denotes the index of the compound and k is the index of the descriptor scale. P_i(calc) denote the values of the property calculated according to Eq. (4.12) with a given set of coefficients $\{a_k\}$. The statistical significance of coefficients a_k is usually characterized by the respective predictor errors, Δa_k.

The quality of the multiple linear (or simple linear) regression can be characterized by several statistical parameters. First, the correlation coefficient of the regression is defined as:

$$R = \sqrt{\frac{\sum_{i=1}^{n} [P_i(\text{calc}) - \bar{P}]^2}{\sum_{i=1}^{n} (P_i - \bar{P})^2}} \tag{4.14}$$

where $\bar{P}$ is the mean value of the dependent variable P. The square of correlation coefficient, R^2 (the coefficient of determination), represents the fraction of the dependent variable P described by the given set of independent variables. In the case of perfect correlation, $R = 1$. A complete absence of correlation between a property P and a set of independent variables (predictors) corresponds to $R = 0$.

The standard error of the multiple linear regression, s, is defined as:

$$s = \sqrt{\sum_{i=1}^{n} \frac{[P_i(\text{calc}) - P_i]^2}{n - p}} \tag{4.15}$$

where p is the number of parameters estimated using the least-squares procedure and n is the number of data points. The difference $(n - p)$ represents the number of residual degrees of freedom. The value of s (or s^2) has been used as a statistical criterion for quick discrimination between the competing multiple regression models, that is, the model with the smallest standard error is usually selected. However, other criteria have also to be accounted for in making the decision about the best regression model. The comparison of the statistical fitness of regression models for different sets of data on a given property P is aided by the normalized standard error of the regression

$$s_0 = \frac{s(n - 1)}{\sqrt{\sum_{i=1}^{n} (P_i - \bar{P})^2}} = \frac{s}{\sigma} \tag{4.16}$$

where σ is the dispersion of the dependent variable, P. The smaller value of s_0 indicates the better regression model.

The selection of the best multiple regression models can be guided by several other statistical criteria. The Fisher F criterion [15] permits to compare the statistical significance of multiple regression models with a different number of regressor variables. The maximum value of the F criterion corresponds to the multiple regression equation with the maximum description of the variance of dependent variable (the property studied). In Table 4.1, the results on the multiple regression treatment are given for data on the aqueous solubility of organic compounds [16]. In the last column of Table 4.1, the values of the F criterion are presented for the best one- to six-parameter multiple regression equations, involving the descriptor on a given line and all descriptors above it. According to this criterion, the QSPR regression equation with four parameters should be the one chosen.

The Student's t test gives information regarding the importance of a single independent variable in a model involving all other independent variables. It is also equivalent to the partial F test, with $t^2 = F$. The only difference is that the t test gives the direction of change of the dependent variable by the change in the predictor variable, which is determined by the sign of the t-test value (cf. Table 4.1). The regressor variables with larger t-test absolute values are considered statistically more significant in the description of the property P.

In most of the QSAR/QSPR work, the equations developed have been used for the prediction of property values for compounds not employed in the development of those equations. Several statistical criteria can be proposed to estimate the reliability of predictions made by a multiple regression equation. First, the data can be split into two parts, one of which ($\{P_1, \ldots, P_k\}$) is used for the development of the model regression equation (fitting sample). The obtained equation is then used for the prediction of data from the other part, $\{P_{k+1}, \ldots, P_n\}$ (validation sample). The sum of squares of errors of predictions

$$\sum_{j=k+1}^{n} [P_j(\text{predicted}) - P_j]^2 \tag{4.17}$$

is a useful characteristic of the stability of the regression equation. Of course, the result will be dependent on the partitioning of the data. A systematic algorithm for the estimation of predictive power of the regression equation is given by the PRESS (*Pre*diction *S*um of *S*quares) statistic. In this statistic, the first line of data is withheld and a multiple regression model developed using

TABLE 4.1 Best Six-Descriptor Correlation Model of Aqueous Solubility log S_w for Data Set of 411 Diverse Structures [16]

n	Descriptor	$a \pm \Delta a$	t Test	R^2	R^2_{cv}	F value
0	Intercept	-0.882 ± 0.138	-6.39			
1	Q_{min}	-16.1 ± 0.7	-24.33	0.348	0.343	218.4
2	N_{el}	-0.113 ± 0.005	-25.12	0.664	0.660	403.4
3	FHDSA(2)	2.55 ± 0.22	11.55	0.776	0.771	469.1
4	ABO(N)	0.781 ± 0.064	12.26	0.842	0.838	541.7
5	^{0}SIC	0.328 ± 0.037	8.91	0.867	0.862	526.9
6	RNCS	-0.0143 ± 0.0022	-6.32	0.879	0.874	487.9

the remaining $n - 1$ data lines. The first line is then replaced and the second line withheld to estimate the regression coefficients without the latter. This procedure is repeated n times, for each line of data in the set. In each case, the value of missing data is estimated from the regression equation developed without its participation and the respective prediction errors (PRESS residuals)$e_i = P_i$(estimated) $- P_i$ (calculated). The PRESS is then defined as:

$$\text{PRESS} = \sum_{i=1}^{n} [P_i(\text{estimated}) - P_i]^2 = \sum_{i=1}^{n} e_i^2 \tag{4.18}$$

For the choice of the best multiple regression model, the model with the smallest PRESS value may be preferred. Stone and Jonathan [17] have suggested a variation of the PRESS statistic, defined as follows (RMSPE = root-mean-square prediction error):

$$\text{RMSPE} = \sqrt{\sum_{i=1}^{n} \frac{e_i^2}{n}} \tag{4.19}$$

This criterion is normalized per single data point and, furthermore, it is directly assessable on the scale of the dependent variable, P.

The PRESS residuals can also be used for the calculation cross-validated coefficient of the determination of a multiple regression, R_{cv}^2, according to the following equation:

$$R_{\text{cv}}^2 = 1 - \frac{\text{PRESS}}{\sum_{i=1}^{n} (P_i - \bar{P})^2} \tag{4.20}$$

where $\bar{P}$ is the mean value of the property correlated. The square root of the cross-validated coefficient of determination defines the cross-validated correlation coefficient, R_{cv}, for a given multiple linear regression. The R_{cv}^2 is expected to be close to the coefficient of determination, R^2, for the multiple regression involving the whole data set (cf. Table 4.1). Substantially smaller R_{cv}^2 values as compared to the R^2 indicate the low predictive power of the respective regression equation. Therefore, the cross-validated correlation coefficient can be considered better statistical characteristics of the multiple regression than the simple correlation coefficient.

Two other criteria have been suggested as the upward adjustment of the standard error of the multiple regression, s [17]. The s^+ criterion is derived to assess the mean value of the average of the squares of prediction errors of n hypothetical data points, conveniently similar to the original data set of n points:

$$s^{+} = s\sqrt{1 + \frac{p + 1}{n}} \tag{4.21}$$

where p is the number of predictor variables in the regression equation. Alternatively,

$$s^{++} = s\Big/\sqrt{1 - \frac{p + 1}{n}} \tag{4.22}$$

which is equivalent to the generalized cross-validation criterion of Wahba [18].

A major challenge in the development of multiple regression equations is connected with the possible multicollinearity of the predictor variables (molecular descriptor scales) [19–21]. In the case of high mutual correlation of predictor variables, the overall statistical characteristics of the regression may be satisfactory, however, the reliability of the predictor coefficients, and thus of the whole regression equation, is low. The multicollinearity can be avoided, at least partly, by examining the correlation coefficients between the different predictor variables in the multiple regression model. In Table 4.2, the correlation coefficients are given for a group of topological indices derived for a set of 482 organic compounds. In most cases, the pair correlation is sufficiently low, except for the pairs *BC*, *DE*, and *EF*.

A simple advice is not to use simultaneously in the QSAR/QSPR multiple linear regression models the descriptor scales that have high correlation coefficients between them. Also, the ridge regression [22] and the principal components regression (PCR) [23–25] techniques are suitable for the treatment of descriptor scales that suffer from multicollinearity. Below, we examine the latter method more closely.

Another possibility to deal with the multicollinearity problem is the use of the preorthogonalization of the descriptor scales. First, it has been proposed to use the orthogonalized quantum chemical indices in QSAR studies [26]. The orthogonalized molecular descriptors have been considered also elsewhere [27–

TABLE 4.2 Pair Correlation Coefficients for Six Common Topological Indices[a]

	A	*B*	*C*	*D*	*E*	*F*
A	1.0000	0.5530	0.3863	0.2572	0.0872	0.0257
B	0.5530	1.0000	0.7588	0.5740	0.1792	0.1027
C	0.3863	0.7588	1.0000	0.1352	−0.2625	0.5038
D	0.2572	0.5740	0.1352	1.0000	0.8454	−0.6141
E	0.0872	0.1792	−0.2625	0.8454	1.0000	−0.9251
F	0.0257	0.1027	0.5038	−0.6141	−0.9251	1.0000

[a] *A*, Wiener index; *B*, Randic index (order 1); *C*, Kier–Hall index (order 1); *D*, average information content (order 1); *E*, average structural information content (order 1); and *F*, average complementary information content (order 1).

32]. The orthogonalization of i descriptors selected in the QSPR model can be carried out by the application of the following equation, which describes the orthogonalization of a vector $|k\rangle$ against vector $|j\rangle$ [33]:

$$|k'\rangle == \frac{r_{jk}|j\rangle - |k\rangle}{\sqrt{1 - r_{jk}^2}} \tag{4.23}$$

where $|k'\rangle$ is the descriptor vector $|k\rangle$ orthogonalized against the descriptor vector $|j\rangle$, while r_{jk} is the correlation coefficient between the nonorthogonal descriptors j and k. After this transformation, descriptor j remains unchanged, as does the correlation coefficient between this descriptor and the property P, since j is the first descriptor against which the second descriptor is orthogonalized. The same procedure has to be repeated for all the remaining descriptors against the first, and then, against the second, third, up to the $(i - 1)$th descriptor.

The correlation coefficient between property P and the orthogonalized descriptor k' is now changed. The new correlation coefficient R_2' can be computed by the following formula:

$$R_2' = \langle P|k'\rangle = \frac{r_{jk}\langle P|j\rangle - \langle P|k\rangle}{\sqrt{1 - r_{jk}^2}} = \frac{r_{jk}R_j - R_k}{\sqrt{1 - r_{jk}^2}} \tag{4.24}$$

where $\langle P|$ is a symbolic bra vector representation of property P, and R_j, R_k, and r_{jk} stand, respectively, for the correlation coefficients between property P and nonorthogonal descriptors j and k and that between the nonorthogonal descriptors j and k. The orthogonalization of all descriptors is achieved by subsequent application of Eq. (4.23) until each of them is orthogonal to all others. In the case of a QSAR/QSPR multiple linear model based on the set of i orthogonalized descriptors d_i $(i = 1, \ldots, N)$, the correlation coefficient between the experimental values of modeled property P and the values estimated by the model P^{est} can then be calculated in a simple way:

$$R = \left[\sum_{i=1}^{I} R_i^2\right]^{1/2} \tag{4.25}$$

where R_i is the correlation coefficient between each orthogonalized descriptor d_i and the modeled property P.

4.3 DERIVATION OF BEST MULTILINEAR REGRESSION MODELS IN LARGE DESCRIPTOR SPACES

4.3.1 Stepwise Regression Methods

The search for the best multiple linear QSAR/QSPR model in the case of a large number of molecular descriptors as candidates for regressor variables is not a trivial task. The calculation of all possible models and the comparison of their statistical characteristics may quickly become impractical with the increase of the number of descriptors and the number of regressor variables in the model. The number of possible correlation models of the nth order on the set of N descriptors is given by the binomial coefficient:

$$C_N^n = \frac{N!}{n!(N - n)!} \tag{4.26}$$

In the case of the set of 1000 descriptors, the number of correlations increases rapidly and it is practically impossible to study all equations of third order:

$$C_{(1000)}^{(2)} = 449{,}500$$

$$C_{(1000)}^{(3)} = 166{,}167{,}000$$

$$C_{(1000)}^{(4)} = 41{,}417{,}124{,}750$$

$$C_{(1000)}^{(5)} = 8{,}250{,}291{,}250{,}200$$

Various so-called stepwise regression techniques have been proposed for the selection of the "best set" of regression predictors among the large number of available descriptors [34–36]. As an example, we present the algorithm of the heuristic method applied in the CODESSA software [37] as follows. Within this method, a preselection of descriptors is accomplished first. All descriptors are checked to ensure that values of each descriptor are available for each structure and that there is a variation in these values. Descriptors for which values are not available for every structure in the data in question or having a constant value for all structures in the data set are discarded. Thereafter, the one-parameter correlation equations for each descriptor are calculated.

To reduce further the number of descriptors in the set, the following criteria are applied and a descriptor is eliminated if:

1. The F value for the one-parameter correlation with the descriptor is below 1.0.
2. The squared correlation coefficient of the one-parameter equation is less than $R^2_{\min}$.
3. The t value for a given parameter is less than t_1 ($R^2_{\min}$ and t_1 are user specified values).
4. The descriptor i is highly intercorrelated (having r_{ij} above a user-specified value r_{full}) with another descriptor j, and this other descriptor has a higher correlation coefficient of the one-parameter equations with the dependent variable.

All the remaining descriptors are then listed in decreasing order according to the correlation coefficient of the corresponding one-parameter correlation equation.

Then the following proceedings are followed:

1. Starting with the top descriptor from the preselected list of descriptors the two-parameter correlations are calculated using the following pairs: the first descriptor with each of the remaining descriptors, the second descriptor with each of the remaining descriptors, and so forth. This procedure is continued until no correlations with an F value above one-third of the maximum F value for a given set has been found.
2. An user-defined number of best pairs showing highest F values in the two-parameter correlations are selected and processed further as the working sets.
3. Each of the remaining descriptors, if not correlated over some user-specified value r_{sig} with the descriptors already included, is added to the selected working set of descriptors. If the resulting correlation has an F value above $F_{working}n/(n + 1)$ (where n is a number of descriptors in the working set plus one), that is, if this correlation is more significant than the working correlation, then this extended set of descriptors is considered for further treatment.
4. After all descriptors have been applied one-by-one and if the maximum number of descriptors_{max} allowed by the user, is not yet achieved, then NS best extended working sets (i.e., the sets with the highest F values) are submitted to the procedure from step 3. Otherwise the procedure is completed and the ND_{max} best correlations found.

The search for the multiparameter regression of the maximum predicting power can also be carried out using the following strategy [38]:

1. All orthogonal pairs of descriptors i and j (with pair correlation coefficient $R_{ij} < R_{min}$) are found in a given descriptor pool. The chance for the absolute orthogonality of two descriptors is negligible and a user-specified value R_{min} has been used as a practical limit for two descriptors being approximately orthogonal (e.g., $R_{min} = 0.1$).
2. The two-parameter regression equations involving all orthogonal pairs of descriptors, obtained in step 1, are calculated. The N_c ($\leq$400) significant pairs with highest multilinear regression correlation coefficients were chosen for the next step.
3. For each significant descriptor pair (i, j), obtained in the previous step, a noncollinear descriptor scale, k (with $R_{ik} < R_{nc}$ and $R_{kj} < R_{nc}$), is added, and the corresponding three-parameter regression treatment performed. When the Fisher criterion at a given probability level, F, is smaller for every three-parameter regression studied than that for the best two-

parameter correlation, the latter is declared the final result. Otherwise, the user-specified number N_c of the descriptor triples with the highest regression correlation coefficients are considered in the next step. The noncollinearity limit, R_{nc}, is, of course, a subjective parameter.

4. For each significant descriptor set, obtained in the previous step, an additional noncollinear descriptor scale is added, and the appropriate $(n + 1)$-parameter regression treatment performed. If the Fisher criterion at the given probability level, F, or the cross-validated correlation coefficient, R_{cv}, obtained for any of these correlations is smaller than that for the best correlation of the previous rank, then the latter is designated as the final result and the search is terminated. Otherwise, the N_c descriptor sets with the highest regression correlation coefficients are stored, and the current step is repeated with the number of parameters increased by one $(n = n + 1)$.

The result has therefore the maximum value of the Fisher criterion and the highest value of the cross-validated correlation coefficient. According to these statistical criteria, it can be considered as the best representation of the property in the given descriptor space.

This stepwise regression algorithm and the regressor selection algorithm based on the orthogonalized descriptors [39–42] have been tested using the results of a total screening of all possible models in a given descriptor space [43]. The QSPR modeling of gas-chromatographic (GC) retention times t_R and Dietz response factors (RF) on 152 diverse chemical compounds was carried out using a pool of 296 different molecular descriptors. All possible multiple linear models involving one to five descriptors were screened to find the highest correlation. The search for the best correlations was also carried out using the stepwise regression described above and the selection based on the orthogonalized descriptor scales. The last two methods were also applied to find the best multiple linear models of higher rank, up to seven regressors. The data on determination coefficients of the best correlations with a different number of predictors are given in Table 4.3. The identical best multiple regression models were obtained using different algorithms up to six parameters. Notably, these models coincide with the "absolutely" best models obtained by the screening of all possible models up to the five parameters. There are small discrepancies between the seven-parameter models developed using the stepwise regression method and the orthogonalization of the descriptor scales. However, neither of these methods exhibited a better overall performance. In the case of the GC retention times, a slightly better result was obtained using the stepwise regression, whereas in the case of the Dietz response factor the orthogonalization technique gave a somewhat higher determination coefficient. The overall result is encouraging, showing that both methods for the selection of the "best" descriptors have been reliable at least up to the six-parameter multiple linear regression models.

TABLE 4.3 Determination Coefficients, R^2, for Best Possible One- to Five-Descriptor and Best Stepwise One- to Seven-Descriptor Multiple Linear Regression Models for Gas-Chromatographic Retention Times, t_R, and Dietz Response Factors (RF) of 152 Compounds Derived from Pool of 296 Molecular Descriptors

	Number of Descriptors in QSPR Model						
Property/Method	1	2	3	4	5	6	7
GC Retention Times t_R							
All regression models	0.901	0.935	0.944	0.954	0.964	–	–
Stepwise regression	0.901	0.935	0.944	0.954	0.964	0.968	0.974
Orthogonalized scales	0.901	0.935	0.944	0.954	0.964	0.968	0.973
GC Response Factors (RF)							
All regression models	0.515	0.763	0.826	0.863	0.887	–	–
Stepwise regression	0.515	0.763	0.826	0.863	0.887	0.893	0.900
Orthogonalized scales	0.515	0.763	0.826	0.863	0.887	0.893	0.902

4.3.2 Principal Component Regression

As discussed, a major problem in the search for the best (most informative) multiple linear correlation equation on a large set of the natural descriptors is connected with the mutual collinearity of descriptors. This leads to the instability of the regression coefficients, the overestimation of the standard errors, and to a critical loss of predictive information. One way to avoid this effect is to transform the nonorthogonal set of the natural descriptors into a set of orthogonal formal scales. The latter, being linear combinations of the natural scales, preserve the information content whereas the negative effects of the scales multicollinearity is removed in the subsequent regression treatment.

The orthogonalization of natural descriptor scales departs from the principal component analysis of the descriptor space. The principal component analysis (PCA) is one of the oldest and best-known technique of multivariate analysis [44–47]. It has been extensively used in different areas of chemistry from the resolution of spectra to the toxicity predictions of compounds [48–52]. The central idea of PCA is to reduce the dimensionality of a data set in which there are a large number of interrelated variables, while retaining as much as possible of the variation present in the data set. This reduction is achieved by transforming the original variables to a new set of variables, the principal components (PC), which are noncorrelated. The PCs are ordered in such a way that the first few, with descending importance, retain most of the variation of all original variables. Principal component analysis thus yields a solution of a set of abstract eigenvectors and an associated set of abstract eigenvalues. Each eigenvector represents an abstract factor, which can be used subsequently as a scale in multiple regression analysis.

In PCA, the initial data matrix (**D**) is (approximately) represented by the inner product of two matrices:

$$\mathbf{D} = \mathbf{RC} \tag{4.27}$$

The row matrix **R**, named the score matrix, has the dimensionality $n \times r$, where n is the number of compounds in the initial data set and r is the number of PCs. The column matrix **C**, named the loadings matrix, has the dimensionality $r \times c$, where c is the number of observable properties in the initial data set. The natural input data can be pretreated to be centered and normalized according to the user's choice.

Various cut-off criteria can be chosen by the user to minimize the number of PCs [46].

Absolute value of the residual matrix. The value depends on whether the natural descriptor scales or prenormalized and precentered descriptor scales were used as an input for the PCA treatment.

Real error. If one has an estimate of the experimental error, then the PCA-calculated real error

$$\mathrm{RE} = \sqrt{\left\{ \sum_{j=n+1}^{c} \varepsilon_j / [r(c - n)] \right\}} \tag{4.28}$$

should be smaller than the user-provided value of the experimental error.

Imbedded error. If one has the estimate of the experimental error, then the PCA-calculated imbedded error function

$$\mathrm{IE} = \sqrt{\left\{ n \sum_{j=n+1}^{c} \varepsilon_j / [rc(c - n)] \right\}} \tag{4.29}$$

should be smaller than the user-provided value of the experimental error.

Factor indicator function

$$\mathrm{IND} = \mathrm{RE}/(c - n)^2 \tag{4.30}$$

reaches a minimum when the correct number of principal components are employed.

Cumulative percentage of the variation is defined as:

$$t_k = \frac{1}{n} \sum_{j=1}^{k} l_j \tag{4.31}$$

where the summation is performed over the k first PCs, l_j is the variance of the jth PC, and n is the number of columns (descriptors) in initial data matrix.

Kaiser's rule. Only the j first PCs with the variance $l_j > 1$ are considered in PCA.

Broken stick model. The first p PCs with $l_k > l'_k$ are considered, where

$$l'_k = \left(\sum_{j=1}^{p} 1/j \right) \Big/ p \tag{4.32}$$

with the notations as above.

Cattell's rule. If the improvement of the total variance of the descriptors remains practically constant when a PC added, then this and subsequent PCs are considered statistically insignificant and the PC expansion is cut off. Consequently, only the (first) k PCs with $l_k - l_{k-1} > \varepsilon$ are considered in PCA, where ε is a small user-determined number.

The process of finding the best correlation with the given number of (formal PC) scales is very fast as only one (the best) correlation has to be considered from the previous step. In addition, a steady correlation improvement results from the consecutive addition of new PC scales. The regression coefficients for formal PC scales already considered in the correlation of the lower order do not change.

In cases where initial descriptor scales have missing values, the nonlinear partial least-squares (NIPALS) algorithm [4] for finding the principal components would be preferred. As a result of the NIPALS treatment, the principal component scales for further multiple linear regression treatment are found, and the best correlations for successive number of PCs can be calculated.

From the standpoint of practical predictions, the use of abstract factors (principal components) is difficult as these have a complex dependence on the natural and chemically or physically meaningful molecular descriptors or the measured properties. Thus it is advisable to proceed with a rotational transformation of the abstract factor space in a way that the new main factors would coincide with some of the original descriptors or properties (test vectors). The corresponding target transformation PCA or target testing (TT) methodology is described in detail elsewhere [44,46,47].

Two general requirements concerning the number of points needed for a test vector and the range spanned by the test values should be considered before carrying out TT treatment. First, the number of points in a test vector must be greater than the number of factors employed in the transformation (*greater-than-n rule*). Results become more reliable as more points are incorporated into test vector. The second requirement for TT concerns the range of values spanned by the test points. Each target vector should span, as completely as possible, the entire range of values for the property being analyzed. Keeping in mind that interpolation is safer than extrapolation, one should enter all the extreme values available on initial data, hoping that the vector will span the complete range of the real vector.

4.3.3 Partial Least Squares

Partial least squares (PLS) is a generalized regression method that allows to build a predictive model between two matrices of data [4,53,54]. It has been extensively used for the development of CoMFA (comparative molecular field analysis) regressions [55, and references therein]. In principle, PLS is based on two PCA models that are rotated to maximize the overlap between the target property (**Y**) and the predictor (**X**) matrices. Several properties, $\mathbf{y}^{(k)}$ can thus be handled simultaneously. According to the PLS procedure, both the property and predictor matrices are decomposed into a product of the respective scores and loadings and to additional error terms as follows [4]:

$$\mathbf{Y} = \mathbf{UQ} + \mathbf{F} \tag{4.33}$$

$$\mathbf{X} = \mathbf{TP} + \mathbf{E} \tag{4.34}$$

The **X** and **Y** matrices are previously mean centered and scaled. Starting from some estimate for a vector $\mathbf{u} = \mathbf{y}^{(k)}$ in matrix **U**, the first approximation to the score column vector **t** is calculated by the following procedure:

$$\mathbf{t} = \frac{\mathbf{Xw}}{\mathbf{w}^{\mathrm{T}}\mathbf{w}} \tag{4.35}$$

where

$$\mathbf{w}^{\mathrm{T}} = \frac{\mathbf{w}_0^{\mathrm{T}}}{\|\mathbf{w}_0^{\mathrm{T}}\|} \tag{4.36}$$

is the normalized row vector of the following weights:

$$\mathbf{w}_0^{\mathrm{T}} = \frac{\mathbf{u}^{\mathrm{T}}\mathbf{X}}{\mathbf{u}^{\mathrm{T}}\mathbf{u}} \tag{4.37}$$

In the last equations (4.35–4.37), subscript **T** denotes the transposed vector. The next estimate of **u** vector is obtained by involving the property matrix **Y** as follows:

$$\mathbf{u} = \frac{\mathbf{Yq}}{\mathbf{q}^{\mathrm{T}}\mathbf{q}} \tag{4.38}$$

where

$$\mathbf{q}^{\mathrm{T}} = \frac{\mathbf{q}_0^{\mathrm{T}}}{\|\mathbf{q}_0^{\mathrm{T}}\|} \tag{4.39}$$

and

$$\mathbf{q}_0^{\mathrm{T}} = \frac{\mathbf{t}^{\mathrm{T}}\mathbf{Y}}{\mathbf{t}^{\mathrm{T}}\mathbf{t}} \tag{4.40}$$

The new **u** vector is inserted into Eq. (3.37) and the next estimate of the score column vector **t** is calculated using Eq. (3.35). This procedure is repeated iteratively until the convergence of vector **t** is achieved. Notably, if the property matrix consists of only one column, that is, there is just one property involved in the PLS treatment, then $q = 1$ and no iterations are needed. The final loadings $\mathbf{p}^{\mathrm{T}}$ and scores **t** of the first PLS component for **X** matrix are then calculated using the following relationships:

$$(\mathbf{p}^{\mathrm{T}})^{(1)} = \frac{\mathbf{p}^{\mathrm{T}}}{\|\mathbf{p}^{\mathrm{T}}\|} \tag{4.41}$$

$$\mathbf{t}^{(1)} = \mathbf{t}\|\mathbf{p}^{\mathrm{T}}\| \tag{4.42}$$

where

$$\mathbf{p}^{\mathrm{T}} = \frac{\mathbf{t}^{\mathrm{T}}\mathbf{X}}{\mathbf{t}^{\mathrm{T}}\mathbf{t}} \tag{4.43}$$

The score vector **t** consists of variables for the first PLS component that is correlated with the property matrix, **Y**. The next PLS component is found by using the same algorithm with the error matrices **E** and **F** as input. The overall PLS procedure is carried out until a satisfactory level of the description of property data is achieved.

The advantage of the PLS treatment of data is that the obtained model involves orthogonal scales that eliminates the problems arising from the collinearity of descriptors. Also, the PLS models are more robust in a sense that the obtained model scales and coefficients are less sensitive to the addition of new data than the multiple linear regression or PCR models. The main disadvantage of PLS is connected with the difficulties in the interpretation of more abstract scales. In addition, it has been pointed out that generally more data are needed for the development of reliable models than in the case of traditional regression methods. A number of variations of PLS has been developed to improve the performance of the method, such as the intermediate least squares [56], continuum regression [57] and continuum powering [58], and principal covariate regression [59].

4.4 NONLINEAR METHODS

The dependence between a molecular property or process studied and molecular descriptors may be intrinsically nonlinear (cf. Fig. 1.2). Such cases are usually

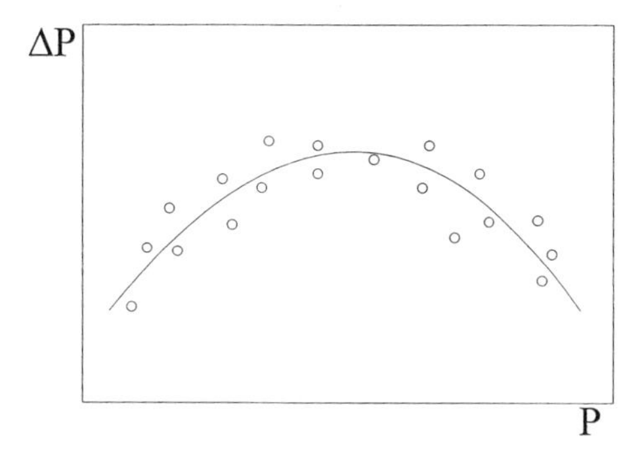

Figure 4.4 Parabolic dependence of linear regression model deviations ΔP on property values P.

characterized by a systematic distribution of the deviations of data predicted by linear models from the experimental data. An example is given in Figure 4.4, where the model deviations have a parabolic dependence on the property P values. Obviously, the square of predictor would be more suitable to describe the respective structure property relationship. The most straightforward method to account for the nonlinearity of the model is the appropriate mathematical transformation of descriptors. Various logarithmic, exponential, and other nonlinear mathematical functions can be applied for that purpose. For instance, the linear least-squares treatment of normal boiling points of a set of 137 non-hydrogen-bonded organic compounds using the roots of gravitation index

$$G_I = \sum_{i,j} \frac{m_i m_j}{r_{ij}^2} \tag{4.44}$$

gives significantly better results when either the square root or the cube root of the descriptor is applied (Table 4.4) [60].

The parabolic dependence of a molecular property on a descriptor is often accounted for just by addition of the respective square term in the QSAR/QSPR equation. For instance, the anaesthetic activity of aliphatic ethers as measured

TABLE 4.4 Correlation Coefficients, R, of One-Parameter Correlations of Normal Boiling Points for Set of Structures Without H Bonding (137 compounds)

Descriptor	R
Square root of the gravitation index, $\sqrt{G_I}$	0.9647
Cube root of the gravitation index, $\sqrt[3]{G_I}$	0.9626
Gravitation index, G_I	0.9497

by ED_{50} for anesthesia of mice is described by the following QSAR equation [61]:

$$\log\left(\frac{1}{C}\right) = (2.29 \pm 0.14) + (0.90 \pm 0.24)\log P - (0.18 \pm 0.09)(\log P)^2$$

$$R^2 = 0.874 \qquad s = 0.135 \qquad n = 28$$

The presence of a square term in the regression equation indicates the presence of an extreme point on the dependence. Thus, whereas an ordinary linear equation extrapolates the maximum or minimum values infinitely along the predictor axis, the square equation may predict a maximum (or minimum) value of the property studied within the real limits of variation of the predictor. Such prediction may be very useful in using the QSPR equations for the molecular design of novel compounds with the best chemical, physical, or pharmacological performance.

In cases when the property studied, P, is determined by several descriptors, the effect of one descriptor may be influenced by some other descriptor. This may happen if several molecular fragments have a combined, nonadditive effect on the property. A possible way to account for such nonlinear effects is the use of the cross terms between the respective fragment descriptors. For example, the $\log K$ of the following reaction series

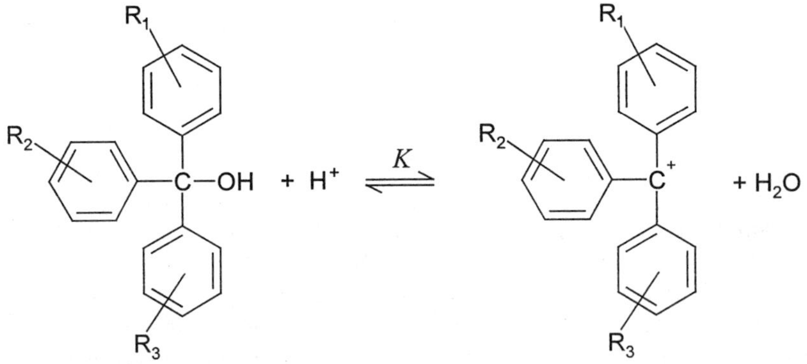

has been described by the following linear free energy relationship (LFER) equation [62]:

$$\log K = \log K_0 + \rho \sum_{i=1}^{3} \sigma_i^+ + q \sum_{i \neq j} \sigma_i^+ \sigma_j^+ \tag{4.45}$$

where σ_i^+ are the resonance constants of substituents in each of three phenyl rings attached to the acidic carbon center.

An analogous QSAR equation describing the chymotrypsin binding Michaelis constant, K_m, was obtained for the L isomers of multiply substituted congeners of compound

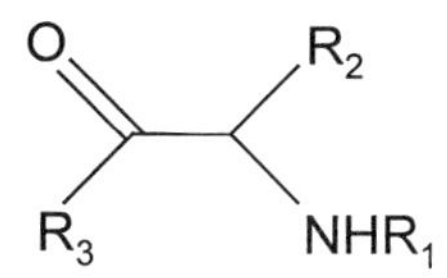

as follows:

$$\log\left(\frac{1}{K_m}\right) = (1.64 \pm 0.46) + (0.77 \pm 0.11)\mathrm{MR}_1 + (1.13 \pm 0.11)\mathrm{MR}_2$$
$$+ (0.47 \pm 0.11)\mathrm{MR}_1 - (0.56 \pm 0.25)I + (1.35 \pm 0.22)\sigma_3^*$$
$$- (0.055 \pm 0.01)\mathrm{MR}_1\mathrm{MR}_2\mathrm{MR}_3$$
$$R^2 = 0.955 \qquad s = 0.333 \qquad n = 84$$

where MR_1, MR_2, and MR_3 are the molecular refractions of the respective substituents, σ_3^* is the Taft induction constant for the substituent R_3, and the indicator variable $I = 1$ if $R_2 = \text{—}CH(CH_3)_2$ or $R_2 = \text{—}CH_2CH(CH_3)_2$.

The nonlinear dependence of the property studied on a molecular descriptor can be accounted for by the introduction of the respective polynomial expansions. The individual members of Legendre or Chebyshev polynomials are orthogonal in the interval $[-1, 1]$. Therefore, the centronormalized descriptor scales $\check{D}_i$ obtained as

$$\check{D}_i = \frac{D_i - \check{D}}{\sigma_D} \tag{4.46}$$

where $\check{D}$ is the mean value of the descriptor and σ_D is its dispersion, would be suitable for the construction of orthogonal higher order scales from the given descriptor, D. For instance, the Legendre polynomials can be constructed proceeding from the following generating formula:

$$P_i(\check{D}) = \frac{1}{2^n n!}\frac{d^n}{d\check{D}^n}(\check{D}^2 - 1) \tag{4.47}$$

The following first polynomials of a series may be used as new descriptors in the QSAR/QSPR treatment:

$$P_0 = 1;\ P_1 = \check{D};\ P_2 = \frac{1}{2}(3\check{D}^2 - 1);\ \ldots \tag{4.48}$$

All above-listed nonlinear representations of the dependence between the property studied and the molecular descriptors are still formally treated by linear least squares, and the nonlinear behavior is accounted for only by the respective transformation of the descriptor scales. However, the regression coefficients may also be involved in a mathematical model in a nonlinear form. In such cases, the coefficients have to be found using a direct nonlinear regression treatment of the data according the respective equations [34]. Let us proceed with the general nonlinear model of the form

$$Y_i = f(c_1, c_2, \ldots, c_k; D_{1i}, D_{2i}, \ldots, D_{li}) + \varepsilon_i = f(\mathbf{c}, \mathbf{D_i}) + \varepsilon_i \qquad (4.49)$$

where $\mathbf{c} = (c_1, c_2, \ldots, c_k)$ is the vector of k regression parameters and $\mathbf{D_i} = (D_{1i}, D_{2i}, \ldots, D_{li})$ includes the l descriptors (independent variables) related to the observable (dependent) variable, Y_i. The least-squares estimate of the vector $\mathbf{c}$ corresponds to the minimum of the error sum of squares defined as:

$$S(\mathbf{c}) = \sum_{i=1}^{n} [Y_i - f(\mathbf{c}, \mathbf{D_i})]^2 \qquad (4.50)$$

where the summation is carried out over all n observables in the data set. To find the least-squares estimate of $\hat{\mathbf{c}}$, the last equation is differentiated with respect to these parameters. Thus, k normal equations are obtained of the following form:

$$\sum_{i=1}^{n} [Y_i - f(\mathbf{c}, \mathbf{D_i})] \left[\frac{\partial f(\mathbf{c}, \mathbf{D_i})}{\partial c_j}\right]_{\mathbf{c}=\hat{\mathbf{c}}} \qquad j = 1, k \qquad (4.51)$$

which can be solved with respect of $\mathbf{c}$. In contrast to linear least squares, the normal equations (3.51) are nonlinear, and therefore it is usually difficult to solve them directly. In most cases, iterative methods have been used for this purpose. In addition, with more parameters involved in the model, multiple solutions exist, corresponding to stationary values of the error sum of squares, $S(\mathbf{c})$. A variety of methods has been developed to follow a local minimum of the function $S(\mathbf{c})$. For instance, the model function $f(\mathbf{c}, \mathbf{D})$ can be linearized with respect of $\mathbf{D}$ by using the Taylor expansion around some estimate $\mathbf{D}^{(0)}$, curtailed at the first derivatives. The optimum parameters $\mathbf{D}^{(\mathrm{opt})}$ are then obtained by solving iteratively the corresponding approximate normal equations [63]. Various optimization methods such as the steepest descent [64], Levenberg-Marquardt's method [65], quasi-Newton, conjugate-gradient, and other algorithms [66–68] are also applicable to find the local best estimates for the parameters, $\mathbf{c}$, of a nonlinear regression. The search for the globally best solution of a nonlinear regression is more complicated. The simulated annealing techniques [69–71] and genetic algorithms [72–75] can be used for the search of global minima on complex surfaces given by the function $S(\mathbf{c})$.

The intrinsically nonlinear QSAR/QSPR models can be developed as the feed-forward artificial neural networks [76–78]. Feed-forward neural networks have been most useful for approaching the classification or pattern recognition of data. A typical artificial neural network (ANN) is built from hierarchical units. Each of these units generates an output as a function of its inputs that may be the outputs of other units or the external data. The units are usually divided into three layers, corresponding to the input, hidden layer, and ouput (Fig. 4.5). With this respect, the units are sequential, each unit can take the output of any or all units from the previous layer, and its output can serve as input to each subsequent unit. In a simple ANN, a weighted sum of input values a_i is calculated in every unit j as:

$$S_j = b_j + \sum_{i=1}^{j=1} a_i w_i \tag{4.52}$$

where w_i are the weights for each respective input and b_j is the so-called bias of the model. The output value a_j is generated as some function of S_j. A commonly used form of this function is sigmoid, expressed as:

$$a_j = \sigma(S_j) = \frac{1}{1 + \exp(S_j)} \tag{4.53}$$

The network is trained to give the best estimates for some set of data (training set) on overall output by adjusting the weights w_i in each layer. The respective error function to be minimized has the following form:

$$E = \frac{1}{2} \sum_{n=1}^{N} (P_n - a_n)^2 \tag{4.54}$$

where P_n are the experimental property values to be predicted for a training set of N elements. A simple steepest descent algorithm can be used for the adjustment of weights by back-propagation of error function. The respective correction term is calculated thus as:

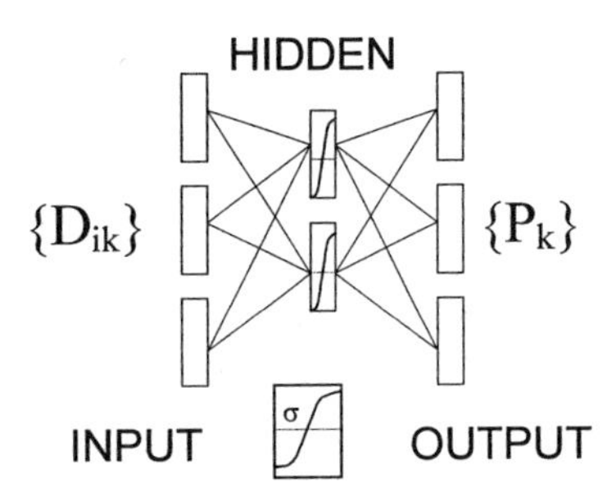

Figure 4.5 Structure of typical artificial neural network.

$$\Delta w_j = -\varepsilon \left(\frac{\partial E}{\partial w_j} \right) \tag{4.55}$$

where ε denotes a user-defined learning rate of the neural network model. To avoid the "overshooting" of the correction, the learning rate has usually a value less than 1. The learning rate can also be changed in the process of teaching the neural network. Other methods, including conjugate gradients [79] and variable metric [80], can be used for the calculation of weights of ANN. The genetic algorithms and simulated annealing have also been used for the training of weights. In the case of genetic algorithms, the weights in a population of ANN are altered using exchange and mutations. Poor performers are eliminated, and the final network includes supposedly the best weights for a given model. In simulated annealing, the weights are changed randomly and the corrections that do not increase the error function are sustained.

The nonlinear QSAR approach using the Chebyshev polynomial expansion and neural networks has been applied for the prediction of genotoxicity of compounds [81]. The structure–activity relationships were developed for a data set of experimental mutagenic potencies of aromatic amines toward *Salmonella typhimurium* TA98+S9 microsomial preparation expressed by the logarithm of the number of revertants per nanomole, log R. Three-layer neural network, made from six input units, five hidden units, and one output unit, was chosen to represent nonlinear QSAR model ($6 \times 5 \times 1$) for log R. The neural network was trained with the back-propagation algorithm on a training set consisting of 50 structures [82]. To stop the training before overfitting, a validation set (consisting of 45 structures) was used. The best neural network model for log R had $R^2 = 0.895$ and involved the following six descriptors: the molecular moment of inertia (I_C), Kier shape index ${}^3\chi$, the number of hydrogen acceptor sites in the molecule, the maximum valence of a nitrogen atom, the partial negative surface area (PNSA1), and γ polarizability of the molecule. Statistically, this model was significantly better than the best six-parameter linear regression model expressed as [83]:

$$\begin{aligned} \log R = &-(0.16 \pm 2.70) + (0.85 \pm 0.15)N_R + (1.26 \pm 0.24)10^{-4}\gamma \\ &- (3.40 \pm 0.68)10^{-2}\ \mathrm{HASA_1} + (1.71 \pm 0.35)10^{-2}\ \mathrm{HDSA} \\ &+ (0.90 \pm 0.18)E_{\mathrm{tot}}(\mathrm{C{-}C}) + (0.90 \pm 0.18)E_{\mathrm{tot}}(\mathrm{C{-}N}) \end{aligned}$$

$$R^2 = 0.8344 \qquad R^2_{\mathrm{cv}} = 0.8056 \qquad s^2 = 0.658 \qquad F = 73.91$$

where N_R denotes the number of rings in the compounds, γ is the γ polarizability, HASA1 and HDSA are the hydrogen bond acceptor and hydrogen bond donor surface areas, respectively. The two last descriptors are the quantum chemically calculated maximum total interaction energies for a C—C bond and for a C—N bond in the molecule, respectively.

The back-propagation neural networks have also been used for QSPR modeling of autoignition temperatures of 233 organic compounds [84]. Using a training set of 85 compounds with 13 different functional groups, the autoignition temperature values for the rest of the validation set of 148 compounds were successfully predicted within the experimental error of ±30°C. The comparison of ANN results with those obtained using multiple linear regression and partial least squares showed better prediction accuracy of neural networks.

REFERENCES

1. B. Kowalski, R. Gerlach, and H. Wold, in *Systems under Indirect Observation* (K. Jörenskog and H. Wold, eds.), pp. 191–206, North-Holland, Amsterdam, 1982.
2. S. Wold, A. Ruhe, H. Wold, and W. Dunn, *SIAM J. Sci. Stat. Comput.* **5,** 735 (1984).
3. S. Wold, *Chemometrics: Mathematics and Statistics in Chemistry*. Reidel, Dordrecht, The Netherlands, 1984.
4. P. Geladi and B.R. Kowalski, *Anal. Chim. Acta* **185,** 1 (1986).
5. A. Höskuldsson, *J. Chemometrics* **2,** 211 (1988).
6. J.A. Burns and G.M. Whiteside, *Chem. Rev.* **93,** 2583 (1993).
7. L.S. Anker and P.C. Jurs, *Anal. Chem.* **64,** 1157 (1992).
8. T. Aoyama, Y. Suzuki, and H. Ichikawa, *J. Med. Chem.* **33,** 2583 (1990).
9. T.A. Andrea, *J. Med. Chem.* **34,** 2824 (1991).
10. A.R. Katritzky, V.S. Lobanov, and M. Karelson, *CODESSA Reference Manual*. University of Florida, Gainesville, 1994.
11. V.V. Nalimov, *The Application of Mathematical Statistics to Chemical Analysis*. Addison-Wesley, Reading, MA, 1962.
12. R. Caulcutt and R. Boddy, *Statistics for Analytical Chemists*. Chapman & Hall, New York, 1983.
13. J.C. Miller and J.N. Miller, *Statistics for Analytical Chemistry*. Ellis Horwood, New York, 1988.
14. P.C. Meier and R.E. Zund, *Statistical Methods in Analytical Chemistry*. Wiley, New York, 1993.
15. R.H. Myers, *Classical and Modern Regression with Applications*. PWS-KENT Publ. Boston, 1989.
16. A.R. Katritzky, Y. Wang, S. Sild, T. Tamm, and M. Karelson, *J. Chem. Inf. Comput. Sci.* **38,** 720 (1998).
17. M.J. Stone and P. Jonathan, *J. Chemometrics* **7,** 455 (1993).
18. G. Wahba, in *Application of Statistics* (P.R. Krishnaiah, ed.), p. 507. North-Holland, Amsterdam, 1976.
19. J.K. Seydel and K.-J. Schafer, *Pharm. Ther.* **15,** 131 (1982).
20. W.J. Dunn, *Chemometrics Intell. Lab. Syst.* **6,** 181 (1982).
21. M. Randić, *J. Mol. Struct. (THEOCHEM)* **233,** 45 (1991).
22. A.E. Hoerl and R.W. Kennard, *Technometrics* **12,** 55 (1970).

23. P.M. Fredericks, J.B. Lee, P.R. Osborn, and D.A.J. Swinkels, *Appl. Spectrosc.* **39,** 303 (1985).
24. J.M. Sutter, J.H. Kalivas, and P.M. Lang, *J. Chemometrics* **6,** 217 (1992).
25. J. Sun, *J. Chemometrics* **9,** 21 (1995).
26. I. Lukovits, *J. Med. Chem.* **26,** 1104 (1983).
27. M. Randić, *New J. Chem.* **15,** 517 (1991).
28. M. Randić, *J. Chem. Inf. Comput. Sci.* **31,** 317 (1991).
29. M. Randić and N. Trinajstić, *New J. Chem.* **18,** 179 (1994).
30. D. Amić, D. Davidović-Amić, and N. Trinajstić, *J. Chem. Inf. Comput. Sci.* **35,** 136 (1995).
31. B. Lučić, S. Nikolić, N. Trinajstić, and D. Juretić, *J. Chem. Inf. Comput. Sci.* **35,** 532 (1995).
32. M. Šoškić, D. Plavšić, and N. Trinajstić, *J. Chem. Inf. Comput. Sci.* **36,** 146 (1996).
33. A. Szabo and N. Ostlund, *Modern Quantum Chemistry*, pp. 15–21. McGraw-Hill, New York, 1989.
34. N.R. Draper and H. Smith, *Applied Regression Analysis*. Wiley, New York, 1981.
35. P.T. Pope and J.T. Webster, *Technometrics* **14,** 327 (1972).
36. R.B. Bendel and A.A. Afifi, *J. Am. Stat. Assoc.* **72,** 46 (1977).
37. A.R. Katritzky, V. Lobanov, and M. Karelson, *Chem. Soc. Rev.* **24,** 279 (1995).
38. A.R. Katritzky, V.S. Lobanov, M. Karelson, R. Murugan, M.P. Grendze, and J.E. Toomey, *Rev. Roum. Chim.* **41,** 851 (1996).
39. B. Lučić, S. Nikolić, N. Trinajstić, and D. Juretić, *J. Chem. Inf. Comput. Sci.* **35,** 532 (1995).
40. B. Lučić, S. Nikolić, N. Trinajstić, D. Juretić, and A. Jurić, *Croat. Chem. Acta* **68,** 435 (1995).
41. D. Amić, D. Davidović-Amić, D. Bešlo, B. Lučić, and N. Trinajstić, *J. Chem. Inf. Comput. Sci.* **37,** 581 (1997).
42. B. Lučić and N. Trinajstić, *SAR QSAR Environ. Res.* **7,** 45 (1997).
43. B. Lučić, N. Trinajstić, S. Sild, M. Karelson, and A.R. Katritzky, *J. Chem. Inf. Comput. Sci.* (1999) (in press).
44. E.R. Malinowski and D.G. Howery, *Factor Analysis in Chemistry*. Wiley-Interscience, New York, 1980.
45. O. Strouf, *Chemical Pattern Recognition*. Wiley, New York, 1986.
46. I.T. Jollcliffe, *Principal Component Analysis*. Springer-Verlag, Berlin, 1986.
47. M. Meloun, M. Militky, and M. Forina, *Chemometrics in Analytical Chemistry*. Ellis Horwood, New York, 1992.
48. S.D. Brown, T.Q. Barker, R.J. Larivee, S.L. Monfre, and H.R. Wilk, *Anal. Chem.* **60,** 252R (1988).
49. S.D. Brown, *Anal. Chem.* **62,** 84R (1990).
50. S.D. Brown, R.S. Bear, Jr., and T.B. Blank, *Anal. Chem.* **64,** 22R (1992).
51. S.D. Brown, T.B. Blank, S.T. Sum, and L.G. Weyer, *Anal. Chem.* **66,** 315R (1994).
52. S.D. Brown, S.T. Sum, F. Despagne, and B.K. Lavine, *Anal. Chem.* **68,** 21R (1996).
53. S. Wold, P. Geladi, K. Esbensen, and J. Öhman, *J. Chemometrics* **1,** 41 (1987).
54. F. Lindgren, P. Geladi, S. Rännar, and S. Wold, *J. Chemometrics* **8,** 349 (1994).

55. T. Oprea and C.L. Waller, *Rev. Comput. Chem.* **11,** 127 (1997).
56. I.E. Frank, *Chemometrics Intell. Lab. Syst.* **1,** 233 (1987).
57. M. Stone and R.J. Brooks, *J. R. Stat. Soc. B* **52,** 237 (1990).
58. A. Lorber, L.E. Wangen, and B.R. Kowalski, *J. Chemometrics* **1,** 19 (1987).
59. S. de Jong and H.A.L. Kiers, *Chemometrics Intell. Lab. Syst.* **14,** 155 (1992).
60. A.R. Katritzky, L. Mu, V.S. Lobanov, and M. Karelson, *J. Phys. Chem.* **100,** 10400 (1996).
61. W.R. Glave and C. Hansch, *J. Pharm. Sci.* **61,** 589 (1972).
62. C.D. Ritchie, W.F. Sager, and E.S. Lewis, *J. Am. Chem. Soc.* **84,** 2349 (1962).
63. O. Hartley, *Technometrics* **3,** 269 (1961).
64. O.L. Davies, ed., *Design and Analysis of Industrial Experiments*. Oliver & Boyd, Edinburgh, 1954.
65. D.W. Marquardt, *J. Soc. Ind. Appl. Math.* **2,** 431 (1963); K. Levenberg, *Quart. Appl. Math.* **2,** 164 (1944).
66. R. Fletcher, *Practical Methods of Optimization*, 2nd ed. Wiley, Chichester, 1987.
67. G.A.F. Seber and C.J. Wild, *Nonlinear Regression*. Wiley, New York, 1989.
68. S. Huet and M.-A. Gruet, *Statistical Tools For Nonlinear Regression; A Practical Guide*. Springer-Verlag, Berlin, 1996.
69. I.O. Bohachevsky, M.E. Johnson, and M.L. Stein, *Technometrics* **28,** 209 (1986).
70. J.H. Kalivas, *Chemometrics Intell. Lab. Syst.* **14,** 1 (1992).
71. U. Hörchner and J.H. Kalivas, *J. Chemometrics* **9,** 283 (1995).
72. L. Davis, ed., *Genetic Algorithms and Simulated Annealing*. Morgan Kaufmann Publishers, San Francisco, 1987.
73. C.B. Lucasius and G. Kateman, *Chemometrics Intell. Lab. Syst.* **19,** 1 (1993).
74. D.B. Hibbert, *Chemometrics Intell. Lab. Syst.* **19,** 277 (1993).
75. D.E. Clark and D.R. Westhead, *J. Comput.-Aided Mol. Des.* **10,** 337 (1996).
76. J.A. Burns and G.M. Whiteside, *Chem. Rev.* **93,** 2583 (1993).
77. B.J. Wythoff, *Chemometrics Intell. Lab. Syst.* **19,** 115 (1993).
78. J. Zupan and J. Gasteiger, *Neural Networks for Chemists*. VCH, Weinheim, 1993.
79. S.M. Muskal and S.-H. Kim, *J. Mol. Biol.* **225,** 715 (1992).
80. V. Kvasnička and J. Pospichal, *J. Mol. Struct. (THEOCHEM)* **235,** 227 (1991).
81. M. Karelson, S. Sild, and U. Maran, *Mol. Simul.* (submitted for publication).
82. T.M. Mitchell, *Machine Learning*. McGraw-Hill, New York, 1997.
83. U. Maran, A.R. Katritzky, and M. Karelson, *Quant. Struct-Act. Relat.* **18,** 3 (1999).
84. J. Tetteh, E. Metcalfe, and S.L. Howells, *Chemometrics Intell. Lab. Syst.* **32,** 177 (1996).

APPENDIX

COMPUTER SOFTWARE FOR CALCULATION OF MOLECULAR DESCRIPTORS AND DERIVATION OF QSAR/QSPR

In this appendix, the software on molecular descriptors provided with this book is described. The software is comprised of the following three modules:

Database on empirical structural descriptors for molecular fragments (substituents)

Program for the calculation of theoretical molecular descriptors

Program for the development of (multiple) linear QSAR/QSPR models

The first module, the database on empirical structural descriptors, includes a selection of substituent inductive, resonance, and steric constants for a variable set of structures. The data correspond to the substituents given in the Tables 2.2, 2.5, and 2.7, respectively.

The set of inductive constants includes the following scales:

σ_m Hammett's σ constant for meta-substituted phenyls
σ_p Hammett's σ constant for para-substituted phenyls
σ^* Taft's σ constant for aliphatic substituents
σ_I dual-substituent parameter σ_I constant
F Swain–Lupton field-effect parameter

The set of resonance constants includes the following scales:

σ^+ Yukawa–Tsuno resonance constant for the π-electron-donating (+R) substituents
σ^- Yukawa–Tsuno resonance constant for the π-electron-accepting (−R) substituents

σ_R dual-substituent parameter σ_R constant
R Swain–Lupton resonance effect parameter
R^+ Swain–Lupton resonance effect parameter for the π-electron-donating (+R) substituents
R^- Swain–Lupton resonance effect parameter or the π-electron-accepting (−R) substituents

The set of steric constants includes the following scales:

E_s Taft's steric constant
E_s^C Taft's steric constant, corrected for the hyperconjugation of alkyl substituents
E_s^0 Taft's steric constant, corrected for the hyperconjugation of alkyl substituents
MR Molecular refraction

The second module of the provided software allows to calculate a number of theoretical descriptors for molecular structures. Several classes of molecular descriptors are available, including the constitutional, topological, geometrical, and quantum mechanical descriptors. To calculate the descriptors for the first three classes, the molecular structure has to be given using one of the predetermined alpha-numerical formats. The quantum chemical descriptors can be extracted and calculated from the information available in the standard output of the MOPAC program system [1].

The selection of constitutional descriptors calculated for a molecular structure include the following:

Number of atoms in the molecule
Number of carbon atoms in the molecule
Number of hydrogen atoms in the molecule
Relative number of atoms of certain type in the molecule
Number of rings in the molecule
Number of single, double, triple, and aromatic bonds in the molecule
Relative number of the bonds of a certain type in the molecule
Molecular weight

The following topological descriptors can be calculated using the attached software provided:

Wiener index W
Randić's indices of different orders (1–3) ${}^i\chi$
Kier–Hall indices of different orders (1–3) ${}^i\chi^v$
Kier–Hall shape indices ${}^i\kappa$

Information content indices of different orders ${}^{k}IC$, ${}^{k}CIC$, ${}^{k}SIC$, ${}^{k}BIC$

The following geometrical descriptors can be calculated:

Moments of inertia of a molecule
Molecular surface area
Molecular volume
Molecular topographical indices

From the output of the semiempirical quantum chemical calculations, the following molecular descriptors are extracted and calculated:

Maximum and minimum partial charges on atoms
Dipole moment of the compound
Frontier molecular orbital energies ε_{HOMO} and ε_{LUMO} and the difference $\varepsilon_{HOMO} - \varepsilon_{LUMO}$
Maximum and minimum bond orders (for bonded atoms) in the molecule
Fukui electrophilicity, nucleophilicity, and the one-electron reactivity indices

The QSAR/QSPR equations relating the user's experimental data with molecular descriptors can be derived using either the multiple least-squares treatment with a predefined set of descriptors or the forward selection of descriptor scales on a large descriptor space. Both treatments can be carried out on a spreadsheet constructed by the user for his or her own data and the data on molecular descriptors available in the attached software provided.

The construction of a spreadsheet begins with entering the program front page, which includes the list of applicable modules (see Scheme 1).

First, the substituent scales can be selected from the appropriate table for the subsequent use in the multiple linear least-squares treatment or in the forward-scale-selection derivation of the best QSAR/QSPR equation (see Scheme 2).

Thereafter, the user can interactively input the experimental property data to be correlated for a preselected list of substituents or compounds (for solvent effect scales) into the spreadsheet (see Scheme 3).

Alternatively, the information about the molecular structures can be read in within the module for the derivation of the theoretical molecular descriptors. Two formats of numerical information about the three-dimensional structure of a molecule are acceptable. First, the *.mol* files by Molecular Design, Ltd., and second, the MOPAC output files with the extension *.mno*. Together with the names of structures, the property values to be correlated should also be inserted into the spreadsheet (see Scheme 4).

After that, the user can select the theoretical molecular descriptors to be calculated. The numerical values of these descriptors are included automatically into the spreadsheet (see Scheme 5).

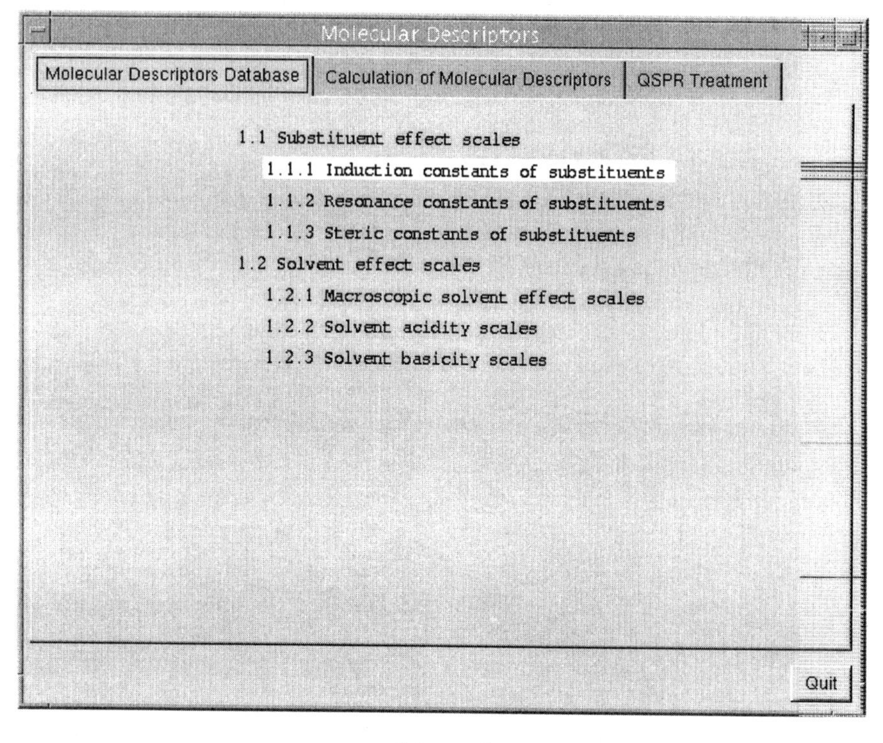

Scheme 1

After the formation of the spreadsheet, using either the empirical or theoretical descriptors, two possibilities are available for the statistical (multiple) linear least-squares treatment of data. First, a small preselected set of descriptors can be used in an individual regression. Second, from a larger set of descriptors, the descriptor scales that give the best statistically justified QSAR/QSPR equation can be selected. The method of forward selection of scales is applied for this purpose [2] (see Scheme 6).

The output of both multiple linear regression treatments includes the following data:

Regression coefficients with the errors
Root mean standard error of the regression
Squared correlation coefficient of the regression
Squared cross-validated correlation coefficient of the regression

The output data can be either printed out or saved in the user-specified file.

The software provided with this text is applicable on the following platforms: Microsoft Windows 95, Solaris 2.6, and AIX 4.2, or later versions of

Induction constants of substituents

File

	σ_m	σ_p	σ^*	σ_I	F
1. H	0.0000	0.0000	0.4900	0.0000	0.0300
2. CH3	−0.0700	−0.1700	0.0000	−0.0400	0.0100
3. CH2CH3	−0.0700	−0.1500	−0.1000	−0.0100	0.0000
4. CH2CH2CH3	−0.0600	−0.1300	−0.1200	−0.0100	0.0100
5. CH(CH3)2	−0.0400	−0.1500	−0.1900	0.0100	0.0400
6. cyclopropyl	−0.0700	−0.2100	0.0400	0.0100	0.0200
7. (CH2)3CH3	−0.0800	−0.1600	−0.1300	−0.0400	−0.0100
8. CH2CH(CH3)2	−0.0700	−0.1200	−0.1300	−0.0300	−0.0100
9. CH(CH3)C2H5	−0.0800	−0.1200	−0.2100	−0.0300	−0.0200
10. C(CH3)3	−0.1000	−0.2000	−0.3000	−0.0700	−0.0200
11. cyclobutyl	−0.0500	−0.1400	−0.1500		0.0200
12. (CH2)4CH3	−0.0800	−0.1500	−0.1600	−0.0300	−0.0100
13. CH2C(CH3)3	−0.0500	−0.1700	−0.1700	−0.0700	0.0300
14. C(C2H5)(CH3)2	−0.0600	−0.1800	−0.3300	−0.0800	0.0300
15. cyclopentyl	−0.0500	0.1400	−0.2000		0.0200
16. cyclohexyl	−0.0500	−0.1500	−0.1800	−0.0200	0.0300

CH3 | Hammett sigma-constant for meta substituted phenyls

Scheme 2

Data sheet: no title

File

	Property	d1	d2	d3
1. H	−1.0000	0.0000	0.0000	0.0000
2. CH3	−1.1700	−0.1700	−0.0700	−0.1700
3. CH2CH3	−1.1700	−0.1500	−0.0700	−0.1500
4. CH2CH2CH3	−1.1200	−0.1300	−0.0600	−0.1300
5. CH(CH3)2	−1.0400	−0.1500	−0.0400	−0.1500
6. cyclopropyl	−1.1400	−0.2100	−0.0700	−0.2100
7. (CH2)3CH3	−1.3800	−0.1600	−0.0800	−0.1600
8. CH2CH(CH3)2	−1.2800	−0.1200	−0.0700	−0.1200
9. CH(CH3)C2H5	−1.3800	−0.1200	−0.0800	−0.1200
10. C(CH3)3	−1.4000	−0.2000	−0.1000	−0.2000
11. cyclobutyl	−1.2500	−0.1400	−0.0500	−0.1400
12. (CH2)4CH3	−1.3800	−0.1500	−0.0800	−0.1500
13. CH2C(CH3)3	−1.1500	−0.1700	−0.0500	−0.1700
14. C(C2H5)(CH3)2	−1.2600	−0.1800	−0.0600	−0.1800
15. cyclopentyl	−1.1500	0.1400	−0.0500	0.1400
16. cyclohexyl	−1.4500	−0.1500	−0.0500	−0.1500

CH2CH2CH3 | Hammett sigma-constant for meta substituted phenyls

Scheme 3

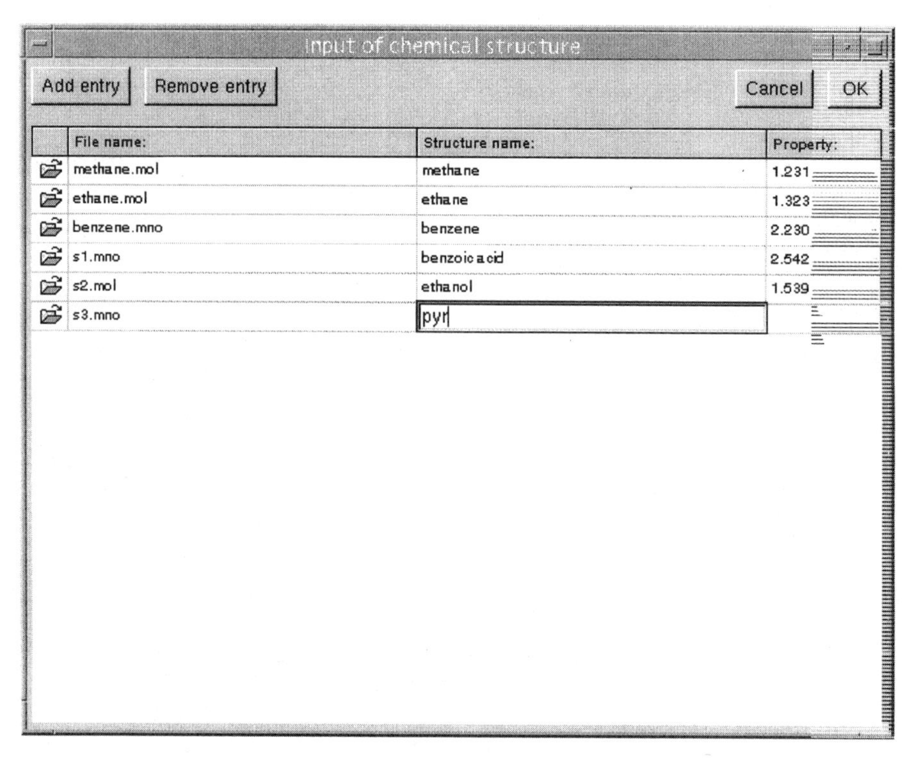

Scheme 4

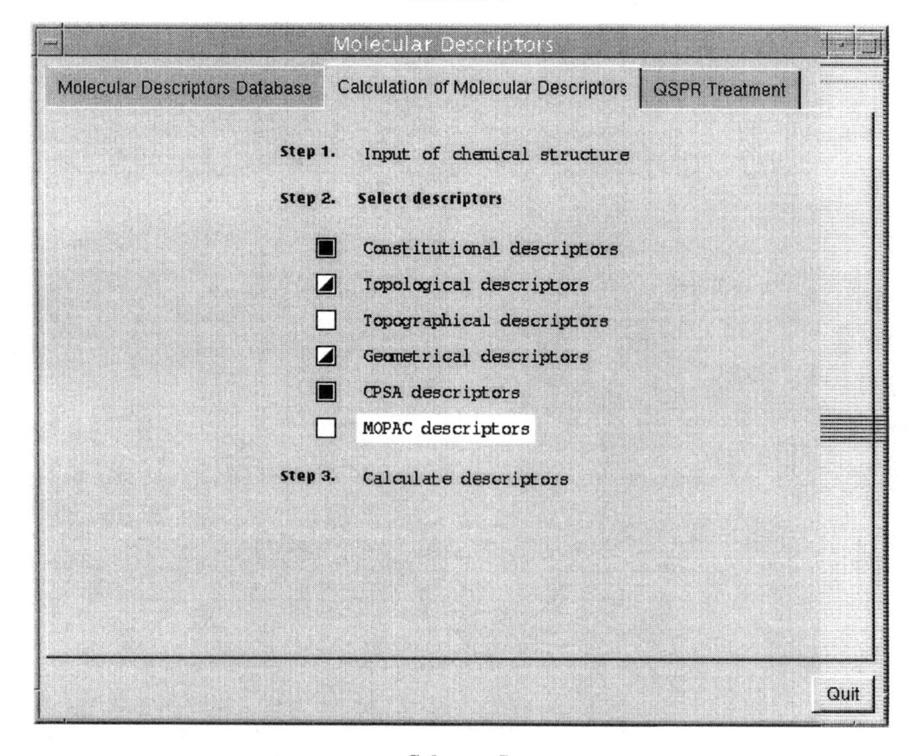

Scheme 5

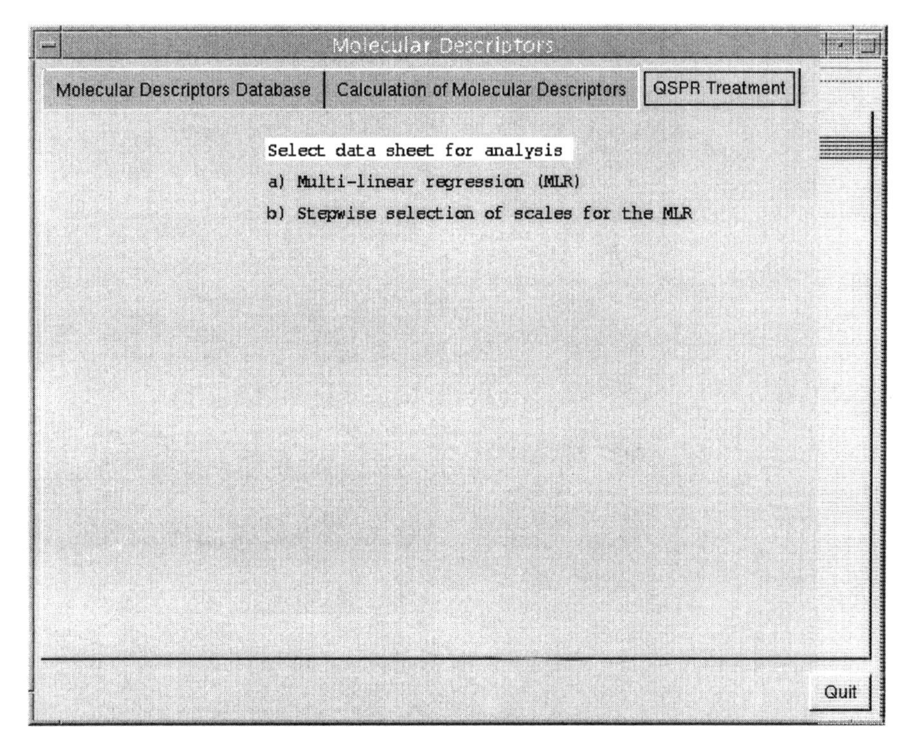

Scheme 6

these systems. The version suitable for the user can be loaded from the CD-ROM provided with this book.

REFERENCES

1. J.J.P. Stewart, *MOPAC 6.0*, QCPE No. 455. Indiana University, Bloomington, IN, 1989.
2. N.R. Draper and H. Smith, *Applied Regression Analysis*, pp. 169ff. Wiley, New York, 1966.

INDEX